Basic
Organic
Chemistry
Part 5: Industrial Products

J. M. TEDDER
University of St. Andrews

A. NECHVATAL
University of Dundee

A. H. JUBB
I.C.I. Ltd., Industrial Laboratory, Runcorn

Basic

Organic

Chemistry

Part 5: Industrial Products

1975
JOHN WILEY & SONS
London   New York   Sydney   Toronto

Library of Congress Cataloging in Publication Data: (Revised)

Tedder, John Michael.
Basic organic chemistry.

Pt. 5 has title: Industrial Products.
Pts. 3-4 by J. M. Tedder, A. Nechvatal, A. W. Murray, and
J. Carnduff; pt. 5 by J. M. Tedder, A. Nechvatal, and A. H. Jubb.
1. Chemistry, Organic. I. Nechvatal, A., joint author. II. Title.
QD253.T38   547   66-17112

ISBN 0 471 85014 4 (Cloth)
ISBN 0 472 85016 0 (Paper)

Printed in Great Britain by William Clowes & Sons Limited
London, Colchester and Beccles

Contributors

GEORGE R. BROWN, M.Sc.,
Chemistry Department, Pharmaceuticals Division, Imperial Chemical Industries Limited, Alderley Park, Macclesfield, Cheshire

JOHN CATTANACH, B.Sc., Ph.D.,
Epsom Division, Group Research and Development Department, The British Petroleum Company Limited, Great Burgh, Epsom, Surrey

ERIC COWLEY, B.Sc., F.R.I.C., M.Chem.A.,
Flavour Development Controller,
Bush Boake Allen Ltd, Ash Grove, Hackney, London, E.8

ANTONY CROSSLEY, Ph.D., B.Sc., A.R.I.C.,
Division Manager,
Edible Oils Division, Unilever Research, Colworth/Welwyn, The Frythe, Welwyn, Hertfordshire

MAX H. DILKE, B.Sc., Ph.D.,
Section Head, Technical Information,
Epsom Division, Research and Development Department, BP Chemicals International Limited, Epsom, Surrey

G. F. DUFFIN, B.Sc., Ph.D., F.R.I.C.,
Managing Director,
Minnesota 3M Research Limited, Pinnacles, Harlow, Essex

T. Edmonds,
The British Petroleum Company Limited, BP Research Centre, Sunbury-on-Thames, Middlesex

Norman F. Elmore, Ph.D.,
Chemistry Department, Pharmaceuticals Division, Imperial Chemical Industries Limited, Alderley Park, Macclesfield, Cheshire

William D. Fordham, B.Sc., Ph.D., F.R.I.C.,
Senior Research Chemist,
Research/Development Department, Bush Boake Allen Ltd, Carpenters Road, Stratford, London E.15

Edward J. Gasson, B.Sc., A.K.C.,
Chemicals Branch Manager,
Research and Development Department, BP Chemicals International Limited, Grangemouth, Scotland

Howard S. Green, B.Sc., Ph.D.,
Project Leader,
Research and Development Department, BP Chemicals International Limited, Salt End, Hedon, Hull

Alexander T. Greer, B.Sc.,
Chemistry Department, Pharmaceuticals Division, Imperial Chemical Industries Limited, Alderley Park, Macclesfield, Cheshire

Owen A. Gurton, B.Sc., B.Pharm., Ph.D.,
Research and Development Manager,
Nobel's Explosives Company Limited, Nobel House, Stevenston, Ayrshire

Reginald H. Hall, A.R.C.S., B.Sc., D.I.C., Ph.D., F.R.I.C.,
Technical Administrator,
Chemicals Branch, Epsom Division, Research and Development Department, BP Chemicals International Limited, Epsom, Surrey

James R. Holker, B.Sc., Ph.D., F.R.I.C.,
Principal Research Officer,
The Shirley Institute, Didsbury, Manchester

KEITH JONES, B.Sc., Ph.D.,
Manager,
Organic Chemicals Section, Unilever Research, Port Sunlight Laboratory, Unilever Limited, Port Sunlight, Wirral, Cheshire

MICHAEL E. B. JONES, B.Sc.,
Research Specialist,
Corporate Laboratory, Imperial Chemical Industries Limited, Runcorn, Cheshire

BRIAN D. JOYNER, F.R.I.C.,
Product Development Manager,
I.S.C. Chemicals Ltd., Avonmouth, Bristol

A. H. JUBB, B.Sc., Ph.D.,
Head of Business Group,
Corporate Laboratory, Imperial Chemical Industries Limited, Runcorn, Cheshire

I. T. KAY, B.Sc., Ph.D.,
Senior Research Officer,
Plant Protection Limited, Jealott's Hill Research Station, Bracknell, Berkshire

M. J. LAWRENSON, B.Sc.,
The British Petroleum Company Limited, Britannic House, Moor Lane, London EC2Y 9BU

PHILIP LUSMAN, B.Sc.,
Research and Development Department, BP Chemicals International Limited, Grangemouth, Scotland

HOWARD G. NAYLOR, B.Sc.,
Process Technologist,
Central Technical Division, BP Chemicals International Limited, Devonshire House, Piccadilly, London W1X 6AY

J. PENNINGTON, B.A.Chem., M.A.Chem.,
Research and Development Department, BP Chemicals International Limited, Salt End, Hedon, Hull

B. K. SNELL, B.Sc., Ph.D.,
Section Manager,
Plant Protection Limited, Jealott's Hill Research Station, Bracknell, Berkshire

C. V. STEAD, B.Sc., Ph.D.,
Senior Research Chemist,
Research Department, Organics Division, Imperial Chemical Industries Limited, Hexagon House, Blackley, Manchester

C. D. S. TOMLIN, B.Sc., M.A., Ph.D.,
Section Manager,
Plant Protection Limited, Jealott's Hill Research Station, Bracknell, Berkshire

DAVID M. WHITEHEAD, B.Sc., M.Inst.P.,
The British Petroleum Company Limited, BP Research Centre, Sunbury-on-Thames, Middlesex

BARRIE WOOD, Ph.D., B.Sc., A.R.I.C.,
Technical Information Officer,
Epsom Division, Research and Development Department, BP Chemicals International Limited, Epsom, Surrey

BERTRAM YEOMANS, B.Sc., A.R.I.C.,
Project Leader,
Research and Development Department, BP Chemicals International Limited, Salt End, Hedon, Hull

Preface to Part 5

The first volume in this series was taken directly from lecture notes. The lecture notes had been prepared because at that time all the available text-books still developed organic chemistry by the then traditional approach of 'preparations and properties'. The lectures and subsequent book sought to enhance the students' understanding of the subject rather than to list a vast array of facts which required to be learnt by rote. In the six years since Part 1 was written a complete change in the outlook of teaching organic chemistry has taken place. The mechanistic approach which was then the exception has now become the norm. Like all revolutions, the mechanistic approach can be carried to excess and we tried to emphasize some of the dangers in the Preface to Part 2. One almost unavoidable result of the universal adoption of the mechanistic approach has been the abandonment of any reference to industrial processes unless they serve to illustrate some particular type of reaction or chemical phenomenon. Since many industrial processes involve highly complex reaction mechanisms, often of a heterogeneous nature and sometimes not fully understood, the present-day student can leave University with a very hazy knowledge of what the chemical industry is all about. In some respects the omission of industrial processes from the undergraduate course is desirable. University teachers are at present a much-criticized company, but by and large they do know what they are talking about when they lecture, even if their delivery is sometimes faulty. In general, university lecturers do not know what the latest developments in industry are, and in contrast to their academic material any lectures they give on industrial processes may be inaccurate and out of date. On the other hand, it is clearly most undesirable that the new graduate should be totally ignorant of application of his subject.

Part 5 of Basic Organic Chemistry is an attempt to solve this problem.

The book commences with an introduction that attempts to show how chemical problems in industry are necessarily associated with economic and sociological considerations. The second Chapter deals with basic raw materials on which the chemical industry depends. This is followed by four Chapters that cover the basic processes involving preparation of the chemical intermediates used for the manufacture of plastics, fibres, etc.; these initial Chapters are primarily concerned with the petrochemical industry. Chapter 7 is an attempt to show the various factors that are brought into consideration when one process is adopted in preference to another.

From Chapter 8 onwards we are more concerned with the products themselves than with the processes by which they are made. Chapters 8 and 9 deal with polymers first as plastics, then as fibres, and these are followed by Chapters on dyestuffs, pharmaceuticals, organic compounds for agriculture and detergents. The next three Chapters deal with types of organic molecules more familiar to the student from Parts 1 to 4: Chapter 14 deals with fuels and explosives; and Chapters 15 and 16 involve a great deal of natural product chemistry, being concerned with food-stuffs, perfumes and flavours. Finally, Chapter 17 gives a brief account of the organic chemicals used in photography and of the fluorocarbon derivatives used as aerosols and refrigerants.

There are thus many aspects of applied organic chemistry that are not dealt with here but we do believe all the major industries are represented.

This book is most likely to find its first appeal to students reading applied chemistry. We certainly hope that it will be useful to such students, but we must emphasize that the book is aimed at the Honours Chemist reading so-called 'pure chemistry'. We believe that the material covered in this book should form part of his chemical general knowledge. In our view a graduate who has no knowledge of the procedures of the chemical industry is like a would-be actor who has neither play nor stage. As with the whole of this series, this book is intended to be read. We hope the student will find the story of the application of his science as interesting and absorbing as we have.

J.M.T.
A.N.
A.H.J.

Acknowledgements

The Editors and individual Authors express their appreciation to many friends and colleagues who have given them advice. The Editors also acknowledge the help they have had from British Petroleum Company Limited, BP Chemicals International Limited, Bush Boake Allen Ltd., Imperial Chemical Industries Limited, I.S.C. Chemicals Ltd., Minnesota 3M Research Ltd., Unilever Research and the Shirley Institute. Thanks must also go to Dr. M. Sparke of BP Chemicals International Limited who kindly acted as link man. The willing and often enthusiastic support we have had from industry has played a very big part in making this book possible.

Contents

CHAPTER 1

Introduction and History of the Chemical Industry

(A. H. JUBB, ICI Corporate Laboratory, Runcorn)

In Parts 1, 2 and 3 we were building in outline an intellectual edifice called organic chemistry. Reactions were classified on a mechanistic basis and compounds were classified on the basis of their reactivity and/or their structure. Part 4 was concerned with applying the concepts developed in the earlier Parts to the compounds found in nature. Compounds were classified by their biosynthetic origin, but the emphasis remained on a mechanistic understanding of the reactions these compounds undergo. This final volume is concerned with synthetic organic chemicals, that is, with compounds prepared in the laboratory or in the factory because they are useful. In Part 4 we mentioned the synthesis of complex natural products as occasionally representing no more than a dilettante desire to emulate nature though we emphasized that usually this was done for far more practical reasons. The compounds and syntheses described in the present volume have the common characteristic that they are useful, be it as a perfume to make a person more attractive, as an explosive to blast in a quarry, or as a fibre to be woven into a fabric for clothing. The importance of a compound in this volume rests not on its chemical significance but in its value on application.

The value of any commodity depends entirely on circumstances; to a traveller in the desert a litre of water could be worth more than all the gold in Fort Knox. This becomes very apparent when pharmaceutical chemicals are being considered. If a particular compound provides the only known cure to a fatal disease its cost is relatively unimportant. On the other hand, a new dyestuff, although providing shades previously unavailable, will not prove important if it is so costly that articles dyed with it become much

1

more expensive than the same articles dyed with similar though
less attractive colours.

The chemistry considered in this volume therefore has an extra
dimension, namely economics. The basic chemical criteria of
mechanistic and thermodynamic feasibility remain paramount.
If a process is thermodynamically impossible then there can be no
point in investigating it, no matter how economically attractive
it may be. However, when there are several routes to a desired
compound economy of operation is what matters, not chemical
elegance. An additional factor, the sociological effect of a process,
is becoming increasingly important, and a process that is both
economically attractive and chemically elegant may have to be
abandoned because of noxious by-products or adverse effects on
the people employed in its manufacture.

In so far as it is one of the principal generators of a country's
wealth, industry must aim to achieve a value for its output greater
than the cost of the input. The input consists of diverse items such
as the cost of the raw materials, steam, electricity, cooling water
(these are termed 'variable costs' in that they vary directly with
the level of production) together with the cost of labour, super-
vision, maintenance, overheads, plant depreciation (money set
aside to replace the plant when it becomes obsolete) and a financial
return on the capital employed in constructing and running the
plant (these items are termed 'fixed costs' as they are independent
of whether the plant is producing at 50% or 90% of its rated
capacity). The output is clearly the monetary value of the product
sold: often referred to as the 'sales realization'. The actual cost of
producing a product can be built up as shown in Table 1.1.

This type of treatment enables an industrialist to begin to
compare different projects on a common basis, although the final
analysis is more complicated and, *inter alia*, requires the build-up
of a total cash flow picture, i.e. cash outflow and cash inflow, over
the life of a project. In the latter treatment allowance is usually
made for the fact that one pound sterling today has a greater
value than one pound sterling in one year's time (Discounted
Cash Flow).

The scale of production has a very important influence on the
cost of producing a product. It has been established that plant
equipment size and cost correlate reasonably well on a logarithmic
relationship known as the '0.6 power law'. Thus:

$$\text{Cost of new plant} = r^{0.6}C$$

where r is the ratio of plant capacities and C the cost of previous

Table 1.1. Manufacturing costs for ethylene (1973 costs) based on naphtha; plant capacity 450,000 tons per annum

Capital costs	10^6£
Total plant cost, including outside battery limits[a]	35.0
Working capital	2.0
Total capital investment	37.0

Manufacturing costs	£/ton of ethylene
1. Variable costs:	
feedstock (naphtha at £20/ton)	78.7
catalyst and chemicals	2.1
less by-product credits	53.0
Net raw materials cost	27.8
utilities (fuel oil, fuel gas, electricity, water)	13.5
Total variable costs	41.3
2. Fixed costs:	
labour and supervision	1.3
maintenance	3.1
tax and insurance	0.8
overheads	3.1
depreciation (on plant capital, excluding working)	7.8
Return on total capital (15%)	12.3
Total cost, including return on capital	69.7

Note: These figures are indicative orders-of-cost and will vary according to plant location.

[a] Battery limits: A geographic boundary defining the area of a specific project for the purpose of construction contracts. It usually embraces the manufacturing area of a proposed plant and includes all process equipment but excludes storage facilities, utilities, administrative buildings and site preparation, unless otherwise specified.

plant. This indicates that a doubling of the capacity of a proposed plant would lead to an increase in cost of about 50%.

This 'law' must be treated with some caution: it applies only on the relatively large scale and over a limited range from the base-case and where a particular plant item is itself increased in size (even then the 0.6 power-index can vary) but not when it is simply replicated; in the last case the cost would be double or treble as appropriate. The concept of a '0.6 power law' is basically derived from the fact that the cost of, say, a reaction vessel is a function of the cost of the material of construction, which is a function of area (i.e. for a spherical reactor, surface area $\propto$ diameter^2) whilst plant capacity is a function of the volume contained by this area of metal (i.e. for a spherical reactor, volume $\propto$ diameter^3). The actual effect of varying the scale of production of ethylene is shown in Figure 1.1.

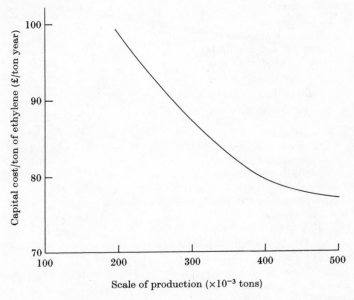

Figure 1.1. Naphtha-based ethylene production: effect of plant capacity on on-site capital costs (1973 basis).
(On-site costs take into account the plant items in the manufacturing area of the plant and exclude provisions of storage, utilities and administrative buildings. The costs of the latter items are usually termed off-site costs or off-battery limit costs.)

For many organic chemicals, there are usually several processes by which a Company could manufacture a particular product. The choice of the production process will be dictated by many factors such as feedstock availability, experience of a particular type of technology, integration into a group of processes, size of the plant, ability to sell any appropriate by-products, etc. The impact of these factors will vary from Company to Company and will play a part in the overall economics so that generally it cannot be said that any one process producing a product is universally superior to the alternative processes.

History of the Chemical Industry

One of the fastest growing industrial sectors in a developed economy is the chemical industry. This industry, which is best defined as a group of related sub-industries, has a major share in the generation of wealth and its prosperity is an essential requisite of continued national growth in developed countries.

The oldest part of the chemical industry had its origin in the nineteenth century and consisted largely of inorganic products based on natural raw materials such as sulphur and salt. Initially, the organic products were largely based on the fermentation of agricultural products. Up to 1860, fermentation processes were used in the production of food and beverage products, of which alcohol and vinegar are the best-known examples. Lactic acid became an established fermentation product before 1900, followed by glycerol, acetone, butan-1-ol, citric acid and bacterial and fungal enzymes, owing, at least in part, to World War I. During the period 1940–1960 there was a rapid expansion in the number of fermentation products that could be produced. These included the more complex molecules, such as antibiotics, vitamins, amino-acids and steroids. At the beginning of the twentieth century there were rapid developments in the production of organic chemicals based on coal; this period was dominated by Germany but other major coal-producing countries, such as the U.S.A., U.K. and France, established coal-tar industries during the period 1914–1920.

Partly as a result of research carried out during World War I, the 1920's saw the creation of the petrochemical industry in the U.S.A. based on the products of the oil industry. World War II brought about a rapid expansion in this industry, largely due to the particular need for strategic materials such as rubber and gasoline (petrol). Thus, it has been estimated that, in 1925, 100 tons of oil-based chemical products were manufactured in the U.S.A. and by 1930 the output had reached 45,000 tons; this level was further increased to 1×10^6 tons in 1940 and the rapid growth during the post-war period raised the output to 49×10^6 tons in 1967.

In the rest of the world, virtually no petrochemical industries existed until after World War II. The growth of the oil industry and particularly the construction of local refineries led to the availability of feedstocks on which petrochemicals could be based. Until 1955, the majority of countries restricted production to the lower olefins and their products. In these countries the absence of natural gas dictated the choice of liquid hydrocarbon fractions for petrochemical feedstocks and, in many cases, the availability of aromatic hydrocarbons from coal sources made their production from oil uneconomic at that time.

European refineries were designed to meet the market demands for the various major fractions, the pattern of which differed from that in the U.S.A. where gasoline was in great demand. Whereas in 1950 the U.S.A. had over 270 cars per 1000 population, the corresponding European figure was 23. Additionally, the greater horse power of American cars contributed to this difference in gasoline demand. This led European countries to produce more of the heavier fractions and less gasoline than in the U.S.A. (Table 1.2).

Table 1.2.   Main oil product yields (% wt.)

	U.S.A.		West Europe	
	1959	1972	1959	1972
Gasolines	41	43	19	17
Middle distillates	27	29	26	32
Fuel oil	11	6	38	37
Others	21	22	17	14

Table 1.3.   Free-world production of organic chemicals (1st stage derivative chemicals; million tons) (Spaght, *Chem. and Ind.*, 12th August, 1967)

Year	U.S.A.	W. Europe	Others	Total	Oil-based (%)
1950	2.95	0.12	0.03	3.1	40
1960	10.05	2.32	0.73	13.1	65
1965	17.0	6.9	3.1	27.0	75
1985 (estd.)	80.0	88.0	32.0	200.0	98
2000 (estd.)	210	240	150	600	>99

As a result, the American petrochemical industry was built up on two principal feedstocks, namely, natural gas condensates (i.e. ethane, propane) and catalytic cracker gases (i.e. ethylene and propylene[a]), originating from the large number of cracking plants. In Europe, owing to the absence of suitable natural gas supplies and to the relatively small number of catalytic crackers, the production of the lower olefins was based on heavier feedstocks,

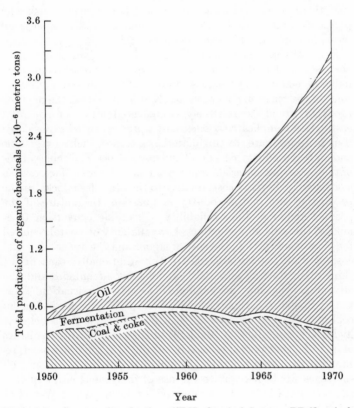

Figure 1.2.   Sources of production of U.K. chemicals [source: BP Chemicals International Limited].

naphtha being dominant. In recent years, the continued growth of refineries and the increased mobility of crude oil has led to rapid growth of petrochemicals production in other countries (Table 1.3).

The manner in which organic chemicals have become increasingly based on oil in the U.K., with fermentation and coal products

[a] Systematic name: propene.

occupying a relatively minor role, is shown graphically in Figure 1.2.

General Considerations

With these historical facts in mind it is pertinent to ask why petrochemicals have become so dominant at the expense of fermentation and coal-based products.

In the absence of economic considerations, the versatility of the chemist is such that virtually all the organic chemicals required by industry could be synthesized from carbon and water. Clearly, some of the reaction pathways involved in producing the required products would be relatively complicated and, consequently, costly. Less complicated and less costly processes can be considered if more appropriate starting materials are used, such as vegetable and animal products, coal or oil and natural gas. The choice of one of these starting materials will depend on a variety of factors such as its availability and cost, relative to the other feedstocks, at the production site in the country in question, the nature of the product required, the availability of suitable conversion technology, the scale of production, the availability of capital required for plant construction, the cost of labour and the availability and cost of water, steam and electricity. Additionally, each of the three main starting materials has inherent advantages and disadvantages that may predominate in a particular situation. Thus, the hydrogen to carbon ratio of oil (1.8:1) is superior to that of coal (0.8:1) when considering their desirability as feedstocks for hydrocarbon production; and large tankers enable large quantities of oil to be transported relatively cheaply. On the other hand, oil supplies can be vulnerable to political pressures; specific oil shortages would necessitate the use of indigenous coal supplies, as exemplified by the German activities during World War II or the current American efforts aimed at coal-gasification in the face of depleted reserves of indigenous natural gas.

The general trend is to avoid the use of agricultural raw materials wherever alternative technology permits, for two principal reasons. First, the hazards inherent in agricultural production lead to unpredictable variation in supplies. Secondly, as the standard of living of the growers increases, prices tend to rise, even in the case of by-products or waste materials which may nominally have zero value but whose collection and transportation costs may become too high. Additionally, agricultural raw materials generally give an inferior yield of synthetic product relative to an equivalent weight of oil. On the other hand, in a limited number of cases it can be the best or only source of particular, usually complicated, organic chemicals.

In the early days of the chemical industry, organic chemicals were produced by fermentation because of the availability of suitable agriculturally based feedstocks and technology, and the lack of economic alternatives. However, fermentation processes have limited applicability, being generally applied to aqueous systems with both reactants and products in low concentration. This leads to a relatively complicated purification of the product. The existence of a thriving coal-mining industry and the need for coke, for production of steel and other strategic materials required by the major powers involved in World War I, provided the driving force for the coal-based era. However, oil has a higher calorific value than coal (1.5:1), does not produce ash, and has a higher density and better burning characteristics. Consequently at a later stage, there was a significant conversion from coal to oil in many industries, including shipping. Research carried out in the U.S.A. during the period 1916–1918, and the growth of the oil industry, largely due to the requirements of an increasing number of cars, led to the emergence of the petrochemical industry. This movement gathered momentum as other countries followed the American pattern and the current availability of relatively cheap oil, due to low cost of world-wide transportation by large tankers and pipelines, accounts for the dominance of oil. Additionally, as in the case of agricultural raw materials, rising living standards have caused coal to increase in price relative to oil because coal mining is a more labour-intensive industry. The petrochemical industry has also benefited from the engineering and scientific developments of the oil industry and the economic savings that follow from using continuously operated, very large-scale chemical plants.

The current large tonnages of chemicals produced are such that, even when economic factors are ignored, production from vegetable products would lead to shortages of, for example, food materials. In fact, the production of synthetic fibres and rubber has obviated the need to use land which can be utilized to satisfy the world's growing food requirements.

At the present time there is appreciable publicity about impending energy shortages which is largely focussed on oil and natural gas supplies. As will be seen in Chapter 2, there is a finite supply but one which will not run out in the next few years. Clearly, local supply positions and political pressures, particularly in relation to the dominant oil-reserve position of the Middle East, will cause problems and inevitably lead to increasing costs of oil. This in turn will provide the incentive for countries to exploit other energy sources, i.e. fossil, nuclear and solar energy, and to attempt to seek ways of utilizing energy more efficiently. Against this uncertain backcloth future generations may experience some moves to a greater dependence on coal- and vegetable-based products.

However, it should be noted that, at the present time, only about 4% of total oil consumption is used in producing the organic chemicals listed in Table 1.3. Spaght has suggested that the corresponding 1985 and 2000 figures will be 5% and 12%, respectively.

CHAPTER 2

Sources of Raw Materials

(A. H. JUBB, ICI Corporate Laboratory, Runcorn)

In Part 1 (Chapters 16 and 17) we briefly discussed the occurrence of organic compounds in nature and gave a short account of the use of coal and petroleum as the raw materials from which the primary products of the chemical industry are derived. The present Chapter looks at the sources of raw materials in much greater detail, paying particular attention to their scale and relative importance.

Agricultural and Forestry-based Raw Materials

Supplies of minerals, coal and oil are limited, whereas agricultural and forestry products are continuously renewable. However, the world's food and timber requirements are such that these materials are not available in unlimited quantities for use as chemical raw materials. It is because of the increasing food demands of a growing world population that there has been active research on the production of synthetic protein concentrates from oil or natural gas. Owing to the tremendous growth in the production of chemicals based on oil, production by means of the processing of agricultural and forestry raw materials represents only a small fraction of the current total output of chemicals. However, such processing plays an important and, in some cases, an indispensable part.

Fats and Oils

Fats and oils are obtained from vegetable and animal sources and their main constituents are mixed triglycerides of the C_{16} and C_{18} acids (see Part 4, pp. 150–155). Some solid vegetable fats such as palm-kernel oil and coconut oil contain large quantities of C_6–C_{14} saturated fatty acids whilst others contain mainly saturated C_{16} and C_{18} and unsaturated C_{18} fatty acids. Liquid vegetable oils, such as cottonseed oil or soya oil contain mainly unsaturated C_{18} acids (oleic, linoleic). Animal fats contain mainly saturated C_{16} and C_{18} acids with small amounts of unsaturated C_{18} acids. Fish oils contain highly unsaturated acids. The world production of the various types is given in Table 2.1.

Table 2.1.   Oils and fats (oil or fat equivalent): 1970 world-production forecast (tons $\times 10^{-6}$) (Source: *World Agriculture*, 1970)

Edible vegetable oils	18.90
Palm oils	4.20
Industrial oils	1.48
Animal fats	12.51
Marine oils	1.26
Total	38.35

Hydrolysis of the various triglycerides is used to provide natural glycerol and fatty acids. Glycerol is also produced by synthetic methods in several countries (see p. 96). The world glycerol production data are listed in Table 2.2. Fatty acids are used in a variety of applications, including soaps, detergents (Chapter 13), coatings and disinfectants. World soap consumption has remained static over the period 1960–1971, whilst the consumption of synthetic detergents has increased by about 160% over the same period (Table 2.3).

Table 2.2.   World glycerol production (tons $\times 10^{-3}$) [Source: *Education in Chemistry*, **9**, No. 1 (1972)]

Year	U.S.A. Natural	U.S.A. Synthetic	Europe total	Japan total	World total	World synthetic (%)
1961	63	64	95	16	255	30
1963	64	73	105	16	285	33
1965	66	91	109	18	310	33
1967	70	96	115	18	330	39
1968	73	91	125	18	335	42
1969	160		122		340	45
1970	150					

Table 2.3. Total world production of soaps, washing and cleaning agents (tons $\times 10^{-6}$) (Source: *European Chemical News*, 16th March, 1973)

	1960	1968	1970	1971[a]
Soaps	6.88	6.49	6.09	6.00
Synthetic detergents	3.43	7.37	8.35	8.90
Scouring agents	0.46	0.58	0.65 ⎤	
Others	0.14	1.01	1.28 ⎦	2.20
Total	10.91	15.45	16.37	17.10
Synthetic detergents (%)	32	48	51	52

[a] Preliminary data

Certain of the natural fatty acids are used as raw materials for the production of intermediates which are in turn used for the manufacture of thermoplastics. Thus, ozonolysis of oleic acid yields pelargonic and azelaic acids:

$$CH_3(CH_2)_7CH{=}CH(CH_2)_7COOH \xrightarrow{\text{Oxidn.}}$$
$$\text{Oleic acid}$$

$$CH_3(CH_2)_7COOH + HOOC(CH_2)_7COOH$$
$$\text{Pelargonic acid} \qquad \text{Azelaic acid}$$

Pelargonic acid is used in the production of ester lubricants, and azelaic acid is used in ester lubricants, low-temperature plasticizers and in the production of nylon 6.9. The world production of azelaic acid is believed to be about 1×10^4 tons p.a.

Nylon 11 ('Rilsan') was introduced in 1950 as a speciality nylon and is produced in France and Brazil. It is obtained from castor oil, which is initially treated with methanol to form a mixture of fatty acid methyl esters. These are separated by distillation and the methyl ricinoleate is then pyrolyzed to give heptanal and methyl undec-10-enoate:

$$CH_3(CH_2)_5CH(OH)CH_2CH{=}CH(CH_2)_7COOCH_3 \longrightarrow$$
$$\text{Methyl ricinoleate}$$

$$CH_3(CH_2)_5CHO + CH_2{=}CH(CH_2)_8COOCH_3$$
$$\text{Heptanal} \qquad \text{Methyl undec-10-enoate}$$

The latter ester is hydrolyzed and the acid treated with HBr in the presence of a peroxide catalyst, to give about a 96% yield of the 11-bromo-acid.

$$ROOR \longrightarrow 2RO\cdot \qquad RO\cdot + HBr \longrightarrow ROH + Br\cdot$$

$$Br\cdot + CH_2{=}CH(CH_2)_8COOH \longrightarrow BrCH_2\dot{C}H(CH_2)_8COOH$$

$$BrCH_2\dot{C}H(CH_2)_8COOH + HBr \longrightarrow BrCH_2CH_2(CH_2)_8COOH + Br\cdot$$

The amino-acid is then formed by reaction with ammonia, the bromine atom being readily replaced.

Fusion of sodium ricinoleate with 70% sodium hydroxide yields octan-2-ol and sebacic acid. The latter is used to produce nylon 6.10, U.S.A. consumption of which was estimated to be about 10,000 tons in 1972.

$$CH_3(CH_2)_5CH(OH)CH_2CH{=}CH(CH_2)_7COOH \xrightarrow{\text{NaOH}}$$
$$\text{Ricinoleic acid}$$

$$CH_3(CH_2)_5\dot{C}H(OH)CH_3 + NaOOC(CH_2)_8COONa$$
$$\text{Octan-2-ol} \qquad\qquad \text{Sodium sebacate}$$

Wood-based Raw Materials

Forest trees are the largest, continuously renewed raw material in the world. The main proportion of wood production is consumed in a variety of markets, construction applications being dominant. Thus, the 1970 world production of industrial round-wood (solid volume of roundwood lumber without bark) was 6.8×10^9 tons (1.2×10^9 cubic metres); additionally, it is estimated that about 1.0×10^9 cubic metres were used as fuel world-wide during 1970, mainly in the underdeveloped countries.

The wide variety of products obtainable from wood is illustrated in Figure 2.1; they are discussed individually in the paragraphs below.

The production of woodpulp (ca. 0.1×10^9 tons during 1970) is the largest 'chemical use' of wood. The production of mechanical woodpulp is not a truly chemical process, being essentially a mechanical process to separate the cellulosic fibres which are then used to produce the cheaper types of paper. Pure pulp is obtained by removal of such materials as lignin, which bonds together the cellulose fibres in the wood, thus providing the cellulose in pure form. The cellulose so obtained is also the basic raw material used in the production of rayon fibres (regenerated cellulose) (see Chapter 9). There are several methods used for the production of pure woodpulp, of which the sulphate and the sulphite processes are most widely used. The waste liquors from the latter process contain hexoses and can be fermented to yield ethanol:

$$C_6H_{12}O_6 \longrightarrow 2C_2H_5OH + 2CO_2$$

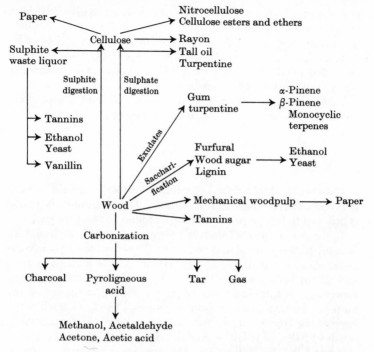

Figure 2.1. Chemicals obtained from wood.

Depending on the type of digestion, a 100 ton/day pulp works could produce $5-12.5 \times 10^3$ litres of alcohol each day by this method. Vanillin (Part 4, p. 144) is obtained by alkaline hydrolysis of lignin-containing materials such as are found in waste sulphite liquor (see Chapter 6).

$$\text{Vanillin}$$

Vanillin

Wood saccharification, whereby waste wood is hydrolyzed with concentrated hydrochloric acid or dilute sulphuric acid, can be used to produce lignin residues together with liquors containing sugars (ca. 4% solutions). The latter can be utilized as a fermentation raw material for the production of ethanol and yeast. The

Scholler process (using dilute H_2SO_4) is reputed to give a 40–50% sugar yield from waste wood and 100 lb. of the latter, on a dry-basis, can produce 2.4–2.5 gallons of 100% ethanol and 4–4.5 lb. of yeast (protein content 50%). Also, wood sugar molasses (concentrated wood sugar solution) can be used in a similar manner to blackstrap molasses as a livestock feed.

Furfural can be obtained as a by-product from the pentoses formed in a variety of wood-processing processes:

$$C_5H_{10}O_5 \longrightarrow \quad \underset{\text{Furfural}}{\left\langle\!\!\!\!\bigcirc_{O}\right\rangle\!\!-\!CHO} \quad + 3H_2O$$

It can also be obtained by hydrolysis of certain vegetable wastes (oat hulls, maize ears, sugarcane stalks). The Dominican Republic is believed to have a production rate around 25,000 tons p.a.

There are various substances present in wood that are not an integral part of the cellular structure. These include such diverse materials as resins, terpenes and tannins. The bulk of turpentine and all the tall oil are obtained as by-products in sulphate pulp processing, turpentine being recovered during steaming of the pulp (ca. 10 kg/ton pulp). Gum is obtained as exudate when the bark of the tree is cut, as with natural rubber; the turpentine is removed from the resulting rosin by distillation. The turpentine is separated, by distillation, into three major components, namely α-pinene, β-pinene and monocyclic terpenes (Part 4, p. 217). The tall oil is composed of fatty acids similar to oleic acid, and rosin acids related to abietic acid. These various materials are frequently referred to as 'Naval Stores', as the original rather crude products of the industry were used for caulking and waterproofing wooden vessels.

In addition to yielding tannins (Part 4, p. 116) used in leather tanning, the wood-tissues, bark and leaves contain essential oils such as cedarwood and sassafras oils.

Essential oils (see Chapter 16), used in the perfumery industry, are volatile materials obtained from vegetable sources, usually isolated by solvent-extraction or steam-distillation. It is stated that about 150 of the 3000 known essential oils have been commercialized. Animal secretions (civet, musk, ambergris) and many plant exudates are used as fixatives (i.e. used to control the rate of evaporation to give a constant level of fragrance). Many of the oils are now made synthetically, and blends of natural and synthetic products are used. In 1969 the U.S. total production of aroma chemicals was 24×10^3 tons, valued at about £22×10^6.

One of the main forestry products is natural rubber (see Chapter 8) obtained as an exudate from the species *Hevea brasiliensis* which originated from the Amazon region of South America. The trees grow rapidly to produce significant amounts of rubber within 6–7 years; high-yielding trees will produce 1200–1500 lb. per acre. Although synthetic rubber has been satisfying an increasing proportion of the world's rubber requirements, some 2.88×10^6 tons of natural rubber are produced annually, a figure that is forecast to increase to 3.7×10^6 tons by 1980.

Table 2.4. World rubber consumption (tons $\times 10^{-3}$)

Year	All rubber	Natural	Natural (%)
1950	2302	1722	43
1955	3099	1887	38
1960	4327	2065	32
1965	6115	2375	28
1970	8955	2900	24.5
1975 (estd.)	11,630	3300	22
1980 (estd.)	14,400	3700	20

Wood carbonization is used to produce charcoal in 20–30% yield. During the carbonization process, acetic acid, methanol and acetone can be obtained as by-products from the pyroligneous liquor (250 gallons from 4000 lb. of hardwood). At the beginning of the twentieth century this was an important source of these materials, but more economic processes emerged to replace this source. Thus, it was reported that there were about 100 such recovery plants operating in the U.S.A. in 1920, the last of which ceased operation in 1969.

In 1967 the estimated output from the gum and wood chemicals sector of the U.S. chemical industry was valued at 211×10^6, which was about 0.5% of the value of the total output of chemicals and allied products (42.2×10^9). In 1969 the former was estimated at 275×10^6, still less than 1% of the latter. The U.S.A. is a major world supplier of gum and wood chemicals, along with Canada, Russia and the Scandinavian countries.

Sugar

In 1970 world sugar production was about 70×10^6 tons, obtained from cane and beet sugar. The waste mother-liquors remaining after removal of the crystallized sugar from evaporated cane sugar solution (containing 55% of sugar, 35–40% of which is sucrose, together with 15–20% of invert sugar, i.e. glucose and fructose) is known as blackstrap molasses. High-test molasses is raw sugar-cane juice in which the sugar has been largely hydrolyzed by acid to give invert sugar (70–80% of sugar, containing 22–27% of sucrose and 50–55% of invert sugar). Sugar-beet molasses contains 48–52% of sucrose and virtually no invert sugar. These sugar-containing solutions can be fermented. Ethanol is recovered in 90% yield, based on sugar content, and is purified by distillation.

Starch

Starch is the principal energy reserve in the plant kingdom (Part 4, Chapter 2). Various types of starch products (grain, potatoes, etc.) can be used for the production of ethanol by fermentation. In the case of maize, cooking is first carried out followed by saccharification to convert the starch into fermentable sugars. This can be effected by dilute mineral acid or by the enzyme diastase.

$$(C_6H_{10}O_5)_n + nH_2O \xrightarrow{\text{Diastase}} nC_6H_{12}O_6$$
$$\text{Starch}$$

$$C_6H_{12}O_6 \xrightarrow{\text{Zymase}} 2C_2H_5OH + 2CO_2$$

The production of ethanol from these various agricultural and forestry products has declined as a proportion of the total production. Thus, in the U.S.A. 90% was based on fermentation during

Table 2.5. U.S.A. Production of ethanol (1970)

Source	% of total alcohol spirits (>90% proof)		
	1961	1967	1970
Synthetic	79.0	79.2	83.0
Fermentation:			
grain and grain products	14.0	13.5	12.0
molasses	3.2	2.2	0.4
fruit juice	2.5	3.9	3.6
Sulphite liquors	1.0	1.1	1.0
Cellulose pulp	0.3	0.1	—
Total production (gallons $\times 10^{-6}$)	321.4	351.8	371.1

1935 but only 9% by 1963 (see Table 2.5). It is estimated that in 1970 the bulk of the ethanol production in Asia (excluding China and Japan) and Latin America, rated at about 725,000 tons, was fermentation-based.

Citric acid is obtained from beet or blackstrap molasses by fermentation with a micro-organism *Aspergillus niger*:

$$C_{12}H_{22}O_{11} + H_2O \xrightarrow[\text{A. niger}]{\text{Air}} \begin{array}{c} CH_2COOH \\ | \\ HOCCOOH \\ | \\ CH_2COOH \end{array}$$

Citric acid

Lactic acid is obtained by controlled fermentation of a variety of carbohydrates such as sucrose, glucose or lactose from maize or potato starch hydrolyzates.

$$\underset{\text{Sucrose}}{C_{12}H_{22}O_{11}} + H_2O \longrightarrow \underset{\text{Glucose}}{C_6H_{12}O_6} + \underset{\text{Fructose}}{C_6H_{12}O_6}$$

$$C_6H_{12}O_6 \xrightarrow{\text{Bacteria}} 2CH_3CH(OH)COOH$$

Lactic acid

Monosodium glutamate [the (−)-form is used as a flavour enhancer; see Chapter 16] can be obtained by fermentation of by-product liquors from sugar-beet production, maize gluten or carbohydrate solutions. In the last case the following reactions occur:

$$C_6H_{12}O_6 + 2O_2 \longrightarrow \begin{array}{c} COCOOH \\ | \\ CH_2CH_2COOH \end{array} + CO_2 + 3H_2O$$

α-Oxoglutaric acid

$$\downarrow NH_3$$

$$\begin{array}{c} H_2NCHCOOH \\ | \\ CH_2CH_2COONa \end{array} \xleftarrow{\text{NaOH}} \begin{array}{c} H_2NCHCOOH \\ | \\ CH_2CH_2COOH \end{array} + H_2O$$

Monosodium
glutamate

Coal as a Source of Chemicals

Coal was formed by anaerobic decomposition of vegetable matter under the influence of high pressure and temperature. All the transition stages can be found, from the lignin-type coals to lignite (brown coal), bituminous coal and anthracite (hard coal); the last of these has the highest carbon content.

In addition to its direct use as a fuel, coal can be processed in three basic ways to produce, *inter alia*, raw materials for the chemical industry, namely by carbonization, gasification or hydrogenation.

There is no rigid distinction between carbonization and gasification except in intent. The former aims at maximum production of coke and liquid and gaseous products, whilst the latter aims at substantial conversion into appropriate gaseous compounds. Carbonization processes have, however, been used to produce mainly coke together with gaseous and liquid by-products, or mainly town-gas with the co-production of coke and liquid products.

Carbonization

In these processes, coal is heated in the absence of air to produce coke as a residue together with volatile matter. The nature of the coal and the heating conditions dictate the nature of the coke and the volatile matter; either low-temperature (<600°C) or high-temperature (>800°C) conditions can be used, but in both cases the principal products are as depicted in Figure 2.2.

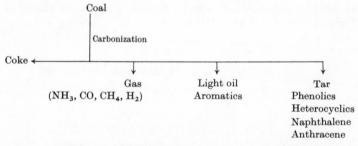

Figure 2.2.   Products produced by coal carbonization.

The high-temperature processes produce between 9 and 14 gallons of tar per ton of coal, depending on the type of carbonizing plant and the operating conditions. The average yield is about 10 gallons, which is approximately 5% by weight of the coal carbonized. Low-temperature processes yield 15–22 gallons of tar per ton of coal carbonized. Figure 2.3 shows the main fractions obtained by distillation of the tar. These fractions can be further processed to yield the individual components in the required purity.

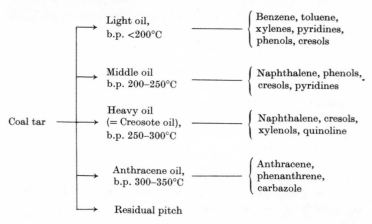

Figure 2.3. Products of distillation of coal tar.

Metallurgical coke is produced by high-temperature carbonization and the aim is to maximize the yield of suitable coke whilst the gas and tar are regarded as by-products. The tar from this type of process contains a high proportion of aromatic compounds [about 10% of naphthalene and also phenol (1%) and pyridine bases (0.1%)]. Recent developments in steel-making have led to more effective use of coke (1250 lb. per ton of pig-iron in 1969 compared with 1550 lb. in 1960 and 3000 lb. in 1917); and, owing to the increased steel production, the overall output of tar from this source has remained reasonably constant.

The primary aim of the gas industry is to manufacture town gas, but inevitably coke is obtained as a by-product when coal is the raw material. Coal used to be carbonized in metal retorts at temperatures around 800°C to produce low gas yields. Later processes using higher temperatures led to higher gas yields with the co-production of coke unsuitable for many domestic appliances. It was found that more suitable smokeless fuels could be obtained by using low-temperature carbonization processes (500–600°C). Several such processes have been developed, each of which can produce its specific tar composition. Generally these tars are lower in aromatics than is gas-works tar but they contain some 10–12% of tar acids. The latter contain more di- and tri-hydric phenols than do the tars from the high-temperature processes and have remained the sole source of catechol.

Gas-works production in the U.K. has decreased drastically in recent years with, initially, the advent of naphtha steam-reforming and more recently the discovery of natural gas in the North Sea (see Table 2.6).

Table 2.6. U.K. production of coal tar (tons $\times 10^{-6}$)

Year	Steel-works	Gas-works	Low-temperature carbonization
1955	1.100	1.830	0.040
1970	0.935	0.342	0.164

The declining importance of coal tar as a source of aromatic compounds can be seen in Table 2.7.

Table 2.7. U.S. production of coal tar products

	1950	1953	1967	1969	1971[a]
Coal tar (10^6 gallons)	749	901	780	769	n.a.
Benzene (10^6 gallons)	192	210	91	102	85
% of total benzene	91	77	9	8	8
Toluene (10^6 gallons)	84	40	19.4	19.6	15
% of total toluene	50	26	3	1	2
Xylenes (10^6 gallons)	72	10	5.5	5.2	n.a.
% of total xylenes	14	9	1	8	n.a.
Naphthalene (10^6 lb.)	289	276	521	496	350
% of total naphthalene	100	100	58	58	50

[a] n.a. = not available.

Coke can be readily converted into calcium carbide in a high-temperature electric arc furnace:

$$CaO + 3C \longrightarrow CaC_2 + CO$$

On treatment with water the carbide readily yields acetylene:

$$CaC_2 + H_2O \longrightarrow CaO + C_2H_2$$

Because of the versatility of acetylene chemistry, this provides coal-based synthetic routes to many organic chemicals. However, owing mainly to the high energy costs, the carbide route has been largely displaced by those based on oil during the time when acetylene uses have been declining (see Chapter 3).

Gasification

In a stream of air, air/steam, oxygen-enriched air, oxygen or oxygen/steam, coke or coal can be largely converted into gaseous products. The use of air/steam leads to producer gas, according to the following equations which operate at different parts of the fuel bed.

$$C + \tfrac{1}{2}O_2 \rightleftharpoons CO \qquad \Delta H = -121 \text{ kJ mol}^{-1} \qquad (1)$$
Partial combustion

$$C + O_2 \rightleftharpoons CO_2 \qquad \Delta H = -405 \text{ kJ mol}^{-1} \qquad (2)$$
Complete combustion

$$C + H_2O \rightleftharpoons CO + H_2 \qquad \Delta H = +120 \text{ kJ mol}^{-1} \qquad (3)$$

$$C + 2H_2 \rightleftharpoons CH_4 \qquad \Delta H = -89 \text{ kJ mol}^{-1} \qquad (4)$$

$$C + CO_2 \rightleftharpoons 2CO \qquad \Delta H = +164 \text{ kJ mol}^{-1} \qquad (5)$$

$$CO + H_2O \rightleftharpoons CO_2 + H_2 \qquad \Delta H = -43 \text{ kJ mol}^{-1} \qquad (6)$$

It can be seen that increasing pressure will affect reactions (3), (4) and (5), and, in fact, favours the formation of methane. This phenomenon is exploited in the Lurgi process where pressure gasification of coal or coke is carried out in the presence of oxygen and water.

Water gas is obtained by reaction of coal or coke with water, as in equations (3) and (7). As both reactions are endothermic, the solid bed is heated periodically whilst steam-blowing.

$$C + 2H_2O \rightleftharpoons CO_2 + 2H_2 \qquad \Delta H = +78 \text{ kJ mol}^{-1} \qquad (7)$$

High temperatures favour CO formation: the equilibrium composition of the dry gas is CO_2 33.1%, CO 0.2% at 400°C and CO_2 0.3%, CO 49.5% at 1000°C. These processes formed the basis of synthesis gas production before the 1950's but were in many cases later replaced by those based on natural gas and oil.

Hydrogenation

Although Berthelot had converted coal into oil by using nascent hydrogen in 1869, it was Bergius in 1913 who used molecular hydrogen to convert coal catalytically into oil. BASF carried out development work which led to the widespread use of the process in Germany before and during the 1939–1945 War. In fact, before 1939 there were thirteen hydrogenation plants operating in Germany for the hydrogenation of coal or tar fractions. ICI carried out coal hydrogenation at Billingham before 1939 but this plant was then converted to creosote-hydrogenation for the production of high-grade gasoline during World War II.

The pulverized coal is slurried with oil and hydrogenated in the presence of an iron catalyst at 480–500°C and 25×10^6 Nm^{-2}. Whereas coal carbonization yields only 8–10% of tar, based on the coal weight, hydrogenation yields about 75% of crude oil. In recent years these plants have become obsolete.

However, with the prospects of diminishing natural gas resources in the U.S.A. there has been significant activity in the development of oil-gasification and coal-gasification processes based in some cases on these traditional techniques. The latter include the Lurgi pressure coal-gasification process, which is already used to produce gas of low thermal capacity. Four other important processes have reached the pilot-plant stage: the 'Bi-gas' process of Bituminous Coal Research, the 'CO_2 Acceptor' process of Consolidation Coal, the 'Hygas' process of the Institute of Gas Technology and the 'Synthane' process developed by U.S. Bureau of Mines.

There has also been significant activity in the development of coal-liquefication processes and four stand out in terms of degree of development. (1) The Char Oil Energy Development (COED) process (FMC Corporation); this consists of fluid-bed coal-pyrolysis to yield a synthetic crude oil, a char product and a gas stream. (2) The 'H Coal' process (Hydrocarbon Research Inc.); the process is based on hydrogenation of coal–oil slurries with a Co/Ni/Mo catalyst in ebullating-bed reactors. (3) The 'Low Ash Coal Development' (Pittsburgh and Midway Coal Co.; a subsidiary of Gulf Oil Co.); the process involves high-pressure coal extraction. (4) The 'Consol Synthetic Fuels' process (CSF) (Consolidated Coal Co. and Office of Coal Research); coal is solvent-extracted and the extract hydrogenated; the resulting crude oil

Table 2.8. U.K. coal production and utilization (tons $\times 10^{-6}$)

	1956	1960	1971
Coal produced	222.0	193.6	144.7
Coal consumed by gas supply industry	27.8	22.3	1.8
Coke ovens	29.3	28.5	23.2

can then be processed to gasoline by conventional cracking and reforming techniques; in 1967, Esso announced their intention to research in this field, but there are no details of their activities.

In 1971 the world coal production was estimated as 21×10^8 tons (U.K. 1.4×10^8 tons, see Table 2.8). This current annual production should be compared with the estimate of world coal reserves of 7.5×10^{12} tons of mineable coal (50% of that actually present; 20% is attributed to the U.S.A.).

Oil

Crude Oils and Their Characteristics

Crude oils, which practically always have natural gas associated with them, vary widely in composition, not only between one oilfield and another, but in some cases between adjoining wells. Essentially, they consist of paraffinic (alkanes), naphthenic (cycloalkanes) and aromatic (mainly benzene derivatives) hydrocarbons.

Originally crude oils were classified as paraffinic, naphthenic (based on the main constituents present in the lighter fractions), or asphaltic (based on the main constituents being high-molecular-weight hydrocarbons, with low hydrogen content, in the heavier fractions). An improved classification consisted of the three types, paraffinic, aromatic and asphaltic, naphthenes being present in varying amounts. From a chemical viewpoint it is probably preferable to classify oils broadly as paraffinic, naphthenic or aromatic depending on the nature of the predominating hydrocarbons. It should be realized, however, that many of the molecules contain mixed structural features, i.e. rings have paraffinic side-

Table 2.9. Fractions obtained from the distillation of crude oil

Fraction	Boiling range (°C)	Range of no. of C atoms in the constituent hydrocarbons
Gas	<20	C_4
Gasoline (naphtha)	20–200	C_4–C_{12}
Kerosene ⎫ (middle distillates)	ca. 175–270	C_9–C_{16}
Gas oil ⎭	ca. 200–400	C_{15}–C_{25}
Lubricating oil	ca. 300 (in vacuo)	C_{20}–C_{30}
Fuel oil	—	
Residue	—	

chains, and saturated and unsaturated rings occur in the same molecule.

Small amounts of organic sulphur, nitrogen and oxygen compounds are usually present in crude oils.

Distillation of the crude oil effects primary separation to produce liquid products boiling over a wide temperature range. The principal fractions are roughly categorized as in Table 2.9. In most cases, the gasoline composition corresponds to the composition of the crude oil, i.e. a highly naphthenic crude gives a highly naphthenic gasoline. The aromatics and naphthenes content of a particular straight-run gasoline fraction (isolated by distillation and not by other thermal or chemical processing) varies as the

Table 2.10. Composition of a Nigerian light naphtha (b.p. 65–164°C). Initial composition in the C_4–C_9 range: paraffins 31.4%, naphthenes 48.4%, aromatics 6.4% (Source: Phillips-Imperial Petroleum Ltd.)

Component	%(w/w)	B.p. (°C)
Isobutane	0.1	−11.7
n-Butane	0.4	−0.5
Isopentane	1.6	27.8
n-Pentane	1.8	36.1
2,2-Dimethylbutane	0.2	49.7
Cyclopentane	0.7	49.3
2,3-Dimethylbutane	0.5	58.0
2-Methylpentane	2.0	60.3
3-Methylpentane	1.5	63.3
n-Hexane	2.5	68.7
Methylcyclopentane	4.3	71.8
2,2-Dimethylpentane	0.2	79.2
2,4-Dimethylpentane	0.3	80.5
Cyclohexane	3.8	80.7
3,3-Dimethylpentane	0.1	86.1
1,1-Dimethylcyclopentane	0.7	87.8
2,3-Dimethylpentane	}1.8	89.8
2-Methylhexane		90.0
cis-1,3-Dimethylcyclopentane	1.7	91.7
3-Methylhexane	1.4	91.8
trans-1,3-Dimethylcyclopentane	1.5	90.7
trans-1,2-Dimethylcyclopentane	2.4	91.9
3-Ethylpentane	0.2	93.5
n-Heptane	2.2	98.4
cis-1,2-Dimethylcyclopentane	0.3	99.5
Methylcyclohexane	8.9	100.9

Table 2.10. *Continued*

Component	%(w/w)	B.p. (°C)
1,1,3-Trimethylcyclopentane	0.9	104.9
2,2-Dimethylhexane	0.1	106.8
Ethylcyclopentane	0.6	103.5
2,5-Dimethylhexane	0.2	109.1
2,4-Dimethylhexane	0.3	109.4
1-*trans*-2-*cis*-4-Trimethylcyclopentane	0.9	109.3
1-*trans*-2-*cis*-3-Trimethylcyclopentane	} 1.5	110.2
3,3-Dimethylhexane		112.0
n-Octane	1.8	125.7
n-Nonane	1.5	150.8
Other C_8 paraffins	4.4	
Other C_9 paraffins	5.8	
Other C_{10} paraffins	2.1	
Other C_8 naphthenes	16.7	
Other C_9 naphthenes	4.0	
Benzene	0.3	80.1
Toluene	1.0	110.6
Ethylbenzene	0.5	136.2
m- and *p*-Xylene	1.9	139.1, 138.3
Isopropylbenzene	0.2	152.4
o-Xylene	0.7	144.4
n-Propylbenzene	0.2	159.2
3- and 4-Ethyltoluene	0.5	161.3, 162.0
Mesitylene	0.4	164.7
2-Ethyltoluene	0.2	165.1
Pseudocumene	0.6	169.3
Hemimellitene	<0.1	176.1
C_{10} aromatics (unknown structure)	0.2	

boiling range increases. This type of fraction is more commonly referred to as naphtha, particularly within the petrochemical industry, and the distillation end-point of this fraction can vary from about 160°C to 190°C (Table 2.10).

It should be noted that, generally, straight-run gasoline fractions are not directly suitable for use in internal combustion engines owing to their low octane ratings (see Chapter 14).

As already indicated, the relative amounts of the principal fractions varies according to the source of the crude oil, as do the relative amounts of the hydrocarbon types present within the naphtha fraction (Table 2.11).

Table 2.11. Yields (wt.-%) of primary products from different crude oils (P = paraffins, A = aromatics, N = naphthenes)

Source	Naphtha	Middle distillates	Fuel oil
Ekofisk (North Sea)	19 P 54, N 31, A 15	34	47
Nigerian Light	25 P 40, N 52, A 8	37	38
Monagas (Venezuela)	1 P 8, N 89, A 3	19	80
Hassi Messaoud (Algeria)	30 P 65, N 27, A 8	40	30

Oil Refinery Processes

If the refining of crude oil were confined simply to distillation or other physical separation processes, which overall retain the chemical composition of the original crude, the production of natural gasoline would be insufficient to provide the world's gasoline requirements, and the octane rating would be inadequate. Several chemical processes are used which modify the nature of the constituents in the crude oil. The main processes are cracking, reforming, alkylation, polymerization and isomerization.

(a) *Cracking*. This process involves rupture of C—C and C—H bonds in the hydrocarbon chains, producing intermediate radicals which then produce smaller paraffin and olefin molecules. The three principal processes 'thermal', 'catalytic' and 'hydrocracking'

$$CH_3(CH_2)_xCH_2CH_2(CH_2)_yCH_3 \longrightarrow$$
$$CH_3(CH_2)_{x-1}CH\!\!=\!\!CH_2 + CH_3(CH_2)_yCH_3$$

are described in Chapter 3.

(b) *Reforming* (see Chapter 3). Whereas the cracking processes convert heavy fractions into lighter ones by causing molecular rupture, reforming processes cause molecular rearrangement. They are used to improve the quality of straight-run and cracked gasoline fractions rather than to increase gasoline yield. In catalytic reforming, the main reactions taking place are: (i) Dehydrogenation of naphthenes to aromatics with the co-production of hydrogen. (ii) Ring isomerization, i.e. conversion of alkylcyclopentanes into cyclohexanes, followed by aromatization. (iii) Dehydrocyclization of paraffins to aromatics. (iv) Isomerization of n-paraffins to isoparaffins.

Additionally, hydrocracking reactions, such as dealkylation of ring side-chains and olefin hydrogenation, can occur.

(c) *Alkylation.* In the oil industry the term alkylation refers to the reaction between an olefin (e.g. ethylene or isobutene) and a paraffin (e.g. isobutane), producing higher branched-chain molecules of high octane rating. The products are often used as blending components for premium gasoline. In the first alkylation plants sulphuric acid was used as catalyst and the exothermic reactions were carried out in the liquid phase at 5–20°C. Anhydrous hydrofluoric acid (operating at 25–45°C) and aluminium chloride can also be used as catalysts. The paraffin is used in excess in order to minimize reaction between two olefin molecules (polymerization).

$$CH_2{=}CH_2 + (CH_3)_2CHCH_3 \xrightarrow{\quad AlCl_3 \quad} (CH_3)_2CHCH(CH_3)_2$$

Ethylene   Isobutane   2,3-Dimethylbutane

$$(CH_3)_2C{=}CH_2 + (CH_3)_2CHCH_3 \xrightarrow{\quad H_2SO_4 \quad} (CH_3)_2CHCH_2C(CH_3)_3$$

Isobutene   Isobutane   2,2,4-Trimethylpentane

(d) *Polymerization.* In the oil industry, the term polymerization refers to the catalytic self-reaction of a gaseous olefin to yield liquid oligomers which are themselves olefins. The original units relied upon thermal reactions (500°C) of an olefin-rich feedstock, but improved yields and better reaction control were subsequently achieved by using catalysts such as kieselguhr impregnated with phosphoric acid. With the latter, olefin conversion of 90% can be achieved at temperatures of 150–220°C and $3.5–5.0 \times 10^6$ Nm^{-2} pressure. Claims have also been made for the use of other catalysts such as sulphuric acid, aluminium chloride and aluminium silicate.

$$(CH_3)_2C{=}CH_2 \xrightarrow{\quad H^+ \quad} (CH_3)_3CCH{=}C(CH_3)_2 \xrightarrow{\quad Isobutene \quad} \text{'Trimer'}$$

Isobutene   2,4,4-Trimethyl-
pent-2-ene

This type of process has become less important in recent years as the products are of poorer quality than those produced in the alternative processes, and it is also believed that a high olefin content in a fuel contributes significantly to air pollution.

(e) *Isomerization.* In the oil industry the term isomerization refers to the rearrangement of *n*-paraffin molecules to their branched isomers and, as already pointed out, takes place to a minor extent in some of the processes mentioned previously, e.g. catalytic reforming. The process is applied to *n*-butane, *n*-pentane and *n*-hexane and use is made of aluminium chloride or platinum catalysts; the latter are less corrosive than the former. 'Pentafining' is a process specifically applied to isomerization of mixtures of pentane, hexane and heptane to their branched isomers; the reaction is carried out at 300–400°C over a supported platinum catalyst in the presence of hydrogen.

The basic mechanism of the reaction is believed to be as follows but it should be noted that the presence of traces of HCl and alkenes are essential to start the reaction:

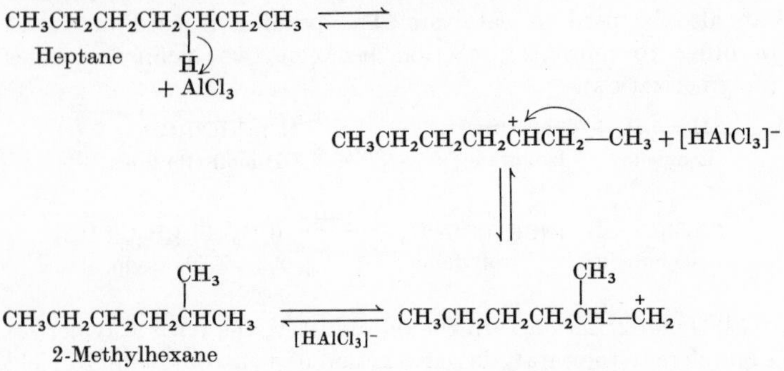

$$CH_3CH_2CH_2CH_2CHCH_2CH_3 \rightleftharpoons$$
Heptane $\quad H$
$\quad + AlCl_3$

$$CH_3CH_2CH_2CH_2\overset{+}{C}HCH_2{-}CH_3 + [HAlCl_3]^-$$

$$\Updownarrow$$

$$\underset{\text{2-Methylhexane}}{CH_3CH_2CH_2CH_2\overset{\displaystyle CH_3}{\overset{|}{C}}HCH_3} \underset{[HAlCl_3]^-}{\rightleftharpoons} CH_3CH_2CH_2CH_2\overset{\displaystyle CH_3}{\overset{|}{C}}H{-}\overset{+}{C}H_2$$

Table 2.12. U.S. refinery processes (as at January, 1969)

Process	Capacity (tons $\times 10^{-6}$ p.a.)
Catalytic cracking	290
Catalytic reforming	127
Hydrocracking	25
Hydrotreating	190
Alkylation: using H_2SO_4	23
using HF	9
Polymerization, using H_3PO_4 } Isomerization, using $AlCl_3$ }	12

The relative importance of these various processes can be seen from Table 2.12.

In addition to the various processes already outlined, there are a variety of other processes which are essential to the refining industry, although some of them have an accessory relationship to the primary aim of converting crude oils into their saleable products. These include hydrogen treatment (to remove impurities such as sulphur-, oxygen- and nitrogen-containing compounds), solvent refining and 'sweetening'. Further flexibility in refinery practice is provided by the use of molecular-sieve technology, whereby normal paraffins can be separated from the branched and cyclic paraffins present in the various streams. One of the main incentives for the removal of normal paraffins from fractions such as gas oil, arises from the requirements of the petrochemical industry for high purity C_{12}–C_{18} normal paraffins as detergent and plasticizer feedstocks.

World Oil Production, Consumption and Reserves

The world production of crude oil was 2.6×10^9 tons in 1972 and was virtually in balance with world consumption estimated at 2.59×10^9 tons. Consumption has been growing at an average rate of 7.8% per annum over the period 1966–1971 and it has been estimated that present annual consumption will treble by the year 2000.

In the light of the consumption figures, it is interesting to review the estimated world reserves of crude oil. Estimates of proved reserves are published regularly and, although annual consumption has been increasing regularly, the net estimates of total proved reserves have risen steadily since records were kept and currently stand at 90.9×10^9 tons, 53% of which are in the Middle East.

The proved reserves probably represent only a small proportion of the world's actual oil resources since they do not include additional 'possible' reserves in existing oilfields or any oil that may result from future exploration. These categories of reserve, unlike proved reserves, cannot be given with any degree of certainty, being based mainly on enlightened opinion. However, in 1958 after a detailed survey, a U.S. geologist estimated that the amount of oil ultimately recoverable might exceed 0.19×10^{12} metric tons. Later estimates range up to 0.45×10^{12} metric tons but cannot be regarded as final.

In addition to these assessed sources of crude oil, potential supplies exist in the form of shale oil, tar sands and coal. The extensive deposits of shale oil in the Green River Formation

(U.S.A.) have been estimated to provide about 10×10^9 metric tons of recoverable oil. The Athabaska tar sands of North Alberta also provide a vast oil potential. In summary, the 1966 U.S. Geological survey estimated the total possible reserves from all sources to be 1.37×10^{12} metric tons and forecast that about 62% of this will be recovered eventually.

Natural Gas

The term 'natural gas' is applied to all varieties of gas, obtained from underground strata, in which the paraffinic hydrocarbons predominate. The gas usually occurs in three types of reservoir: (i) Dry-gas reservoirs, where the gas is accompanied by only small amounts of liquid hydrocarbons when reduced to atmospheric pressure and temperature. It is estimated that about half the world's reserves occur in the absence of oil. (ii) Condensate reservoirs in which the gas is accompanied by larger amounts of liquid hydrocarbons. (iii) Oil reservoirs in which the gas is present as a 'gas-cap' above the oil surface or solution gas obtained as the pressure is reduced on removal of the oil from the well.

The gaseous hydrocarbon constituents are mainly methane and ethane, with varying minor amounts of the heavier paraffins. When small amounts of normally liquid hydrocarbons (pentane, hexane and heptane) are present the gas is referred to as 'wet gas'. Olefins are almost always absent. Other possible constituents comprise nitrogen, carbon dioxide and H_2S, and in certain cases helium is present in recoverable concentrations.

The gas composition varies widely from gas field to gas field (Table 2.13) and, in fact, the composition of a particular well may change during its lifetime.

The presence of large amounts of ethane and propane in natural gas in many parts of the U.S.A. together with the existence of a highly developed system of gas processing and distribution (via 225,000 miles of trunk pipelines) enabled these gases to be used as a major feedstock for ethylene production. In Western Europe, natural gas was mainly methane and therefore could not be used in this way. The gas strikes in the North Sea off the Lincolnshire coast were also virtually pure methane. However, the more recent oil strikes in more northerly regions appear to have significant amounts of associated gas containing C_2–C_5 species.

The estimated world reserves of natural gas are 1.6×10^{15} cu. ft. In this context it should be noted that the U.S.A., which has reserves estimated at 265×10^{12} cu. ft., is currently using natural

Table 2.13. Variation in composition of natural gas at the well-head (volume %)

	Lacq (France)	Slochteren (Holland)	Hassi R'Mel (Algeria)	Sarmas (Roumania)	Los Angeles (U.S.A.)	Dachava (U.S.S.R.)
Methane	69.4	81.9	83.5	99.2	77.5	98.0
Ethane	2.8	2.7	7.0	—		0.7
Propane	1.5	0.38	2.0	—		—
Butane	0.7	0.13	0.8	—		—
Pentane	0.3	0.05	0.3	—	16.0	—
Hexane and above	0.3	0.03	0.1	—		—
H_2S	15.2	—	—	—	—	—
CO_2	9.5	0.8	0.2	—	6.5	0.1
N_2	0.3	14.0	6.1	0.8	—	1.2

gas at the rate of 22×10^{12} cu. ft. p.a. This situation is causing concern and leading to development of coal-gasification processes and also to a move to ease restrictions on imports of crude oil fractions, particularly of naphtha feedstocks for ethylene crackers. The high rate of natural-gas consumption is due, in part, to its use as a primary energy source. Thus, in 1972, 34% of the U.S.A. primary energy consumption was based on natural gas compared with 12% for the U.K. and 1.0% for Japan.

The term 'liquefied natural gas' (LNG) is usually used to describe methane that has been liquefied for transportation. 'Liquefied petroleum gas' (LPG) is used to describe variable mixtures of predominantly propane and butane. It can originate from natural-gas processing, petrochemical plants or refinery off-gases. The last are sometimes called 'liquid refinery gases' (LRG). LPG from petrochemical or refinery sources will generally contain valuable amounts of C_3 and C_4 olefins, unlike that from natural-gas sources.

CHAPTER 3

Preparation of Primary Petrochemicals

In the preceding Chapter the importance of oil as the primary source of starting materials for the chemical industry was emphasized, and a brief description was given of the nature and treatment of crude oil. The present Chapter is concerned with a more detailed consideration of the preparation of useful chemical intermediates from petroleum feedstocks.

1. Cracking

(J. CATTANACH, Research and Development Department, BP Chemicals International Limited, Epsom)

Introduction

Petrochemical intermediates such as ethylene and propylene* are primarily produced by the thermal cracking of light hydrocarbon feedstocks in the presence of steam. Commercial units are therefore generally referred to as steam crackers. Ethane and propane recovered from natural gas account for over 90% of the feedstocks used in the U.S.A. while Western Europe and Japan depend almost entirely on light distillate fractions, often called naphtha (maximum boiling range 35–200°C) produced from crude oil. Cracking of gas oil (maximum boiling range 175–360°C) is likely to increase in the U.S.A. and technology has already been developed for that of crude oil.

Thermal cracking was first introduced as a commercial process in 1912 to increase the yield of middle distillate fractions (150–340°C) from crude oil. Other thermal cracking processes developed include visbreaking, which involves limited molecular degradation

* Systematic name: propene.

34

under mild thermal cracking conditions to improve the viscosity of heavy distillate fractions (>250°C), and delayed coking or fluid coking processes in which the thermal cracking conditions are so severe that the feedstock is completely converted into coke, middle distillate, gasoline (50–200°C) and gaseous products. The gasoline fraction produced by thermal cracking is of too low a quality for modern engines, and thermal cracking as a petroleum-refining process has been replaced by catalytic cracking and hydrocracking which cause both extensive molecular degradation and rapid intermolecular and intramolecular rearrangements. Catalytic processes operate at lower temperatures and are more selective, giving higher yields of the required light distillate fractions and high-quality gasoline.

Catalytic cracking is the major source of propylene in the U.S.A. but otherwise catalytic cracking and hydrocracking have limited application for manufacture of chemicals except for the production of feedstocks for thermal cracking. The availability of olefinic by-products from refinery operations failed to meet the growing demand of the chemicals industry and steam cracking was there-fore established as a separate process for producing the required olefinic intermediates. Present consumption of ethylene and propylene in the free world's chemicals industry is estimated at 22×10^6 and 11×10^6 metric tons p.a., respectively. Steam crackers based on liquid feedstocks also provide substantial quantities of other chemical intermediates, such as butadiene, butenes, isoprene and aromatics. Modern naphtha crackers have ethylene capacities in the order of $250-500 \times 10^3$ tons p.a. and they require upwards of 1 million tons p.a. of feedstock.

Steam crackers are designed to operate at conditions that make full use of the basic chemical and physical principles favouring the formation of the required olefins.

Thermochemistry

The thermal decomposition of paraffins and olefins can be represented by molecular equations such as:

$$
\begin{array}{lll}
A & \rightleftharpoons B + C & \Delta H^\ominus \text{ (kJ/mol}^{-1}) \\
C_3H_8 & \rightleftharpoons CH_3CH{=}CH_2 + H_2 & +126 \\
C_3H_6 & \rightleftharpoons HC{\equiv}CCH_3 + H_2 & +163 \\
C_3H_6 & \rightleftharpoons H_2C{=}C{=}CH_2 + H_2 & +172 \\
C_2H_6 & \rightleftharpoons C_2H_4 + H_2 & +138 \\
C_2H_4 & \rightleftharpoons C_2H_2 + H_2 & +176
\end{array}
$$

The equilibrium constant K_p expressed in terms of partial pressures is given by the expression:

$$K_p = (p_B p_C / p_A)_{equil.}$$

The equilibrium constant varies with temperature in accordance with the van't Hoff equation:

$$d(\ln K_p)/dt = \Delta H / RT^2$$

Dehydrogenation reactions are highly endothermic, so their equilibrium constants increase rapidly with increasing temperature (Figure 3.1). Equilibrium therefore favours a higher proportion of the unsaturated species at the higher temperatures. Potential hydrocarbon conversion reactions can be predicted in a qualitative way by comparing relative stabilities as measured by

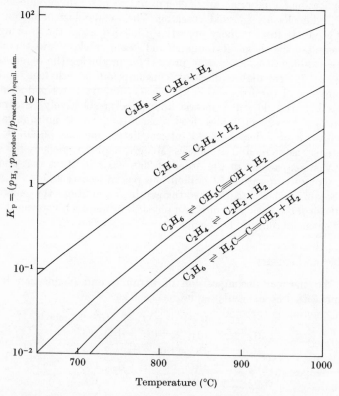

Figure 3.1. Equilibrium constants for hydrogenation reactions.

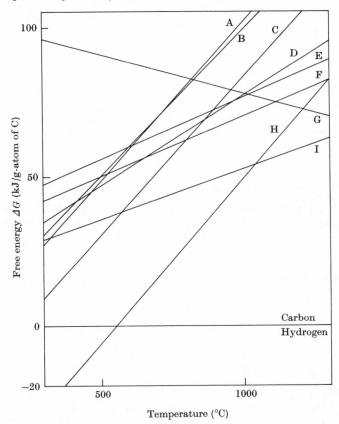

Figure 3.2. Free energy of formation of hydrocarbons (ΔG) (kJ/g-atom of C).
A, Hexane. B, Cyclohexane. C, Ethane. D, Propylene. E, Buta-1,3-diene. F, Ethylene. G, Acetylene. H, Methane. I, Benzene.

the free energy of formation per gram-atom of carbon as shown in Figure 3.2.

A reaction is thermodynamically possible only if it is accompanied by a decrease in free energy ($\Delta G < 0$) and at equilibrium the free energy of the system is at a minimum. Dehydrogenation of paraffins to olefins is favoured at high temperatures, while at low temperatures the corresponding hydrogenation reaction is favoured:

$$C_2H_6 \rightleftharpoons C_2H_4 + H_2 \qquad \text{Inversion temperature } 720°C$$

At higher temperatures dehydrogenation of ethylene to acetylene is favoured:

$$C_2H_4 \rightleftharpoons C_2H_2 + H_2 \qquad \text{Inversion temperature } 1152°C$$

Steam crackers for ethylene production therefore operate between 750° and 900°C whilst processes for acetylene manufacture operate above 1200°C.

The high stability of potential by-products such as methane, benzene, other aromatic compounds and free carbon has important implications in the design of steam crackers.

Bond Dissociation Energies

Bond dissociation energies for C—C and C—H bonds within typical hydrocarbon molecules are dependent on their immediate molecular environments (Table 3.1). In general C—C bonds are

Table 3.1. Bond dissociation energies (kJ mol^{-1}) for the reaction:
$$R_1—R_2 \rightleftharpoons R_1\cdot + R_2\cdot$$

R_1	H	CH_3	C_2H_5	iso-C_3H_7	tert-C_4H_9	C_6H_5
R_2						
CH_3	435	368	356	351	335	389
C_2H_5	410	356	343	335	322	377
n-C_3H_7	410	356	343	335	322	377
iso-C_3H_7	397	351	335	326	305	368
tert-C_4H_9	381	335	322	305	285	351
C_6H_5	381	389	377	368	351	418
$C_6H_5CH_2$	356	301	289	285	268	331
$CH_2{=}CHCH_2$	356	343	331	339	—	372
$CH_2{=}CH$	431	385	372	364	351	414

weaker than C—H bonds and their bond energy depends on the state of hybridization, the strength of the bond increasing as the *s* character of the bond increases. Multiple bonds have much higher bond energies:

Bond	C—C	C$=$C	C$\equiv$C
D(C—C) (kJ mol^{-1})	335–368	682	962

Single C—C bonds adjacent to multiple bonds (α-bonds) are considerably strengthened by the increased *s* character resulting from electronic overlap with the sp^2 orbital of the double bond.

In consequence the s character of the adjacent β C—C bond is reduced and this bond is considerably weakened, e.g.:

$$D(\text{C—C}) \text{ (kJ mol}^{-1}) \quad \overset{682 \quad 385 \quad 343}{\text{C}=\text{C—C—C}}$$

The presence of an odd electron in a free radical has a similar effect on the adjoining α- and β-bond energies. However, the rate at which hydrocarbon molecules decompose and the nature of the products formed cannot be predicted solely from a consideration of simple fission of the weakest bonds within the molecule.

Reaction Mechanism

The role played by free radicals in determining the reaction rate and resultant products when hydrocarbons are pyrolyzed was first outlined over 35 years ago by F. O. Rice. Extensive research during the past 25 years has added detailed quantitative information and clearly confirmed the basic principles included in Rice's original free-radical mechanism (see Part 2, Chapter 11) which consists of three main parts:

(a) *Initiation.* The primary process in hydrocarbon pyrolysis is the homolysis of the weakest C—C bond to form two radicals in a unimolecular reaction such as:

$$\text{C}_{12}\text{H}_{26} \longrightarrow \text{C}_2\text{H}_5 \cdot + \text{C}_{10}\text{H}_{21} \cdot$$

Fission of C—H bonds to yield H atoms is generally insignificant.

(b) *Propagation.* Small radicals formed in the chain-initiation reaction can abstract a hydrogen atom from a reactant molecule to give a larger radical and a saturated compound, e.g.:

$$\text{C}_2\text{H}_5 \cdot + \text{C}_{12}\text{H}_{26} \longrightarrow \text{C}_2\text{H}_6 + \text{C}_{12}\text{H}_{25} \cdot$$

Larger radicals formed are thermally unstable and extremely short-lived. They decompose by fission of the C—C bond β to the radical site, forming an olefin and a new radical:

$$\text{C}_8\text{H}_{17}\text{CH}_2\text{CH}_2\dot{\text{C}}\text{HCH}_3 \longrightarrow \text{C}_8\text{H}_{17}\text{CH}_2 \cdot + \text{C}_3\text{H}_6$$

This β-fission reaction is repeated until a thermally more stable radical is formed:

$$\text{C}_8\text{H}_{17}\text{CH}_2\text{CH}_2\dot{\text{C}}\text{HCH}_3 \longrightarrow \text{C}_3\text{H}_6 + 4\text{C}_2\text{H}_4 + \text{CH}_3 \cdot$$

This 'unzipping' reaction is responsible for the high ethylene yields obtained from normal paraffins. Isomerization of higher primary radicals to secondary radicals, by 1–4 and 1–5 hydrogen-transfer reactions:

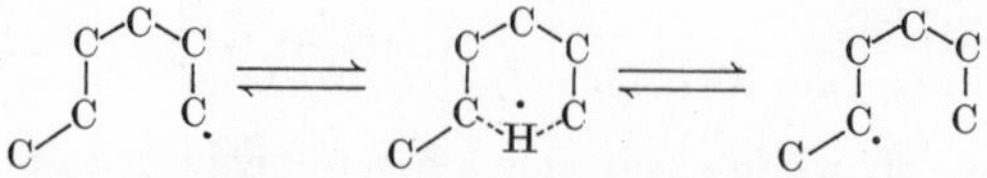

will give rise to propylene on subsequent β-fission:

$$C_3H_7CH_2\dot{C}HCH_3 \longrightarrow C_3H_7 \cdot + C_3H_6$$

Smaller radicals having no C—C bond in the β-position can give rise to hydrogen atoms:

$$C_2H_5 \cdot \longrightarrow C_2H_4 + H \cdot$$

The radicals that are relatively stable at cracking temperatures, in order of decreasing stability, are $H \cdot > CH_3 \cdot > C_2H_5 \cdot > sec\text{-}C_3H_7 \cdot > tert\text{-}C_4H_9 \cdot$. These smaller radicals regenerate a larger radical in the chain propagation step by hydrogen abstraction:

$$CH_3 \cdot + C_{12}H_{26} \longrightarrow CH_4 + C_{12}H_{25} \cdot$$

Hence a chain mechanism is established by which the parent compound can decompose very much faster than by the initiation reaction alone.

The probability of forming a primary, secondary or tertiary radical in the above chain-propagation step is governed by a combination of the number of hydrogen atoms that can be abstracted to give the new radical and the relative abstraction rates (Part 3, p. 313) which depend on the differences in activation energy

$$RS = k_1/k_2 = (A_1/A_2)\ e^{\Delta E/RT}$$

where RS = relative rate, and either $\Delta E = E_{sec} - E_{prim} = 8$ kJ mol^{-1} or $\Delta E = E_{tert} - E_{prim} = 17$ kJ mol^{-1}. The relative abstraction rates of secondary to primary and tertiary to primary hydrogen atoms at 825°C are 1.7 and 6.4, respectively. The presence of methyl substituents in branched paraffins favours tertiary radical formation and reduces the number of primary free radicals which would otherwise decompose to ethylene. Branched paraffins therefore give higher yields of C_3 and higher olefins than do normal paraffins.

(c) *Termination.* The chain reaction is terminated by removal of the chain propagation radicals through recombination or disproportionation:

$$(H \cdot + H \cdot \xrightarrow{\text{Wall}} H_2; \text{ important in ethane pyrolysis})$$

$$CH_3 \cdot + CH_3 \cdot \longrightarrow C_2H_6$$

$$CH_3 \cdot + C_2H_5 \cdot \longrightarrow C_3H_8 \text{ or } CH_4 + C_2H_4$$

$$C_2H_5 \cdot + C_2H_5 \cdot \longrightarrow C_4H_{10} \text{ or } C_2H_6 + C_2H_4$$

This free-radical mechanism may be used to predict the principal primary products and their approximate distribution. It is generally assumed that after a small induction period the concentration of radicals in the system will have attained a steady-state value, such that the rate of their formation and disappearance are equal. The actual concentration of radicals in the system is relatively small, normally about 10^{-9} to 10^{-13} mol l^{-1}. Hydrocarbons decompose mainly as a result of the chain mechanism and the average chain length λ, which is a relative measure of the importance of the two decomposition pathways, is generally long (>10), where

$$\lambda = \frac{\text{Rate of decomposition by chain reactions}}{\text{Rate of decomposition by initiation reactions}}$$

Reaction products are therefore almost completely derived from reactions occurring in the chain-propagation process. These depend on the nature and number of radicals that can be produced in the chain-initiation reaction which in turn depends upon the molecular structure of the parent hydrocarbon.

Application to Specific Hydrocarbon Types

Simplified overall reaction schemes for ethane and *n*-pentane can be represented as follows:

Ethane

Initiation	$C_2H_6 \longrightarrow CH_3 \cdot + CH_3 \cdot$
Propagation	$CH_3 \cdot + C_2H_6 \longrightarrow CH_4 + C_2H_5 \cdot$ (transfer)
	$C_2H_5 \cdot \longrightarrow C_2H_4 + H \cdot$
	$H \cdot + C_2H_6 \longrightarrow H_2 + C_2H_5 \cdot$
Termination	$H \cdot + H \cdot \longrightarrow H_2$ at wall
	$C_2H_5 \cdot + C_2H_5 \cdot \longrightarrow C_4H_{10}$

n-Pentane

Initiation $\qquad\qquad\qquad\qquad$ $C_5H_{12} \longrightarrow CH_3\cdot + C_4H_9\cdot$

$\qquad\qquad\qquad\qquad\qquad\quad$ $C_5H_{12} \longrightarrow C_2H_5\cdot + C_3H_7\cdot$

Propagation $\qquad$ $R\cdot + C_5H_{12} \longrightarrow RH + C_5H_{11}\cdot$

$\qquad$ where $R\cdot = H\cdot$, $CH_3\cdot$, $C_2H_5\cdot$

$\qquad\qquad$ and $C_5H_{11} = C_5H_{11}\text{-}1^{ry}$, $C_5H_{11}\text{-}2^{ry}$ and $C_5H_{11}\text{-}3^{ry}$

$\qquad\qquad\qquad\qquad$ (a) $\qquad\quad$ (b) $\qquad\qquad$ (c)

(a) Chain 1 $\quad$ $\overset{\frown\frown\frown}{\dot{C}\text{—}C}\text{—}C\text{—}C\text{—}C \longrightarrow C_2H_4 + C_3H_7\cdot$

$\qquad\qquad\qquad\qquad$ $C_3H_7\cdot \longrightarrow C_2H_4 + CH_3\cdot$

(b) Chain 2 $\quad$ $C\text{—}\overset{\frown\frown\frown}{\dot{C}\text{—}C}\text{—}C\text{—}C \longrightarrow C_3H_6 + C_2H_5\cdot$

$\qquad\qquad\qquad\qquad$ $C_2H_5\cdot \longrightarrow C_2H_4 + H\cdot$

(c) Chain 3 $\quad$ $C\text{—}C\text{—}\overset{\frown\frown\frown}{\dot{C}\text{—}C}\text{—}C \longrightarrow C_4H_8\text{-}1^{ry}$ and $CH_3\cdot$

$\qquad$ where $\overset{\frown\frown\frown}{\dot{C}\text{—}C}\text{—}C$ indicates β-fission.

Termination $\qquad$ $CH_3\cdot + CH_3\cdot \longrightarrow C_2H_6$

$\qquad\qquad\qquad$ $CH_3\cdot + C_2H_5\cdot \longrightarrow C_3H_8$

Naphthenes (cycloparaffins) are more stable than the corresponding paraffin at cracking conditions. Alkylnaphthenes tend to dealkylate and some dehydrogenation to aromatics can occur. The major products formed from cyclohexane are butadiene, ethylene and hydrogen:

$$cyclo\text{-}C_6H_{12} \longrightarrow C_4H_6 + C_2H_4 + H_2$$

Aromatic hydrocarbons are stable and are therefore generally unchanged under thermal cracking conditions. Large alkyl substituents crack within themselves by the normal free-radical mechanism. The α-bond adjacent to the aromatic ring is very strong, so methyl groups formerly present and those produced by dealkylation are largely retained. The major product formed from aromatics are condensation products, such as tar and coke:

$$\text{Aromatics} \longrightarrow \text{Polynuclear aromatics} \longrightarrow \text{Coke}$$

The above discussion only relates to the thermal cracking of individual hydrocarbons at relatively low conversion levels. When a hydrocarbon mixture is cracked, the radicals generated by the least stable components accelerate the decomposition of more stable components. The cracking rate of the least stable component is simultaneously reduced owing to the depletion of free radicals. Reaction velocity constants for typical components present during naphtha cracking may be ordered as follows:

Normal α-olefins > multi-branched paraffins > normal paraffins
iso-paraffins > C_6-naphthenes > C_5-naphthenes > aromatics

Naphthenes (cycloparaffins), which are relatively stable alone, are therefore degraded to a significant extent during naphtha cracking.

Secondary Reactions

As the conversion level is increased the partial pressure of the reactant decreases and the partial pressures of the primary products increase to a maximum level at which the rate of formation equals the rate of disappearance. Beyond this point, the primary products react faster than they are being formed, so their concentration goes through a maximum (Figure 3.3).

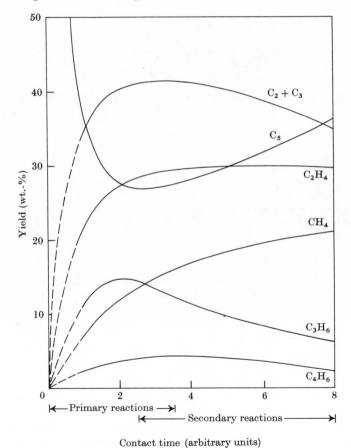

Figure 3.3. Effect of cracking severity on product yields from a light naphtha.

Ethylene is relatively stable but it can be removed by reactions such as:

$$C_2H_4 + C_2H_4 \longrightarrow C_2H_3\cdot + C_2H_5\cdot$$
$$C_2H_5\cdot + C_2H_4 \longrightarrow C_2H_6 + C_2H_3\cdot$$
$$C_2H_3\cdot + C_2H_3\cdot \longrightarrow C_4H_6$$

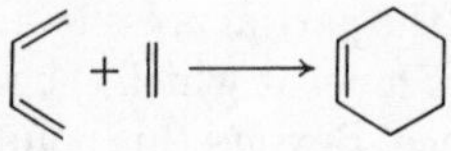

Propylene is thermally less stable and the main reactions by which it decomposes are:

$$C_3H_6 \longrightarrow C_2H_3\cdot + CH_3\cdot$$
$$CH_3\cdot + C_3H_6 \longrightarrow CH_4 + C_3H_5\cdot$$
$$C_2H_3\cdot + C_3H_6 \longrightarrow C_2H_4 + C_3H_5\cdot$$
$$n(C_3H_5\cdot) \longrightarrow (C_3H_5)_n \longrightarrow \text{Aromatics} + H_2$$
$$H\cdot + C_3H_6 \longrightarrow C_3H_7\cdot$$
$$C_3H_7\cdot \longrightarrow C_2H_4 + CH_3\cdot$$

Butadiene is thermally stable but can undergo Diels–Alder-type reactions:

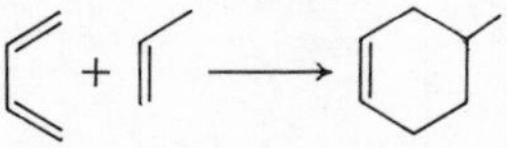

Condensation of olefins, diolefins and aromatics gives rise to polynuclear aromatics which then react further to form coke. Secondary reactions leading to coke are very prevalent in the stationary skin-layer adjacent to the reactor surface. The coke which is formed has to be removed periodically by air oxidation.

Reaction Kinetics

Paraffins of high molecular weight decompose faster than those of low molecular weight and, since the activation energies for paraffins having 5 or more carbon atoms ($n > m > 5$) are similar, the relative ratio of the rate constants is equivalent to the ratio of their respective A factors:

$$\frac{k_n}{k_m} = \frac{A_n\, e^{-E_n/RT}}{A_m\, e^{-E_m/RT}} \approx \frac{A_n}{A_m}$$

In the initial stages of the reaction, the rate of disappearance is generally independent of both the pressure and the surface-to-volume ratio of the reactor. The reaction therefore appears to be unimolecular and to follow first-order kinetics (Part 3, p. 255):

$$k = \frac{1}{t} \ln \frac{1}{1-x}$$

The yield of cracked products is therefore directly proportional to the run length during the initial stages of conversion and the cracking rate almost doubles for each 20° rise in temperature.

In practice the reaction rates are not completely independent of pressure and the reaction order is greater than 1.0 and closer to 1.3. Rice and Herzfeld showed that the overall reaction order could vary depending on the chain-termination reactions involved. If the chain-terminating reaction involves radicals initiating the chain mechanism the overall order is 1.5; if it involves a heavier radical which could otherwise undergo β-fission it is 1.0; and if it involves two of the heavier radicals it is 0.5. Intermediate orders indicate that several termination reactions are occurring simultaneously.

Feedstocks

Potential olefin yields are mainly controlled by the nature of the feedstock. The distribution of feedstocks used in world-wide ethylene production is as follows:

Feedstock	Ethane	Propane	Butane	Naphtha	Gas oil
Distribution (%)	36	11	3	47	3

Paraffins are the preferred components, so feedstocks having higher hydrogen contents generally give higher olefin yields (Table 3.2).

Table 3.2. Typical product yields (wt.-%) for various cracker feedstocks in a single-pass operation[a]

Feedstock	Ethane	Propane	Butanes	Naphtha	Gas oil	Crude oil
Cracking temp. (°C)	830	845	830	870	870	—
Hydrogen	3.6	1.2	0.7	1.0	0.6	0.8
Methane	7.1	21.1	17.2	19.2	8.6	11.1
Acetylene	0.3	0.3	0.1	0.6	0.5	1.0
Ethylene	46.0	30.3	21.6	32.2	21.3	23.2
Ethane	(38.3)	3.7	4.4	4.7	2.7	2.8
Propylene	2.0	19.7	28.2	14.6	13.4	11.6
Propane	0.4	(18.2)	0.8	0.4	0.4	0.9
C_4's	2.3	4.0	(21.9)	7.6	9.1	6.0
Gasoline	—	1.5	5.1	15.0	24.4	16.5
Fuel oil	—	—	—	4.0	19.0	26.1

[a] The overall ethylene yield in commercial ethane cracking units is increased to 77 wt.-% by recycling the unconverted ethane and thereby increasing ethane conversion to 100 wt.-%. Ethane resulting from liquid feedstock cracking is also frequently recovered and cracked in a smaller furnace to boost ethylene yields.

Hydrocarbon-type compositional data for a typical naphtha feedstock are given in Table 3.3, which also shows that acyclic paraffins, especially normal paraffins, are the best olefin precursors.

Table 3.3.   Composition of a typical naphtha feedstock

Hydrocarbon type	Wt.-% of feedstock	Range of carbon nos.	Selectivity coefficient[a]
Normal paraffins	30	4–12	0.60
Branched paraffins	30	4–12	0.40
Naphthenes	25	5–12	0.25
Aromatics	15	6–9	0.05

[a] Potential ethylene plus propylene yield (wt.-%) which can be derived from a given hydrocarbon type when the mixture is steam cracked is given by the product of the selectivity coefficient and the wt.-% of the hydrocarbon type. In the above case the predicted total basic olefin yield (ethylene plus propylene) is 37.0 wt.-%.

Process Variables

Conditions for successful operation of steam crackers for high olefin production are short residence time (0.1–0.4 sec), high temperature (750–900°C), low overall pressure ($<3 \times 10^5$ Nm^{-2}), low hydrocarbon partial pressure (steam dilution 25–100 wt.-%) and rapid quenching to minimize secondary reactions.

The cracking severity is controlled mainly by two independent variables, namely contact time and temperature, which are adjusted to give the required product distribution from a given feedstock.

Contact Time

Reactions giving rise to the required olefins can generally be represented by molecular equations, such as:

$$\underset{1-x}{C_2H_6} \rightleftharpoons \underset{x}{C_2H_4} + \underset{x}{H_2}$$

where x = fraction decomposed at some intermediate time t_1. The approach to equilibrium EA is related to the equilibrium constant K_p at t_∞ by:

$$\text{EA} = \frac{n_{C_2H_4} \cdot n_{H_2}}{n_{C_2H_6}} \cdot P/K_p \qquad 0 < \text{EA} \leqslant 1.0$$

where n = mole fraction at time t_1 and P = total hydrocarbon pressure in the system. The ethylene yield should therefore increase as the approach to equilibrium increases (i.e. $x \to x_\infty$ as EA $\to 1.0$). However, as the ethylene concentration increases so will the rate of undesirable secondary reactions. The yield of desirable products at any one temperature therefore goes through a maximum with increasing contact time (Figure 3.3). The contact time at any one temperature is normally chosen to give a specific EA value consistent with optimum product yields. Typical EA values range from 0.3 to 0.90, lower values being used at higher conversions (see Figure 3.3). The corresponding contact times are in the order of 0.1–1.0 seconds.

Cracking Temperature

The overall activation energy for hydrocarbon pyrolysis is about 222 kJ mol^{-1}, so the total rate of conversion increases rapidly with increasing temperature. The activation energies for initiation, propagation and termination reactions are about 351, 46 and 0 kJ mol^{-1}, respectively. As the temperature is increased, the rate of initiation reactions and hence the rate of production of free radicals increase rapidly. The rates of termination reactions which are responsible for the removal of free radicals are temperature-independent. Increasing the temperature will thus increase the concentration of free radicals in the system and therefore the rate of favourable reactions leading to the required basic olefins.

Initiation reactions are highly endothermic, so the corresponding equilibrium constant increases rapidly with temperature, thereby favouring free-radical formation. Conversely, termination reactions are highly exothermic, so the corresponding equilibrium constant decreases rapidly with increasing temperature. Chain-propagation reactions are essentially thermally neutral and so the corresponding equilibrium constants are virtually independent of temperature. In general, higher concentrations of free radicals will therefore be favoured at higher temperatures. Equilibrium also favours increased ethylene yields at higher temperatures because of the increase in K_p:

$$P_{C_2H_4} = \frac{P_{C_2H_6}}{P_{H_2}} \cdot K_p$$

Modern ethylene plants operate between 750°C (low severity) and 900°C (high severity) according to the required ratio of ethylene to higher olefin products. Figure 3.4 shows how the cracked product spectrum is affected by increasing the cracking tempera-

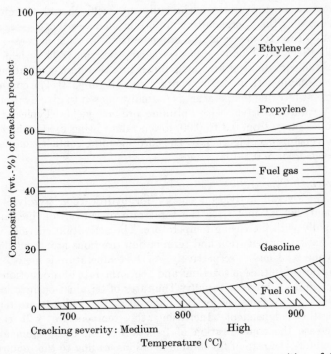

Figure 3.4. Effect of cracking temperature on the composition of the cracked products from a light naphtha cracker.

ture, while at the same time reducing the contact time to maintain high olefin yields.

Steam Dilution

The steam diluent reduces the hydrocarbon partial pressure, thereby favouring initiation and β-fission reactions at the expense of termination and other bimolecular reactions involving reactive chain carriers. The approach to equilibrium during dehydrogenation reactions is given by:

$$EA = \frac{x^2}{1-x} \cdot \frac{P}{K_p}$$

and the associated ethylene yield by:

$$x = \left[\frac{1}{1 + \dfrac{P}{EA}} \right]^{\frac{1}{2}}$$

Olefin yields are therefore favoured by low hydrocarbon partial pressures. The actual steam-dilution level employed is dictated by the type of feedstock and the cracking severity. Typical dilution levels fall within the following ranges:

Feedstock	Ethane	Propane	Naphtha	Gas oil
Dilution level (wt./wt.)	0.25–0.40	0.30–0.50	0.40–0.80	0.80–1.00

Steam dilution also prevents tar formation by reducing the dew point of heavy products, renders the coil surface passive, thereby reducing potential coking reactions, and improves the transfer of heat to the reactants. Steam also helps to remove coke by means of the water–gas reaction:

$$C + H_2O \longrightarrow CO + H_2$$

Thermal Cracker Design

Direct Heating

Tubular furnaces occupy the dominant position for olefin production from gaseous, naphtha and gas oil feedstocks, and modern plants utilize this design. A general flow diagram is shown in Figure 3.5 (p. 50). The furnace consists of two parts, the convection section which vapourizes the feed and preheats the feed and steam diluent to reaction temperature and the radiant section which supplies the necessary heat of reaction and any additional sensible heat. Multiple furnaces generally containing between two and eight cracking coils per furnace are employed so that individual furnaces can be taken off stream for decoking without unduly upsetting the overall plant throughput. The configuration and dimensions of the coils in the radiant section (where cracking occurs) vary according to the type of design and the products required. They are normally 30–160 m in length and 50–120 mm in diameter. The temperature profile of the coil is controlled by the firing of gas burners located on the furnace walls.

The cracked product is rapidly quenched, either directly by an oil or water spray or indirectly by transfer-line heat exchangers, to a temperature between 300° and 400°C to minimize secondary reactions. The heat recovered is used to generate high-pressure steam. Transfer-line heat exchangers are generally followed by a direct oil-spray quench.

Heavy oil, tar and quench oil separate in the base of the primary fractionator and the gaseous hydrocarbons, gasoline fraction and dilution steam are removed as an overhead fraction. The gaseous

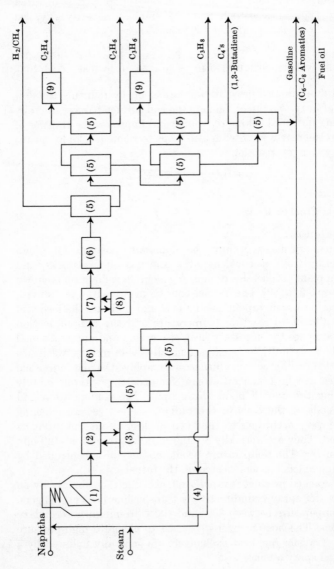

Figure 3.5. Naphtha cracker and product recovery system. 1, Furnace. 2, Quench. 3, Oil fractionator. 4, Steam generator. 5, Distillation column. 6, Cooler. 7, Compressor. 8, Acid wash. 9, Selective hydrogenator.

hydrocarbon product is then separated from the gasoline and dilution steam in a separate column.

The cracked gases are then compressed to about $3.5 \times 10^6 \, Nm^{-2}$ in four or five stages. Acid gases, such as hydrogen sulphide and carbon dioxide, are removed by a caustic wash at one of the interstages. Acetylenes are normally removed by selective hydrogenation of the mixed or individual product streams. They can also be removed and recovered as a by-product by solvent extraction.

The compressed gases consisting of hydrogen and hydrocarbons are dried over alumina or molecular sieves and cooled to about $-70°C$ before being fed to the demethanizer. Compressed methane, ethylene and propylene are used as refrigerants in the purification system. Ethylene and propylene are then purified and recovered by low-temperature, high-pressure distillation. Ethane and propane can be recycled and cracked in separate furnaces. Butadiene may be extracted from the C_4-stream. The C_5-200°C or gasoline fraction contains significant quantities of C_6–C_8 aromatics which can be extracted (see Chapter 5); otherwise the dienes present are selectively hydrogenated and the fraction used for motor gasoline.

Other Cracking Processes

Linear Alpha Olefin Production

Paraffin waxes consisting of C_{14}–C_{34} normal paraffins are the preferred feedstock for α-olefin production. At the mild conditions employed (ca. 550°C; contact time = 3 sec) the required C_6–C_{20} α-olefins are relatively stable and are formed by reactions, such as:

$$\dot{C}H_3 + C_{10}H_{21}CH_2CH_2CH_2C_7H_{15} \longrightarrow CH_4 + C_{10}H_{21}\dot{C}HCH_2CH_2C_7H_{15}$$
$$C_{10}H_{21}\dot{C}HCH_2CH_2C_7H_{15} \longrightarrow C_{10}H_{21}CH{=\!=}CH_2 + C_7H_{15}CH_2.$$

Consequently a distribution of α-olefins is obtained in which the yield of individual olefins decreases with increasing molecular weight.

Indirect Heating

Tubular furnaces are normally limited to feedstocks with end points lower than 350°C so as to avoid excessive coke formation. Novel methods have been developed in which the heat necessary for cracking is supplied by a heat-transfer agent which has been preheated to a much higher temperature than the reaction temperature. Heat-transfer agents employed include solids, such as coke or refractory solids, liquids such as molten metal or molten

salts, and steam superheated to 2000°C. Heavy feedstocks including crude oils can be cracked by this technique. Coke is
removed from the cracking zone together with the heat-transfer
agent, and the heat generated by combustion of the coke in a
separate reactor is used to reheat the transfer agent. Cracking
temperatures range from 780° to 1500°C according to whether
ethylene or acetylene is to be produced, and the corresponding
residence times are 1.0 to 0.1 sec, respectively. Ethylene yields
from crude oils are in the order of 20–30 wt.-%.

The Wulff process is primarily designed for acetylene production
and uses a cyclic operation. A honeycombed network of ceramic
tiles is initially preheated by fuel-gas combustion. The hydrocarbon feed diluted with steam is admitted in the second cycle and
cracked on the hot ceramic tiles.

Autothermic Cracking

In autothermic cracking the heat necessary for cracking is
supplied by combustion of part of the feed and the overall reaction
is self-sustaining. Combustion can be achieved in the form of a
submerged flame but now it is normally conducted in the lower
section of the reactor, a forced circulation of solid inorganic
particles being used to complete combustion. Temperatures
generated in the reaction zone range from 860° to 970°C and the
residence time is about 0.2 sec.

Electrical Processes

Considerable interest has been shown in recent years in the
concept of directly expending electrical energy for the production
of chemical intermediates by utilizing the energy in a plasma.
Temperatures in excess of 8000°C can be generated and any
hydrocarbon molecule would be dissociated into atoms. The
products obtained will depend on the rate of quenching but
inevitably they will be small molecules. Acetylene formation is
normally favoured and direct production from methane, naphtha,
crude oils or even coal particles is technically feasible.

Catalytic Cracking

Propylene is recovered as a by-product in the catalytic cracking
of gas oils to gasoline. Catalytic cracking is a low-pressure process
which normally operates at 450–600°C with a strongly acidic
catalyst containing a crystalline aluminosilicate (molecular sieve).
Cracking of paraffinic components proceeds by a carbonium ion
mechanism which has a β-fission step similar to that found for free

radicals. A small carbonium ion reacts with the paraffin to give a larger carbonium ion which then decomposes to give an olefin and a primary carbonium ion.

$$C_3H_5^+ + C_{12}H_{26} \longrightarrow C_3H_6 + C_{12}H_{25}^+$$
$$C_{12}H_{25}^+ \longrightarrow C_3H_6 + C_9H_{19}^+$$

The primary carbonium ion produced rapidly isomerizes to the more stable secondary form:

$$C_5H_{11}CH_2CH_2CH_2CH_2^+ \longrightarrow C_5H_{11}CH_2CH_2\overset{+}{C}HCH_3$$

β-Fission of this secondary carbonium ion will yield propylene:

$$C_5H_{11}CH_2CH_2\overset{+}{C}HCH_3 \longrightarrow C_5H_{11}CH_2^+ + C_3H_6$$

These reactions proceed until the smallest fragments formed contain three carbon atoms and the resulting carbonium ion regenerates the chain. Propylene yields range between 2 and 4 wt.-% on feed.

Lower olefins are also recovered to a lesser extent from Fisher–Tropsch, thermal-cracking and coal-gasification units.

Hydrocracking

Hydrocracking has come into prominence in recent years as an alternative to catalytic cracking for the production of paraffins of lower molecular weight from heavy crude-oil fractions. Propane, butane and naphtha range products have potential importance as feedstock for olefin production. The process is operated under pressure (15–20×10^6 Nm^{-2}) at temperatures up to $450°C$ with an excess of hydrogen in the presence of a dual-functional catalyst, such as a dispersed metal on an acidic zeolite.

2. Reforming

(T. EDMONDS, Petroleum Division, BP Research Centre, Sunbury-on-Thames)

Introduction

Catalytic reforming is the basic refinery process used to improve the octane number of light distillates for burning in modern high-compression internal combustion engines. The octane number of a fuel is a functional property.* It is, nevertheless, related to the chemical composition of the components of the distillate. Figure 3.6 shows the general trends in the octane number with the number of carbon atoms in a molecule for various component types in the feedstock. The light distillate feedstock, boiling between 30°C and

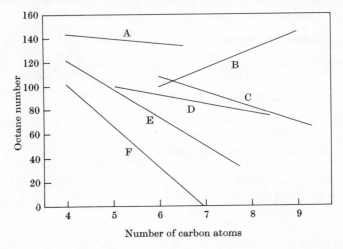

Figure 3.6. Octane numbers of various components of light-distillate fuels, showing the effect of the number of carbon atoms in the molecule of different classes of compound.
A, Olefins. B, Aromatics. C, Naphthenes. D, Dimethyl-paraffins. E, Methyl-paraffins. F, n-Paraffins.

200°C, is a complex mixture of paraffins, cycloparaffins (naphthenes) and aromatic molecules with five to eleven carbon atoms. Its exact composition will depend upon its source. Typical compositions and octane numbers of light distillates from Kuwait, Iranian Light, and Libyan crude oils are summarized in Table 3.4. In finished gasoline, octane numbers in excess of 90 are required; clearly the octane numbers of these feedstocks are too low. Figure 3.6 indicates that, to raise the octane number, the propor-

* The octane number is determined in a single-cylinder engine of variable compression ratio by comparison with a mixture of 'iso-octane' (2,2,4-trimethylpentane) and n-heptane. It is the volume percentage of iso-octane in the blend giving the same intensity of knock as the sample fuel under the same operating conditions. Iso-octane is allocated a value of 100 on an arbitrary 'octane number scale', while n-heptane has a value 0.

Table 3.4. Properties of light distillates from Kuwait, Iranian Light and Libyan crude oils

Boiling range	30–65°C			65–145°C			145–190°C		
Crude source	Kuwait	Iranian Light	Libyan	Kuwait	Iranian Light	Libyan	Kuwait	Iranian Light	Libyan
Yield (wt.-% on crude)	2.9	2.7	2.3	10.02	12.71	10.15	6.83	7.65	6.7
Total sulphur (wt. ppm)	200	910	17	240	880	5	960	1200	50
Hydrocarbon type analysis (wt.-%)									
Paraffins	97.5	98	94	72	56	55	64	45.5	51
Naphthenes	2.5	2	5.2	19	31	40.4	18	35	39.5
Aromatics	0	0	0.8	9	13	4.6	18	19.5	9.5
Octane number	71.5	71.5	72	44	50	51.5	30	35	22

tion of aromatics, branched-chain paraffins and olefins must be increased while the overall molecular weight of the feed is decreased. The objective is achieved in catalytic reforming by operating under conditions to maximize the concentration of aromatics in the product. The yields attained make the separation of individual aromatic components economically attractive for chemical feedstocks, particularly when the boiling range and composition of the light distillate are carefully selected.

The individual reactions that occur over a reforming catalyst reflect the complexity of the feedstock. They may be divided into three main classes: dehydrogenation reactions [(a), (b)], isomerization reactions [(c), (d)] and hydrocracking reactions [(g), (h)]. Reactions (e) and (f) combine dehydrogenation and isomerization.

(a) Dehydrogenation of cyclohexanes to aromatics:

$$\text{cyclo-}C_6H_{12} \longrightarrow C_6H_6 + 3H_2$$

(b) Dehydrogenation of paraffins to olefins:

$$C_6H_{14} \longrightarrow C_6H_{12} + H_2$$

(c) Isomerization of alkylcyclopentanes to cyclohexanes:

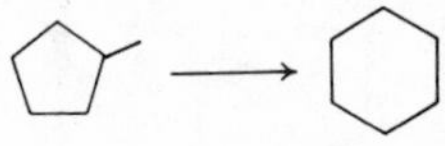

(d) Isomerization of n-paraffins to iso-paraffins:

$$n\text{-Hexane} \longrightarrow 2\text{- and 3-methylpentane, dimethylbutanes}$$

(e) Dehydrocyclization of paraffins to aromatics:

$$n\text{-}C_6H_{14} \longrightarrow C_6H_6 + 4H_2$$

(f) Dehydroisomerization of alkylcyclopentanes to aromatics:

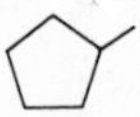

(g) Hydrocracking of paraffins:

$$C_6H_{14} + H_2 \longrightarrow 2C_3H_8$$

(h) Dealkylation of *gem*-dialkyls formed in dehydrocyclization, as illustrated.

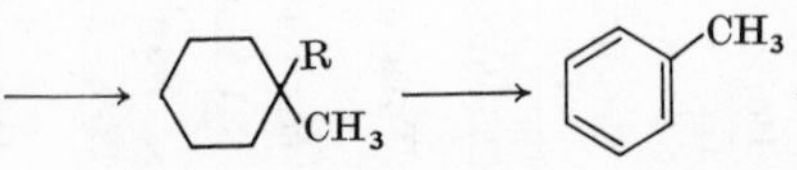

Most molecular types in the feedstock will undergo more than one of these reactions (a)–(h) during processing.

The catalyst is normally platinum supported on high-purity alumina. It provides two kinds of site: those of the metal for hydrogenation/dehydrogenation reactions and acid sites on the support for isomerization. Hydrocracking reactions can take place on either site. The quantity of platinum on the catalyst is between 0.3 and 1.0 wt.-%, and this small concentration must be distributed so that most of the platinum atoms are available for reaction with feedstock molecules. It is usually introduced as chloroplatinic acid which leaves 0.5–1.0 wt.-% of chlorine as acid sites on the large surface area of the alumina.

The platinum on the catalyst is seriously deactivated by sulphur compounds and therefore the feedstock is desulphurized to < 3 ppm by wt. of sulphur before the reforming (sulphur levels in light distillate are quoted in Table 3.4).

The process is operated at 500–525°C and at pressures 1.0–4.0×10^6 Nm^{-2}. To ensure that the active surface of the catalyst does not become rapidly deactivated by carbon deposits under these very severe conditions, it is necessary to carry out the reactions in the presence of hydrogen. In practice, a hydrogen-rich gaseous phase is circulated with vapourized feedstock through fixed beds of catalyst. After reaction the products are cooled, liquids are separated and removed from the system, and the gases (mainly hydrogen) are recirculated to the reaction zone.

The hydrocarbon composition of the feed and product are

Table 3.5. Hydrocarbon composition (wt.-%) of a reformer feed and product (octane no. of the feed = 60 and of the reformate 100)

No. of carbon atoms	Paraffins		Naphthenes		Aromatics	
	Feed	Reformate	Feed	Reformate	Feed	Reformate
C$_3$	0	Trace				
C$_4$	0.1	2.0				
C$_5$	1.9	6.8	0.1	0.3		
C$_6$	6.4	9.3	2.5	0.6	0.4	4.4
C$_7$	12.1	8.2	6.6	0.5	3.6	17.6
C$_8$	15.6	4.4	6.8	0.3	5.6	23.0
C$_9$	13.3	1.3	5.6	0.1	4.3	16.4
C$_{10}$	8.8	0.2	3.0	Trace	1.0	4.1
C$_{11}$	1.8	0	0.5	0	0	0.5
Total	60.0	32.2	25.1	1.8	14.9	66.0

compared in Table 3.5. The results show that the major objectives of enhancing the concentration of aromatics in the product while reducing the overall molecular weight are successfully achieved with the concomitant increase in octane number.

Thermodynamic Considerations

The conditions employed to attain the maximum yield of aromatics are determined by the thermodynamics of the reactions. Some thermodynamic equilibrium constants for C_6 components and their heats of reaction at 500°C are given in Table 3.6.

Table 3.6. Thermodynamic data for catalytic reforming reactions

Reaction	K	ΔH (kJ mol^{-1})
Cyclohexane $\rightleftharpoons$ Benzene + 3H$_2$	6×10^5	+221
Methylcyclopentane $\rightleftharpoons$ Cyclohexane	8.6×10^{-2}	−15.9
n-Hexane $\rightleftharpoons$ Benzene + 4H$_2$	7.8×10^4	+266
n-Hexane $\rightleftharpoons$ 2-Methylpentane	1.1	−5.9
n-Hexane $\rightleftharpoons$ 3-Methylpentane	7.6×10^{-1}	−4.6
n-Hexane $\rightleftharpoons$ Hex-1-ene + H$_2$	3.7×10^{-2}	+130

The equilibrium constants in Table 3.6 apply when the partial pressures of the various components are expressed in atmospheres. For components of higher molecular weight the thermodynamic data are increasingly favourable for aromatization of naphthenes and dehydrocyclization of paraffins but less favourable for isomerization of paraffins and alkylcyclopentanes.

The formation of benzene from both cyclohexane and n-hexane involves strongly endothermic reactions and the production of hydrogen. Theoretically it is advantageous to operate at as high a temperature and as low a hydrogen partial pressure as possible to maximize the yield of aromatics; practically, however, deactivation of the catalyst by deposition of carbon sets a limit on both these parameters. The formation of olefins is also favoured by high temperature and low hydrogen partial pressures; although this reaction does not take place to a large extent under typical reforming conditions it is nevertheless important because olefins have been identified as intermediates in some reactions. The equilibrium between methylcyclopentane and cyclohexane favours the former, but it is disturbed by rapid dehydrogenation of cyclohexane to benzene. For the equilibrium between n-hexane and the methylpentanes, 2-methylpentane is the favoured isomer.

The thermodynamic data also suggest that 30–35% of dimethyl-butanes should be present in the isomers at 500°C.

The approach to equilibrium for individual components of the feed has been extensively studied over typical reforming catalysts and various authors agree that (i) cyclohexanes are rapidly converted into an equilibrium mixture of aromatics, (ii) when cracked products are ignored, equilibrium amounts of benzene are formed from methylcyclopentane, and (iii) the equilibrium concentrations of dimethyl isomers are not observed in paraffin distributions.

However, it should be remembered that results on pure compounds may not exactly predict the behaviour of mixtures because of the influence of one compound on another at the catalyst surface.

Reaction Mechanisms

No detailed mechanism has yet been postulated for catalytic reforming. The reactions of cyclohexane, methylcyclopentane, cyclohexene and methylcyclopentene have been studied over one catalyst exhibiting isomerization sites alone, a second catalyst exhibiting dehydrogenation sites alone and a third dual-function catalyst. The results led to the following scheme for reactions in reforming of C_6 hydrocarbons. The vertical reaction paths in the scheme take place on the metal hydrogenation-dehydrogenation sites, and the horizontal reaction paths on the acid sites of the support.

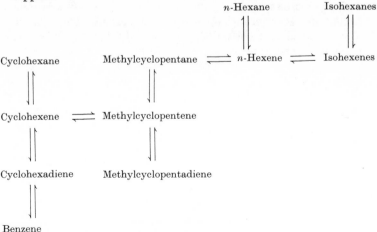

Reaction paths for C_6 molecules in reforming.

According to this scheme the formation of benzene from cyclohexane only requires dehydrogenation sites and therefore occurs over both the dehydrogenation catalyst and the dual-function catalyst. On the other hand, the dehydroisomerization of methylcyclopentane to benzene requires four stages: dehydrogenation to methylcyclopentene, isomerization to cyclohexene and two final dehydrogenation steps. This reaction only occurs with the dual-function catalyst; furthermore, it was subsequently shown to take place even when the dual-function catalyst is a physical mixture of the two components with the different sites existing on two different catalyst particles. Isomerization reactions on acid sites are believed to take place via a carbonium ion mechanism. The isomerization step appears to be rate-determining for the dehydroisomerization of methylcyclopentane, and this can be explained by the fact that the carbonium ion mechanism requires the formation of an unstable primary carbonium ion. Reaction intermediates, methylcyclopentene and cyclohexene, were identified in the products of this reaction at low contact times and conversions.

The stages in the dehydrocyclization of n-hexane are even more complex, requiring C_5 cyclization to methylcyclopentane followed by dehydroisomerization to benzene, i.e. a total of 6 steps. This has led to the postulate of direct C_6 cyclization over platinum sites. The latter mechanism has been proved over a single-function catalyst on which the acid sites were poisoned. However, the rate of this reaction and hence its contribution over a dual-function catalyst is smaller than in C_5 cyclization.

The proposed scheme does not take into account hydrocracking, coking and polymerization, which are minor reactions. On the other hand, it does unify the production of aromatics from both paraffins and naphthenes.

A new generation of reforming catalysts has recently been marketed. These employ two metal components on the alumina support. The mechanism of reaction does not appear to differ from that on the standard catalyst, but they have the ability to operate at conditions closer to those that are thermodynamically ideal without loss of activity.

3. Direct Oxidation of Butane and Naphtha

(H. S. GREEN, Research and Development Department,
BP Chemicals International Limited, Hull)

Introduction

Acetic acid was first manufactured commercially by the fermentation of ethanol to produce vinegar and by the distillation of wood (see Chapter 2). During World War I, escalation of the demand for acetone, for which acetic acid was an intermediate, gave impetus to the development of the first synthetic routes to the acid which were based on the oxidation of acetaldehyde. This method is still of major importance although the acetaldehyde is produced mainly from ethylene (Wacker process) rather than acetylene or ethanol as in earlier processes (cf. Fig. 3.7).

$$C_2H_4 + \tfrac{1}{2}O_2 \xrightarrow{\text{Pd catalyst}} CH_3CHO$$

Other major routes to acetic acid are the carbonylation of methanol (BASF and Monsanto):

$$CH_3OH + CO \xrightarrow[\text{catalyst}]{\text{Rh or CO}} CH_3COOH$$

and of increasing commercial importance since the 1950's is the direct liquid-phase oxidation of butane (Celanese and Hüls) and of naphtha (BP Chemicals).

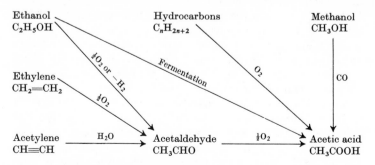

Figure 3.7. Principal commercial routes to acetic acid.

Of the feedstocks required for the process, *n*-butane is available in large quantities in the U.S.A. from natural gas, whilst in Europe low-boiling naphthas are more readily available as products of petroleum refining. A typical suitable naphtha is a fraction (b.p. 15–100°C) comprising mainly mixed C_5–C_7 hydrocarbons (see Table 3.7).

Table 3.7. Composition of naphtha used for oxidation to acetic acid

Component	Wt.-%
n-Pentane	15–25
n-Hexane	10–20
Isopentane	10–20
2- and 3-Methylpentanes	10–20
Aromatics and cyclic hydrocarbons (naphthenes)	10–20
Other C_4–C_8 paraffins	5–10

Description of the Process

Liquid oxidation of both butane and naphtha is usually carried out at 150–200°C and 3–9×10^6 Nm^{-2}. The reaction is conducted in pressure vessels made from stainless steel because of the corrosive nature of the products. Air, or oxygen-enriched air, is dispersed through the liquid reactants as a stream of small bubbles. The reaction is highly exothermic (1.5×10^6 J per Kg of hydrocarbon reacted), and to remove the heat liberated and hence control the reaction temperature the reaction liquid is circulated over heat-exchange tubes fed with water. As heat is absorbed, the water is converted into steam which is used to supply heat for distillation of the products.

Products are removed either directly from the reactor or as a condensate from the vapour issuing from the head of the reactor; the remaining permanent gases, comprising carbon oxides, nitrogen and a trace of oxygen are vented. Unchanged hydrocarbon which separates from the overhead condensate is returned to the reactor. Material removed directly from the reactor has a much lower hydrocarbon content and separation is unnecessary. The aqueous product contains formic, acetic and propionic acid together with the intermediate products of oxidation, mainly ketones and esters, some unchanged hydrocarbon and some high-boiling material. Neutral products, separated by distillation, can be recycled to the reactor to produce more acid or further purified to yield additional products such as acetone or butan-2-one. Acid products are separated from water and purified in a complex series of distillations.

Product ratios vary widely with feedstock and operating conditions. In general, the ratio of acetic to formic acid is 4–8:1 and for naphtha oxidation the ratio of acetic to propionic acid is 5–7:1. Little propionic acid is produced from butane. Depending upon the feedstock, 1 ton of hydrocarbon yields from 1.0–1.4 tons of acid products.

Reaction Mechanisms

Oxidation of paraffin hydrocarbons in the liquid phase produces acids of lower carbon number by cleavage of carbon—carbon bonds. Thus in theory n-butane can yield two C_2 fragments or one C_1 and one C_3 fragment, leading to the formation of two molecules of acetic acid or one each of formic and propionic acid. More combinations are, of course, possible from hydrocarbons of higher molecular weight.

The oxidation proceeds by a free-radical chain mechanism which can be considered to take place in two well-defined stages. Alkylperoxy-radicals and hydroperoxides are formed in the first stage, and then decompose in a variety of ways to produce other species. If the decomposition of alkylperoxy-radicals and hydroperoxides is minimized as, for instance, in low-temperature ($<100°C$), low-conversion ($<1\%$) systems then the reactions can be represented by the following sequences:

$$\text{Initiation} \quad \text{Radical source} \longrightarrow \text{R} \cdot$$
$$(\text{RH} + O_2 \longrightarrow \text{R} \cdot + HO_2 \cdot) \tag{1}$$
$$\text{Propagation} \quad \text{R} \cdot + O_2 \longrightarrow RO_2 \cdot \tag{2}$$
$$RO_2 \cdot + \text{RH} \longrightarrow RO_2H + \text{R} \cdot \tag{3}$$
$$\text{Termination} \quad 2\text{R} \cdot \longrightarrow \left.\begin{matrix} \\ \\ \\ \end{matrix}\right\} \tag{4}$$
$$\text{R} \cdot + RO_2 \cdot \longrightarrow \left.\right\} \text{Stable products} \tag{5}$$
$$2RO_2 \cdot \longrightarrow \left.\right\} \tag{6}$$

Abstraction of hydrogen (3), which is believed to be the rate-controlling step, adheres to the normal free-radical attack pattern, i.e. tertiary > secondary > primary (cf. Part 1, p. 17).

In commercial oxidations, which are of necessity carried out to high conversion at high rates (high temperature), alkylperoxy-radicals, dialkyl peroxides and hydroperoxides undergo further reaction and the process becomes a very complex series of interrelated reactions in which hydrocarbons, acid products and a wide variety of oxygenated intermediates all participate.

The mechanisms by which acids can be produced are diverse. For example, hydroperoxide formed by reaction (3) can undergo decomposition to produce a secondary alkoxy radical. This

$$CH_3CH_2\underset{\underset{OOH}{|}}{C}HCH_3 \longrightarrow CH_3CH_2\underset{\underset{O\cdot}{|}}{C}HCH_3 + \cdot OH \qquad (7)$$

subsequently degrades to a lower alkyl radical and acetaldehyde. The acetaldehyde so formed is further oxidized to acetic acid.

$$CH_3CH_2\underset{\underset{O\cdot}{|}}{C}HCH_3 \longrightarrow CH_3CH_2\cdot + CH_3CHO \qquad (8)$$

Propionic acid can be produced by an analogous mechanism from *n*-pentane or *n*-hexane. Formaldehyde is a product of decomposition of primary alkoxy-radicals and is further oxidized to formic acid.

$$CH_3CH_2CH_2CH_2O\cdot \longrightarrow CH_3CH_2CH_2\cdot + HCHO \qquad (9)$$

Ketones can be obtained from alkoxy-radicals by disproportionation (reaction 10) or by the decomposition of alkoxy-radicals derived from branched-chain hydrocarbons (reaction 11):

$$2CH_3CH_2\underset{\underset{O\cdot}{|}}{C}HCH_3 \longrightarrow CH_3CH_2\underset{\underset{O}{\|}}{C}CH_3 + CH_3CH_2\underset{\underset{OH}{|}}{C}HCH_3 \qquad (10)$$

$$CH_3CH_2\overset{\overset{CH_3}{|}}{\underset{\underset{O\cdot}{|}}{C}}CH_3 \longrightarrow CH_3CH_2\cdot + CH_3COCH_3 \qquad (11)$$

Carbon oxides, the other principal products of the oxidation, arise from the decomposition of acyl and acylperoxy-radicals:

$$CH_3CO\cdot \longrightarrow CH_3\cdot + CO \qquad (12)$$

$$CH_3COOO\cdot \longrightarrow CH_3O\cdot + CO_2 \qquad (13)$$

The above reaction sequences serve to illustrate only a few of the reactions by which the main products of the oxidation are formed. Even in the simplest case, when the feedstock is substantially pure butane, many more reactions are possible. For a naphtha feedstock the number is further multiplied and a wide variety of both acid and neutral products may be obtained in addition to those required. Fortunately most of the unwanted species will undergo further oxidation and hence may be returned to the reaction stage.

The fact that butane and naphtha oxidation processes produce acetic acid in commercially acceptable yield by reactions of such complexity is attributable to the great stability of this acid towards further oxidation.

Production Statistics and End Uses

Acetic Acid

The importance of acetic acid as a major petrochemical is shown by the annual growth rate of its production in the U.S.A.; it rose from 6% between 1950 and 1960 to almost 9% in the following decade. Demand for the acid in the U.S.A. in 1970 was 0.9×10^6 metric tons. It is estimated that the total world production will exceed 2×10^6 metric tons a year by 1975. Of this total approximately 0.8×10^6 metric tons a year (35–40%) will be produced by paraffin oxidation processes.

Acetic acid is used mainly for the manufacture of cellulose acetate via the anhydride (see Chapter 9), of vinyl acetate for emulsion paints and of acetate esters for use as solvents. Other important although smaller outlets include pharmaceuticals, chloroacetic acids and foodstuffs (pickling).

A breakdown of the U.S.A. market for acetic acid in 1970 is given in Table 3.8.

Table 3.8. End uses of acetic acid in the U.S.A. for 1970

End uses	% of market
Cellulose acetate	44
Vinyl acetate monomer	31
Solvent esters	14
Chloroacetic acid	3
Others	8

Formic Acid

The world capacity for formic acid is difficult to estimate since there are a multiplicity of small-scale producers, many of whom derive the acid as a by-product of other processes. In the U.S.A. the capacity in 1971 was 35×10^3 tons p.a.

The main uses of formic acid are in textile dyeing and finishing, leather tanning and chemical synthesis. In Europe there is also a growing market for the acid in the treatment of silage to combat the effects of dampness during storage.

Propionic Acid

Production statistics for propionic acid are not available. The main outlets are chlorinated herbicides, vinyl propionate for emulsion paints (Germany only) and calcium propionate which is used as a bread preservative. The acid is also marketed as a grain preservative.

CHAPTER 4

Primary Petrochemicals, I: Alkanes, Alkenes and Alkynes

(A. H. Jubb, ICI Corporate Laboratory, Runcorn)

Methane and Carbon Monoxide

'Synthesis Gas'

Methane, which is mainly used as an energy source, is obtained as natural gas and from petrochemical plants and refineries. One important use of methane is for the production of 'synthesis gas', i.e. mixtures of H_2 and CO in varying proportions. Two principal methods are used, either steam reforming or partial oxidation; they can also be used with the higher paraffins.

In the steam reforming process, the first stage involves reaction of methane and steam over a supported catalyst:

$$CH_4 + H_2O \rightleftharpoons CO + 3H_2 \qquad \Delta H = +205 \text{ kJ mol}^{-1}$$
$$CH_4 + 2H_2O \rightleftharpoons CO_2 + 4H_2 \qquad \Delta H = +163 \text{ kJ mol}^{-1}$$

The catalysts are generally based on nickel oxide plus a variety of promoters and are contained in externally heated reactor tubes. The reactions are carried out at pressures in excess of 2.5×10^6 Nm^{-2} and temperatures above 700°C. The 'steam ratio' (moles, steam:carbon) employed is 3:1.

Generally the effluent from the reformer contains unchanged hydrocarbon and a secondary reforming step is used over a catalyst operating above 1000°C.

Carbon monoxide can be removed in a 'shift converter' where it reacts catalytically with the excess steam to produce CO_2 and H_2. Catalysts containing Cr, Cu and Zn are probably used at 150–300°C to reduce the CO level to below 0.5%. When necessary, the CO_2 is then removed by absorption on solids or in solution by means of a variety of reagents. Steam reforming has been widely used to produce town gas; in the U.K. over 90% of town gas was obtained in 1968 by steam-reforming naphtha, but this situation has since changed with the discovery of the gas fields in the North Sea.

In the partial oxidation process, which can be catalytic or non-catalytic, methane is treated with a deficiency of oxygen:

$$CH_4 + \tfrac{1}{2}O_2 \longrightarrow CO + 2H_2 \qquad \Delta H = -36 \text{ kJ mol}^{-1}$$
$$CH_4 + O_2 \longrightarrow CO_2 + 2H_2$$

It is likely that the oxygen initially reacts to give $CO_2 + H_2O$ and that this is followed by a slower reaction of these products with the excess of methane to give CO–H_2 mixtures.

Table 4.1 gives some typical figures for 'synthesis gas' obtained by partial oxidation.

Table 4.1. 'Synthesis gas' compositions (vol.-% of dry, crude product) from the Shell partial oxidation process according to feedstock. Each product also contains ca. 1.4% of N_2

Component	Natural gas	Naphtha	Heavy fuel oil
H_2	60.9	51.6	46.1
CO	34.5	41.8	46.9
CO_2	2.8	4.8	4.3
CH_4	0.4	0.4	0.4

Table 4.2. Composition of 'synthesis gas' required for some typical chemical syntheses

End-use synthesis	Synthesis gas required
Ammonia	$3H_2:1N_2$
Methanol	$2H_2:1CO$
Oxo-reactions	$1H_2:1CO$
Phosgene, acetic acid, aldehydes, acrylates	CO
H_2O_2, benzene reduction, nitrobenzene reduction, etc.	H_2

There are several commercial processes using this partial oxidation technique of which the Texaco, Shell and Topsøe processes are the most widely known. Once again, CO conversion and CO_2 removal would be carried out as above where appropriate.

If pure CO is required, the gas is passed through an ammoniacal solution of copper salts which absorbs the CO. The latter is then liberated by heating the solution, washed with water and dried.

If pure hydrogen is required, the CO level can be reduced by catalytic reaction with steam, the resulting CO_2 being removed by absorption techniques. Currently about 90% of the world production of hydrogen is based on oil-derived feedstocks.

'Synthesis gases' are often processed further to suit the particular requirements of the downstream consuming-reactions. Table 4.2 shows some examples.

Methanol

The production of methanol from 'synthesis gas' is an established process:

$$CO + 2H_2 \rightleftharpoons CH_3OH \qquad \varDelta H = -109 \text{ kJ mol}^{-1}$$

Carbon monoxide and hydrogen react at 350–450°C and $25–45 \times 10^6$ Nm^{-2} over Cr–Zn catalysts to give partial gas conversion (<20%) per pass. The methanol is condensed out and the unchanged gases are recycled; ICI have recently introduced an improved highly selective copper-based catalyst which allows the use of pressures in the region of 7×10^6 Nm^{-2} (250°C). Carbon dioxide can also be converted into methanol, but the reaction is used to a lesser extent:

$$CO_2 + 3H_2 \longrightarrow CH_3OH + H_2O \qquad \varDelta H = -58.5 \text{ kJ mol}^{-1}$$

A limited amount of methanol is obtained as a co-product in the liquid-phase oxidation of butane as operated by Celanese at Pampa, Texas, for the production of acetic acid.

Methanol can be oxidized to formaldehyde in the vapour phase

$$CH_3OH + \tfrac{1}{2}O_2 \longrightarrow HCHO + H_2O \qquad \varDelta H = -158 \text{ kJ mol}^{-1}$$

over a silver catalyst at 500–650°C. In recent years catalysts based on iron and molybdenum have been developed and are claimed to give a 92% yield of formaldehyde. Formaldehyde is used in phenol–formaldehyde and urea–formaldehyde resins (see Chapter 8) and as an intermediate for the production of pentaerythritol, hexamethylenetetramine and glycollic acid. Its polymers form a useful thermoplastic material commercialized as 'Delrin' and 'Celcon'.

Fischer–Tropsch Reaction

In 1925, work in Fischer's Institute (Germany) established that hydrocarbons of high molecular weight could be obtained by catalytic (Fe–Co) reaction of CO and H_2 at 250–300°C and

0.1–1×10^6 Nm^{-2}. Further development during 1926–1928 led to improved catalysts based on cobalt, culminating in commercial production in Germany in 1936. By 1941, nine plants were in operation and in 1942 had an annual capacity of 0.74×10^6 tons. The process, named the Fischer–Tropsch process, has been little used apart from its widespread adoption in Germany during the war period.

The one exception is a plant operated by Sasol (South Africa), using a combination of processes developed by Kellogg (U.S.A.) and Lurgi–Ruhrchemie, respectively. The former process uses an iron catalyst circulating in the gas stream (H_2:CO ratio = 6:1), to produce mainly low-boiling hydrocarbons (50–70% straight chain), together with some water-soluble oxygenated compounds (mainly ethanol). The latter operates with a fixed-bed iron catalyst (H_2:CO ratio = 2:1) to produce essentially heavy oils and paraffins which are 90–95% straight-chain (Table 4.3). The process is operated because the lack of indigenous oil supplies has required the South African Government to seek alternative routes to liquid fuels based on coal which is available in vast quantities.

Table 4.3. Products from Fischer–Tropsch reactions (wt.-%)

Product	Catalyst-cycling process[a]	Fixed-bed process[a]
C_3–C_4	7.7	5.6
Petrol (C_5–C_{11})	72.3 (70%)	33.4 (50%)
Medium oil	3.4 (60%)	16.6 (40%)
Semisolid paraffins	3.0	10.3
Medium paraffins (m.p. 57–60°C)	—	11.8
Hard paraffins (m.p. 95–97°C)	—	18.0
Alcohols and Ketones	12.6	4.3
Organic acids	1.0	Trace

[a] Figures in parentheses indicate the olefin concentration in the fraction.

Many theories have been proposed for the reaction mechanism, including carbide intermediates and, without doubt, a complex set of heterogeneous reactions is involved. The overall reactions can be represented:

$$n CO + 2n H_2 \longrightarrow (CH_2)_n + n H_2O \qquad \Delta H = -192 \text{ kJ mol}^{-1}$$

$$2n CO + n H_2 \longrightarrow (CH_2)_n + n CO_2$$

Carbon Monoxide

Other outlets for carbon monoxide include phosgene and formic acid. The former is obtained by reaction of an excess of carbon monoxide with chlorine in a reactor packed with activated charcoal:

$$CO + Cl_2 \longrightarrow COCl_2$$

The main use of phosgene is in the production of isocyanates from amines, polycarbonate resins by reaction with dihydroxy-compounds, and isocyanuric acid by reaction with ammonia. Formic acid is produced by absorbing CO in sodium hydroxide solution, or in methanol at high pressure:

$$CO + NaOH \longrightarrow HCOONa \xrightarrow{H^+} HCOOH$$

$$CO + CH_3OH \longrightarrow HCOOCH_3 \xrightarrow{NH_3} HCONH_2 \xrightarrow[H_2O]{H^+} HCOOH$$

This acid is also obtained, to a growing extent, as a by-product of the hydrocarbon oxidation process to acetic acid (see Chapter 3, p. 61).

Carbon monoxide and water react with olefins at high pressure under acid conditions to yield tertiary carboxylic acids. The reaction is known as the Koch reaction and pivalic acid is obtained in this manner from isobutene:

$$(CH_3)_2C{=}CH_2 + CO + H_2O \xrightarrow{H^+} (CH_3)_3CCOOH$$
Pivalic acid

The reaction is also applied to higher branched olefins, propylene trimer or diisobutene, to give branched-chain acids used as paint driers. The catalytic addition of carbon monoxide and hydrogen to olefinic double bonds is called the 'oxo'-reaction and is discussed in Chapter 6 (p. 153).

Methane: Other Uses

Methane itself is utilized to produce hydrogen cyanide, carbon disulphide and chlorinated methanes.

Hydrogen cyanide can be obtained from methane by several processes. The Andrussow process is based on the reaction of ammonia, methane and air over a platinum-based catalyst at 1000°C:

$$CH_4 + NH_3 + \tfrac{3}{2}O_2 \longrightarrow HCN + 3H_2O$$

A modification of this process initially oxidizes ammonia with air over platinum gauze containing 10% of rhodium at 900°C:

$$4NH_3 + 5O_2 \longrightarrow 4NO + 6H_2O$$

then, after cooling to 600°C, the gaseous products are caused to react with methane:

$$2NO + 2CH_4 \longrightarrow 2HCN + 2H_2O + H_2$$

In the Degussa process, used in Germany, use is made of the reaction discovered by Kuhlmann in 1842, whereby ammonia reacts directly with methane:

$$CH_4 + NH_3 \longrightarrow HCN + 3H_2$$

The reaction is endothermic and is carried out over a platinum catalyst contained in reactor tubes coated with a ceramic material; temperatures around 1200–1300°C are used. The hydrogen can be recovered as a useful by-product.

In the Shawinigan process, which can also be applied to propane or heavier feedstocks, methane is treated with ammonia at 1400°C in a fluidized bed of coke heated electrically. Yields of 85% are claimed based on both methane and ammonia.

Hydrogen cyanide can also be obtained indirectly by reactions starting from carbon monoxide.

$$CO + CH_3OH \xrightarrow[10^6 \ Nm^{-2}]{100°C} \underset{\substack{\text{Methyl} \\ \text{formate}}}{HCOOCH_3} \xrightarrow[40°C/1.5 \times 10^6 \ Nm^{-2}]{NH_3}$$

$$\underset{\text{Formamide}}{HCONH_2} \xrightarrow[200–500°C]{Al_2O_3} HCN + H_2O$$

An increasing amount of HCN is obtained as a by-product in the production of acrylonitrile from propylene (see Chapter 6, p. 145).

Carbon disulphide, traditionally manufactured from charcoal and sulphur, can be obtained by reaction of methane with sulphur at 700°C and 0.2×10^6 Nm^{-2} (Thacker process):

$$CH_4 + 2S_2 \longrightarrow CS_2 + 2H_2S$$
$$CH_4 + S_2 \longrightarrow CS_2 + 2H_2$$
$$CH_4 + 2H_2S \longrightarrow CS_2 + 4H_2$$

The hydrogen sulphide produced is recovered as sulphur by using the Claus process:

$$H_2S + \tfrac{3}{2}O_2 \longrightarrow SO_2 + H_2O; \qquad SO_2 + 2H_2S \longrightarrow 3S + 2H_2O$$

Although methane is less reactive than its homologues to chlorine, it reacts readily at 400–440°C. By adjusting the overall ratio of chlorine to methane it is possible to obtain mainly methyl chloride, at one extreme, or mainly carbon tetrachloride at the other extreme. Thus, at a chlorine-to-methane molar ratio of 0.6:1, it is reported that the following mole ratios of products are obtained: methyl chloride 6.0, methylene dichloride 3.0, chloroform 1.0, carbon tetrachloride 0.25. The methyl chloride can be isolated by distillation, and further chlorination of the other products can be carried out by photo-activation in the liquid phase. Carbon tetrachloride has the largest market.

$$CH_4 + Cl_2 \longrightarrow HCl + CH_3Cl \xrightarrow{Cl_2} CH_2Cl_2 + HCl$$

$$\xrightarrow{Cl_2} CHCl_3 + HCl \xrightarrow{Cl_2} CCl_4 + HCl$$

Methyl chloride is often produced by reaction of methanol with hydrogen chloride.

Acetylene

Acetylene can be obtained from calcium carbide and as a co-product from ethylene crackers (Chapter 2). Additionally, it can be obtained as a primary product from hydrocarbons by several processes (Chapter 3) which make use of the fact that, although acetylene is thermodynamically unstable at normal temperatures, it becomes less so at high temperatures; in fact at 1200°C it is the least unstable of the hydrocarbons (Figure 3.2, Chapter 3) and therefore can be synthesized from the latter.

The main impetus for the industrial development of acetylene chemistry arose from the work of Reppe in Germany. This led to a significant production of a variety of materials based on acetylene during World War II. It is reported that the peak of acetylene production (440×10^6 m^3) was reached in Germany during 1943–1944, 86% being produced by the carbide process. Up-to-date statistics on total acetylene production are not readily available but Table 4.4 shows the situation in 1968.

In 1970, the acetylene obtained in the U.S.A. from petrochemical fractions (0.21×10^6 tons) was used to produce vinyl chloride (40%), vinyl acetate (30%), acrylates (22%), various acetylenic chemicals and chlorinated solvents. Non-chemical use

Table 4.4. Acetylene production (tons × 10⁻³)

Country	Total	From petrochemical feedstock
Germany	318	187
Austria	14	Nil
Spain	57	Nil
France	173	19
Italy	265	195
U.S.A.	465^a	Not reported
Japan	503	46

^{a} 0.21×10^6 tons based on petrochemical feedstocks in 1970.

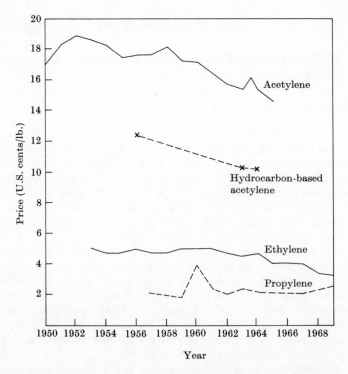

Figure 4.1. Approximate average annual U.S.A. prices of hydrocarbons. The acetylene price is the average for chemical and non-chemical uses. The propylene price is the average for refinery, chemical and polymer grades. (Source: U.S. Bureau of Census; U.S. Tariff Commission.)

of acetylene, mainly for welding and metal-cutting, represents an important but relatively small proportion of total consumption.

Acetylene has to be utilized in a dilute form at pressures exceeding 0.2×10^6 Nm^{-2}, and pressure vessels are designed to withstand pressure surges of twelve times the operating pressure; this is to contain any acetylene explosion that might occur. Generally, although there are potential hazards associated with the use of acetylene due to the ease of decomposition, the reactions of acetylene are relatively easy to carry out and purification of the products is also relatively uncomplicated.

However, acetylene-based syntheses have been steadily replaced by those based on ethylene owing to the lower cost of the latter in recent years (see Figure 4.1), so that there are few areas where acetylene remains the dominant feedstock. In this context, it is interesting that during 1944 Germany produced about half its ethylene requirements by partial hydrogenation of acetylene, a situation forced by necessity. Figure 4.2 (p. 76) indicates the versatility of acetylene chemistry, but the following rather more detailed discussion indicates those areas where acetylene continues to play a significant role.

Vinyl chloride (see Chapter 7) is readily obtained in 85–90% yield by vapour-phase reaction of acetylene with HCl over a HgCl$_2$ catalyst supported on carbon:

$$CH{\equiv}CH + HCl \longrightarrow CH_2{=}CHCl \qquad \Delta H = -184 \text{ kJ mol}^{-1}$$

Vinyl acetate is produced by the vapour-phase reaction between acetylene and acetic acid at 170–220°C over a catalyst comprising zinc or cadmium acetate on active charcoal.

$$CH{\equiv}CH + CH_3COOH \longrightarrow CH_2{=}CHOOCCH_3$$

Acetylene can be dimerized to vinylacetylene which, on hydrochlorination, produces chloroprene, the monomer that is used to produce neoprene rubber (see Chapter 8):

$$2CH{\equiv}CH \longrightarrow \underset{\text{Vinylacetylene}}{CH{\equiv}CCH{=}CH_2} \xrightarrow{\text{HCl}} \underset{\text{Chloroprene}}{CH_2{=}CClCH{=}CH_2}$$

Acetaldehyde is readily obtained by catalytic hydration of acetylene with sulphuric acid in the presence of mercuric sulphate as catalyst (Part 1, p. 108):

$$CH{\equiv}CH + H_2O \longrightarrow CH_3CHO$$

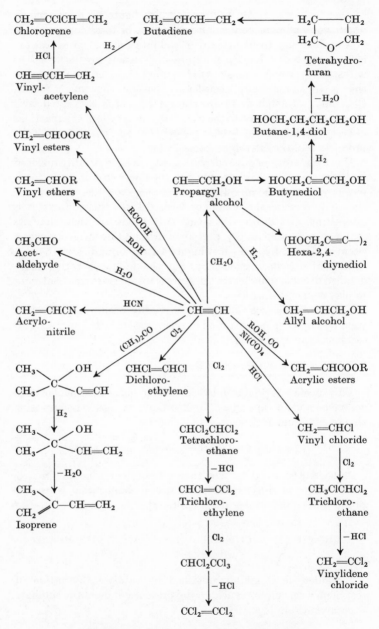

Figure 4.2. Some general reactions of acetylene.

A second route to acetaldehyde, developed by Reppe, was based on the reaction of methanol with acetylene to give methyl vinyl ether which was then hydrolyzed with dilute phosphoric acid. This route is no longer used and the hydration route is obsolete in most countries. However, the ether and its higher homologues continue to be made on a limited scale by the first stage.

Although there are several propylene-based processes available for producing acrylic acid and acrylates, those based on acetylene remain important, e.g.:

$$HC\equiv CH + CO + ROH \xrightarrow[40-50°C]{Ni(CO)_4} CH_2=CHCOOR$$

The older version utilizes nickel carbonyl in aqueous HCl, which supplies some of the CO in addition to acting as a catalyst. BASF have developed a process using nickel bromide (0.1%) as catalyst in tetrahydrofuran as solvent; acrylic acid is obtained and is subsequently esterified with the appropriate alcohol.

Acrylonitrile can be produced from acetylene but this process has been virtually superseded by the propylene-based routes (see Chapter 6, p. 145):

$$CH\equiv CH + HCN \xrightarrow{CuCl/NH_4Cl} CH_2=CHCN$$

Reaction of acetylene with formaldehyde gives propargyl alcohol and, subsequently, butyne-1,4-diol:

$$CH\equiv CH \xrightarrow{HCHO} CH\equiv CCH_2OH \xrightarrow{HCHO} HOCH_2C\equiv CCH_2OH$$

$$\text{Propargyl} \qquad\qquad\qquad \text{Butyne-1,4-diol}$$
$$\text{alcohol}$$

The latter product can be readily hydrogenated to butane-1,4-diol which, on dehydrogenation over a copper catalyst, readily produces γ-butyrolactone. On reaction with methylamine at 200°C under pressure, the latter yields 1-methylpyrrolidone; this is used as an extraction solvent for both acetylene and aromatic hydrocarbons.

$$HOCH_2C\equiv CCH_2OH \xrightarrow{H_2} HOCH_2CH_2CH_2CH_2OH \xrightarrow{-H_2}$$

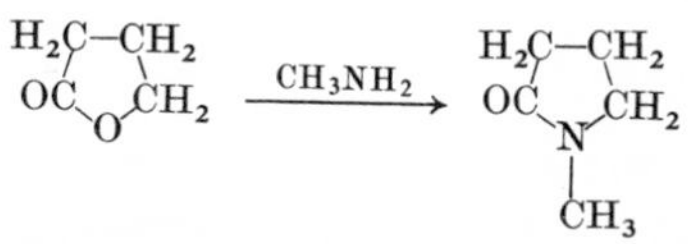

γ-Butyrolactone   1-Methylpyrrolidone

SNAM Progetti have developed a process for isoprene production based on the catalysed reaction of acetone with acetylene. The intermediate 2-methylbutyn-2-ol is selectively hydrogenated to methylbutenol which is subsequently dehydrated to isoprene.

$$CH{\equiv}CH + \begin{array}{c}CH_3\\CH_3\end{array}{>}CO \xrightarrow[2\times10^6\ Nm^{-2}]{10-40°C} \begin{array}{c}CH_3\\CH_3\end{array}{>}\overset{OH}{\underset{|}{C}}{-}C{\equiv}CH \xrightarrow[Pd/Rh/BaSO_4]{H_2}$$

$$\begin{array}{c}CH_3\\CH_3\end{array}{>}\overset{OH}{\underset{|}{C}}{-}CH{=}CH_2 \xrightarrow[250-300°C]{Al_2O_3} \begin{array}{c}CH_3\\CH_2\end{array}{>}C{-}CH{=}CH_2$$
                                                                                                 Isoprene

Addition of chlorine to acetylene in tetrachloroethane at 80°C in the presence of antimony pentachloride or ferric chloride yields 1,1,2,2-tetrachloroethane, $(CHCl_2)_2$. The latter readily loses HCl on treatment with lime at 50°C, or on pyrolysis at 220–320°C over barium and copper chloride supported on carbon, to yield trichloroethylene:

$$CHCl_2CHCl_2 \longrightarrow CHCl{=}CCl_2 + HCl$$

Trichloroethylene is used as a solvent and for metal degreasing; it has to be stabilized by the addition of acetylenic alcohols. It is also produced via chlorination of ethylene dichloride, a process that has come into greater prominence with the advent of oxychlorination techniques (Chapter 6).

Further addition of chlorine yields pentachloroethane which on dehydrochlorination with lime provides perchloroethylene:

$$CHCl{=}CCl_2 + Cl_2 \longrightarrow CHCl_2CCl_3 + HCl$$
                                                                 Pentachloroethane

$$CHCl_2CCl_3 \xrightarrow{Ca(OH)_2} CCl_2{=}CCl_2$$
                                                        Perchloroethylene

Perchloroethylene is primarily used as a dry-cleaning solvent and once again there has been a gradual move to base its production on ethylene.

Ethylene

Ethylene is the most important single raw material used as a chemical feedstock; its world-wide consumption has grown from 6.5×10^5 tons in 1950 to 1.9×10^7 tons in 1970 (Figure 4.3) and

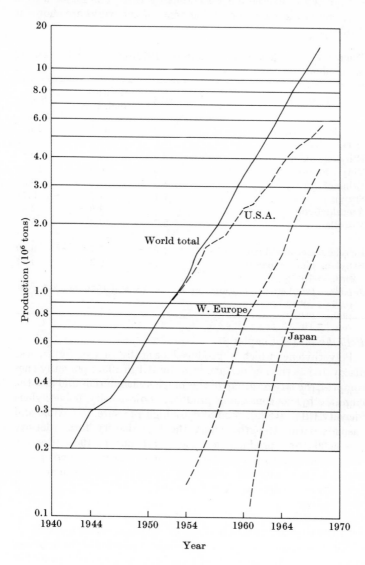

Figure 4.3. World production of ethylene.

is expected to exceed 5.5×10^7 tons by 1980. The major uses of ethylene in the main consuming areas of the world are shown in Table 4.5.

Table 4.5. Ethylene end-use pattern (1972) (% of total consumed per product)

Product	U.K.	W. Europe (incl. U.K.)	U.S.A.
L.D. Polyethylene	39.5	39.5	28.2
H.D. Polyethylene	6.3	13.5	11.6
Ethylene oxide	19.6	13.1	20.1
Vinyl chloride	16.8	16.9	15.4
Ethanol	8.0	2.7	6.3
Styrene	7.1	7.5	9.7
Acetaldehyde	—	3.8	2.8
Miscellaneous	2.7	3.0	5.9
Ethylene consumed (tons $\times 10^{-6}$)	1.04	7.70	9.10
Ethylene plant capacity (tons $\times 10^{-6}$)	1.18	9.35	9.24

Japanese ethylene production was 4×10^6 tons in 1970 and 3.5×10^6 tons in 1971.

Polyethylene (cf. Chapter 8)

Polyethylene, which is produced basically in two forms, was discovered as the low-density form by ICI in 1932; polyethylenes comprise the largest group of thermoplastic materials and also the largest ethylene-consuming products. Low-density polyethylene (density 0.910–0.925), produced by high pressure, is flexible and has good tear-strength, whilst the high-density form (density 0.94–0.96), first produced at modest pressure in 1953 after the work of Ziegler, is more crystalline, more resistant to creep and has a higher softening point (ca. 120°C). World demand for polyethylene is illustrated in Figure 4.4.

Ethylene Oxide

Ethylene oxide used to be obtained by reaction of ethylene with hypochlorous acid followed by dehydrochlorination of the resulting chlorohydrin with lime, ethylene dichloride and dichlorodiethyl ether being obtained as by-products:

$$CH_2{=}CH_2 + HOCl \longrightarrow ClCH_2CH_2OH \xrightarrow{Ca(OH)_2} H_2C{-}CH_2$$
$$\underset{O}{\diagdown\diagup}$$

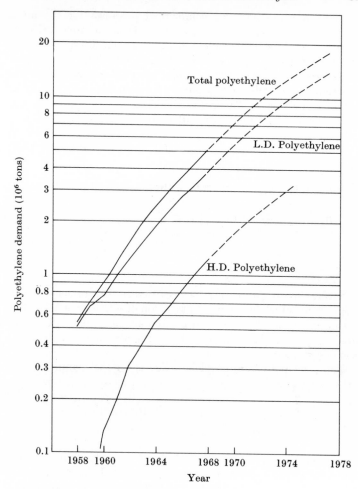

Figure 4.4. World demand for polyethylene (excluding China).

Although an 85% yield of ethylene oxide can be obtained, based on ethylene, and plant capital costs are relatively modest, the high consumption of chlorine (2 tons per ton of oxide) has led to the process being displaced by the direct oxidation route. This move has been accelerated by the ability to use the chlorohydrin plants for the production of propylene oxide. By mid-1969 about 90% of U.S.A. plant capacity was based on oxidation processes.

There are at least four commercial processes for the direct oxidation of ethylene to ethylene oxide, all using a supported silver catalyst. The Union Carbide and the Scientific Design processes use air-oxidation, whilst that developed by Shell utilizes oxygen; units producing 100×10^3 tons p.a. are now commonplace. The reaction is generally carried out with ethylene concentrations of <3% by volume at 200–290°C and pressures in the range 1.0–2.5×10^6 Nm^{-2}. Yields of the order of 65–70% are obtained.

$$CH_2{=}CH_2 + \tfrac{1}{2}O_2 \longrightarrow H_2C\overset{\displaystyle}{\underset{O}{-}}CH_2 \qquad \varDelta H = -146 \text{ kJ mol}^{-1}$$

$$CH_2{=}CH_2 + 3O_2 \longrightarrow 2CO_2 + 2H_2O \qquad \varDelta H = -1{,}322 \text{ kJ mol}^{-1}$$

$$CH_2{=}CH_2 \xrightarrow{\;k_1\;} H_2C\overset{\displaystyle}{\underset{O}{-}}CH_2$$
$$\underset{k_3}{\searrow} \quad \underset{k_2}{\swarrow}$$
$$CO_2$$

Several investigations have been carried out to establish whether the CO_2-forming step is direct oxidation of ethylene or the subsequent oxidation of ethylene oxide. It was found that $k_2/k_1 \simeq 0.08$ and $k_3/k_1 \simeq 0.4$, showing that parallel oxidation of ethylene was the more significant.

The principal use of ethylene oxide is for the manufacture of ethylene glycol:

$$H_2C\overset{\displaystyle}{\underset{O}{-}}CH_2 + H_2O \longrightarrow HOCH_2CH_2OH$$

This comprises over 55% of the U.S.A. ethylene oxide consumption and over 80% of the Japanese consumption. Ethylene glycol is still mainly used in antifreeze mixtures, but an important and fast-growing market is as a feedstock for polyester fibre production. Other uses include the production of surfactants, ethanolamines, higher ethylene glycols and glycol ethers.

Vinyl Chloride

Union Carbide have claimed the direct chlorination of ethylene to vinyl chloride in 87% yield but there is no evidence that any process based on these claims has been commercialized. In the event, ethylene is converted into vinyl chloride in two stages. The first stage involves the intermediate production of ethylene dichloride which can be carried out by either direct chlorination of ethylene or by oxychlorination of ethylene with HCl/air.

$$CH_2{=}CH_2 + Cl_2 \longrightarrow ClCH_2CH_2Cl + HCl$$
$$CH_2{=}CH_2 + 2HCl + \tfrac{1}{2}O_2 \longrightarrow ClCH_2CH_2Cl + H_2O$$

The ethylene dichloride is then dehydrochlorinated to vinyl chloride, the HCl produced being recycled to the oxychlorination step where appropriate. Details of the chemistry involved in these reactions are given in Chapter 7.

Ethanol (see Chapter 6, p. 159)

There are two synthetic processes used for the production of ethanol from ethylene. In the first ethylene (usually 95% concentrated, but lower concentrations can be used) is absorbed in a series of 95–98% sulphuric acid absorbers at 55–85°C and 1.5–4.0×10^6 Nm^{-2}.

$$CH_2{=}CH_2 + H_2SO_4 \longrightarrow CH_3CH_2OSO_3H$$
Ethyl hydrogen sulphate
$$2CH_2{=}CH_2 + H_2SO_4 \longrightarrow (CH_3CH_2)_2SO_4$$
Diethyl sulphate

The intermediate ethyl hydrogen sulphate and the by-product diethyl sulphate are hydrolyzed with steam.

$$CH_3CH_2OSO_3H + H_2O \longrightarrow CH_3CH_2OH + H_2SO_4$$
$$(CH_3CH_2)_2SO_4 + H_2O \longrightarrow C_2H_5OC_2H_5 + H_2SO_4$$

The alcohol and ether are distilled from the weak sulphuric acid which is then concentrated for re-use in the absorbers. The ether and ethanol are recovered by distillation; the yield of the latter is about 90%.

The second process involves the catalyzed hydration of ethylene over a phosphoric acid catalyst supported on diatomaceous earth at 300°C and 6–7×10^6 Nm^{-2}. The ethylene conversion is only about 5% per pass but ethanol yields of the order of 95% are obtained.

Ethanol is used as a solvent and as a raw material for the production of acetaldehyde, acetic acid and methyl acetate.

Styrene

The intermediate used for styrene production is ethylbenzene which can be obtained from catalytic reforming plants or by the direct ethylation of benzene. The direct ethylation of benzene by ethylene over aluminium chloride as catalyst is described in

Chapter 5 (p. 121). Ethylbenzene is also produced by vapour-phase reaction at 200–250°C and 1.4×10^6 Nm^{-2} of ethylene and benzene over a fixed bed of kieselguhr impregnated with phosphoric acid. Boron trifluoride catalysts are used in Universal Oil Company's 'Alkar' process, and claims have also been made for the use of molecular-sieve catalysts.

The main processes for converting ethylbenzene into styrene involve catalytic dehydrogenation techniques. The process is reversible and the equilibrium favours styrene formation at higher temperatures, but a limit is imposed by the instability of styrene and ethylbenzene. The vapourized feed is heated with superheated steam and dehydrogenated at 630°C over a catalyst such as zinc or chromium oxide. Generally, ethylbenzene conversions of 40–60% are achieved, giving 90–92% yields of styrene. Vacuum-distillation is used to separate the styrene from unchanged ethylbenzene; polymerization inhibitors are added during distillation and during storage.

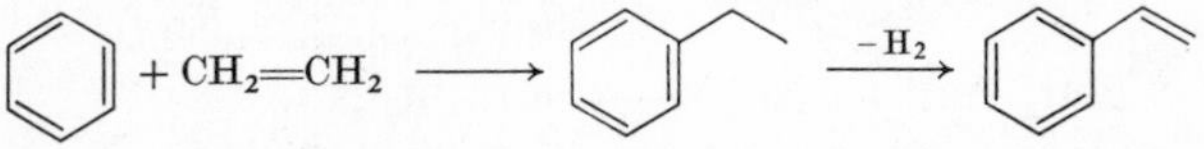

Union Carbide developed a process involving the liquid-phase oxidation of ethylbenzene (manganese acetate catalyst) to phenylmethylcarbinol (α-methylbenzyl alcohol); acetophenone and benzoic acid were co-produced and the ketone could be readily converted into the alcohol. The latter could then be dehydrated readily to styrene at 300°C over a TiO$_2$ catalyst.

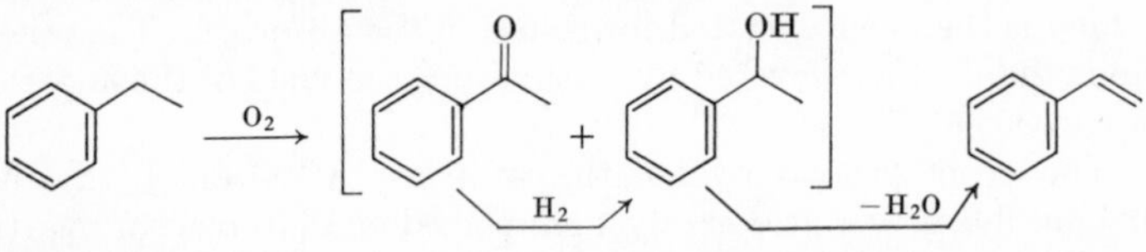

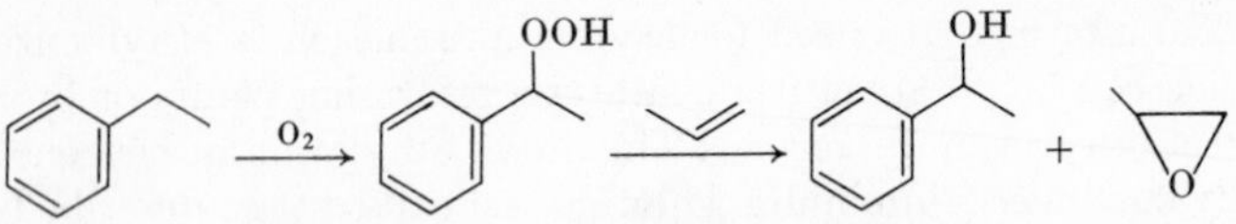

Recently a new process has been developed whereby ethylbenzene is oxidized to its hydroperoxide at 130°C and 0.5×10^6 Nm^{-2}; 13% conversion of ethylbenzene is used to give 85% yield of the hydroperoxide. After concentration (to 35%), the hydroperoxide solution is treated at 110°C with propylene in the presence of tungsten or molybdenum naphthenate to produce the alcohol and propylene oxide (see p. 93). The alcohol is dehydrated to styrene over a TiO_2 catalyst. It is reported that three plants (in the U.S.A., Holland and Spain) have been constructed based on this technology.

Acetaldehyde

Up to the early 1960's the bulk of the U.S. acetaldehyde was obtained from ethanol by oxidation or dehydrogenation:

$$C_2H_5OH + \tfrac{1}{2}O_2 \xrightarrow[\text{Ag catalyst}]{375-500°C} CH_3CHO + H_2O$$

$$C_2H_5OH \xrightarrow[250-300°C]{\text{Cu/Cr catalyst}} CH_3CHO + H_2$$

In Europe, at that time, a significant amount of acetaldehyde was still being made from acetylene.

Phillips had described the reaction of ethylene with palladium chloride in 1894 and further publications appeared by Anderson (1934) and Kharasch (1938). However, it was in 1956 that the commercial potential was appreciated and the reaction developed as the Wacker process. Essentially, ethylene is oxidized in the presence of an aqueous solution of cupric chloride and palladium chloride. Two variants can be used: a one-stage process in which the catalyst is regenerated *in situ* (developed by Farbwerke Hoechst) by oxygen, and a two-stage process whereby air is used to regenerate the catalyst in a separate reactor (developed by Wacker-Chemie). Yields of 95% are reported from ethylene. The basic steps in the process can be summarized as follows:

$$CH_2{=}CH_2 + PdCl_2 + H_2O \longrightarrow CH_3CHO + Pd + 2HCl$$
$$Pd + 2CuCl_2 \longrightarrow PdCl_2 + 2CuCl$$
$$2CuCl + \tfrac{1}{2}O_2 + 2HCl \longrightarrow 2CuCl_2 + H_2O$$

The net reaction is:

$$CH_2{=}CH_2 + \tfrac{1}{2}O_2 \longrightarrow CH_3CHO$$

Thus, the reaction is catalytic with respect to palladium. Other oxidants have been claimed in place of $CuCl_2$, e.g. benzoquinone, PbO_2 and Fe^{3+}, but only the first of these has been used commercially.

The reaction rate increases with $[Pd^{2+}]$ and $[C_2H_4]$ and is inhibited by $[Cl^-]$ and $[H^+]$; the rate expression is

$$-\frac{d/[C_2H_4]}{dt} = \frac{k[PdCl_4^{2-}][C_2H_4]}{[Cl^-][H_3O^+]}$$

This expression also holds for propylene, but-1-ene and but-2-ene. From experiments with deuteriated ethylene (C_2D_4) it was found that $k_{C_2H_4}/k_{C_2D_4} = 1.07$, suggesting that C—H or C—D bond cleavage is not rate-determining. Reaction of ethylene with $PdCl_2$ in D_2O solution produced deuterium-free acetaldehyde. Thus, all the four hydrogen atoms of the latter are present in the original ethylene molecule. The observed results can be accommodated in the mechanism proposed by Henry (Scheme 1).

$$[PdCl_4]^{2-} + C_2H_4 \underset{k_{-1}}{\overset{k_1}{\rightleftharpoons}} [PdCl_3C_2H_4]^- + Cl^-$$

$$k_{-2} \left\Vert\; k_2 \;\right. H_2O$$

$$[PdCl_2(OH_2)C_2H_4]^0 + Cl^-$$

$$k_{-3} \left\Vert\; k_3 \;\right. H_2O$$

$$[PdCl_2(OH)C_2H_4]^- + H_3O^+$$

$$k_{-4} \left\Vert\; k_4(\text{slow}) \right.$$

$$[PdCl(CH_2CH_2OH)] + Cl^-$$

$$\downarrow k_5$$

$$CH_3CHO + Pd^0 + Cl^-$$

SCHEME 1.

The intermediate step can be visualized as shown in Scheme 2.

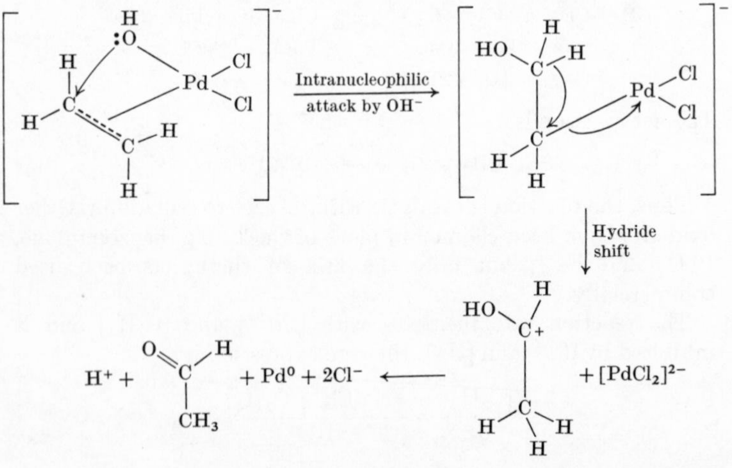

SCHEME 2.

In 1968, 39% of the U.S., 46% of the West European and virtually 100% of the Japanese acetaldehyde production was based on ethylene.

The manufacture of acetic acid, acetic anhydride, butan-1-ol and 2-ethylhexan-1-ol accounts for over 80% of the acetaldehyde consumption. The production of vinyl acetate, pentaerythritol, chloral and pyridine bases consumes the bulk of the remainder. Butan-1-ol and 2-ethylhexan-1-ol are produced from acetaldehyde by the following series of condensations and hydrogenations.

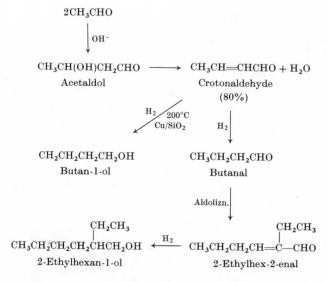

The 1972 U.S. acetaldehyde plant capacity is estimated at 0.7×10^6 tons; the 1970 demand was 0.62×10^6 tons; the end-uses were acetic acid and acetic anhydride 49%, butanols 14%, 2-ethyl-hexanol 12%, others 25%.

Vinyl Acetate

Announcement of the acetaldehyde process based on ethylene was followed by a large number of publications on the reaction between ethylene and palladium chloride in acetic acid systems. Subsequently, several processes were developed and commercialized in U.S.A., Europe and Japan. They are modifications of two basic approaches. First, a liquid-phase reaction, carried out at 120–130°C and 1.0×10^6 Nm^{-2} in the presence of cupric chloride and lithium salts. Oxygen is used for the *in situ* regeneration of the cupric ion which in turn re-oxidizes zerovalent palladium. The presence of water causes the formation of acetaldehyde. Secondly, vapour-phase conditions in which ethylene, oxygen and acetic acid are passed over solid catalysts containing palladium (i.e. carbon/$PdCl_2$/$CuCl_2$ or $PdCl_2$ on Al_2O_3) at 160°C and 5–10×10^5 Nm^{-2}. In each case vinyl acetate yields in excess of 90% have been claimed. The intermediate step involved in these reactions is believed to be as shown in Scheme 3.

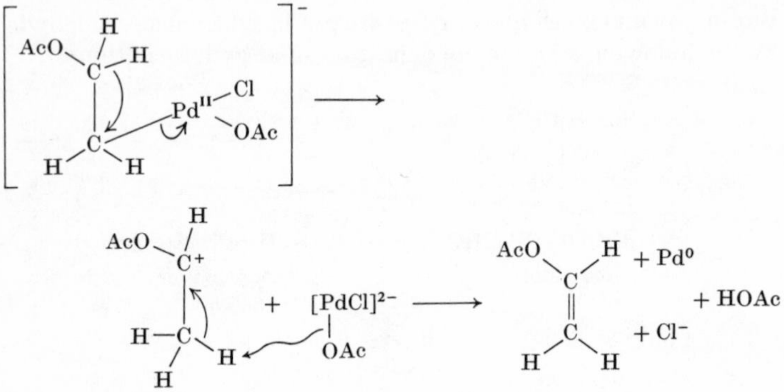

SCHEME 3.

The vinyl acetate yield is critically dependent on [Cl^-] and if this is too high then ethylidene diacetate is formed preferentially (cf. Scheme 4).

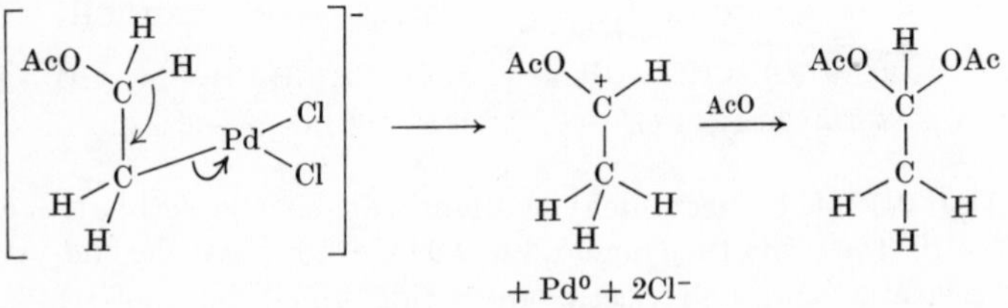

SCHEME 4.

In addition to the ethylene route, vinyl acetate can be produced from acetaldehyde and acetic anhydride (or acetic acid), the intermediate ethylidene diacetate being pyrolyzed in the presence of a strong acid:

$$CH_3CHO + (CH_3CO)_2O \longrightarrow CH_3CH(OCOCH_3)_2 \longrightarrow$$
$$CH_3COOCH{=}CH_2 + CH_3COOH$$

The 1970 U.S.A. vinyl acetate production was 0.36×10^6 tons, of which 90% was polymerized to produce poly(vinyl acetate), poly(vinyl alcohol) and poly(vinyl butyral); the remainder is co-polymerized with other monomers such as vinyl chloride. The poly(vinyl acetate) was mainly used in latexes, resins, emulsion paints and adhesives.

Long-chain Olefins

An extension of the Ziegler polymerization of ethylene can be used to give oligomers of ethylene, mainly in the range C_4–C_{20} (Poisson distribution obtained with a maximum concentration at C_{12}). Finely powdered aluminium is activated in a solvent by ball-milling and treated with hydrogen under pressure in the presence of triethylaluminium. The resulting diethylaluminium hydride is treated under pressure with ethylene to yield triethyl-aluminium:

$$2Al + 3H_2 + 4Al(C_2H_5)_3 \longrightarrow 6AlH(C_2H_5)_2$$
$$6AlH(C_2H_5)_2 + 6C_2H_4 \longrightarrow 6Al(C_2H_5)_3$$

A portion of the triethylaluminium is recycled to the hydride-forming step and the remainder is treated under pressure $(7$–$10 \times 10^6\ Nm^{-2})$ with ethylene at temperatures below $130°C$. Ethylene adds to the alkylaluminium in the following manner:

$$Al(C_2H_5)_3 + nCH_2{=}CH_2 \longrightarrow Al{\Big\langle}\begin{array}{l}(CH_2CH_2)_xH\\(CH_2CH_2)_yH\\(CH_2CH_2)_zH\end{array}$$

At higher temperatures $(130$–$250°C)$, a competing reaction occurs:

$$Al{-}\begin{array}{l}(CH_2CH_2)_xH\\(CH_2CH_2)_yH\\(CH_2CH_2)_zH\end{array} + 3CH_2{=}CH_2 \longrightarrow$$

$$Al(C_2H_5)_3 + \left\{\begin{array}{l}H(CH_2CH_2)_{x-1}CH{=}CH_2\\H(CH_2CH_2)_{y-1}CH{=}CH_2\\H(CH_2CH_2)_{z-1}CH{=}CH_2\end{array}\right.$$

this reaction can be carried out thermally or catalytically (Ni metal at $50°C$). Generally, the non-catalytic method is used and about 95% of straight-chain α-olefins are obtained; some double-bond migration and chain-branching occur in side-reactions. The $Al(C_2H_5)_3$ can be recovered by distillation or extraction.

The olefins obtained are used as feedstocks for plasticizer alcohols $(C_7$–$C_9)$ and detergents $(C_{12}$–$C_{18})$. This process is known as the 'Alfene' process and was originally developed by the Continental Oil Co. and operated by Gulf Oil in the U.S.A.

In a variant on this process, the 'Alfol' process, the mixture of trialkylaluminium compounds is oxidized to the corresponding trialkoxides which are subsequently hydrolyzed by sulphuric acid to yield a mixture of C_4–C_{22} primary alcohols as illustrated.

$$3CH_2{=}CH_2 + \tfrac{3}{2}H_2 + Al \longrightarrow (C_2H_5)_3Al$$

$$(C_2H_5)_3Al + nCH_2{=}CH_2 \longrightarrow \begin{matrix} H(CH_2CH_2)_x \\ H(CH_2CH_2)_y{-}Al \\ H(CH_2CH_2)_z \end{matrix} \xrightarrow{O_2}$$

$$\begin{matrix} H(CH_2CH_2)_xO \\ H(CH_2CH_2)_yO{-}Al \\ H(CH_2CH_2)_zO \end{matrix} \xrightarrow{H_2SO_4} \begin{cases} H(CH_2CH_2)_xOH \\ H(CH_2CH_2)_yOH + \tfrac{1}{2}Al_2(SO_4)_3 \\ H(CH_2CH_2)_zOH \end{cases}$$

The process is operated in the U.S.A. and Germany and the alcohols produced are used as plasticizer feedstocks (C_6–C_{10} converted into phthalate esters) and detergent feedstocks (C_{12}–C_{18}).

Ethylene–Propylene Rubbers

As a result of Natta's work on the polymerization of propylene, copolymers of ethylene and propylene (ca. 50:50) were produced commercially by using Ziegler catalysts. The elastomers (EPR elastomers) obtained have good abrasion-, ozonolysis- and oxidation-resistance and are potentially cheaper than other elastomers. As they contain no unsaturated centres they can only be cured with peroxides, which has some distinct disadvantages. In order to overcome these problems and to allow sulphur-curing, a third monomer, containing two dissimilar double bonds, is incorporated (5–10%). The products are known as 'ethylene–propylene terpolymer' rubbers (EPT). Dicyclopentadiene, hexa-1,4-diene and ethylidenenorbornene are examples of the dienes used. Consumption of such polymers is recorded in Table 4.6.

Table 4.6. EPT rubber consumption (tons $\times 10^3$)

Country	1965	1970	1975 (estd.)
U.S.A.	13	75	150
U.K.	—	3	23
W. Europe	1	23	85
Japan	—	13	55
Total World	14	150	320

Propylene

Propylene is obtained as a co-product from ethylene crackers and also from refinery processes. Typically a naphtha-based ethylene cracker will provide a propylene-to-ethylene ratio in the range (0.4–0.5):1. In the U.S.A. a significant amount of propylene is consumed by refineries for alkylate and polymer gasoline production.

Polypropylene

Unlike ethylene, which can be polymerized at high or low pressure, polypropylene is produced only by Ziegler-type technology. In most of its applications polypropylene competes with high-density polyethylene and is used in injection moulding and extrusion applications. Polypropylene is also used in fibre form as continuous filament, monofilament and staple in addition to slit and fibrillated film. Details of the method of production and the three stereoregular forms are given in Chapter 8.

Acrylonitrile (cf. Chapters 6 and 7)

Acrylonitrile was first manufactured commercially in Germany during the 1930's from acetylene. This route continued to dominate acrylonitrile production until the early 1960's, when its position was rapidly replaced by propylene ammoxidation (see Chapter 6, p. 145):

$$CH_3CH{=}CH_2 + NH_3 + O_2 \longrightarrow CH_2{=}CHCN + 3H_2O$$

In Chapter 7 (p. 175) the various processes for the manufacture of acrylonitrile are compared and the reason for the preference for ammoxidation is discussed.

In 1970 the world production of acrylonitrile was estimated at 1.4×10^6 tons. The 1972 U.S. plant capacity was 5.4×10^5 tons and the demand in 1971 was 4.6×10^5 tons; the end-uses were acrylic and modacrylic fibres 60%, ABS and SA resins 18%, nitrile rubber 6%, others (including exports) 16%.

A limited amount of acrylonitrile is converted into acrylamide, the polymer from which is used, *inter alia*, as a flocculating agent:

$$CH_2{=}CHCN + H_2O \xrightarrow{\text{Dil. } H_2SO_4} CH_2{=}CHCONH_2$$

In Japan, Ajinomoto produce monosodium glutamate (a food flavour enhancer; Chapter 16) from acrylonitrile, in addition to producing it by fermentation:

$$CH_2{=}CHCN + CO + H_2 \xrightarrow{\text{Co catalyst}} OHCCH_2CH_2CN \xrightarrow{\text{HCN/NH}_3}$$

$$NCCH(NH_2)CH_2CH_2CN \xrightarrow{\text{NaOH/H}_2O} NaOOCCH(NH_2)CH_2CH_2COOH$$

A number of Companies have developed processes for the production of adiponitrile by the reductive dimerization of acrylonitrile. One of these, which has been commercialized, involves electrolytic reduction at a lead cathode in a divided cell. Adiponitrile is an intermediate in the preparation of nylon 6,6 (see Chapter 9).

$$2CH_2{=}CHCN + 2H_2O + 2e^- \longrightarrow NC(CH_2)_4CN + 2OH^- \qquad \text{Cathode}$$
$$\text{reaction}$$
$$H_2O \longrightarrow 2H^+ + \tfrac{1}{2}O_2 + 2e^- \qquad \text{Anode reaction}$$

A new route to nylon 6,6 polymer has recently been reported; this could involve the addition of hydrogen chloride to acrylonitrile to yield chloropropionitrile which is then converted into adiponitrile by reactions with iron powder. The iron chloride so formed is converted back to iron and recycled.

Oxo-alcohols (cf. Chapter 6)

n- and Iso-butanol are produced from propylene by hydroformylation, generally via the intermediate aldehydes:

$$CH_3CH{=}CH_2 + CO + H_2 \longrightarrow CH_3CH_2CH_2CHO + (CH_3)_2CHCHO$$

$$\Big\downarrow H_2 \qquad\qquad\qquad \Big\downarrow H_2$$

Butan-1-ol   Isobutyl alcohol

Propan-2-ol (cf. Chapter 6)

Propan-2-ol, prepared by the hydration of propylene, was the first petrochemical to be commercialized (around 1920). As with ethylene, both the sulphuric acid method and the catalytic hydration method can be used (see p. 83).

$$CH_3CH{=}CH_2 \xrightarrow{\text{H}_2O} (CH_3)_2CHOH$$

Acetone

Acetone is a co-product with phenol in the cumene process (see p. 95, Chapter 5, p. 132, and Chapter 7, p. 181), in which propylene is one of the starting materials. To a decreasing extent, acetone is also manufactured from propan-2-ol by dehydrogenation or by oxidation at high temperatures (see Chapter 7, p. 179).

$$(CH_3)_2CHOH \rightleftharpoons (CH_3)_2CO + H_2$$

$$(CH_3)_2CHOH + \tfrac{1}{2}O_2 \xrightarrow[550-650°C]{Cu\ catalyst} (CH_3)_2CO + H_2O \qquad (85-90\%)$$

Shell have developed a liquid-phase oxidation process whereby oxygen is passed into propan-2-ol and hydrogen peroxide at 80–120°C at $(0.2-0.3) \times 10^6$ Nm^{-2}.

$$(CH_3)_2CHOH + O_2 \longrightarrow (CH_3)_2CO + H_2O_2$$

Acetone is also prepared by the direct oxidation of propylene (Chapter 7, p. 182), and is formed as a co-product in the manufacture of glycerol from acrolein (see p. 96).

Propylene Oxide

Currently, the main commercial route to propylene oxide is via propylene chlorohydrination; many redundant ethylene oxide plants have been modified for this purpose. Propylene dichloride is obtained as a by-product.

$$CH_2{=}CHCH_3 + HOCl \longrightarrow \underset{\underset{Cl\ \ OH}{|\ \ \ |}}{CH_2CHCH_3} \xrightarrow{Ca(OH)_2} \underset{\diagdown\!O\!\diagup}{H_2C{-}\!{-}CHCH_3}$$

Throughout the world a large amount of research effort has been directed to development of a commercially viable direct oxidation route to propylene oxide. Although there have been many claims of such processes, only one has yet been commercialized. They include a Celanese process involving use of peracetic acid, and reactions involving combined oxidation of propylene and either acetaldehyde or isobutyraldehyde.

Kellog and Bayer have both carried out investigations into the electrochemical production of propylene oxide. Propylene is passed into an anode compartment of a diaphragm cell containing dilute sodium chloride as anolyte. Hypochlorous acid formed at the anode reacts to give propylene chlorohydrin, which diffuses into the cathode compartment and decomposes in the highly alkaline catholyte, as shown in the following scheme.

Anode reactions: $2Cl^- \longrightarrow Cl_2 + 2e^-$

$$Cl_2 + H_2O \longrightarrow HOCl + H^+ + Cl^-$$

$$HOCl + CH_3CH{=}CH_2 \longrightarrow$$

$$CH_3CHClCH_2OH + CH_3CH(OH)CH_2Cl$$

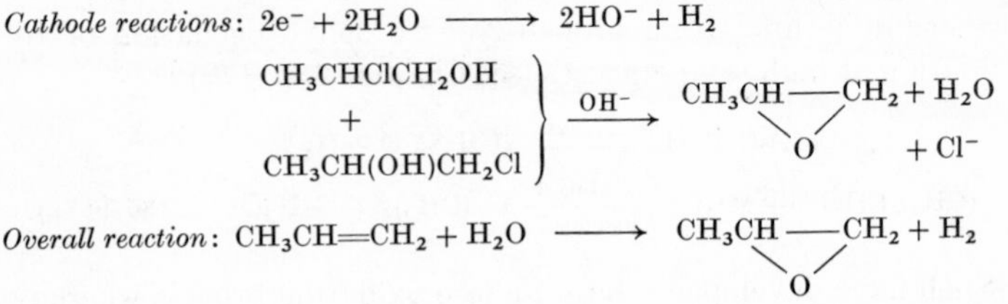

Cathode reactions: $2e^- + 2H_2O \longrightarrow 2HO^- + H_2$

$$\left.\begin{array}{c} CH_3CHClCH_2OH \\ + \\ CH_3CH(OH)CH_2Cl \end{array}\right\} \xrightarrow{OH^-} CH_3CH\!\!\overset{\displaystyle\diagup\diagdown}{\underset{O}{\rule{0pt}{0pt}}}\!\!CH_2 + H_2O + Cl^-$$

Overall reaction: $CH_3CH{=}CH_2 + H_2O \longrightarrow CH_3CH\!\!\overset{\diagup\diagdown}{\underset{O}{\rule{0pt}{0pt}}}\!\!CH_2 + H_2$

However, there is no evidence that the process has been commercialized. The exception that has been commercialized is a process developed jointly by Scientific Design and Atlantic Richfield which is reported to form the basis of three recently constructed plants in the U.S.A., Holland and Spain. This is represented simply as:

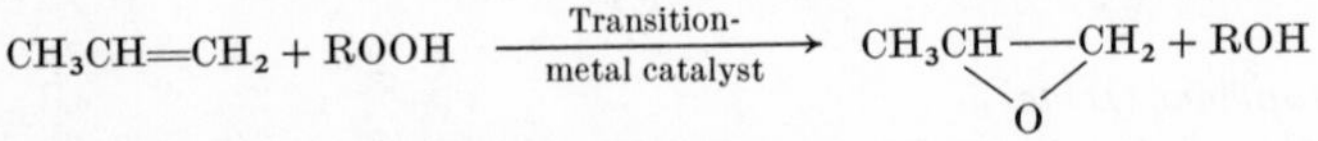

$$CH_3CH{=}CH_2 + ROOH \xrightarrow[\text{metal catalyst}]{\text{Transition-}} CH_3CH\!\!\overset{\diagup\diagdown}{\underset{O}{\rule{0pt}{0pt}}}\!\!CH_2 + ROH$$

where ROOH is $C_6H_5CH(CH_3)OOH$ or Bu^tOOH.

As a result of work on the reaction of t-butyl hydroperoxide with cyclohexene in the presence of catalytic amounts of vanadium salts, Gould *et al.* proposed that the mechanism is neither a free-radical process nor a redox oxidation process but a rate-determining heterolysis of the O—O bond in the complex formed by the metal and the hydroperoxide. In detail, they proposed the following sequence of reactions.

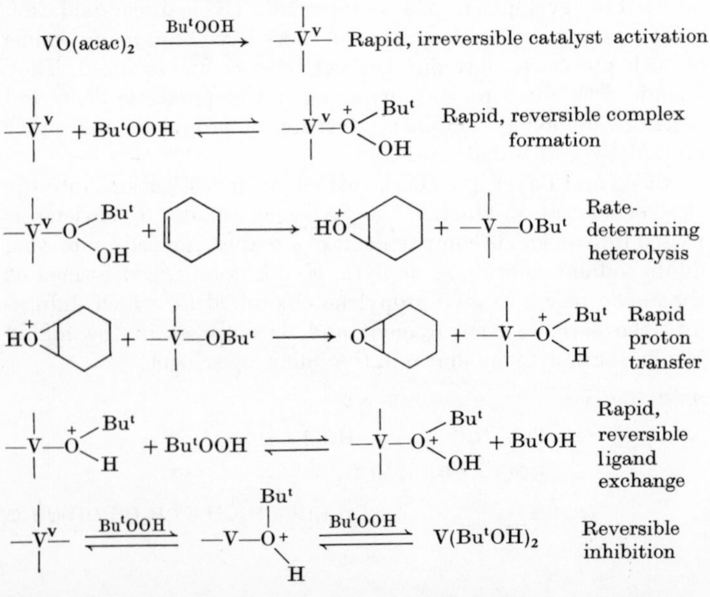

Propylene oxide is used for the production of propylene glycol (for unsaturated polyesters, etc.), urethane foams, surfactants, etc.

Cumene

Benzene is more readily alkylated with propylene than with ethylene (Chapter 5, p. 121) and the reaction can be carried out either in the vapour phase over a supported phosphoric acid catalyst ($250°C$ and 2.5×10^6 Nm^{-2}) or in the liquid phase with aluminium chloride/HCl as catalyst ($70–100°C$ and $1–5 \times 10^5$ Nm^{-2}). Sulphuric acid is also a catalyst for the liquid-phase reaction.

$$CH_3CH{=}CH_2 + H^+ \longrightarrow (CH_3)_2CH^+$$

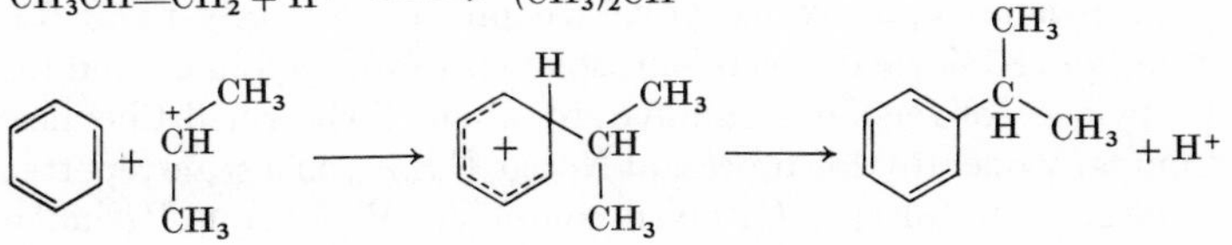

Cumene is mainly used as a raw material for the co-production of phenol and acetone (see Chapter 5, p. 132); a small amount is used, in the form of its hydroperoxide, as a polymerization initiator.

Allyl Chloride

Chlorination of propylene at $450–550°C$ (propylene : $Cl_2 = 4:1$) produces allyl chloride in 80–85% yield; 1,2- and 1,3-dichloropropane are the main by-products (see Part 2, p. 129).

$$CH_2{=}CHCH_3 + Cl_2 \longrightarrow CH_2{=}CHCH_2Cl + HCl$$

The main end-use for allyl chloride is the production of

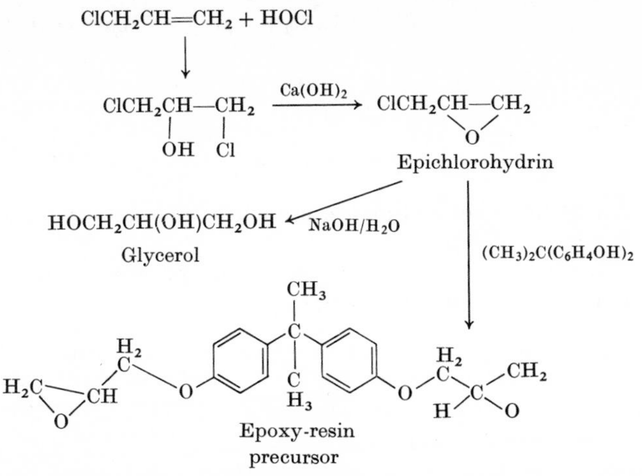

SCHEME 5.

epichlorohydrin, used in the production of epoxy-resins and an intermediate in one of the synthetic routes to glycerol (cf. Scheme 5).

Hydrolysis of allyl chloride yields allyl alcohol, whilst reaction with ammonia can be used to yield allylamine.

$$CH_2{=}CHCH_2Cl \begin{array}{l} \xrightarrow{\text{NaOH}} CH_2{=}CHCH_2OH + NaCl \\[2em] \xrightarrow{\text{NH}_3} CH_2{=}CHCH_2NH_2 + HCl \end{array}$$

Acrolein

Propylene can be oxidized over a copper oxide catalyst at 350°C to give an 85% yield of acrolein; steam is used as a diluent and the propylene conversion is limited to about 20%. Shell Chemical Company operate this process at Norco, U.S.A., on a scale reported to be 20×10^3 tons p.a. Catalysts comprising Bi, Mo, and P (similar to those for acrylonitrile production) have also been claimed for this process. The mechanism is believed to involve the formation of an allylic intermediate at the catalyst surface by abstraction of a hydrogen atom from the methyl group of propylene. Rapid isomerization of the allylic intermediate then occurs before addition of an oxygen atom.

$$CH_2{=}CH\overset{*}{C}H_3 \xrightarrow{-H} CH_2{=}CH\overset{*}{C}H_2{-}Cat \rightleftharpoons Cat{-}CH_2CH{=}\overset{*}{C}H_2$$

$$\downarrow O \qquad\qquad\qquad\qquad\qquad \downarrow O$$

$$CH_2{=}CH\overset{*}{C}H_2O{-}Cat \qquad\qquad Cat{-}O{-}CH_2CH{=}\overset{*}{C}H_2$$

$$\downarrow -H \qquad\qquad\qquad\qquad\qquad \downarrow -H$$

$$CH_2{=}CH\overset{*}{C}HO \qquad\qquad\qquad OHCCH{=}\overset{*}{C}H_2$$

Attack at either end of the propylene molecule has been shown by tracer experiments.

Acrolein is the starting material for the Shell Chemical Company's route to glycerol:

$$CH_2{=}CHCHO + (CH_3)_2CHOH \longrightarrow CH_2{=}CHCH_2OH + (CH_3)_2CO$$
Acrolein   Propan-2-ol   Allyl alcohol   Acetone

$$CH_2{=}CHCH_2OH \longrightarrow HOCH_2CH(OH)CH_2OH$$
Glycerol

The acrolein is treated with an excess of propan-2-ol in the vapour phase (400°C) over a catalyst containing magnesium and zinc to give a 77–80% yield of allyl alcohol. The latter is then converted into glycerol by reaction with aqueous hydrogen peroxide in the presence of a tungstic oxide catalyst. Glycerol yield is 85–90% based on allyl alcohol.

Acrolein can be oxidized to acrylic acid in 70% yield with air in the presence of steam over a Co, Bi, or Mo catalyst (420–480°C). Propylene (10% in air and steam) can be oxidized directly to acrylic acid over a Co or Mo catalyst. Acrolein readily forms a dimer which can be converted into hexane-1,2,6-triol by hydrolysis followed by hydrogenation.

$$2CH_2{=}CHCHO \longrightarrow \begin{array}{c} \overset{H_2}{C} \\ HC \overset{}{\diagup} \diagdown CH_2 \\ \| \qquad | \\ HC \diagdown_{O} \diagup CHCHO \end{array} \xrightarrow[2,\ H_2]{1,\ H_2O}$$

$$HOCH_2CH(OH)CH_2CH_2CH_2CH_2OH$$

Methionine (Part 4, pp. 367–368), used as an animal feed supplement, can be readily produced from acrolein by reaction with methanethiol.

$$CH_2{=}CHCHO + CH_3SH \longrightarrow CH_3SCH_2CH_2CHO$$

$$\Big\downarrow \begin{array}{l} +HCN \\ +(NH_4)_2CO_3 \end{array}$$

$$\underset{\text{Methionine}}{CH_3SCH_2CH_2\overset{\overset{\displaystyle NH_2}{|}}{C}HCOOH} \xleftarrow{\ H^+\ } CH_3SCH_2CH_2 \begin{array}{c} H \\ \diagdown \overset{N}{\diagup} \diagdown \\ C \qquad CO \\ \diagup \diagdown \quad \diagup \\ OC {-} NH \end{array} + NH_3 + H_2O$$

Isoprene

Isoprene is manufactured from propylene by the Goodyear–Scientific Design process, which begins by dimerizing propylene at 150–200°C and 20×10^6 Nm^{-2} with tripropylaluminium and nickel to yield 2-methylpent-1-ene:

$$\underset{\text{2-Methylpent-1-ene}}{2CH_3CH{=}CH_2 \longrightarrow CH_2{=}C(CH_3)CH_2CH_2CH_3}$$

The latter is then isomerized at 150–300°C over a fixed-bed, supported phosphoric acid catalyst to 2-methylpent-2-ene:

$$\underset{\text{2-Methylpent-1-ene (2MP-1)}}{CH_2{=}C(CH_3)CH_2CH_3} \longrightarrow \underset{\text{2-Methylpent-2-ene (2MP-2)}}{(CH_3)_2C{=}CHCH_2CH_3}$$

The thermodynamic equilibrium (227°C) is reported to be 30.6% of 2MP-1 and 69.4% of 2MP-2. The 2-methylpent-2-ene is then 'cracked' at 650–750°C in the presence of small amounts of HBr and steam to yield isoprene (65% yield) (some cyclopentadiene and piperylene are also formed and are removed by distillation):

$$(CH_3)_2C{=}CH_2CH_2CH_3 \longrightarrow CH_2{=}C(CH_3)CH{=}CH_2 + CH_4$$
$$\text{Isoprene}$$

The world isoprene capacity was reported to be around 5×10^5 tons p.a. in 1972 and Goodyear have a 50×10^3 tons p.a. unit based on the above technology. It is used as the monomer for production of stereospecific polyisoprene whose properties closely resemble those of natural rubber (see Chapter 8). In 1970 world polyisoprene consumption was estimated at 320×10^3 tons.

Propylene Dimerization

In recent years there has been appreciable patent activity covering the dimerization of propylene by catalysts based on liganded nickel complexes and an alkylaluminium halide. It is believed that, for example, $(PBu_3^n)_2NiCl_2$ and $C_2H_5AlCl_2$ produce the catalytically active species (1) at elevated temperature or in the presence of propylene.

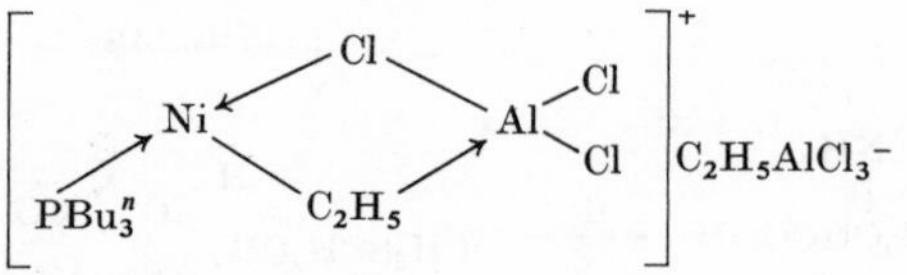

(1)

Propylene dimerization is then believed to occur by a rapid olefin–alkyl exchange together with a rate-determining insertion reaction, as in Scheme 6.

The skeletal distribution obtained depends on the temperature (low temperatures favour formation of intermediate (2)), the nature of the aluminium species and the nature of the ligand. Thus, Ph_3P leads to a relatively high proportion of straight-chain hexene, whilst tricyclohexylphosphine produces only small amounts (ca. 2%), the main product being 2,3-dimethylbutenes.

To date there is no evidence that this reaction has been commercialized.

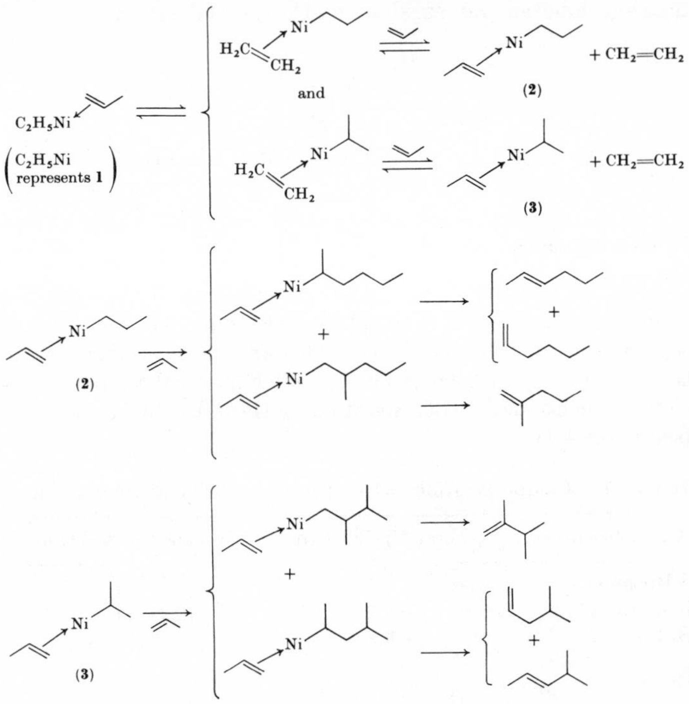

SCHEME 6.

Propylene Trimer and Tetramers

These highly branched oligomers of propylene are produced in refineries in units constructed to make polymer gasoline from C_3 and C_4 olefins. The latter type of product has declined in importance. The trimer is mainly used for the production of nonylphenol and 'iso-decanol'. The tetramer markets have declined in recent years with the cut-back in ABS detergents (see Chapter 13).

Disproportionation of Propylene

The Phillips Petroleum Co. have developed a process ('Triol' process) whereby propylene is converted into ethylene and *n*-butenes by reaction at 150–210°C and 10^6 Nm^{-2} with cobalt and molybdenum catalysts.

It is believed that rapid equilibration occurs between reactant and the disproportionation products, probably via formation of a quasi-cyclobutane intermediate on the metal catalyst:

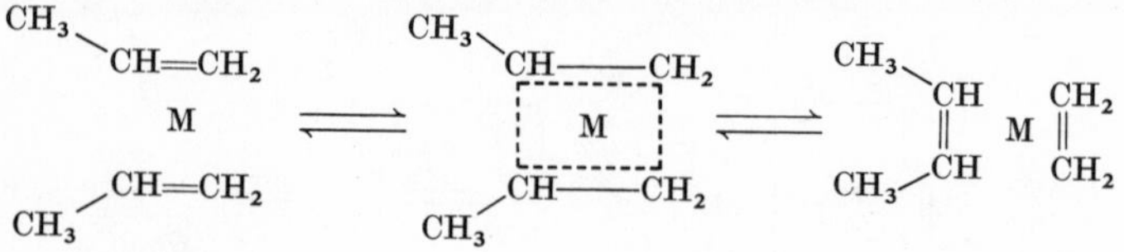

C_4 Hydrocarbons

Although ethane (b.p. $-88.6°C$), ethylene (b.p. $-103.6°C$), propane (b.p. $-42°C$) and propylene (b.p. $-47.6°C$) can be separated by distillation, the C_4 hydrocarbons (separable from the C_2–C_3 species by distillation) have boiling points that are too close to allow such a method of separation for the individual components (see Table 4.7).

Table 4.7. C_4-Composition (wt.-%) from naphtha steam cracking

C_4 hydrocarbon	B.p.(°C)/760 mm	Typical composition (%)
n-Butane	0.5	7
Isobutane	-11.7	1
But-1-ene	-6.3	18
But-2-ene $\begin{cases} cis\text{-} \\ trans\text{-} \end{cases}$	$\left.\begin{matrix} 3.7 \\ 0.9 \end{matrix}\right\}$	7
Isobutene	-6.9	37
Butadiene	-4.4	30

Butadiene can be recovered by extractive distillation with furfural or 1-methylpyrrolidone. The butenes can be concentrated to around 90–95% by extractive distillation with acetonitrile to separate them from butanes. Isobutene is extremely reactive and can be removed from C_4 mixtures by absorption in 50% sulphuric acid.

Butadiene

Butadiene is an important starting material in the synthetic rubber industry (Chapter 8). The production of butadiene in the major areas in 1970 is shown in Table 4.8. A large proportion of this (96% in Europe, 79% in U.S.A.) was used for the production of styrene–butadiene rubber (SBR; 3.5×10^6 tons consumed world-wide in 1970), stereoregular polybutadiene rubber (7.3×10^5 tons consumed world-wide in 1970) and nitrile–butadiene rubber (1.9×10^5 tons consumed in 1970, world-wide). Butadiene is also

Table 4.8. Butadiene production and plant capacities ($tons \times 10^6$) in the major producing countries

Country or area	Butadiene capacity (1972)	Butadiene production (1972)
W. Europe	1.260	0.940
U.K.	0.244	0.200
U.S.A.	1.800	1.553
Japan	0.687	0.579[a]

[a] 0.563 in 1971.

used in styrene–butadiene resins and in acrylonitrile–butadiene–styrene resins.

Before World War II, Reppe discovered that nickel catalysts of the type $Ni(PPh_3)_2Cl_2$ produced cyclo-octatetraene and benzene from acetylene. In 1954, Reed showed that by using similar catalysts $[Ni(PPh_3)_2(CO)_2$ activated by acetylene] butadiene could be oligomerized to cyclo-octa-1,5-diene (**4**).

(*cis,cis*-COD)

4

Further extensive work, particularly by Wilke and his co-workers, led to a series of reactions based on nickel(II) catalysts reduced with aluminium alkyls in the presence of dienes (see Scheme 7); the presence of phosphine or phosphite ligands plays an important role.

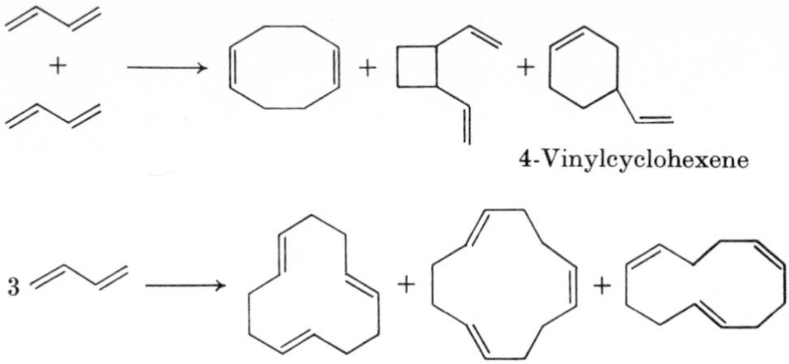

4-Vinylcyclohexene

Cyclododeca-1,5,9-triene (CDT)

(three isomers)

SCHEME 7 (part of).

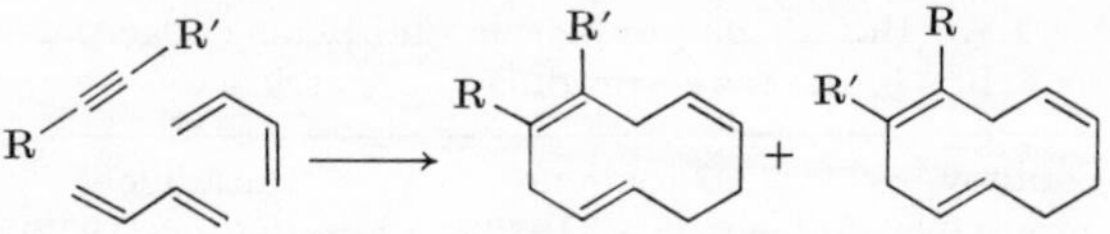

Dialkylcyclododeca-1,4,8-trienes

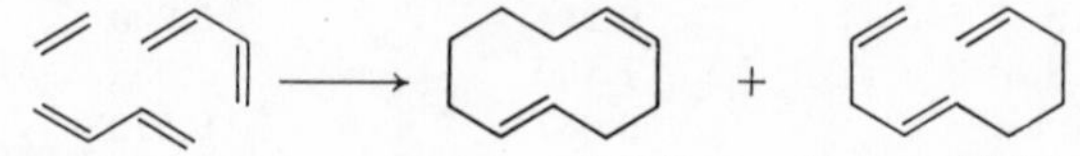

Cyclodeca-1,5-diene   Deca-1,4,9-triene

SCHEME 7 (contd.)

The choice of catalyst can allow the butadiene to be converted in high yield into a particular product. It was found that reduction of Ni(acac)$_2$ with aluminium alkyls in the presence of ligands (CO, PR$_3$, 1,5-COD, 1,5,9-CDT) gave the co-ordinatively unsaturated Ni–L complex which was extremely reactive towards butadiene (in which it was soluble), catalysing its conversion into CDT.

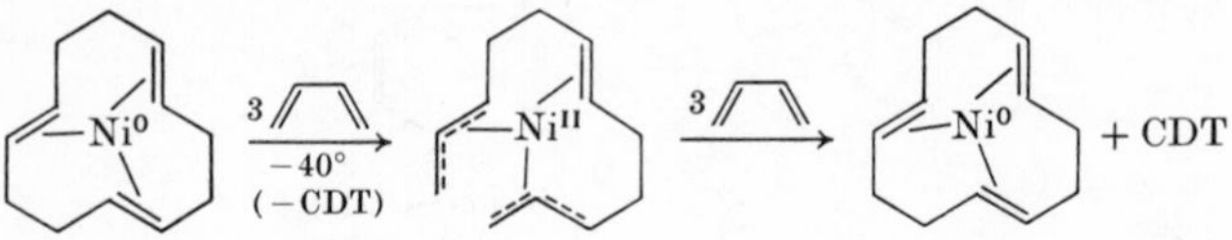

trans,trans,trans-CDT is obtained in this reaction but the *trans,trans,cis*-form is the main product when TiCl$_4$ and alkyl-aluminium halides are used as catalysts. The latter catalyst is more effective in catalysing formation of CDT than are the nickel-based systems.

If one of the co-ordination positions around the nickel atom is 'blocked' with a strong ligand, then only two butadiene molecules can co-ordinate to form a bis-π-allyl complex (see Scheme 8).

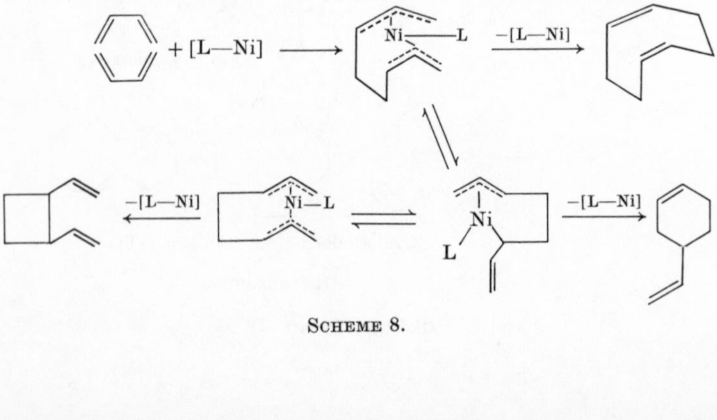

SCHEME 8.

The use of P(OR)₃ ligands favours COD formation, whilst PR₃ ligands lead to a mixture of COD and vinyl cyclohexene.

Butadiene and ethylene react according to the mechanism of Scheme 9 to yield cyclodeca-1,5-diene and deca-1,4,9-triene.

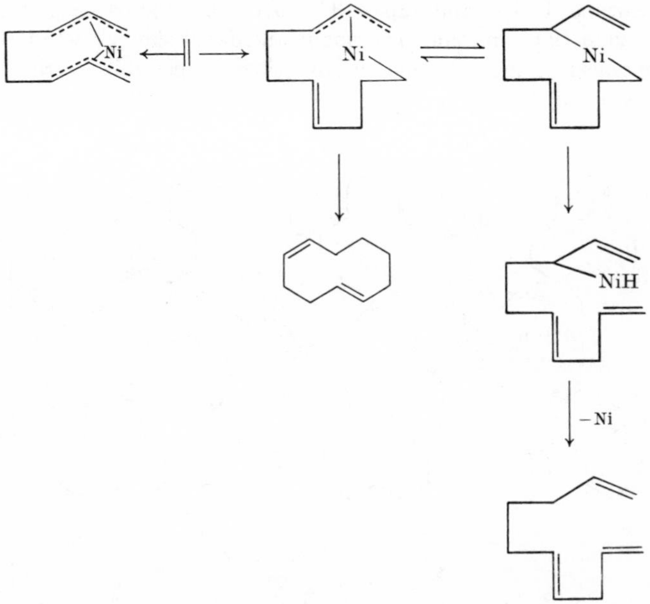

SCHEME 9.

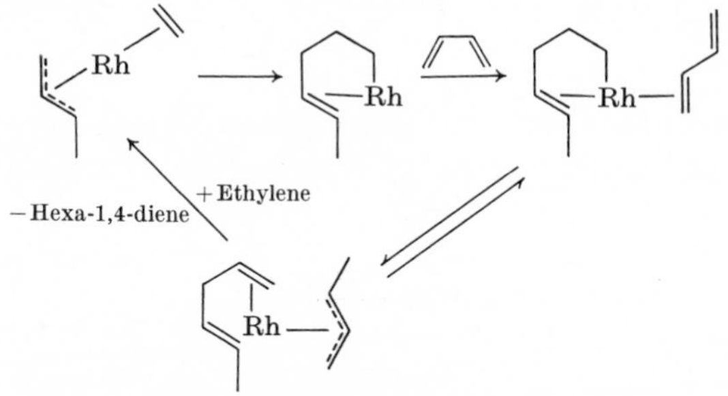

In the presence of rhodium chloride, ethylene and butadiene react readily at 50°C to give a 90% yield of hexa-1,4- and -2,4-diene. The proportion of the latter increases under more vigorous conditions. Similar products are obtained on using nickel or iron compounds in the presence of alkylaluminium compounds.

Hexa-1,4-diene is produced in the U.S.A. and Japan as a termonomer used in ethylene–propylene termonomer rubbers.

Cyclo-octa-1,5-diene and cyclododeca-1,5,9-triene are produced in Europe, U.S.A. and Japan. The latter is the starting material for nylon-12 production in Germany and Japan, lauryl lactam (the monomer) being obtainable by several possible variations (Scheme 10).

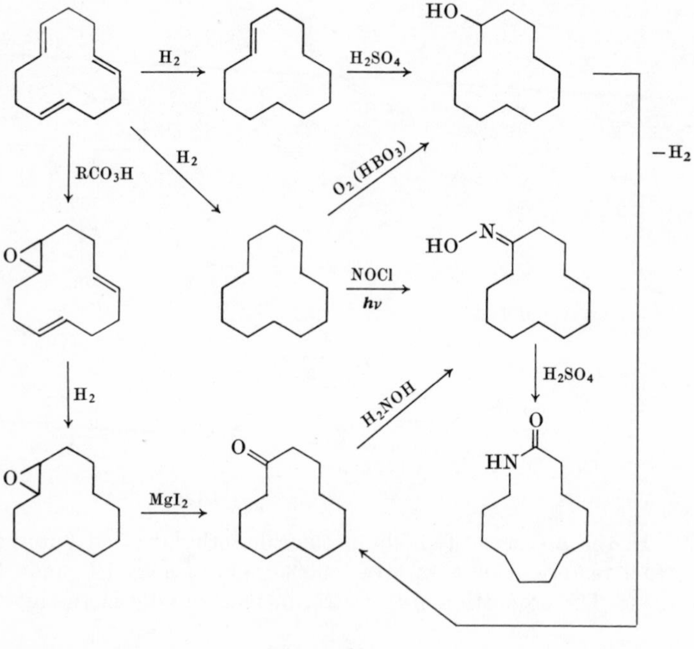

SCHEME 10.

Cyclo-octadiene can be carbonylated in the presence of PdL_2Cl_2 and EtOH to yield the ester (5). The corresponding acid

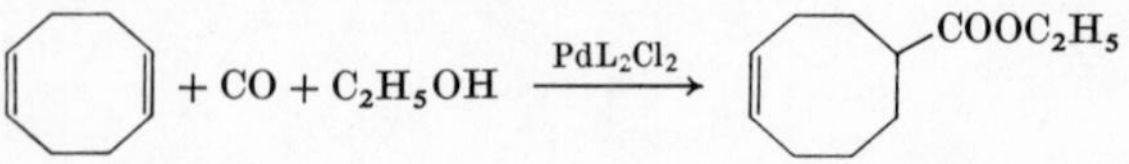

(5)

can be converted into azelaic acid in high yield (>80%) by treatment with aqueous sodium hydroxide at high temperatures (320°C). These reactions have not as yet been commercialized.

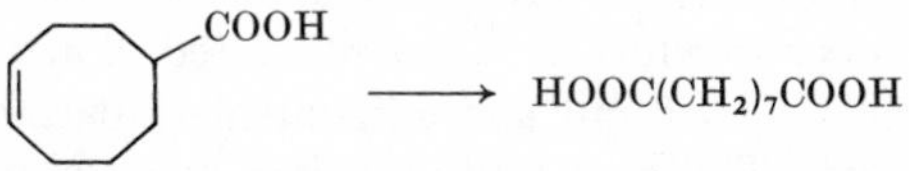

$$\text{(structure)} \quad \longrightarrow \quad HOOC(CH_2)_7COOH$$

Homogeneous gas-phase chlorination of butadiene at 330–400°C produces 3,4-dichlorobut-1-ene and 1,4-dichlorobut-2-ene (3:2 molar ratio). The 1,4-dichlorobut-2-ene can be converted into its allylic isomer 3,4-dichlorobut-1-ene by heating with cuprous chloride, and, being the lower-boiling isomer, it can be continuously removed by distillation. Dehydrochlorination of the 3,4-dichlorobut-1-ene can be carried out by heating with caustic soda solution, and chloroprene is thus obtained in over 90% yield.

$$CH_2{=}CHCCl{=}CH_2$$
Chloroprene

$$-HCl \nearrow$$

$$CH_2{=}CHCH{=}CH_2 \quad \longrightarrow \quad CH_2{=}CHCHClCH_2Cl$$
3,4-Dichlorobut-1-ene

$$\searrow$$

$$ClCH_2CH{=}CHCH_2Cl$$
1,4-Dichlorobut-2-ene

The dichlorobutenes can be treated with sodium cyanide in the presence of cuprous chloride, to yield 1,4-dicyanobut-2-ene. On hydrogenation in the vapour phase adiponitrile is formed which on further hydrogenation yields hexamethylenediamine (HMD).

$$\left.\begin{array}{l} ClCH_2CH{=}CHCH_2Cl \\ CH_2{=}CHCHClCH_2Cl \end{array}\right\} \xrightarrow[CuCl]{NaCN} NCCH_2CH{=}CHCH_2CN \xrightarrow{H_2}$$

$$NCCH_2CH_2CH_2CH_2CN \xrightarrow{H_2} HMD$$

du Pont are believed to be still operating a plant based on this process.

Recently, du Pont and Esso have announced that they have developed new adiponitrile processes based on butadiene, but they provide little actual detail of their processes. The du Pont

route involves hydrocyanation of butadiene in the presence of a
zerovalent nickel catalyst. Initially, HCN is added to butadiene
to form a mixture of isomeric unsaturated mononitriles. These are
separated and the non-linear compounds re-isomerized to the
equilibrium mixture which is again separated. Further addition
of HCN is then carried out and adiponitrile separated from the
dinitrile mixture. The Esso route involves reaction of butadiene
with iodine and copper cyanide, to give the cuprous iodide complex
of dehydroadiponitrile. Hydrolysis with aqueous HCN solution
furnishes dehydroadiponitrile in high yield; this reaction also
regenerates the iodine and cuprous cyanide for recycle.

Butadiene reacts with SO_2 under pressure to yield a cyclic
sulphone which, on reduction, yields tetramethylene sulphone
('sulpholane'), b.p. 285°C, which is completely miscible with water:

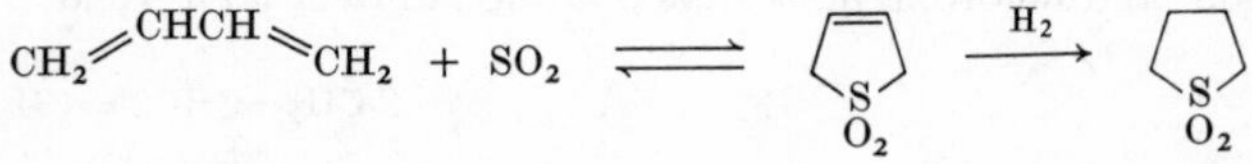

Sulpholane

Sulpholane is used as a solvent for CO_2 removal and as an extractive
solvent in aromatics plants.

Isobutene

Isobutene is important in the production of butyl rubber (95%
isobutene; 95×10^3 tons consumption in U.S.A. in 1970; 255×10^3
tons world-wide in 1970) (cf. Chapter 8).

Isobutene polymers, based on isobutene and *n*-butenes, are
viscous liquid copolymers used as viscosity improvers for lubricating oils and as sealing compounds.

Isobutene is readily converted into mixed dimers and trimers
by absorption in 60% sulphuric acid at 20°C, followed by heating
to 90°C (Part 1, p. 65). 80% of dimers and 20% of trimers are
obtained in 90% overall yield.

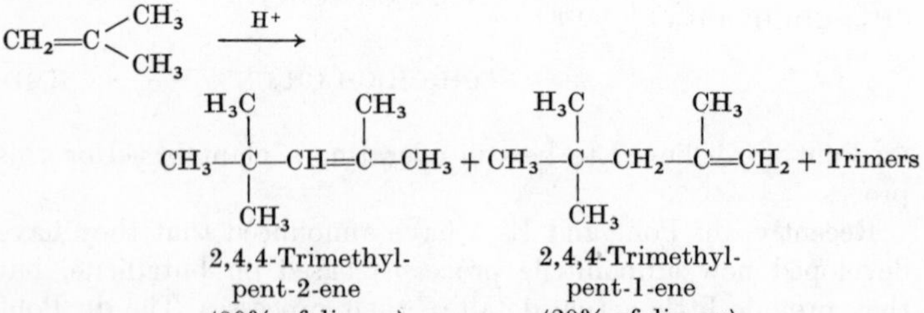

2,4,4-Trimethyl-   2,4,4-Trimethyl-
pent-2-ene   pent-1-ene
(80% of dimers)   (20% of dimers)

The dimers are used to alkylate phenols for detergent applications and also carbonylated to produce C_9 alcohols.

Mixed branched heptenes are produced by reaction of isobutene with propylene in the presence of $AlCl_3$ or phosphoric acid and used for the production of iso-octanol.

At least four processes have been developed for the production of isoprene from isobutene, all based on the Prins reaction (reaction between an olefin and an aldehyde). In the process developed and commercialized in the U.S.S.R., pure isobutene is treated with formaldehyde in 1.5% H_2SO_4 to give 4,4-dimethyl-1,3-dioxan (DMD) in 75–80% yield. The latter is then cracked to isoprene over a calcium phosphate-based catalyst at 300–400°C to give an 80–85% yield of isoprene. The process developed in France by Institut Française du Pétrole is very similar but can employ a mixed C_4 stream; in the process, particularly with dilute isobutene streams, by-products (polyols) are formed in significant amounts.

$$\underset{CH_3}{\overset{CH_3}{>}}C{=}CH_2 + 2HCHO \xrightarrow{H^+} \underset{CH_3}{\overset{CH_3}{>}}C\underset{O\text{---}CH_2}{\overset{CH_2\text{---}CH_2}{<}}O \longrightarrow$$

DMD

$$CH_2{=}\overset{\overset{\displaystyle CH_3}{|}}{C}{-}CH{=}CH_2 \ + \ HCHO + H_2O$$

Isoprene

Variations on this process have been established in several countries, particularly in Japan (by Sumitomo) and Germany (Bayer). The basis of the latter's process was work done in Ludwigshafen, Germany, in 1938–1939; it uses a solid ion-exchange resin to effect the Prins reactions and a fluid-bed supported phosphoric acid catalyst to crack the DMD.

A more recent development is the Marathon (U.S.A.) process in which chloromethyl methyl ether replaces formaldehyde. Reaction with isobutene is effected in the presence of $TiCl_4$:

$$(CH_3)_2C{=}CH_2 + CH_3OCH_2Cl \xrightarrow{TiCl_4} (CH_3)_2CClCH_2CH_2OCH_3$$

The condensation product is then pyrolyzed to give isoprene, together with methanol and HCl which are then recycled for production of the chloromethyl ether:

$$(CH_3)_2CClCH_2CH_2OCH_3 \longrightarrow CH_2{=}C(CH_3)CH{=}CH_2 + CH_3OH + HCl$$
Isoprene

Marathon claim that the overall yield is 80% based on isobutene and they have licensed the process to Ashland Oil in Venezuela.

Isobutene is hydrated under mild conditions (60% H_2SO_4 at 20°C), to give *tert*-butyl alcohol (cf. Chapter 6, p. 159). Liquid-phase chlorination of isobutene produces methallyl chloride in 88% yield.

The Escambia Chemical Corporation developed a process for the production of methacrylic acid based on the oxidation of isobutene with nitric acid and nitrogen dioxide.

$$CH_2{=}C(CH_3)_2 + O_2 \longrightarrow (CH_3)_2C(OH)COOH \longrightarrow$$

$$CH_2{=}C(CH_3)COOH$$
$$\text{Methacrylic acid}$$

Isobutene is also used for alkylation of phenols to yield various antioxidants, butylated *p*-cresol (BHT) being one of the main products.

n-*Butenes and* n-*Butane*

n-Butenes and *n*-butane can be dehydrogenated to butadiene in endothermic reactions. The thermodynamic equilibria are such that high temperatures and low pressures are desirable to maximize production of the diene.

$$CH_3CH_2CH_2CH_3 \underset{+H_2}{\overset{-H_2}{\rightleftharpoons}} \begin{matrix} CH_2{=}CHCH_2CH_3 \\ + \\ CH_3CH{=}CHCH_3 \end{matrix} \underset{+H_2}{\overset{-H_2}{\rightleftharpoons}} CH_2{=}CHCH{=}CH_2$$

(with $-H_2 / +H_2$ pathway connecting $CH_3CH_2CH_2CH_3$ directly to $CH_2{=}CHCH{=}CH_2$)

In the Houdry one-stage dehydrogenation of *n*-butane, temperatures of 590–675°C at 0.5×10^6 Nm^{-2} are used. The *n*-butane is fed through reactors containing shallow beds of pelleted chromia–alumina mixed with an inert material. Each bed is on-line for about 5–10 minutes before the feed is switched to the next reactor to allow the catalyst in the first reactor to be regenerated. This is done by purging with steam and then burning-off deposited carbon in an air stream. Butadiene is separated from butane and butenes by liquid–liquid extraction or extractive distillation; the butane and butenes are then recycled to the reactor. Overall butadiene yields of about 60% are obtained. The dehydrogenation of *n*-butenes is generally carried out in the presence of steam to lower the hydrocarbon partial pressure and minimize carbon deposition on the catalyst.

On hydration (Chapter 6), but-1-ene and but-2-ene both yield butan-2-ol. Because of the ease of polymer formation the direct hydration method is not used; 80–85% sulphuric acid is used under mild conditions. Dehydrogenation of butan-2-ol proceeds at 350–400°C over a metallic catalyst to give ethyl methyl ketone (butan-2-one), used mainly as a solvent. Ethyl methyl ketone can also be made from but-1-ene and but-2-ene by reaction with palladium chloride (cf. acetaldehyde). But-1-ene reacts about one-quarter as fast as ethylene and at 0.9–2×10^6 Nm^{-2}/90–120°C, gives an 85–88% yield of the ketone.

Pure but-1-ene is used increasingly in the production of co-polymers with other α-olefins and epoxidation of but-1-ene provides butene oxide which is converted into butane-1,2-diol for use in polymeric plasticizers and stabilization of chlorinated solvents.

The oxidation of naphtha has been discussed in Chapter 3. A number of processes particularly involving the C_4 fraction have been developed.

CHAPTER 5

Primary Petrochemicals, II: Aromatic Hydrocarbons

Introduction

(H. G. NAYLOR, Central Technical Division, BP Chemicals International Limited, London)

The majority of the large-tonnage products that contain benzenoid nuclei can be synthesized from benzene, toluene or individual C_8

Table 5.1. Major aromatic compounds: their manufacture and uses

Compound	Production method	Major uses
Benzene	1. Extraction 2. Dealkylation of toluene	Polystyrene (via ethyl-benzene), phenol (via cumene), cyclohexane, detergents
Toluene	Extraction	Upgrading of gasoline, benzene, toluene di-isocyanate[a]
Ethylbenzene	1. Extraction 2. Alkylation of benzene	Polystyrene
p-Xylene	1. Extraction 2. Isomerization of C_8 aromatics	Polyester fibres and, to a minor extent, films
o-Xylene	1. Extraction 2. Isomerization of C_8 aromatics	Phthalate plasticizers, alkyl and polyester resins
m-Xylene	Extraction	Limited use, isophthalic acid
Cumene	Alkylation of benzene	Phenol (acetone as by-product)

[a] A mixture of 2,4- and 2,6-tolylene di-isocyanate used in the manufacture of polyurethane foam.

aromatic hydrocarbons. The major source of aromatic hydrocarbons was formerly coal tar but with the advent of the internal combustion engine and an increasing need for gasoline (petrol), petroleum rapidly became the dominant source. However, benzene is still extracted in worthwhile quantities from coke-oven by-products (see Chapter 2).

The primary aromatic compounds that are either extracted from petroleum or synthesized in high purity are listed in Table 5.1, together with the methods of production and the products that are manufactured from them.

The tonnages of the individual aromatics produced within Western Europe for 1970 are listed in Table 5.2, together with the plant capacities.

Table 5.2. Western Europe: aromatics production for 1970 and total plant capacity for 1970

Aromatic compound	Production 1970 (tons × 10^{-3})	Plant capacity 1970 (tons × 10^{-3})
Benzene	2700	3600
Ethylbenzene	1400	1800
Cumene	900	1000
o-Xylene	250	300
p-Xylene	300	400

Manufacturing Routes to Aromatic Hydrocarbons

(H. G. NAYLOR, Central Technical Division, BP Chemicals International Limited, London)

The concentration of any one compound in crude oil is inevitably very low, and therefore concentration techniques are employed before extraction. Distillation of petroleum to give an approximate C_6–C_8 cut is the first obvious step, but even then the concentration of desired aromatic hydrocarbons is somewhat low. This narrow fraction is therefore catalytically reformed to increase the aromatic content, which has beneficial effects not only in the petrochemical industry but also in the production of high-quality motor gasolines. The compositions of a typical catalytic reformer feed and product are listed in Table 5.3, and the consequent increases in the research octane number are included. The octane number of a gasoline is a measure of its anti-knock properties, and the higher the value, the higher is the quality (see Chapter 3, p. 54).

Gasoline produced in thermal cracking (see Chapter 3) is also a major source of aromatic hydrocarbons of increasing importance. The appropriate fraction is hydrogenated to saturate olefins and diolefins before solvent extraction.

Table 5.3. Composition (wt.-%) of catalytic reformates and their octane ratings

Compound	Feed	RON 90[a]	RON 96[a]	RON 100[a]
Aromatics:				
Benzene	0.8	4.3	5.2	6.9
Toluene	3.4	20.1	23.1	26.4
Ethylbenzene	1.1	3.2	3.7	3.8
p-Xylene	0.9	4.0	4.5	5.1
m-Xylene	1.8	8.5	8.7	10.2
o-Xylene	1.2	4.2	4.7	5.9
C_9	0	6.4	7.7	8.2
C_{10} and higher	0	0	0	0
Total aromatics	9.2	50.7	57.6	66.5
Total paraffins	66.5	44.3	41.4	33.1
Total naphthenes	24.3	5.0	1.0	0.4

[a] Research Octane Number without the addition of lead alkyl.

The subsequent processing of the aromatics concentrate is dependent on the particular products required. An idealized flow-sheet to provide all the important individual compounds is shown in Figure 5.1, and the various processes are described in the following Sections.

Liquid–Liquid Extraction

Azeotropic mixtures are formed between the aromatic hydrocarbons and some of the saturated compounds present in catalytic reformate. Therefore conventional fractionation cannot yield a pure aromatic stream and liquid–liquid extraction has been adopted in the majority of cases.

The aromatic portion of the reformate can be regarded as a solute in solution with the saturated compounds (the solvent). If a further solvent is added that is immiscible with the saturated compounds, then the aromatic compounds are distributed between the two solvents at equilibrium according to the partition coefficients μ:

$$C_1/C_2 = \mu$$

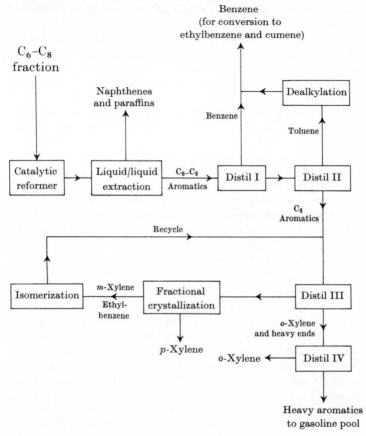

Figure 5.1. Process scheme for recovery of the important individual aromatics.

where C_1 is the concentration of an aromatic hydrocarbon in the saturated hydrocarbons and C_2 is its concentration in the added solvent. However, the selectivity of the added solvent to aromatic layer is the prime factor in a process of this nature, and the partition function tells us nothing about the slight miscibility of one solvent within the other. Therefore the ratio of the concentrations of the aromatic hydrocarbons and the saturated hydrocarbons in the two liquid phases is of most importance and consequently it is more appropriate to use a selectivity ratio β:

$$\beta = [\chi_A/\chi_S]_I / [\chi_A/\chi_S]_{II}$$

where χ_A and χ_S are the molar fractions of aromatic and saturated compounds, respectively, in the two liquid phases I and II. If I denotes the added solvent, then the greater the value of β the higher will be the selectivity to aromatic compound. The situation is now analogous to fractionation, and further increases in purity can be gained by repeating the extraction until the desired purity is attained.

Present-day commercial practice employs a continuous, rather than the 'batch' process, involving a number of mixers and settlers. The solvent is passed down a column up which the reformate is flowing, and thorough mixing within the column is ensured by means of rotating discs. The solvent leaving the bottom of the column is rich in aromatic hydrocarbons and contains only a small quantity of paraffins and naphthenes (cycloalkanes), which are removed by extractive stripping before the aromatic compounds are distilled. The solvent from the bottom of the column is recycled to the process. Typical solvents that can be used are sulpholane, 1-methylpyrrolidone–ethylene glycol and dimethyl sulphoxide. The process details vary according to the solvent used.

Production of o- and p-Xylene

Usually the tonnage of individual xylenes required is in excess of the mixed xylenes produced from liquid–liquid extraction and subsequent distillation. Therefore there is very often a departure from the idealized flow-scheme shown in Figure 5.1 and o- and p-xylene are produced directly from catalytic reformate. A cut is taken that contains predominantly the C_8 aromatics. o-Xylene is the only isomer that can satisfactorily be separated by distillation (see Table 5.4). It is advisable to ensure that there are no

Table 5.4. Boiling points and melting points of the C_8 aromatic isomers.

Compound	B.p. (°C)	M.p. (°C)
o-Xylene	144.41	−25.19
m-Xylene	139.10	−47.88
p-Xylene	138.35	+13.25
Ethylbenzene	136.15	−95.01

paraffins or naphthenes (cycloalkanes) in the feed to the o-xylene columns as they can form azeotropic mixtures with the aromatic hydrocarbons.

The major industrial method of producing *p*-xylene is by fractional crystallization (melting points of C_8 aromatics are given in Table 5.4).

The quantity of *p*-xylene that can be crystallized from a given quantity of feedstock (i.e. the extraction efficiency) will be governed by the phase equilibria for that feedstock. A typical feedstock to a crystallization unit, if we assume prior removal of *o*-xylene, contains approximately 50 wt.-% of *m*-xylene, 20 wt.-%

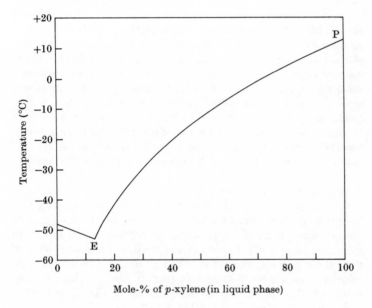

Figure 5.2. *m*-Xylene/*p*-xylene eutectic. (For P–E see text.)

each of *p*-xylene and ethylbenzene, and a residual quantity of *o*-xylene (2–4 wt.-%). Strictly, the phase equilibria for the quaternary system should be considered, but the principles can be effectively demonstrated if it is accepted that, with a feedstock of the composition given above, it is the *m*-xylene/*p*-xylene eutectic shown in Figure 5.2 that is the governing factor. As the temperature is reduced from, say, 25°C where the feedstock is completely liquid, no *p*-xylene will crystallize until the line PE is met. At this point pure *p*-xylene will commence to crystallize such that the composition–temperature relation of the liquid follows the line PE until the point E is reached; then an intimate mixture of *m*-xylene and *p*-xylene crystallizes and the whole solution becomes solid. When pure *p*-xylene is required, cooling is stopped before the eutectic composition E is reached; the pure *p*-xylene crystals can then be either filtered off or separated in a centrifuge. The C_8 aromatic isomer system is very nearly ideal, and the presence of ethylbenzene and *o*-xylene alters the eutectic composition only from ca. 13% to 10% per mole of *p*-xylene; it does, however, depress the freezing point of the eutectic mixture from ca. −65°C to −86°C.

In commercial practice the feedstock is cooled to ca. −70°C and the mixture is then centrifuged to recover the crystals of *p*-xylene. As some of the other isomers wet the crystals and also are retained in the spaces in crystal agglomerates, the recovered crystals are warmed slightly to provide a small amount of pure *p*-xylene as a wash liquid, and the mixture is centrifuged again. The process is repeated until the recovered *p*-xylene is of the correct purity (>99.2 wt.-%). The wash liquid obtained during each centrifuge stage is recycled to avoid undue loss of *p*-xylene, but the maximum theoretical quantity that can be extracted is fixed at ca. 60% of that present in the feedstock by the eutectic point of the system.

Isomerization of the C_8 Aromatics

The market for *m*-xylene is very small, and unfortunately this isomer is in the greatest concentration in the feedstock. The mother-liquor from the crystallization process, which is rich in *m*-xylene, is isomerized towards the equilibrium composition of the xylene isomers, whereafter it can be recycled to the *o*-xylene columns (see Figure 5.1) and further *o*- and *p*-xylene can be extracted. The equilibrium concentrations of the C_8 aromatics with temperature are shown in Figure 5.3.

The three xylene isomers isomerize easily but ethylbenzene is slow to reach its equilibrium content. It is thought that the reaction mechanism involves the smooth displacement of the methyl groups around the aromatic nucleus and, therefore, that direct conversion of *o*-xylene into *p*-xylene cannot occur (cf. Scheme 1).

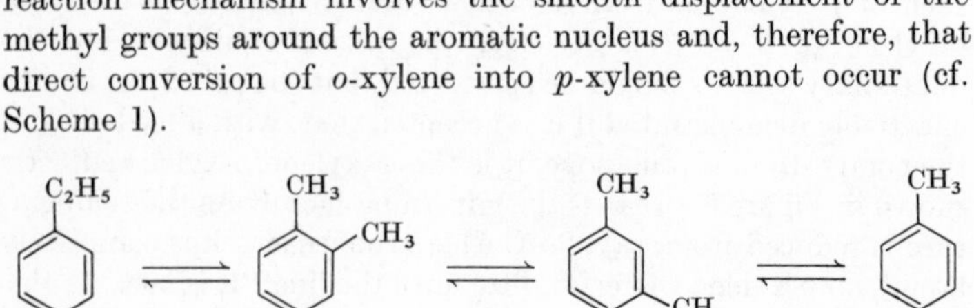

SCHEME 1.

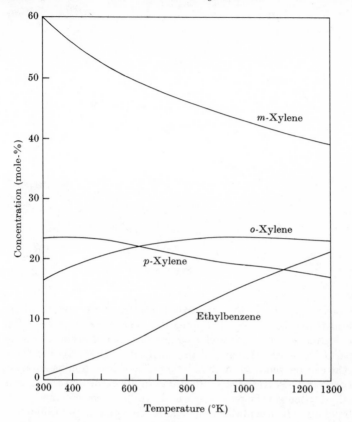

Figure 5.3. Equilibrium concentrations of C_8 hydrocarbons at various temperatures.

Evidence has also been presented suggesting that isomerization over a zeolite catalyst occurs via transalkylation to a trimethylbenzene, e.g. as in Scheme 2 (p. 118).

It is probable that both mechanisms occur in the isomerization of the xylene isomers. Isomerization of ethylbenzene, however, which occurs only in the presence of hydrogen, can be explained as shown in Scheme 3 (p. 118).

Isomerization processes have been developed for both the liquid and the vapour phase, but nearly all commercial processes use the latter. There are two main catalyst systems, silica–alumina and platinum on silica–alumina, and the choice between them

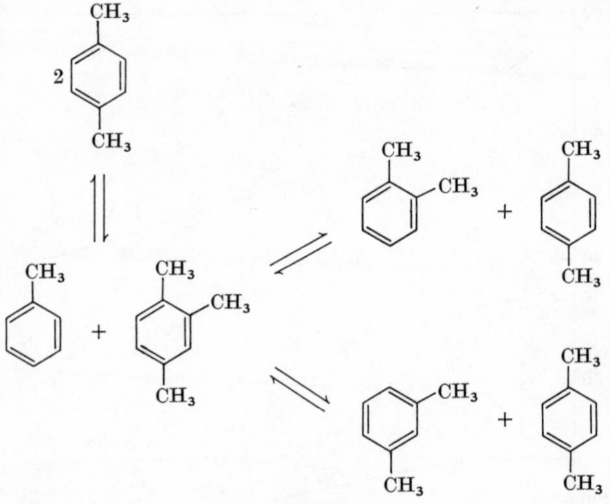

SCHEME 2.

largely depends on the quantity of ethylbenzene present in the feed to the isomerizer. If appreciable quantities are present the noble-metal catalyst is preferred as this is able to isomerize some part of the ethylbenzene and also cracks it to lighter hydrocarbons that can be removed by distillation. The silica–alumina catalyst, however, has little, if any, effect on ethylbenzene and the level of ethylbenzene in the recycle stream to the o-xylene columns would build up to unacceptable proportions. The other major differences between the two systems are: (i) hydrogen must be present with

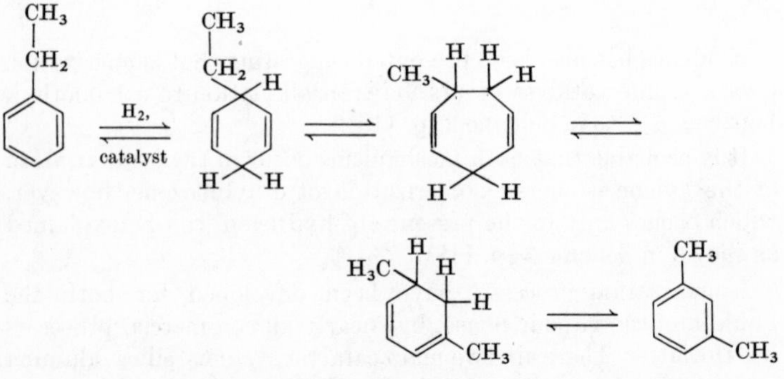

SCHEME 3.

the feed to the noble-metal catalyst to allow the isomerization of ethylbenzene; (ii) the silica–alumina catalyst requires frequent regeneration by oxidative burn-off while the platinum-on-alumina can operate for a year or more without regeneration. Typical operating conditions for a noble-metal catalyst are a temperature of about 460°C at 2.5×10^6 Nm^{-2} and a feed-to-hydrogen molar ratio of 1:10.

Specific Extraction Processes

Two further processes for the extraction of p-xylene are now in commercial operation. In one the feedstock is passed in the liquid phase over a bed of Y-type molecular sieve when the other isomers, which are less strongly held, pass through as the raffinate stream.

The p-xylene is desorbed by a specific eluent which can be distilled off and recycled to the adsorber. As p-xylene is required at high purity (>99.2 wt.-%) the eluent is also used to purge the feedstock from the void spaces in the adsorber. The process is operated in a continuous manner by employing the moving-bed principle in which the feed is passed up the adsorber down which the sieve is travelling. The adsorption of p-xylene, purging of the voids, and desorption of the p-xylene are accomplished in successive zones within the bed and the various streams are either admitted or removed from the side of the adsorber. In practice the moving bed is simulated by moving the feed and eluent inlets, and the product outlets up and down the bed in discontinuous steps; the greater the number of these points the greater the approach to a true moving bed. It is claimed that over 95% of the p-xylene present in the feed can be recovered at 99.3% purity.

In the second process the complexing reaction of HF–BF$_3$ with an aromatic molecule is used.

$$ArH + HF + BF_3 \rightleftharpoons ArH_2^+BF_4^-$$

The ionic complex formed can then be extracted into an excess HF layer for separation. The stability of the complex formed is very sensitive to the structure of the aromatic molecule, and the stabilities are in the ratio m-:o-:p-xylene as 20:2:1, with the ethylbenzene complex much less stable. By careful adjustment of the concentration of BF$_3$, it is possible to effect complete extraction of the m-xylene; p-xylene can then be separated from the remaining isomers by superfractionation and distillation. The HF–BF$_3$ will also act as an isomerization catalyst at elevated reaction temperatures.

Hydrodealkylation of Toluene to Benzene

The hydrodealkylation of toluene can be carried out either thermally or catalytically, and there is no clear advantage with either system.

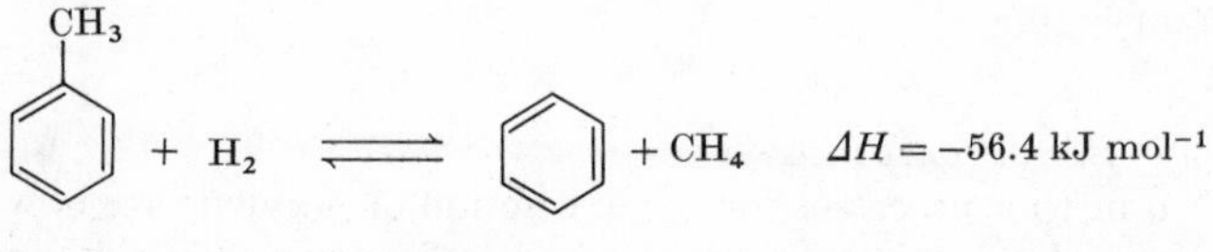

$$\text{C}_6\text{H}_5\text{CH}_3 + \text{H}_2 \rightleftharpoons \text{C}_6\text{H}_6 + \text{CH}_4 \qquad \Delta H = -56.4 \text{ kJ mol}^{-1}$$

In the thermal process toluene and hydrogen in a molar ratio of ca. 1:3 are passed into a reactor at 650–750°C and 4–6×10^6 Nm^{-2}. The benzene is recovered by distillation from the liquid effluent. A portion of the recycle hydrogen is vented to improve the conversion into benzene by reducing the concentration of methane. This vent gas is scrubbed to recover entrained product.

The stoichiometry of the reaction indicates that increasing total pressure has no effect on the thermodynamic equilibrium, but it does increase the rate of reaction. It also improves the selectivity to benzene by reducing side reactions such as the formation of polynuclear aromatics and coke. A further side reaction is the formation of biphenyls, e.g.:

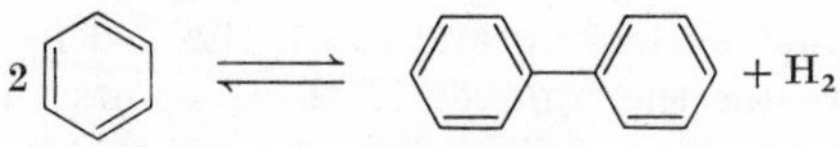

Thus a high partial pressure of hydrogen is maintained to shift the equilibrium towards benzene. Under the conditions mentioned, ca. 3% of biphenyls are formed at equilibrium; these compounds are recycled to the reactor to suppress further formation.

A radical chain mechanism has been postulated:

$$(\text{R}\cdot + \text{H}_2 \rightleftharpoons \text{RH} + \text{H}\cdot \qquad \text{Initiation})$$

$$\text{H}\cdot + \text{C}_6\text{H}_5\text{CH}_3 \rightleftharpoons (\text{C}_6\text{H}_6\text{CH}_3)\cdot \rightleftharpoons \text{C}_6\text{H}_6 + \text{CH}_3\cdot$$

$$\text{CH}_3\cdot + \text{H}_2 \rightleftharpoons \text{CH}_4 + \text{H}\cdot$$

When catalysts are employed they must have a low acid function to avoid ionic reactions which lead to coke formation; chromic oxide on alumina and zeolites are examples. Operating temperatures are somewhat lower than for the thermal reaction. Little is known about the reaction mechanism.

Production of Ethylbenzene and Cumene by Alkylation

Ethylbenzene and cumene are manufactured by reaction of benzene with ethylene and propylene, respectively. This reaction, which is catalysed by a Lewis acid, is known as a Friedel–Crafts synthesis, e.g.:

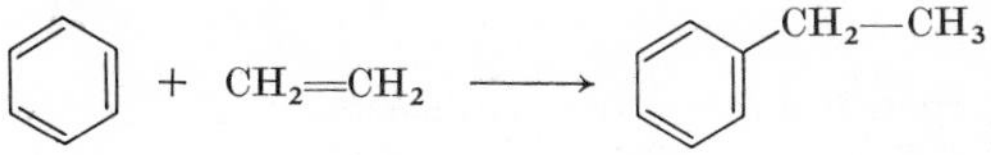

$$\Delta H = -104 \text{ kJ mol}^{-1}$$

A number of catalyst systems have been developed, but that most frequently used is aluminium chloride in a liquid-phase reaction. Thermodynamically the reaction goes to completion at temperatures below 500°C, but side reactions do occur leading to polysubstituted aromatics.

The reaction mechanism involves the necessary presence of HCl and this can be produced by the introduction of either water or ethyl chloride. In the presence of aluminium chloride and hydrogen chloride protonation of the double bond of the olefin occurs, and then this electrophilic species interacts with the π-electrons of the aromatic nucleus (Part 2, pp. 109 and 212).

$$CH_2\!=\!CH_2 + HCl + AlCl_3 \longrightarrow [C_2H_5]^+[AlCl_4]^-$$
$$[C_2H_5]^+[AlCl_4]^- + C_6H_6 \longrightarrow [C_6H_6C_2H_5]^+[AlCl_4]^-$$
$$[C_6H_6C_2H_5]^+[AlCl_4]^- \overset{\cdot}{\longrightarrow} C_6H_5C_2H_5 + HCl + AlCl_3$$

At the reaction temperature aluminium chloride exists in its dimeric form which is ineffective as a catalyst. However, it forms a complex with the benzene, and further aluminium chloride will dissolve in the complex in the catalytically active, monomeric form.

In the commercial process the liquid effluent from the reactor is allowed to settle and the aluminium chloride is recycled to feed. The ethylbenzene or cumene is recovered by distillation in high purity and the heavy ends which contain the polysubstituted benzenes are further treated with aluminium chloride–benzene complex under conditions such that transalkylation occurs, e.g.

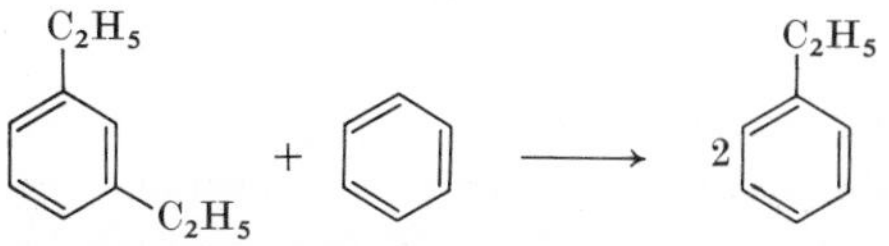

as shown. The product is returned for recovery of the ethyl-benzene.

The reaction conditions for alkylation with ethylene are more severe than are required with propylene, and therefore feedstock for cumene manufacture can contain ethylene which will remain unchanged. The presence of propylene in ethylbenzene production, however, will lead to unwanted products.

Benzene

(A. H. JUBB, ICI Corporate Laboratory, Runcorn)

The largest outlet for benzene is ethylbenzene (see p. 121) which is an intermediate in the preparation of styrene (Chapter 4. p. 83) and this in turn is the building unit of a major plastic (Chapter 8, p. 225). Although phenol is manufactured in a number of ways (see p. 132 and Chapter 7, p. 170) these are almost all based on benzene, so that phenol production is the second major use of benzene. Table 5.5 shows the use of benzene in 1971. In the U.S.A.

Table 5.5.   Benzene end-use pattern in U.S.A. in 1971

End-use	Amount consumed (tons $\times$ 10^{-6})	% of total
Ethylbenzene (for styrene)	1.650	44.2
Phenol	0.717	19.2
Cyclohexane	0.584	15.6
Maleic anhydride	0.144	3.9
Detergent alkylate	0.140	3.8
Aniline	0.132	3.5
Dichlorobenzene	0.042	1.1
DDT	0.019	0.5
Other chemical uses	0.301	8.2
	3.729	100.0

during 1971, 92% of the benzene produced was obtained from petrochemical sources and 8% from coal.

Cyclohexane

Benzene is readily hydrogenated to cyclohexane either in the liquid phase or in the vapour phase over nickel catalysts. The former processes generally employ temperatures around 150–200°C with the appropriate pressure to maintain liquid-phase conditions. The main variation in the various processes is the method used to dissipate the heat of reaction.

$$C_6H_6 + 3H_2 \longrightarrow C_6H_{12} \qquad \Delta H = -206 \text{ kJ mol}^{-1}$$

Cyclohexane is used mainly in the production of adipic acid, hexamethylenediamine, and caprolactam.

Oxidation of cyclohexane with air at 150°C and $1 \times 10^6 \text{ Nm}^{-2}$ in the presence of cobalt naphthenate (1 ppm) is carried out to a cyclohexane conversion of about 8% to produce a 70% yield of cyclohexanol (A) and cyclohexanone (K) together with a wide variety of by-products. After removal of unchanged cyclohexane, the crude products are oxidized with 50% nitric acid at 85°C in the presence of a copper–vanadium catalyst to yield approximately 1 lb. of adipic acid per lb. of crude product. The adipic acid is purified by recrystallization from water; a mixture of adipic, glutaric and succinic acid can be recovered from the mother-liquors.

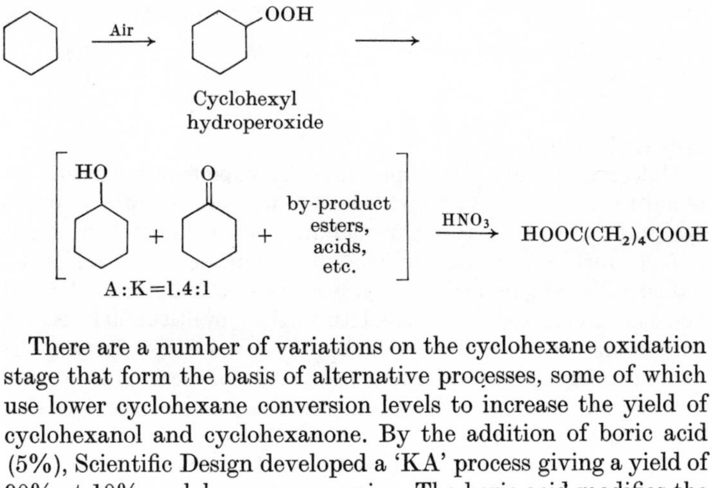

Cyclohexyl
hydroperoxide

A:K=1.4:1

There are a number of variations on the cyclohexane oxidation stage that form the basis of alternative processes, some of which use lower cyclohexane conversion levels to increase the yield of cyclohexanol and cyclohexanone. By the addition of boric acid (5%), Scientific Design developed a 'KA' process giving a yield of 90% at 10% cyclohexane conversion. The boric acid modifies the decomposition of the intermediate hydroperoxide and also readily esterifies the cyclohexanol, leading to higher A:K ratios (10:1).

There are claims in the literature for the direct oxidation of cyclohexane to adipic acid, including oxidation in the presence of acetic acid with cobalt catalysts, but there is little evidence that they have been commercialized.

The Celanese Corporation have developed a variant of the normal 'KA' route whereby cyclohexane is air-oxidized to about

25% conversion. After removal of unchanged cyclohexane, the cyclohexanol and cyclohexanone are removed by distillation and oxidized by nitric acid to adipic acid. The residual complex mixture of by-products from the first oxidation stage is esterified with pentane-1,5-diol to a complex mixture of polyesters. The latter are then hydrogenolyzed over copper chromite to give a mixture of diols from which hexane-1,6-diol is separated by distillation. Vapour-phase amination of the diol is carried out by reaction with ammonia over a Raney nickel catalyst to yield hexamethylenediamine (HMD). Pentane-1,5-diol, obtained in the process from glutaric acid formed during cyclohexane oxidation, is recycled to the esterification stage. Hexamethylenediamine is prepared from adipic acid in a two-stage process. Adipic acid is first converted into adiponitrile by treatment with an excess of ammonia in the liquid or vapour phase, in the presence of a dehydrating catalyst (boron phosphate); the nitrile is then catalytically hydrogenated.

$$HOOC(CH_2)_4COOH \xrightarrow[\text{Boron phosphate} \atop (350-450°)]{NH_3} NC(CH_2)_4CN \xrightarrow[\text{Ni}]{H_2} H_2N(CH_2)_6NH_2$$

Adipic acid	Adiponitrile	HMD
	(85%)	(94%)

Maleic Anhydride

Maleic anhydride can be produced by vapour-phase oxidation of most materials containing four or five unbranched carbon atoms (paraffins excepted) over a vanadium pentoxide catalyst; these include furfural, crotonaldehyde and n-butane. However, most industrial production is based on benzene (see Chapter 4). Air and benzene are mixed and passed through a multitubular reactor containing vanadium pentoxide supported on an inert carrier. The contact time is of the order of 0.1 sec at atmospheric pressure and 400–450°C. The reaction is highly exothermic and the liberated heat is removed by fused salt which circulates between the catalyst tubes; the heat content of the salt is used to generate steam. The exit gases are cooled in a waste-heat boiler and finally by cooling water. They are then absorbed in water, whereby the anhydride is converted into maleic acid. The latter is then dehydrated back to the anhydride by water-removal under vacuum-distillation conditions or by azeotropic distillation with hydrocarbons. It is finally purified by vacuum distillation. There are variants of the basic process, mainly in the recovery section, and some manufacturers use a switch condenser to recover over 85% of the product as the anhydride. The remainder can then be recovered by absorption in water; 60–65% yields are obtained based on benzene.

$$2 \; \bigcirc \; + \; 9O_2 \; \xrightarrow{V_2O_5} \; 2 \; \begin{array}{c} CHCO \\ \| \\ CHCO \end{array} \!\!\! \diagdown \!\!\! O \; + 4H_2O + 4CO_2$$

The 1971 U.S.A. maleic anhydride plant capacities were around 0.14×10^6 tons p.a. with a demand estimated at 0.1×10^6 tons. The uses were in polyester resins 50%, fumaric acid 15%, pesticides (see Chapter 12, p. 434) 10%, alkyd resins (see Chapter 8, p. 231) 5%; the remaining uses include plasticizers, lubricants and malic acid.

Alkylbenzenesulphonates

The preparation of alkylbenzenesulphonates for use as detergents is described in Chapter 13 (p. 463):

$$\text{Olefin} + C_6H_6 \; \xrightarrow[\text{AlCl}_3]{\text{HF or}} \; R\!-\!C_6H_5 \; \xrightarrow[\text{H}_2\text{SO}_4]{\text{SO}_3} \; RC_6H_4SO_3H$$

When R is a branched chain, a so-called 'hard' detergent is obtained and because such compounds are not biodegradable they have been displaced by 'soft' linear alkylbenzenesulphonates.

Aniline

Aniline is an important raw material for the production of dyes, pharmaceuticals and rubber auxiliaries. It is prepared by the same process as is used in the laboratory (Part 1, p. 136): benzene is nitrated with a mixture of nitric and sulphuric acid, the water formed in the reaction being azeotropically removed with the excess of benzene. Traditionally, nitrobenzene was reduced in a batch process in agitated reactors in which iron turnings, water and 30% hydrochloric acid reacted to give aniline in 90–95% yield. Generally, these plants had annual capacities below 10×10^3 tons. They have now been superseded by more economic continuous vapour-phase hydrogenation processes, with one of several catalysts, at 270°C in plants with annual capacities as high as 50×10^3 tons. The processes used include a fixed-bed version with a nickel sulphide catalyst deposited on alumina and liquid-phase processes with supported palladium, nickel or copper catalysts. A process involving ammonolysis of chlorobenzene is now obsolete.

The main use for aniline (>50%) is in the rubber industry for the production of antioxidants and vulcanization accelerators, followed by use in production of isocyanates for rigid-foam applications. Other important end-products include dyes, photographic and pharmaceutical chemicals.

Chlorobenzenes

Chlorination of benzene at 40–60°C can be carried out to give primarily monochlorobenzene (benzene:Cl_2 molar ratio 1:0.6) or, because of its relatively increasing importance, by the addition of $AlCl_3$ whereupon dichlorobenzenes become the major products (a lower benzene:Cl_2 ratio is used).

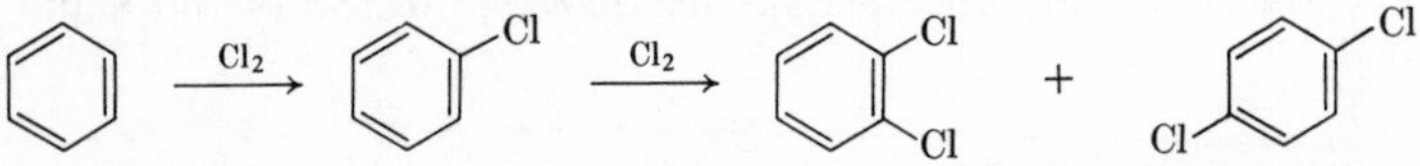

o-Dichlorobenzene is used for the synthesis of various organic intermediates, such as 3,4-dichloroaniline, used in the dyestuffs industry, and for the production of pesticides; it is also used as a solvent. The *para*-isomer is used as a deodorant and for moth control.

DDT

The insecticide DDT (dichlorodiphenyltrichloroethane) requires chlorobenzene as a starting material (see Chapter 12, p. 425).

Xylenes

(J. PENNINGTON, Research and Development Department, BP Chemicals International Limited, Hedon, Hull)

Phthalic Anhydride from o-*Xylene*

As recently as 1960, only 10% of the world production of phthalic anhydride was based on *o*-xylene, naphthalene being the major starting material. By 1970 *o*-xylene had outstripped naphthalene, accounting for about 55% of the 1.5×10^6 tons of phthalic anhydride produced in that year. This trend is continuing, with virtually all new production capacity based on *o*-xylene as feedstock.

In all but one plant, *o*-xylene is converted into phthalic anhydride by gas-phase oxidation over a fixed-bed catalyst. The exception is a unit of ca. 15×10^3 tons p.a. operated by Rhone Progil (France), who use a liquid-phase process showing similarities to the Amoco process for terephthalic acid (below). Two processes, those of von Heyden and BASF, together account for the major proportion of new production capacity. They are similar enough for a single process description to suffice. The reaction can be expressed by the equation:

$$\Delta H = -1290 \text{ kJ mol}^{-1}$$

Filtered air is compressed and preheated, and *o*-xylene (95% minimum purity) is then vapourized by injection into the hot-air stream. The quantity of *o*-xylene in the air must be controlled to be less than 1% molar so as to avoid explosive mixtures.

Reaction takes place in multitubular reactors, containing a vanadium oxide-based catalyst, at 350–400°C. The heat of reaction is transferred, via molten salts circulating around the tubes, to coils in which high-pressure steam is generated. New catalysts show high selectivity (over 75% of theory) at high gas throughputs, while engineering developments have permitted the construction of larger reactors containing up to 15,000 tubes. Hence a high output per reactor is possible, in spite of the limitation on *o*-xylene concentration.

After partial cooling, the gaseous products enter a bank of 'switch-condensers' operating in sequence. Each condenser passes through a cooling phase, during which phthalic anhydride condenses as crystals on finned tubes. This is followed by a heating phase to remelt the crude product, which is drawn off to intermediate storage. Residual gases from the condensers are scrubbed with water, to remove any remaining organic impurities, and discharged through a stack.

The crude product, containing about 99% phthalic anhydride even at this stage, is subjected to thermal treatment followed by continuous vacuum-distillation in a train of 2 or 3 columns. The purity of the final product is 99.8% or better. This material may be stored, transported and used in the molten form, or resolidified as flakes as required.

The largest single outlet for phthalic anhydride, consuming about half of the world production, is the manufacture of phthalate ester plasticizers for use in PVC (polyvinyl chloride) (see Chapter 8, p. 223). The greater part of the remainder is incorporated into alkyd resins (surface coatings and paints) or unsaturated polyester resins (for glass-fibre reinforced mouldings). The former are a versatile group of resins and comprise essentially dibasic acids, polyols (particularly glycerol or pentaerythritol) and monobasic fatty acids, which limit ester cross-linking, confer hydrocarbon solubility and, if unsaturated, provide drying properties.

Terephthalic Acid and Dimethyl Terephthalate from p-*Xylene*

Terephthalic acid (TPA) and dimethyl terephthalate (DMT) have the same major end-use, the production of polyester (polyethylene terephthalate) fibres, such as ICI's Terylene (Chapter 9, p. 284).

Until the mid-60's, all terephthalic acid was converted into the dimethyl ester for final purification, owing to difficulties resulting from the very low volatility and solubility of the acid. Since that time, technological improvements have resulted in increasing capacity for high-purity 'fibre-grade' acid, which in 1970 represented about 20% of the total world production of ca. 1.75×10^6 tons.

It is believed that the original process for the manufacture of terephthalic acid, the oxidation of p-xylene with aqueous nitric acid, is now employed by only one company (du Pont); however, the scale is not inconsiderable at over 0.2×10^6 tons p.a. The rapid expansion in, and profitability of, the polyester fibre market resulted in a proliferation of processes (all liquid-phase). However, two now dominate, sharing about 80% of world production capacity. These are the Amoco (or Mid-Century) and the Witten process.

Direct oxidation of p-xylene to p-toluic acid in the presence of small amounts of cobalt (or manganese) salts occurs relatively readily, but further oxidation to terephthalic acid is extremely slow. The problem has been tackled in three ways. First, certain promoters accelerate the oxidation of p-toluic acid. These include bromide ion (Amoco) and organic ketones and aldehydes. Then Teijin (Japan) have demonstrated that much higher cobalt concentrations give acceptable rates. Economic feasibility would depend on efficient recovery and recycle of the catalyst. Finally, p-toluic acid may be converted into more readily oxidizable derivatives, in particular the methyl ester (Witten). A brief mention should also be made of the Henkel processes, operated only in Japan, wherein potassium benzoate is disproportionated, or potassium (ortho)phthalate is isomerized, to the terephthalate salt.

In the Amoco process p-xylene (99% min.), solvent acetic acid and a solution of the cobalt–manganese catalyst and bromide

Direct oxidation

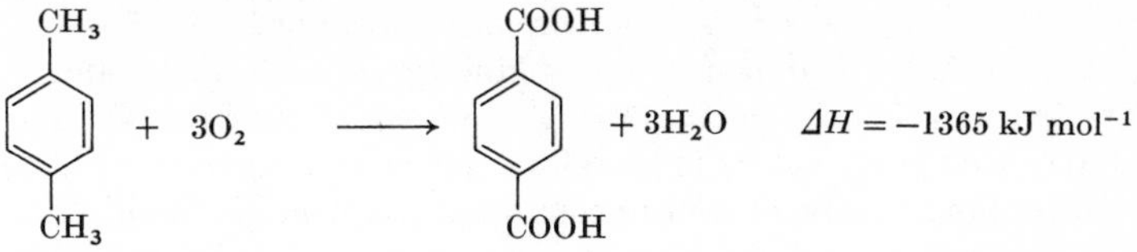

$$\text{(p-xylene)} + 3O_2 \longrightarrow \text{(terephthalic acid)} + 3H_2O \qquad \Delta H = -1365 \text{ kJ mol}^{-1}$$

Witten process

$$\text{(p-xylene)} + \tfrac{3}{2}O_2 \longrightarrow \text{(p-toluic acid)} + \tfrac{3}{2}H_2O \qquad \Delta H = -691 \text{ kJ mol}^{-1}$$

$$+ CH_3OH \longrightarrow + H_2O \qquad \Delta H = -29 \text{ kJ mol}^{-1}$$

$$+ \tfrac{3}{2}O_2 \longrightarrow + \tfrac{3}{2}H_2O \qquad \Delta H = -674 \text{ kJ mol}^{-1}$$

$$+ CH_3OH \longrightarrow + H_2O \qquad \Delta H = -29 \text{ kJ mol}^{-1}$$

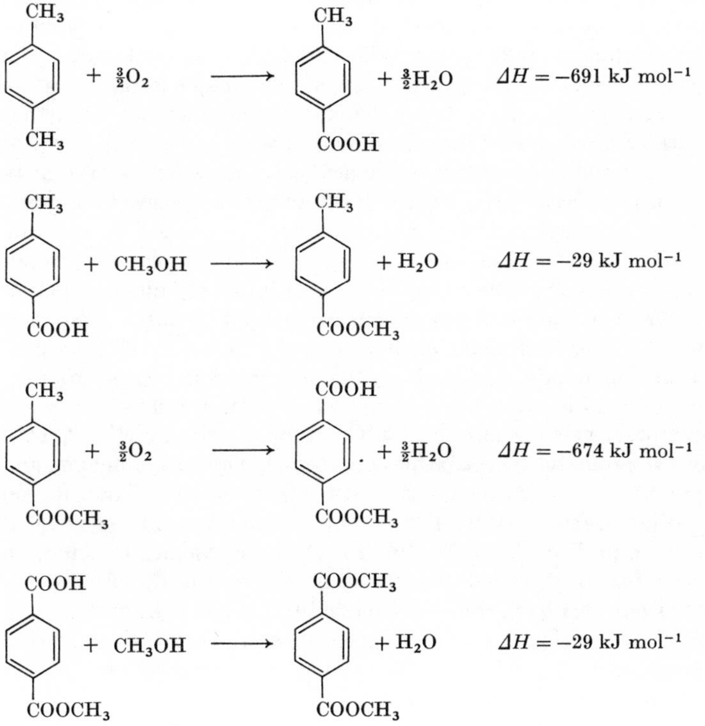

promoter are fed continuously to one or more (parallel or series) reactors, through which air is passed. The heat of reaction may be removed, and the temperature controlled at 175–230°C, by allowing the reaction mixture to boil under reflux at a predetermined pressure, usually within the range 1.5–3.0×10^6 Nm^{-2}. If required, the water content of the mixture may be regulated by removing water from the reflux condensate before returning it to the reactor. The oxygen content of the evolved gas is maintained at a low level, by controlling the air flow, to avoid the possibility of forming an explosive gas mixture at the reactor head.

Terephthalic acid has a low solubility in acetic acid and is precipitated continuously during the oxidation process. The slurry from the reactors is therefore cooled and centrifuged. The mother-liquor (solvent and unconverted *p*-xylene) is distilled, to remove water and high-boiling residues, and recycled. The residues, containing the catalysts, may be recycled in part, treated to recover catalyst components or discarded. The solid from the centrifuges is leached with hot acetic acid and dried to give technical-grade terephthalic acid (TPA) (ca. 99% pure). The yield is well over 90% of theoretical. The major impurity is the intermediate *p*-carboxybenzaldehyde. The product may be further purified or converted into the dimethyl ester (see below).

The bromide promoter/acetic acid solvent system is extremely corrosive, necessitating the use of titanium or expensive high-grade alloys in much of the plant. However, the process shows wide applicability in the oxidation of alkyl-aromatics in general, allowing single plants to be used as multiproduct units if required.

Direct oxidation of *p*-xylene is carried out by three companies who operate individual plants, on scales of $4-7 \times 10^4$ tons p.a., using cobalt catalysts, with organic promoters in lieu of bromide. Acetic acid is again the solvent, but oxidation temperatures are comparatively low (around 130°C). Mobil use ethyl methyl ketone as the promoter (Olin Mathieson process), while acetaldehyde and paraldehyde play the same role in the Eastman and Toyo Rayon processes, respectively. Furthermore, Mobil use oxygen (99.5%) at high pressure (ca. 3.5×10^6 Nm^{-2}) as the oxidant, to achieve a very high rate of oxidation. In each case, part of the promoter is oxidized, via peroxidic intermediates (which maintain a high Co^{III}/Co^{II} ratio), to give by-product acetic acid. Unconverted promoter is recycled with the solvent. In other respects, process details are similar to those for the Amoco process.

Technical-grade terephthalic acid is usually esterified with a large excess of methanol at high temperatures and pressures (over 200°C, ca. 15×10^6 Nm^{-2}). Amphoteric metal oxides (SnO, ZnO, Sb$_2$O$_3$, etc.) are typically used as catalysts. Purification of the ester is conventionally by distillation, occasionally including crystallization stages.

Amoco's alternative purification process entails hydrogenation of a solution of the technical-grade material (10–15% by wt.) in water or acetic acid, at 250–280°C, under autogenous pressure. A low partial pressure of hydrogen is used to avoid saturation of the aromatic ring. The catalyst is supported palladium. p-Carboxy-benzaldehyde and a number of coloured impurities are reduced to more soluble species. On cooling, the solution gives crystals of fibre-grade acid, which are washed and dried. In the Mobil process, technical-grade terephthalic acid is hydrogenated and filtered in the gas phase, entrained in superheated steam; the fibre-grade product is recovered by direct condensation to give suspended solid particles, which are separated from the entrainer in cyclones.

In the first stage of the Witten process, p-xylene is continuously oxidized with air to p-toluic acid at about 150°C. No solvent is used, and the catalyst is a hydrocarbon-soluble cobalt compound (e.g. naphthenate or fatty-acid salt) at low concentration, probably less than 0.1% by wt. As usual, the oxygen concentration in the evolved gas must be controlled. Heat removal and temperature control are effected by allowing the mixture to boil under reflux (at ca. 4–8×10^5 Nm^{-2}), possibly supplemented by circulation of the reaction mixture through external coolers. The conversion of the p-xylene probably exceeds 70% per pass. After removal of the p-xylene (for recycle), the p-toluic acid is esterified with methanol at a temperature in excess of 200°C (pressure over 2.5×10^6 Nm^{-2}); no catalyst is required. The methyl p-toluate can now be oxidized to monomethyl terephthalate at about 200°C and 1.5–2.5×10^6 Nm^{-2} in the presence of the cobalt catalyst carried over from the first oxidation step. The reactor product is again esterified with methanol, to give a mixture containing dimethyl terephthalate. The crude mixture is subjected to vacuum-fractionation in a distillation train, to give fibre-grade dimethyl terephthalate. Unconverted methyl p-toluate and other intermediates can be recycled, while high-boiling residues are discarded. The fibre-grade ester may be stored in the molten state or converted into solid flakes.

It has been demonstrated that p-xylene and methyl p-toluate can be oxidized simultaneously. Esterification of the mixed product gives methyl p-toluate and dimethyl terephthalate, the former being recycled. The commercial status of this simplified process is unknown.

Isophthalic Acid from m-*Xylene*
Isophthalic acid is produced in pure form (over 99%) or in admixture with terephthalic acid. World consumption was no more than 5–6×10^4 metric tons in 1970 but is growing steadily. Manufacture is largely by the Amoco process (see terephthalic acid). The major outlet is in unsaturated polyester resins.

Cumene

(R. H. HALL, Research and Development Department, BP Chemicals International Limited, Epsom)

Manufacture of Phenol and Acetone

The conversion of cumene into its hydroperoxide and decomposition of the latter to phenol and acetone were discovered by Hock and Lang during a study of the autoxidation of hydrocarbons; details were published in 1944. Dry cumene (1) was shaken with oxygen at 85°C under ultraviolet irradiation to give the hydroperoxide (2) which was isolated as the crystalline sodium salt. Cautious acidification of the latter, extraction into ether, and distillation under low pressure gave pure (2). The rate of production of the hydroperoxide in the oxidation was slow, a yield of about 7% being obtained in 24 h. Treatment of the hydroperoxide with boiling aqueous 10% sulphuric acid for 1.5 h. gave phenol and acetone. The phenol was isolated as its benzoate, the overall yield of the latter from the hydroperoxide being about 75%.

$$
\underset{(1)}{\underset{\overset{|}{CH_3}}{\overset{\overset{CH_3}{|}}{C_6H_5-CH}}} \xrightarrow{O_2} \underset{(2)}{\underset{\overset{|}{CH_3}}{\overset{\overset{CH_3}{|}}{C_6H_5-C-OOH}}} \xrightarrow{H^+} C_6H_5OH + (CH_3)_2CO
$$

$$
\downarrow
$$

$$
C_6H_5OOCC_6H_5
$$

Several industrial companies immediately became interested in this process as the basis of a new industrial route to phenol and acetone, notably the Distillers Company Ltd. in the U.K. and the Hercules Powder Co. in the U.S.A. Pooling of the technology developed by these two companies led to the Distillers–Hercules phenol process which, since BP Company Limited acquired the chemical assets of Distillers Co. Ltd. in 1967, has become known as the BP–Hercules process.

Cumene is first made by alkylation of benzene (in excess) with propylene (see p. 121) either in the liquid phase at about 100°C with aluminium chloride as catalyst or in the vapour phase at about 250°C and $3.45 \times 10^6 \ Nm^{-2}$ over a catalyst comprising phosphoric acid on kieselguhr. In both processes, the broad features of which were known before development of the cumene–phenol process, the conversion of benzene is limited to minimize the formation of di-, tri- and higher-alkylated products; the unconverted benzene and products higher-boiling than cumene are separated and recycled to the alkylation stage in the case of the liquid-phase process.

$$C_6H_6 + CH_3CH{=}CH_2 \xrightarrow[\text{H}_3\text{PO}_4; \text{ vapour phase}]{\text{AlCl}_3; \text{ liquid phase}} C_6H_5CH(CH_3)_2$$

The oxidation of the cumene is usually carried out in two or more oxidizers in series. These are usually towers, although stirred autoclaves are also used. Air is the usual oxidizing agent but oxygen is sometimes employed. The reaction temperature is 95–130°C and is usually highest in the first oxidizer of a series, falling progressively as the hydroperoxide concentration increases. By employing these reaction temperatures, the need for the ultraviolet irradiation used by Hock and Lang is avoided. The oxidation is carried out in an anhydrous system or in the presence of small amounts of water. Some alkali is generally fed to the oxidizers to maintain a pH of about 7. The final concentration of hydroperoxide is limited to less than about 40% (preferably 20–30%) in order to maintain a high selectivity. $\alpha\alpha$-Dimethylbenzyl alcohol, $C_6H_5C(CH_3)_2OH$, and acetophenone are formed as by-products in small amounts.

Oxidation of cumene under the above conditions is a free-radical chain process:

Initiation: $RH + \cdot X \longrightarrow R\cdot + HX$ $[R = C_6H_5C(CH_3)_2]$

where $X\cdot$ arises from $\begin{cases} ROOH \longrightarrow RO\cdot + \cdot OH \\ RH + RO\cdot \longrightarrow ROH + R\cdot \\ RH + \cdot OH \longrightarrow R\cdot + H_2O \end{cases}$

Propagation: $R\cdot + O_2 \longrightarrow RO_2\cdot$
$RO_2\cdot + RH \longrightarrow ROOH + R\cdot$

Termination: $2RO_2\cdot \longrightarrow ROOR + O_2$ or other stable products
$2RO_2\cdot \longrightarrow 2RO\cdot + O_2$
$RO\cdot \longrightarrow PhCOMe + \cdot CH_3$
$\downarrow$
$HCHO + HCOOH$

The product from the final oxidizer is concentrated by vacuum-distillation to about 80% hydroperoxide concentration and fed to the cleavage reaction. The cumene recovered from the concentration is purified and recycled to the oxidation stage.

Cleavage of the hydroperoxide is normally carried out in a continuous manner by feeding the concentrate and the acid catalyst (usually sulphuric acid) separately and continuously into a vessel or series of vessels containing previously decomposed hydroperoxide and allowing the products to overflow into a receiver. The reaction is run at about 50–90°C and the very considerable heat of reaction (226 kJ mol⁻¹) is removed by heat exchange, e.g. by cooling coils in the reactor or by allowing the acetone to reflux.

The cleavage reaction is ionic, the mechanism probably being as illustrated.

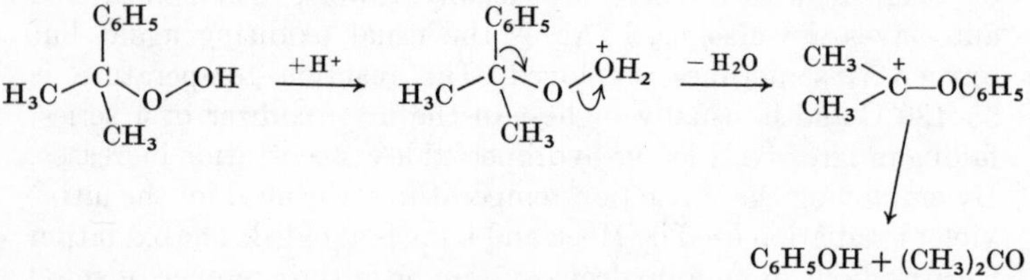

$$C_6H_5OH + (CH_3)_2CO$$

In addition to the main reaction some side reactions occur, leading to formation of α-methylstyrene (**4**) and its dimers cumylphenol (**5**), *p,p′*-isopropylidenediphenol (**6**) (commonly called 2,2-diphenylolpropane in industry) and mesityl oxide (**7**), together with water, some tar, and traces of numerous other compounds. The acetophenone present in the crude hydroperoxide concentrate passes unchanged through the cleavage reaction, as does some of the α,α-dimethylbenzyl alcohol (**3**).

$$C_6H_5C(CH_3)_2OH \longrightarrow C_6H_5C(CH_3){=}CH_2 + H_2O$$

          (**3**)   (**4**)

                                                          ↘

                                                     Dimers

$$C_6H_5OH + C_6H_5C(CH_3)_2OH \longrightarrow C_6H_5C(CH_3)_2{-}C_6H_4OH$$

                  (**3**)   (**5**)

$$2C_6H_5OH + (CH_3)_2CO \longrightarrow (CH_3)_2C{<}^{C_6H_4OH}_{C_6H_4OH} + H_2O$$

                                                   (**6**)

$$2(CH_3)_2CO \longrightarrow (CH_3)_2C{=}CHCOCH_3 + H_2O$$

                                   (**7**)

The cleavage product is neutralized and the various products are recovered by distillation. The acetone and water and almost all the hydrocarbons (cumene and α-methylstyrene) are taken overhead in one or two stills and thus separated from the phenol and high-boiling substances. The acetone is purified by further fractionation. The cumene is purified and recycled to the oxidation, and the α-methylstyrene is either separated as a pure product or hydrogenated to cumene for recycle.

The phenol-containing stream from the first distillation is further distilled in a crude phenol column to separate crude phenol overhead from acetophenone and higher-boiling materials. The base product is burnt, treated to recover cumylphenol, or cracked at high temperatures (e.g. 300°C) when it gives phenol and α-methylstyrene; any αα-dimethylbenzyl alcohol present forms α-methylstyrene. These are recycled to an earlier still. Aceto-phenone is occasionally isolated from the cracker distillate.

The crude phenol obtained as distillate from the crude phenol column is finally purified to remove traces of hydrocarbons, mesityl oxide, etc. This can be done by simple distillation or by extractive distillation with water or some other suitable liquid. The overall yields from cumene to phenol and acetone can be over 90%; the quantity of acetone produced is about 0.6 ton per ton of phenol. The phenol obtained is of very high purity and passes very stringent specifications.

CHAPTER 6

Some Specific Processes

Processes operated on the plant scale often differ greatly from experiments carried out in the laboratory. On the plant very much higher pressures and temperatures are available and heterogeneous catalysts are frequently employed. The purpose of the present Chapter is to discuss some of the processes carried out in large petrochemical installations. The topics chosen are far from comprehensive but are intended to give a chemist acquainted only with laboratory work some idea of large-scale operations.

Oxidation

(B. WOOD, Research and Development Department, BP Chemicals International Limited, Epsom)

Many of the simpler hydrocarbons produced by the petroleum chemicals industry (Chapter 2) are used as raw materials for oxidation processes. Such oxidations range from only small alterations in hydrocarbon structure through partial oxidation to epoxides, alcohols, aldehydes, ketones and/or carboxylic acids, to complete combustion. As well as hydrocarbons themselves, intermediate oxidation products (e.g. alcohols, aldehydes or ketones) may be used as starting materials. Propylene, for example, may be converted first into acrolein, and the latter then oxidized to acrylic acid in a second stage:

$$CH_2{=}CHCH_3 \longrightarrow CH_2{=}CHCHO \longrightarrow CH_2{=}CHCOOH$$

Such oxygenated starting materials may be obtained from processes other than oxidation of hydrocarbons; aldehydes are manufactured by the OXO process (see p. 153) and alcohols are produced by hydration of olefins. However, the present trend is towards development of direct, one-stage oxidation processes based on a cheap hydrocarbon feedstock. This Section is limited to a simple consideration of this kind of process.

By far the most important industrial oxidant for conversion of hydrocarbons is oxygen, often as air, since this is available in unlimited amounts at low cost. Processes based on other oxidants such as ozone, chromic or nitric acid, sulphur or its compounds, or hydrogen peroxide, have largely been displaced by direct oxidation processes employing oxygen (air).

Oxidation reactions are exothermic (see examples in Table 6.1) and one of the more important practical problems of large-scale chemical manufacture is rapid removal of this heat of reaction in order to limit the degree of oxidation and give the optimum yield of desired product, indeed to avoid an uncontrolled temperature rise leading to complete combustion.

Table 6.1. Heat evolved in typical hydrocarbon oxidation reactions at constant pressure

Reaction	ΔH (kJ mol^{-1})
$CH_4 + \frac{1}{2}O_2 \longrightarrow CO + 2H_2$	-36
$CH_4 + O_2 \longrightarrow CH_2O + H_2O$	-282
$CH_4 + 2O_2 \longrightarrow CO_2 + 2H_2O$	-802
$C_2H_6 + \frac{1}{2}O_2 \longrightarrow C_2H_4 + H_2O$	-105
$C_2H_6 + O_2 \longrightarrow CH_3CHO(g) + H_2O$	-322
$C_2H_6 + 3O_2 \longrightarrow 2CO_2 + 3H_2O$	-1425

During recent years the petroleum chemicals industry has developed novel oxidation processes that have extended the product range beyond that of oxygenated compounds. By carrying out catalytic vapour-phase oxidation of propylene in presence of ammonia, or of ethylene in presence of hydrogen cyanide, acrylonitrile is produced directly; the former reaction is called ammoxidation and the latter oxycyanation (see pp. 145, 178).

Other important variants are (i) oxychlorination, or more precisely oxyhydrochlorination (i.e. chlorination with an oxygen–hydrogen chloride reactant mixture; see p. 142), and (ii) oxidative dehydrogenation by which alkanes and/or monoalkenes are catalytically dehydrogenated in the vapour phase in presence of oxygen to give diolefins. For example, but-1-ene and but-2-ene are converted into buta-1,3-diene by this means:

$$CH_3CH_2CH{=}CH_2 \\ \text{or} \quad \xrightarrow{\;\;O_2\;\;} \quad CH_2{=}CHCH{=}CH_2 + H_2O \\ CH_3CH{=}CHCH_3$$

Two other co-oxidations are the reaction of either sulphur dioxide or phosphorus trichloride with a *n*-alkane in the presence of oxygen (air) to give alkanesulphonic acids (sulphoxidation) and alkylphosphonyl dichlorides, respectively:

$$RH + SO_2 + \tfrac{1}{2}O_2 \longrightarrow RSO_3H$$
$$RH + PCl_3 + \tfrac{1}{2}O_2 \longrightarrow RP(O)Cl_2 + HCl$$

Types of Oxidation Process

Oxidation processes used in the petroleum chemicals industry may be classified as liquid- or vapour-phase, with or without a catalyst. In general terms, non-catalytic oxidations of hydrocarbons in both the liquid and the vapour phase proceed by a free-radical mechanism and often result in a range of oxygenated products. On the other hand, catalytic oxidations are more specific, the function of the catalyst being to increase the rate of reaction to the wanted product relative to the rate of undesired side-reactions; this allows lower reaction temperatures to be used with a corresponding improvement in product selectivity.

Although much work has been done on the mechanism of catalysis, it is not possible to predict precisely what catalyst will be needed for a particular process. Hence, the selection and industrial development of a catalyst for a new manufacturing process is still largely based on experimentation and know-how.

Non-catalytic Vapour-phase Oxidation. A noteworthy example of this type of oxidation is a process developed in America for conversion of butane–propane mixtures, called liquefied petroleum gas (LPG), into a range of products including formaldehyde, acetaldehyde, methyl alcohol, *n*-butyl alcohol, isobutyl alcohol and acetic acid. As originally developed this was an air-oxidation process in which an excess of LPG was treated at a temperature of 370°C and a pressure of 7–8×10^5 Nm^{-2}. Later this operation was switched from air to oxygen, which was claimed to increase both plant capacity and product range.

Another industrially important non-catalytic vapour-phase reaction is the manufacture of synthesis gas (a carbon monoxide–hydrogen mixture) by partial oxidation of methane (natural gas) with a restricted amount of oxygen at 1300–1500°C. Steam is often mixed with the oxygen to help control the oxidation (see Chapter 2, p. 23). Synthesis gas is used for production of methanol and also in OXO reactions (see p. 153).

Catalytic Vapour-phase Oxidation. An outstandingly important reaction in this class is the direct oxidation of ethylene to ethylene

$$C_2H_4 + \tfrac{1}{2}O_2 \longrightarrow \underset{\underset{O}{\diagdown\diagup}}{H_2C - CH_2}(g) \qquad \Delta H = -146 \text{ kJ mol}^{-1}$$

oxide, a process which has now displaced the older route via ethylene chlorohydrin. Conversion of an olefin into its 1,2-epoxide is called epoxidation.

$$C_2H_4 + HOCl \longrightarrow \underset{\underset{OH \quad Cl}{|\qquad|}}{CH_2 - CH_2} \xrightarrow{Ca(OH)_2} \underset{\underset{O}{\diagdown\diagup}}{H_2C - CH_2}$$

The catalyst used to epoxidize ethylene is elemental silver on an inert support (e.g. α-alumina), which may be promoted with an alkaline-earth compound such as barium peroxide or calcium oxide. Reaction temperatures of 260–290°C are employed with ethylene–air mixtures of below the lower explosive limit of 3% ethylene; with ethylene–oxygen mixtures, temperatures as low as 230°C can be used. Pressures of $1–3 \times 10^6$ Nm^{-2} are commonly employed. The yield of ethylene oxide is between 50 and 70%, carbon dioxide being the main by-product. The major outlet for ethylene oxide is its hydrolysis to ethylene glycol, the latter being used in antifreeze compositions and in the manufacture of polyethylene terephthalate (see Chapter 9). Other uses of ethylene oxide are as a fumigant and for preparation of polyethylene glycols and non-ionic surface-active agents (Chapter 13).

Other catalytic vapour-phase oxidation processes operated industrially include oxidation of naphthalene or *o*-xylene to phthalic anhydride; of benzene (or butenes) to maleic anhydride (Chapter 5); of propylene to acrolein and/or acrylic acid, and ammoxidation of propylene to acrylonitrile (see p. 145).

Non-catalytic Liquid-phase Oxidation. Industrially one of the most important applications of this type of reaction is the oxidation of C_4–C_8 naphtha fractions to acetic acid, with the co-production of smaller amounts of formic and propionic acid. This process has been described in Chapter 3. However, because non-catalytic liquid-phase oxidation of paraffinic hydrocarbons results in random oxidative attack with formation of a variety of oxygenated products, this type of oxidation is often unattractive for commercial purposes and has only been exploited in the case of lower paraffins which yield relatively simple mixtures of oxygenated compounds. Improved selectivity can, however, be achieved in the case of higher straight-chain paraffins by carrying out oxidation with a restricted supply of oxygen (3–4.5%) at 165–170°C in presence of boric acid, whereupon secondary alcohols become major products. The function of the boric acid is to stabilize the alcohols as their borate esters. This process is known as the Bashkirov reaction and is applied to liquid-phase oxidation of cyclohexane to cylohexanol (see also Chapter 4), an intermediate for the manufacture of adipic acid which is used in production of polyamides (Chapter 9).

Another non-catalytic liquid-phase oxidation of great industrial significance, used in manufacture of phenol, is the conversion of cumene into cumene hydroperoxide, oxidative attack being on the tertiary carbon atom of the side chain (Chapter 5, p. 132).

Catalytic Liquid-phase Oxidation.   Numerous industrial oxidation processes of this type have been developed during recent years. Of particular novelty is the process for direct oxidation of ethylene to acetaldehyde with oxygen (air) in presence of an aqueous redox system comprising palladium and copper chlorides. Reaction is normally effected at 50–140°C and a pressure between 0.6 and 1.5×10^6 Nm^{-2}, and proceeds via decomposition of the ethylene–palladium chloride complex in presence of water, whereby acetaldehyde is formed with consequent reduction of the palladium salt. The latter is then re-oxidized by the redox system. The yield of acetaldehyde

$$CH_2{=}CH_2 + PdCl_2 + H_2O \longrightarrow CH_3CHO + Pd + 2HCl$$
$$2CuCl_2 + Pd \longrightarrow PdCl_2 + 2CuCl$$
$$2CuCl + 2HCl + \tfrac{1}{2}O_2 \longrightarrow 2CuCl_2 + H_2O$$

on ethylene is 95%, and the reaction is exothermic ($\Delta H = -243$ kJ mol^{-1}) (see Chapter 4, p. 85). Higher olefins can also be oxidized by this means; for example, propylene is converted into acetone, and *n*-butenes into ethyl methyl ketone.

An important extension of this process is reaction of ethylene with acetic acid in an anhydrous liquid phase containing copper and palladium salts in presence of oxygen, to produce vinyl acetate. The overall reaction is:

$$C_2H_4 + CH_3COOH + \tfrac{1}{2}O_2 \longrightarrow CH_2{=}CHOCOCH_3 + H_2O$$

Vinyl acetate is used in the manufacture of emulsion paints.

Other applications of catalytic liquid-phase oxidations are the conversion of *p*- and *m*-xylene into terephthalic and isophthalic acid, respectively (Chapter 5, p. 128). Similarly, toluene can be

oxidized in the liquid phase to benzoic acid which can be used
for production of phenol (Chapter 7, p. 173), and of caprolactam
via hexahydrobenzoic acid (Chapter 7, p. 192).

Chlorination

(A. H. JUBB, ICI Corporate Laboratory, Runcorn)

The world production of chlorine in 1970 was around 2×10^7
tons, approximately 80% of which was used in the production of
vinyl chloride, solvents and pesticides. The U.S.A. 1970 demand
alone was 9.6×10^6 tons, 59% of which was used to produce
chlorinated solvents.

Chlorination of organic substrates is carried out in either the
liquid or the vapour phase for two purposes: first, to introduce
chlorine atom(s) into a molecule to provide a product where the
chlorine atoms play an integral part in dictating its properties;
secondly, to introduce the chlorine atom(s) in such a manner that
they can be subsequently displaced or eliminated to introduce
other functionality into the molecule. Chlorination of paraffins
occurs by a free-radical process and in many of the commercial
applications the aim is to produce the monochloro-derivative (a
mixture of isomers). To this end the chlorination is carried out
with a large molar excess of paraffin in order to minimize formation
of polychlorides; the monochlorides have about the same
chlorination rate as the parent paraffin. The nature of directive
effects is discussed in Part 2, pp. 132–137.

Chlorination of olefins can take place by addition mechanisms,
usually at low temperatures (below 250°C), or by an allylic substitution
mechanism at higher temperatures. In the liquid phase, in
the dark, chlorine adds to a double bond by a heterolytic (ionic)
process.

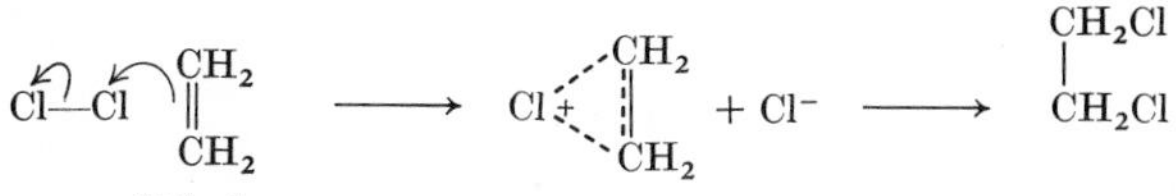

Ethylene

In the presence of light, homolytic (free-radical) addition of
chlorine predominates in the vapour phase (Part 1, p. 59). At higher
temperatures the allylic substitution mechanism prevails because
addition of a chlorine atom to the olefin is reversible whilst
abstraction of a hydrogen atom from an allylic position is irre-

versible. Thus, for propylene (Part 2, p. 129) we have the annexed reactions.

(i) $Cl\cdot + CH_3CH{=}CH_2 \rightleftharpoons CH_3\overset{\cdot}{C}HCH_2Cl$ Reversible addition

$$\downarrow {+Cl_2}$$

$$CH_3CHClCH_2Cl + Cl\cdot$$
Propylene
dichloride

(ii) $Cl\cdot + CH_3CH{=}CH_2 \longrightarrow (CH_2CHCH_2)\cdot + HCl$ Irreversible
Allyl radical   allylic
substitution

$$\downarrow {+Cl_2}$$

$$CH_2ClCH{=}CH_2 + Cl\cdot$$
Allyl chloride

One of the basic economic disadvantages of chlorination is that, in many cases, each reacting chlorine molecule leads to introduction of one chlorine atom into the organic substrate and co-production of a molecule of hydrogen chloride. This ultimately holds even for conversion of ethylene into ethylene dichloride, in that the latter is mainly converted into vinyl chloride. In so far as chlorine currently costs £30–40 per ton it can be seen that only when the HCl can be utilized do chlorination reactions compete with alternative processes. Various Companies have sought alternative ways of recycling the HCl. These included electrolysis of hydrochloric acid and reaction of HCl with Fe_2O_3, the product $FeCl_3$ being oxidized back to Fe_2O_3 with co-production of chlorine. Shell have claimed the development of a viable process based on the Deacon process:

$$4HCl + O_2 \xrightarrow[\text{400–600°C}]{\text{Cu or Fe salts}} 2Cl_2 + 2H_2O$$

The alternative that has received the greatest amount of interest and is now widely used is oxychlorination, whereby the HCl is oxidized *in situ*. The first application of this technique to benzene (see Chapter 7, p. 172) was developed by the German firm of Raschig around 1930 and was subsequently improved by Hooker Chemical Co. (U.S.A.):

$$C_6H_6 + HCl + \tfrac{1}{2}O_2 \longrightarrow C_6H_5Cl + H_2O$$

Table 6.2. Main industrial chlorination processes

Feedstock	Reaction conditions	Main products[g]
Carbon monoxide	Excess of Cl_2 over activated carbon, 0.1×10^6 Nm^{-2}	Phosgene
Methane	350–450°C, varying $CH_4:Cl_2$ ratios	Methyl chloride[a] (0.05) Methylene chloride (0.21) Chloroform (0.10) Carbon tetrachloride[b] (0.20)
Ethane	300–450°C	Ethyl chloride[c] (0.28)
Ethylene	Liquid phase, 40°C, $0.1–1.0 \times 10^6$ Nm^{-2} Vapour phase, $FeCl_3$ catalyst, 0.1×10^6 Nm^{-2}	Ethylene dichloride[d] (3.10)
Ethylene dichloride	Liquid phase, $AlCl_3$ catalyst	1,1,2-Trichloroethane (0.10) 1,1,2,2-Tetrachloro-ethane
Acetylene	Liquid phase in tetra-chloroethane, 80°C, $SbCl_3$ catalyst	1,1,2,2-Tetrachloro-ethane
Trichloroethylene	70–80°C, $FeCl_3$ catalyst	Pentachloroethane
Propylene	$C_3H_8:Cl_2$ ratio 4:1, 500°C	Allyl chloride (0.12 in 1969)
Butadiene	Vapour phase 350–400°C	3,4-Dichlorobut-1-ene 1,4-Dichlorobut-2-ene (see p. 105)
n- and iso-Pentane	300°C, 0.5×10^6 Nm^{-2}, excess of paraffin	Monochloropentanes (hydrolysed to yield pentyl alcohols by Sharples Chemicals Inc., U.S.A., until early 1960's)
Paraffins (1) $C_9–C_{13}$	70–120°C to 50–70% Cl_2 content	
(2) $C_{13}–C_{17}$	70–120°C to 40–60% Cl_2 content	Cutting oils and plasticizer extenders (0.04)[e]
(3) $C_{20}–C_{25}$	70–120°C to 40–60% Cl_2 content	
(4) $C_{12}–C_{18}$	Monochlorination	Monochlorides[f] for benzene alkylation

Table 6.2.

Feedstock	Reaction conditions	Main products[g]
Benzene	40–60°C, Fe powder catalyst or hν, 40–60°C	Monochlorobenzene (0.20) o- and p- Dichlorobenzene (0.06) (depending on Cl_2 input) Hexachlorobenzene
Toluene	PCl_3 or hν, 100–150°C, control Cl_2 input to dictate product	Benzyl chloride (0.04) Benzylidene chloride (0.002) Benzotrichloride (0.006)

[a] Also manufactured by oxychlorination of methane or reaction of methanol with HCl.

[b] Also manufactured by chlorination of CS_2.

[c] Also manufactured by reaction of ethanol with HCl.

[d] Also manufactured by oxychlorination (see Chapter 7, p.168).

[e] Larger quantities of plasticizer extenders are used outside U.S.A.

[f] Monochlorides dehydrochlorinated to yield olefins which are then used to alkylate benzene; used in Germany.

[g] Figures in parentheses are the 1972 U.S.A. consumption data (tons × 10^{-6}).

More recently it has been widely applied to the production of vinyl chloride (see Chapter 7, p. 168):

$$CH_2{=}CH_2 + 2HCl + \tfrac{1}{2}O_2 \longrightarrow ClCH_2CH_2Cl + H_2O \longrightarrow CH_2{=}CHCl$$

The main industrial uses of chlorination reactions are listed in Table 6.2 and, in order to indicate their relative importance, the 1970 U.S.A. production data are also given.

On high-temperature chlorination in the presence of a large excess of chlorine, hydrocarbons undergo carbon bond-cleavage to yield ultimately C_1 and C_2 fragments; these subsequently yield carbon tetrachloride and hexachloroethane. This type of reaction has been termed 'chlorinolysis' by Haas. The American Company, Stauffer, has developed such methods for utilizing the by-products from a vinyl chloride plant to produce carbon tetrachloride and perchloroethylene. Chlorinolysis is used commercially for the production of hexachlorocyclopentadiene from pentane and chlorine and of hexachlorobutadiene from butanes and chlorine.

Ammoxidation

Synthesis of Acrylonitrile from Propylene

(E. J. GASSON, Research and Development Department, BP Chemicals International Limited, Grangemouth)

Over the last decade, the world demand for acrylonitrile (mainly for acrylic fibres and oil-resistant synthetic rubbers and resins) has been increasing, on average, by over 10% per annum. This demand has largely been stimulated by the dramatic fall in the cost of production from propylene and ammonia, as compared with the earlier routes from acetylene and hydrogen cyanide or from ethylene cyanohydrin, which have now been rendered virtually obsolete (see Chapter 7, p. 175). The growth of the propylene–ammonia process has been most rapid in Europe, with an eight-fold increase in capacity since 1965, and in Japan, with a six-fold increase over the same period.

It has been estimated that the world production of acrylonitrile by this process exceeded 1.5×10^6 tons in 1971 and that the capacities in Europe, the Americas and Japan each exceed 6.4×10^5 tons p.a. at the present time (early 1972).

The first publication of the reaction of propylene, ammonia and oxygen to give acrylonitrile was in a patent to Cosby of Allied Chemical in 1949, but this disclosed a yield no higher than 6% and appears to have aroused little or no research interest at that time. By 1950–1951, Hadley and his co-workers of the Distillers Company Limited (DCL) had developed a two-stage vapour-phase catalytic process, in which propylene was first oxidized to acrolein over a supported copper oxide catalyst in the presence of selenium, and the acrolein was then treated with ammonia and air over a supported molybdic oxide catalyst to give acrylonitrile.

Although this process was not commercialized (being super-seded by the single-stage route), the overall yield was good and it seems probable that publication of the process patents served to encourage a number of other industrial research groups to seek a single-stage catalyst.

The first break-through in the single-stage route was the dis-covery of the bismuth molybdate catalyst in 1957, by Idol of Standard Oil, Ohio (Sohio), and this was followed in 1959 by DCL's tin–antimony oxide catalyst. By this time, Hadley (DCL) had suggested that this reaction could be considered as one of a general series, in which a compound containing an 'activated methyl group' is treated with ammonia and air over a suitable catalyst to give the corresponding nitrile. He called this reaction 'ammoxidation', a name that has now been almost universally adopted.

$$RCH_3 + NH_3 + \tfrac{3}{2}O_2 \longrightarrow RCN + 3H_2O$$

Sohio were the first to go into production (in 1960) and, since then, have licensed their process widely throughout the world. Their original catalyst, bismuth phosphomolybdate supported on silica, was later replaced in some plants by an antimony–uranium oxide–silica gel catalyst and it is now reported (1972) that a third catalyst has been developed. The DCL process (later to become the BP/Ugine process) was first commercialized in 1965 by Ugilor in France and by Border Chemicals in Scotland. Four plants are now (1972) using this process, compared with over twenty using the Sohio process or a modification of it.

Although many catalysts have now been patented for this reaction, almost all of those in commercial use fall into one of two categories: (*a*) those based on bismuth molybdate and (*b*) those based on antimony oxide. Catalysts in the latter group contain at least one other metal oxide, e.g. tin, iron or uranium. According to the patent literature, both the above types of catalyst have now been improved by the addition of further promoters. A third group of patented catalysts, consisting of molybdates or tungstates promoted with tellurium dioxide, is at present represented by the cerium–molybdenum–tellurium oxide catalyst used in the Montecatini–Edison plant in Italy.

The volume of published work on the kinetics and mechanism of this reaction is now very large, particularly in relation to the bismuth molybdate catalyst. It is generally (but not universally) agreed that the reaction proceeds with the intermediate formation of acrolein. The conversion of propylene into acrolein is first-order with respect to propylene and zero-order with respect to oxygen, at the normal operating concentrations of the latter. At low oxygen concentration the oxygen-dependence tends towards first order. The second stage, the reaction of acrolein with ammonia and oxygen to give acrylonitrile, is first-order in acrolein and ammonia and zero-order in oxygen. This reaction is, however, very much the faster, and the overall conversion of propylene into acrylonitrile therefore appears to have a zero-order dependence on ammonia as well as on oxygen, when these are present in about stoichiometric ratio to the propylene. The activation energy for the first stage is 79 kJ mol^{-1} and for the second is variously stated to be between 12.5 and 29.3 kJ mol^{-1}.

The primary, and rate-determining, step in the reaction is the adsorption of propylene on the catalyst, with the abstraction of a hydrogen atom from the methyl group, to give a symmetrically π-bonded allylic adsorbate:

$$\text{CH}_2\text{---CH---CH}_2$$

Then follows the abstraction of a second hydrogen atom from either methylene group and the addition of an oxygen anion (supplied from the catalyst, not directly from the gaseous oxygen) to give adsorbed acrolein, which at least partially desorbs to give free acrolein in the gas phase. In the second stage of the reaction, it is not yet certain whether the adsorbed acrolein reacts with molecular ammonia (to give an aldimine) or with an imino- or imido-ion, which has been produced by hydrogen abstraction from the absorbed ammonia. The latter hypothesis has been favoured by some because it can be shown that simultaneous attack of the ammonia does occur over the bismuth molybdate catalyst. However, over a tin–antimony catalyst no such attack on the ammonia takes place, yet the conversion of acrolein into acrylonitrile over this catalyst is still a very fast reaction.

The final stage involves the transfer of oxygen from the gaseous state to the catalyst, to replace that used in the reaction. In fixed-bed reactors, the feed composition and operating conditions are controlled so that this re-oxidation of the catalyst occurs concurrently and in equilibrium with the ammoxidation reaction, so that no change occurs in the catalyst under normal operating conditions. In fluid-bed reactors, it may be necessary to incorporate a catalyst re-oxidation zone, e.g. by feeding a proportion of the air at a point below the rest of the reactants.

The principal by-products are hydrogen cyanide and carbon oxides, but small amounts of acetonitrile, acrolein, acetone and acetaldehyde are also produced.

The main reactions are highly exothermic:

$$\text{C}_3\text{H}_6 + \text{NH}_3 + \tfrac{3}{2}\text{O}_2 \longrightarrow \text{CH}_2\text{=CHCN} + 3\text{H}_2\text{O} \quad \Delta H = -515 \text{ kJ mol}^{-1}$$
$$\text{C}_3\text{H}_6 + 3\text{NH}_3 + 3\text{O}_2 \longrightarrow 3\text{HCN} + 6\text{H}_2\text{O} \quad \Delta H = -946 \text{ kJ mol}^{-1}$$
$$\text{C}_3\text{H}_6 + 3\text{O}_2 \longrightarrow 3\text{CO} + 3\text{H}_2\text{O} \quad \Delta H = -1075 \text{ kJ mol}^{-1}$$
$$\text{C}_3\text{H}_6 + \tfrac{9}{2}\text{O}_2 \longrightarrow 3\text{CO}_2 + 3\text{H}_2\text{O} \quad \Delta H = -1925 \text{ kJ mol}^{-1}$$

Overall, in a commercial reactor, the heat release might be about 700 kJ per gram-mole of propylene fed, and one of the principal constraints in commercial operation is the need to remove this heat of reaction while maintaining the catalyst temperature within the effective operating limits. This is done in the Sohio process by the use of fluidized catalyst with internal cooling coils, and in the BP/Ugine process by the use of fixed-bed multitubular reactors in which the catalyst is contained in tubes of small internal diameter, cooled by heat-transfer salt circulated through the reactor shell, between the tubes. In both types of reactor, the heat of reaction is used to generate high-pressure steam.

The reactors operate in the range 400–470°C, usually at contact times of 2–4 sec for a fixed bed and 10–20 sec for a fluid bed. In commercial operation, the feed composition is usually designed to avoid the flammability region and is in the range propylene 5–8%, ammonia 5–9%, steam 0–30% and the balance air. The ammonia:propylene ratio is usually in the range 1.0–1.1:1. At lower ratios the acrylonitrile yield falls whilst the acrolein yield increases and the latter may cause polymerization problems downstream of the reactors. Unconverted ammonia must be removed from the reactor product gases before they are cooled to the dew-point, in order to prevent polymerization of the by-product acrolein; this is usually done by scrubbing the gas with a hot sulphuric acid–ammonium sulphate solution, from which ammonium sulphate may be recovered by crystallization. However, there is little profit in isolating this by-product and, today, most acrylonitrile plants are operated so as to produce the minimum amount of ammonium sulphate consistent with the maximum acrylonitrile yield.

After neutralization of the ammonia, the reactor product gas is further cooled and scrubbed with water to extract all the organic products and hydrogen cyanide, which are then recovered in a subsequent distillation, to give crude acrylonitrile containing about 70–75% of acrylonitrile, 12–15% of hydrogen cyanide, 2–4% of acrolein, and 1–3% of acetonitrile, with 5–7% of water and minor amounts of other by-products. The subsequent purification stages are the same in principle in the Sohio and the BP/Ugine process, but they differ in the order in which they are carried out. In the BP/Ugine process, the acrolein and other aldehydes are removed by reaction with some of the hydrogen cyanide to give cyanohydrins, which are easily separated from the other products by distillation. Hydrogen cyanide is taken overhead in the next column, and then acetonitrile is separated from acrylonitrile by a hydro-extractive distillation, in which aqueous acetonitrile is taken from the base of the column and aqueous acrylonitrile is removed as distillate. The wet acrylonitrile is dried and further purified in one or more finishing columns.

In the Sohio process, the acetonitrile is separated first, by hydro-extractive distillation, then the overhead product is distilled to remove hydrogen cyanide. The crude acrylonitrile is then 'topped and tailed' to remove acetone and cyanohydrins in subsequent distillations.

The final products are of very high quality. The acrylonitrile is over 99.98% pure (excluding any added stabilizers) and the hydrogen cyanide (BP/Ugine process) is over 99.5% pure.

Reliable figures for the yields from commercial reactors are not easily obtained. Many of the figures quoted in the literature and in patents clearly refer to laboratory-scale reactors (which normally give considerably higher yields than the full-scale plants), and furthermore it is not always clear whether the quoted yields are based on the quantity of reactant fed (i.e. 'yield per pass') or on the quantity converted (i.e. 'ultimate yield' or 'efficiency of conversion'). The latter figure, which is the larger (unless the conversion is quantitative), is very often quoted in spite of the

Table 6.3. Catalysts and reported yields for the ammoxidation of propylene

Original catalyst	Additives	Patentee	Year	Acrylonitrile yield (%)
Bi, P, Mo, Si	—	Sohio	1957	53–59
	Ba, F, Si	Sohio	1962	73
	V	SNAM	1966	67
	Fe, Ni, Mn, Mo, Tl	Sumitomo	1970	84
Sb, Sn	—	Distillers	1959	60
	Cu, Fe, etc.	Distillers	1963	71
Sb, Fe	—	Sohio	1963	63
	Cu, Nb	Sohio	1965	65
	B, P, V	Nitto	1968	73
	V, Te	Nitto	1969	79
Sb, U	—	Sohio	1962	75
	Cu	Distillers	1963	73
	—	Monsanto	1969	80
Te, Mo, Ce, Si	—	Montecatini–Edison	1968	80
Te, Mo, Cu, P	—	Goodrich	1969	65
Te, Mo, Fe	—	SNPA	1969	78
Te, W, Fe, Si	—	Nitto	1970	76

fact that only the yield per pass is relevant to processes such as this, in which there is no recycle of unconverted reactants. Even on the laboratory scale, yields vary with the catalyst and promoters used, and also with the feed composition, operating pressure and contact time.

The list of catalysts and yields in Table 6.3 has been selected from the patent literature. It is not comprehensive but is designed to illustrate the way in which the various types of catalyst have been improved by the addition of further promoters and modifiers. The yields quoted are based on the amount of propylene fed and all refer to laboratory-scale reactors, although they may not all be strictly comparable because of differences in feed composition and contact time. It is probable that few of the newer catalysts have yet been used commercially.

In this Table, all the catalyst elements listed were present as oxides or salts but, for brevity, oxygen has been omitted from the list of catalyst contents.

Pyridine Syntheses
(A. H. JUBB, ICI Corporate Laboratory, Runcorn)

Although pyridine syntheses do not all involve ammoxidation, there is a similarity between some of the syntheses and the preparation of acrylonitrile which justifies placing these topics together. Pyridine usage is relatively small by petrochemical standards, but some pyridine bases are used in important applications. Table 6.4 lists the important bases and their main end-uses. Although references to pyridine synthesis date back over 100 years, all requirements for pyridine bases, with the exception of 5-ethyl-2-methylpyridine, were met from coal-tar sources until around 1954. There is a vast number of references to a variety of methods for manufacturing these various bases, often as mixed products but, in the event, only a relatively small number have actually been commercialized, and none of them yields pyridine as the sole product, picolines always being also produced.

It is estimated that in 1970 the world demand for pyridine itself was about 8500 tons, of which about 1500 tons were provided from natural sources. The 1970 demand for the three picoline bases is estimated at about 7000 tons (about 800 tons from natural sources), whilst that for 5-ethyl-2-methylpyridine is estimated at about 8500 tons. It is believed that the present world plant capacities for synthetic and natural pyridine bases total around 30000 tons p.a.

Table 6.4. Pyridine bases and their end-uses

Base	End-uses
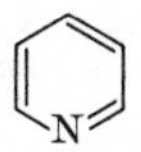 Pyridine B.p. 115°C	Production of bipyridyl herbicides and piperidine. Denaturation of alcohol. Pharmaceuticals (antiseptics such as cetylpyridinium chloride and antihistamines based on aminopyridine).
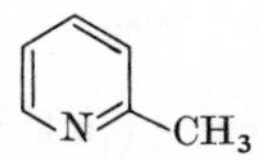 α-Picoline B.p. 129°C	Reaction with CH_2O followed by dehydration yields 2-vinyl-pyridine, used in adhesive applications. Pesticides, coccidiostats and defoliants and production of a fertilizer additive to prolong action of nitrogen in the soil.
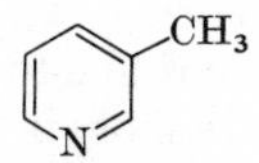 β-Picoline B.p. 144°C	Pharmaceuticals such as 'Niacin' and 'Niacinamide' for vitamin B additives in food and animal feeds.
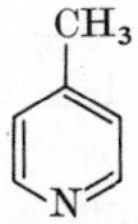 γ-Picoline B.p. 145°C	Isonicotinic hydrazide drugs for use against tuberculosis; surfactants, solvents and pesticides.
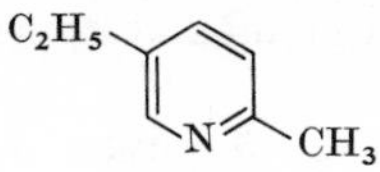 5-Ethyl-2-methylpyridine B.p. 144°C	Pharmaceuticals such as 'Niacin' (pyridine-3-carboxylic acid). 2 - Methyl - 5 - vinylpyridine (used in tyre-cord adhesives and elastomers).

The synthetic production of pyridine and β-picoline on a commercial scale is achieved by reaction of a mixture of acetaldehyde, formaldehyde and ammonia in a catalyst fluidized bed (silica–alumina). The ratio of the two bases formed depends on the amount of formaldehyde present and is usually adjusted to give a pyridine-to-β-picoline weight ratio of about 1.5:1, the combined yields being in the range 60–70%.

When formaldehyde is omitted from the feed, α- and γ-picoline are obtained as the main products (combined yield around 68%),

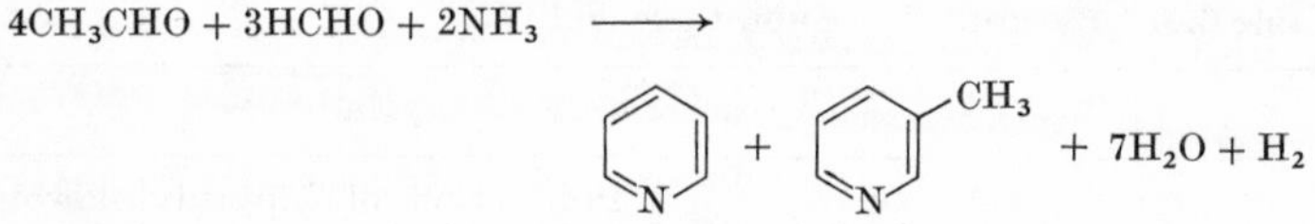

$4CH_3CHO + 3HCHO + 2NH_3 \longrightarrow$

$+$ $+ 7H_2O + H_2$

together with some 5-ethyl-2-methyl- and 3-ethyl-4-methyl-pyridine.

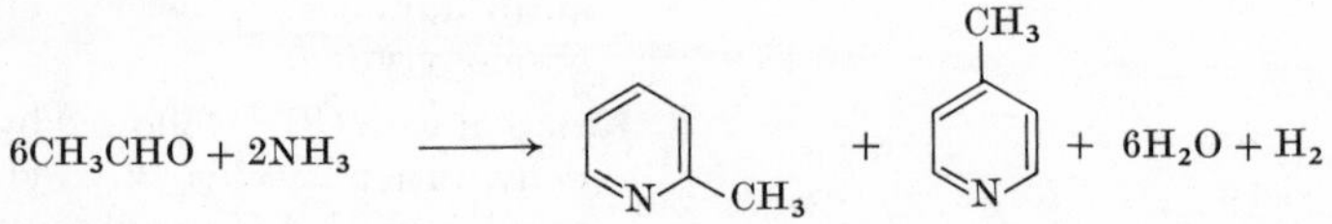

$6CH_3CHO + 2NH_3 \longrightarrow$ $+$ $+ 6H_2O + H_2$

These processes, which are operated in U.S.A., Italy and Japan, are based on the original work of Chichibabin and were first developed commercially by Reilly Tar and Chemical Corporation in the U.S.A. (see Part 3, pp. 125–126). British Petroleum have developed an as yet uncommercialized variant of these processes, and patent claims indicate that pyridine-to-picoline weight ratios of 5.7:1 can be achieved by addition of air to the mixed aldehyde feed; this Company has also patented the use of crotonaldehyde in place of acetaldehyde.

The acrolein route is a two-stage process in which propylene is initially air-oxidized over a fixed-bed bismuth molybdate catalyst at 330°C and 3×10^5 Nm^{-2}. The reactor off-gas, containing acrolein, is then mixed with ammonia and passed to a second reactor containing fluidized silica–alumina catalyst at 450°C. The process gives pyridine-to-β-picoline weight ratios around 1.5:1. It is believed that this process or modifications of it are being operated in Japan and India.

The commercial production of 5-ethyl-2-methylpyridine is carried out by condensing paraldehyde (acetaldehyde trimer) with ammonia in the liquid phase in the presence of an ammonium salt catalyst:

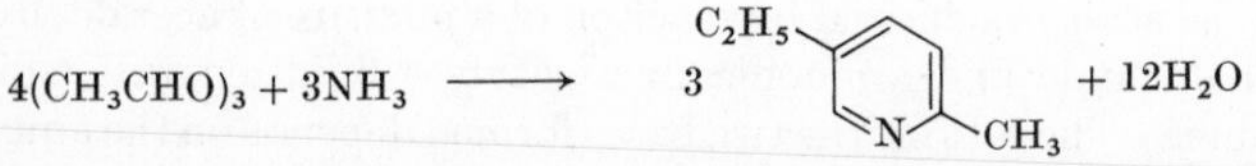

$4(CH_3CHO)_3 + 3NH_3 \longrightarrow 3$ $+ 12H_2O$

The paraldehyde, in acetic acid, is mixed with 30–40% aqueous ammonia, with which it reacts at 5×10^6 Nm^{-2} and 220–260°C. The dialkylpyridine is obtained in 60–70% yield, together with small amounts of α- and γ-picoline. The process is operated in France, Italy, Sweden, Switzerland, U.S.A. and possibly in Russia. Alkylpyridines, particularly the 5-ethyl-2-methyl compound can be thermally hydrodealkylated in the presence of hydrogen at 650–750°C and 0.5–5.0×10^6 Nm^{-2}. Thus, the 5-ethyl-2-methyl compound can be converted into mixtures of pyridine and α-picoline, the ratio of the two products depending on reaction conditions.

Although they have not been commercialized, several additional routes to pyridine have received extensive development study and are worthy of brief mention.

(a) ICI reported the catalytic vapour-phase reaction of tetra-hydrofurfuryl alcohol with ammonia in 60% yield at 55% alcohol conversion.

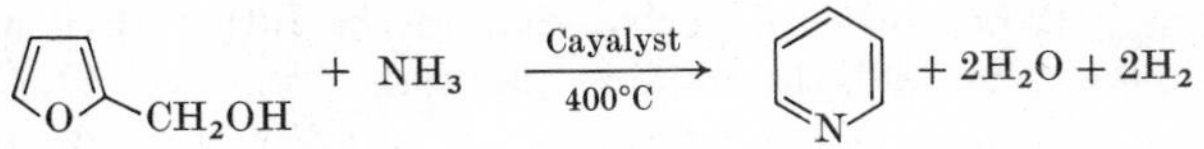

$$\text{[structure]} + NH_3 \xrightarrow[400°C]{\text{Cayalyst}} \text{[structure]} + 2H_2O + 2H_2$$

Quaker Oats developed a similar two-stage process with piperidine as product of the first-stage vapour-phase reaction; this was then dehydrogenated to pyridine in 80–90% yield.

(b) Diels–Alder reaction of acrolein with methyl vinyl ether proceeds to give 2,3-dihydro-2-methoxypyran in 95% yield. On reaction with ammonia in the vapour phase, the dihydropyran derivative gives pyridine in 84% yield.

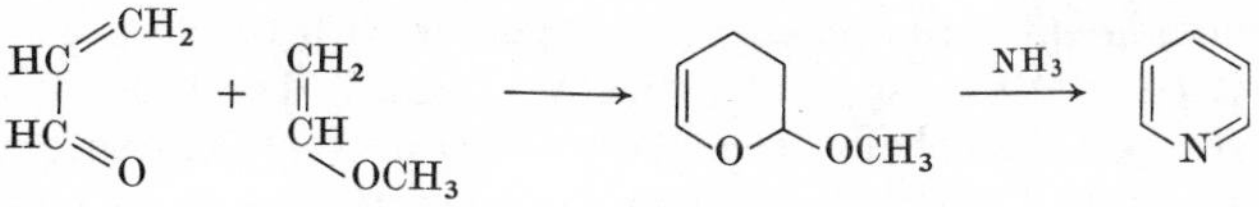

$$\text{[structure]} + \text{[structure]} \longrightarrow \text{[structure]} \xrightarrow{NH_3} \text{[structure]}$$

It is largely due to their relatively high raw-material costs that these processes have not been commercialized.

Hydroformylation or the 'OXO Reaction'

(M. J. LAWRENSON, Research and Development Department, BP Chemicals International Limited, Sunbury-on-Thames)

The catalytic addition of carbon monoxide and hydrogen to an olefinic double bond to give aldehydes is commonly referred to as the 'OXO' reaction although this term has now largely been superseded by the more descriptive term 'hydroformylation'. This homogeneously catalysed liquid-phase reaction gives a mixture of linear and branched aldehydes:

$$RCH=CH_2 \xrightarrow{+H_2+CO} RCH_2CH_2CHO \text{ and } RCH(CH_3)CHO$$

Other isomeric products may be formed if the catalyst induces double-bond isomerization of the parent olefin. Aldehydes are reactive intermediates and can be converted into acids by oxidation, amines by reductive amination and polyols by reaction with formaldehyde. However, most of the aldehydes produced are converted into alcohols and used as such. Straight-chain alcohols are preferred commercially because they are more stable to heat and form derivatives such as esters more readily. Unwanted branched aldehydes such as isobutyraldehyde are used as fuel although it has been suggested that in the future they will be cracked to give back the starting materials.

The conventional hydroformylation catalyst is octacarbonyldicobalt, $Co_2(CO)_8$, which can be formed from any cobalt salt suitably soluble in the olefinic feed. In practice it is formed by reducing cobalt naphthenate or oleate with carbon monoxide. The active catalyst component is believed to be tetracarbonylhydridocobalt, $HCo(CO)_4$, formed in the presence of the hydrogen required for hydroformylation.

The normal operating conditions for cobalt carbonyl as catalyst are 19–21×10^6 Nm^{-2} and 140–$180°C$. A simplified process flow diagram is shown in Figure 6.1. Olefin, hydrogen and carbon monoxide enter the reactor (A), together with recycled gas and recycled catalyst, and aldehydes are formed. The product then passes to the decobalter (B) where chemical treatment with acids forms a cobalt salt that can be extracted with water to recover and recycle the cobalt. Removal of cobalt at this stage is essential to prevent (i) decomposition of catalyst downstream of the hydroformylation reactor, (ii) loss of aldehyde by catalysis of condensation reactions and (iii) poisoning of hydrogenation catalysts used later. The cobalt is then converted back into an oil-soluble form (C) and reintroduced into the reactor. The product of the aldehyde distillation column (D) is passed into the hydrogenator (E). Although significant quantities of alcohol may be formed in the hydroformylation reactor above $140°C$ by cobalt-catalysed hydrogenation of the aldehyde product, current commercial practice is to hydrogenate the product in a separate step, e.g. over a solid nickel catalyst, which is far more efficient. The alcohols so formed are separated by distillation (F). The product distribution

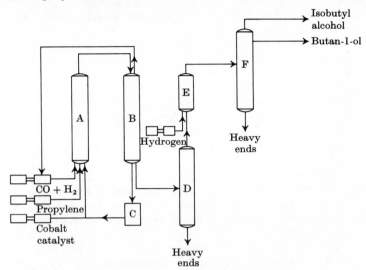

Figure 6.1. Flow diagram for synthesis of butanol.
A, Hydroformylation. B, Decobalting. C, Catalyst recycle. D, Aldehyde distillation. E, Hydrogenation. F, Butanol distillation.

after (A) is aldehydes 78–82%, alcohols 10–12%, formic esters about 2% and high-boiling products (6–8%) produced by aldol condensation of the aldehydes. The ratio of linear to branched aldehydes and alcohols is ca. 3:1 or slightly higher.

The conventional hydroformylation process produces a C_{n+1} alcohol from a C_n olefin. When higher olefins are in limited supply it is more convenient to manufacture higher alcohols from low olefins by the Aldox process which converts a C_n olefin into a C_{2n+2} alcohol. The Aldox process is essentially a conventional hydroformylation process with the cobalt catalyst modified with a metal compound such as an organic zinc salt to catalyse aldol condensation of the initial aldehyde product, e.g. propylene into

$$CH_3CH{=}CH_2 + H_2 + CO \longrightarrow CH_3CH_2CH_2CHO + CH_3CH(CH_3)CHO$$

$$CH_3CH_2CH_2CHO \longrightarrow CH_3CH_2CH_2CH(OH)\underset{\underset{CH_3CH_2}{|}}{C}HCHO \longrightarrow$$

$$CH_3CH_2CH_2CH{=}\underset{\underset{CH_3CH_2}{|}}{C}CHO \xrightarrow[Ni]{H_2} CH_3CH_2CH_2CH_2\underset{\underset{CH_3CH_2}{|}}{C}HCH_2OH$$

2-ethylhexan-1-ol. Both hydroformylation and condensation may be carried out in the same reactor.

The more recent Shell process uses a phosphine-stabilized carbonylcobalt catalyst. When octacarbonyldicobalt is treated with trialkylphosphine, carbon monoxide is evolved and an ionic species $[Co(CO)_3(R_3P)_2][Co(CO)_4]$ is formed. This is readily converted into a hydrocarbon-soluble form $[Co(CO)_3(R_3P)]_2$ which is the catalyst precursor. This complex is stable at pressures as low as 0.1×10^6 Nm^{-2}, whereas under comparable conditions octacarbonyldicobalt decomposes to carbon monoxide and metallic cobalt, which is inactive. It is believed that the active catalyst is a hydridocarbonyl similar to that obtained from octacarbonyldicobalt except that one carbonyl group is replaced by phosphine. Carbonyl complexes of rhodium have also been patented as hydroformylation catalysts, but these are not yet used in commercial operation.

The cobalt–phosphine catalyst system is a far better hydrogenation catalyst than the unsubstituted carbonyl, facilitating the production of alcohols in a one-step process. However, substantial amounts of olefin are competitively hydrogenated to alkane. The linear:branched ratio in the product is considerably higher than that for the unsubstituted carbonyl, such that the overall product analysis is: alkane 20%, branched alcohols 9%, normal alcohols 64% and high-boiling products 7%. The enhanced linear-to-branched ratio of ca. 7:1 is believed to be due to the stereoselectivity of $HCo(CO)_3(Bu_3P)$. It should be noted, however, that owing to competitive hydrogenation to alkane the overall yield of the particularly desirable normal alcohols is not much improved. The particular advantage of the phosphine catalyst is that reaction conditions are milder, 4.0–5.5×10^6 Nm^{-2} of H_2 and CO and 150–195°C, obviating the use of expensive very high-pressure equipment.

Thermodynamics and Mechanism

Extensive thermodynamic data are available only for the reaction involving ethylene which yields one aldehyde, propionaldehyde:

$$CH_2{=}CH_2 + H_2 + CO \longrightarrow CH_3CH_2CHO$$

The heats of formation for ethylene, carbon monoxide, and propionaldehyde are +52, −110 and −204 kJ mol^{-1}, respectively, at 25°C, from which ΔH is −146 kJ mol^{-1}. Hence $\Delta G = -146 + 0.24T$ or −61 kJ mol^{-1} at 25°C, and the equilibrium constant for the reaction is 4.1 N^{-2} m^4. Hydrogenation of the olefin is thermodynamically preferred with $\Delta G = -94.5$ kJ mol^{-1} at 25°C.

The mechanism of octacarbonyldicobalt-catalysed hydroformylation has never been verified by the isolation or detection of either the hydridocarbonyl or other intermediate, but it is generally postulated on the basis of olefin reactions conducted at room temperature that a template reaction occurs whereby addition of the olefin forms a π-complex, hydrogen is transferred to give an alkyl intermediate, and *cis*-insertion of a carbon monoxide ligand then occurs to give an acyl group; the further addition of hydrogen precedes elimination of the aldehyde (see the following).

$$\begin{array}{c} H \quad CHR \\ | \quad \parallel \\ (CO)_4Co + CH_2 \end{array} \rightleftharpoons \begin{array}{c} H----CHR \\ | \quad \parallel \\ (CO)_4Co---CH_2 \end{array} \rightleftharpoons \begin{array}{c} CH_2R \\ | \\ (CO)_4Co-CH_2 \end{array}$$

$$\rightleftharpoons \begin{array}{c} O \\ \parallel \\ C \\ (CO)_3Co----CH_2-CH_2R \end{array} \rightleftharpoons \begin{array}{c} (CO)_3Co-C-CH_2CH_2R \\ \parallel \\ O \end{array}$$

$$\xrightarrow{CO} \begin{array}{c} (CO)_4Co-C-CH_2CH_2R \\ \parallel \\ O \end{array} \xrightarrow{H_2} RCH_2CH_2CHO + HCo(CO)_4$$

An alternative mechanism has been proposed involving $HCo(CO)_3$ as the active species obtained by dissociation of $HCo(CO)_4$:

$$HCo(CO)_4 \rightleftharpoons HCo(CO)_3 + CO$$

but this is doubtful since hydroformylation can occur at very high pressure where the equilibrium should lie wholly to the left-hand side.

When octacarbonyldicobalt is used as a catalyst under conditions such that recovered olefin is little isomerized, approximately the same linear-to-branched ratio is obtained from a terminal olefin and its isomeric internal olefins. Apparently rearrangement of the alkyl group attached to the cobalt occurs with a preference to the formation of a linear alkyl intermediate.

When $[Co(CO)_3(R_3P)]_2$ is used as catalyst precursor it is believed that the catalytic species and intermediates are essentially the same as those postulated for octacarbonyldicobalt except that one carbonyl group has been replaced by phosphine. The hydridocarbonyl $HCo(CO)_3(Bu_3P)$ has been synthesized and shown to

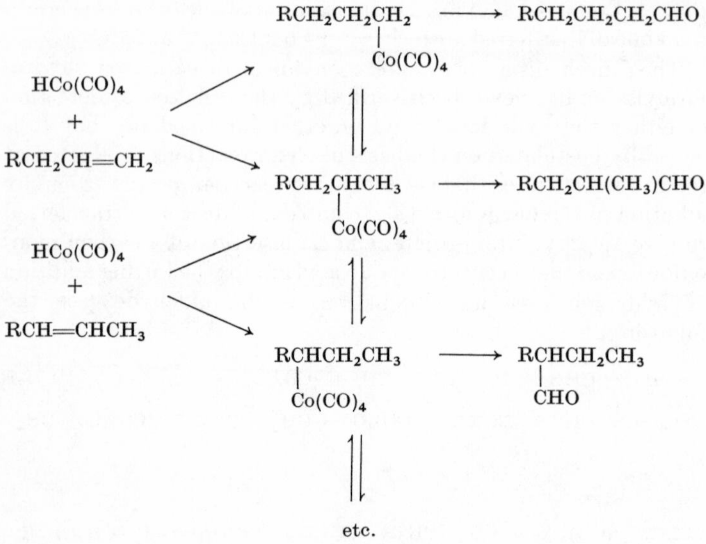

$$RCH_2CH_2CH_2 \longrightarrow RCH_2CH_2CH_2CHO$$
$$\overset{|}{Co(CO)_4}$$

$$HCo(CO)_4$$
$$+$$
$$RCH_2CH=CH_2$$

$$RCH_2CHCH_3 \longrightarrow RCH_2CH(CH_3)CHO$$
$$\overset{|}{Co(CO)_4}$$

$$HCo(CO)_4$$
$$+$$
$$RCH=CHCH_3$$

$$RCHCH_2CH_3 \longrightarrow RCHCH_2CH_3$$
$$\overset{|}{Co(CO)_4} \qquad\qquad \overset{|}{CHO}$$

etc.

produce $CH_3CH_2COCo(CO)_3(Bu_3P)$ in the presence of ethylene under pressure of carbon monoxide and that this reacts with hydrogen to give the corresponding aldehyde and $[Co(CO)_3(Bu_3P)]_2$ in accord with the mechanism shown above.

Applications

Figures for Western European capacities for hydroformylation products in 1971 show the importance of the reaction and relative amounts of each product. Propylene is the most important feedstock. Butanols find uses as solvents (375×10^3 tons p.a.), and butan-1-ol is condensed and hydrogenated to give 2-ethylhexanol (585×10^3 tons p.a.). Esterification of this alcohol with phthalic anhydride gives the dialkyl phthalates used in plasticization of PVC which may contain up to 50% of plasticizer (see Chapter 8). Other plasticizer alcohols (330×10^3 tons p.a.) include 'iso-octanols' obtained by hydroformylating propylene–butene codimers, and C_7–C_9 alcohols from a C_6–C_8 olefin cut derived from wax cracking. Alcohols derived from the hydroformylation of higher olefins such as dodecene are sulphonated to give detergent intermediates (3×10^3 tons p.a. at present, expected to rise to 36×10^3 tons p.a. in 1972). Ethylene is hydroformylated on a much smaller scale to produce propionaldehyde which is converted into propan-1-ol or propionic acid.

Hydration of Olefins

(P. LUSMAN, Research and Development Department, BP Chemicals International Limited, Grangemouth)

The commercial application of olefin hydration is confined almost entirely to the production of ethanol and propan-2-ol;

$$RCH=CH_2 + H_2O \rightleftharpoons RCH(CH_3)OH \qquad (R = H \text{ or alkyl})$$

some butan-2-ol is also made. Two processes, termed indirect and direct hydration, respectively, are used. Though these employ very different conditions, their basic chemistry is similar. Olefin hydration is an acid-catalysed reaction which proceeds via a carbonium ion formed from the olefin and a proton donated by the acid (cf. Part 1, p. 56). The normal orientation rules are followed (Part 2, pp. 77–81), so that for propylene and olefins of higher molecular weight secondary alcohols are formed almost exclusively.

In indirect hydration the source of protons is usually concentrated sulphuric acid. The carbonium ion reacts with sulphate ion to form mono- and di-ethyl sulphate, and these esters are hydrolysed to the alcohol in the second stage of the process.

$$R\overset{+}{C}HCH_3 + HSO_4^- \rightleftharpoons \underset{H_3C}{\overset{R}{>}}CHOSO_2(OH) \rightleftharpoons$$

$$\underset{H_3C}{\overset{R}{>}}CHOSO_3^- + H^+$$

$$\underset{H_3C}{\overset{R}{>}}CHOSO_3^- + R\overset{+}{C}HCH_3 \rightleftharpoons \left(\underset{H_3C}{\overset{R}{>}}CHO\right)_2 SO_2$$

$$\left(\underset{H_3C}{\overset{R}{>}}CHO\right)_2 SO_2 + H_2O \rightleftharpoons RCH(OH)CH_3 + \underset{H_3C}{\overset{R}{>}}CHOSO_2(OH)$$

$$\underset{H_3C}{\overset{R}{>}}CHOSO_2(OH) + H_2O \rightleftharpoons RCH(OH)CH_3 + H_2SO_4$$

To produce ethanol, ethylene is absorbed in 94–98% w/w sulphuric acid at 55–85°C under a pressure of 2.0–3.0×10^6 Nm^{-2}. Ethylene is bubbled through a counter-current flow of acid, e.g. in a column; 99% of the ethylene is absorbed. The product, which contains 0.3–0.4 part of ethylene (as esters) per part by weight of acid, is diluted with 1.0–1.4 parts by weight of water and heated to 60–80°C at atmospheric pressure. This is sufficient to hydrolyse the diethyl sulphate; the monoester is hydrolysed in a second reactor at ca. 100°C. Ethanol is separated from the dilute acid solution by distillation to give an aqueous solution containing 50–65% w/w ethanol. The dilute acid (35–65% w/w) is reconcentrated for re-use in the absorption stage. Propylene is more reactive than ethylene and milder conditions suffice in the absorption stage: 85% sulphuric acid is used and a temperature of 25–30°C. Otherwise the processes are very similar.

In direct hydration the source of protons is concentrated phosphoric acid held in the pores of a silica carrier. The latter is usually in the form of small spheres or cylinders, and the reaction is performed in exactly the same way as a fixed-bed heterogeneous catalytic process. However, the reaction is not on a solid surface as in the majority of such processes, but is still a liquid-phase reaction in the acid held within the pores of the carrier.

Water and ethylene (0.4–0.8 molar ratio $H_2O:C_2H_4$) are passed as vapour over the catalyst at 250–300°C and about $6 \times 10^6 \ Nm^{-2}$. The overall reaction of the carbonium ion is:

$$\overset{+}{R}CHCH_3 + H_2O \rightleftharpoons RCH(OH)CH_3 + H^+$$

Though esters may be intermediates in the reaction and traces of these appear in the product, the bulk of the latter is ethanol and no separate hydrolysis stage is necessary. The condensed product is an aqueous solution containing 10–20% w/w of ethanol. The conversion of ethylene per pass is only 5–6%. Most of the unconverted ethylene is fed back to the reactor, a small quantity being removed from the recycle loop to purge impurities (mainly from polymerization of ethylene) from the system and maintain the ethylene concentration in the recycled hydrocarbon gas at 90–95%. If the ethylene in the purge stream is purified and re-used, the ultimate selectivity on ethylene to ethanol is ca. 95%. The low conversion per pass is a result of the low value of the gas-phase equilibrium constant of the reaction at the reaction temperature. At 270°C the value is $5 \times 10^{-8} \ N^{-1} \ m^2$. Since the reaction is exothermic ($\Delta H = -46 \ kJ \ mol^{-1}$) the value of the equilibrium constant decreases as the temperature, and therefore the reaction rate, increases. Higher water:ethylene ratios and greater reaction pressure both increase the equilibrium ethylene conversion. Both of these increase costs because extra energy is required to heat and vaporize the extra water, and equipment capital costs rise with pressure. Also, for a given temperature and reactant ratio, there is a maximum operating pressure which results from the need to avoid condensation of water on the catalyst; such an occurrence washes acid off the carrier. The usual process conditions are therefore a compromise between these opposing technical and economic forces and are chosen to give the lowest overall operating cost consistent with reliable continuous operation.

The direct propylene hydration process operates under milder conditions but is basically similar to that of ethylene. Temperatures of 150–220°C and pressures up to 3.5×10^6 Nm^{-2} are used. At a given temperature the propylene hydration equilibrium is less favourable for alcohol than is the ethylene hydration equilibrium. This is compensated by the reaction rates being faster at lower temperatures. At 200°C the equilibrium constant is ca. 1×10^{-7} N^{-1} m^2. Because of the lower temperature, the operating pressure has to be lower than for ethanol to prevent condensation of water. The net result is that the conversion of propylene per pass is similar to that of ethylene in the ethanol process.

The indirect process has the advantage of lower pressures and temperatures; the higher conversion of olefin per pass means that more dilute (i.e. ethane/ethylene and propane/propylene), and therefore cheaper, olefin streams can be used. Such streams used in the direct process would necessitate excessive purging of impurities from the recycle loop. The direct process has fewer corrosion problems and therefore maintenance costs are lower; the crude product is purer and therefore easier to work-up to the stringent specifications often applied to the final product. The combination of ease of operation and cheaper running costs has made the direct hydration process the preferred synthetic route to ethanol and propan-2-ol. However, some Companies still consider that indirect propylene hydration retains advantages for small-scale plants on sites where pure propylene is not available.

In the product recovery and purification sections of the plants the by-products that have to be separated are mainly hydrocarbon polymer, aldehydes and ketones, ethers, and higher alcohols. Of these, both diethyl ether and di-isopropyl ether can be recovered for sale, and the polymer can be used as fuel. In the simplest purification procedure the bulk of the water is removed in a primary concentration stage. This is followed by two distillations, the first to remove lower-boiling components and the second to remove higher-boiling compounds and more water. A purer product can be obtained if before the initial concentration the alcohol is extractively distilled with water. Most of the impurities distil overhead, and a dilute aqueous side-stream is taken for concentration.

The products of the final concentration from ethylene and propylene are, respectively, the aqueous azeotropes: 95.5% w/w ethanol (b.p. 78.2°C) and 88% w/w propan-2-ol (b.p. 80.1°C). To obtain the anhydrous alcohols the azeotropes are distilled with an organic solvent which entrains the water; benzene is normally used.

Applications

The first attempts to commercialize a primitive version of the indirect ethylene hydration process were made early in the present century. Coke-oven gas was used as a source of ethylene. By the 1930's the process was well established using petroleum-derived ethylene. Since then ethylene hydration has largely replaced fermentation of carbohydrates as a source of industrial ethanol. Fermentation is more expensive unless there is a cheap local source of carbohydrate or the price is subsidized by government protection of agricultural products. This is true for some parts of Europe, notably France and Italy, where there is still a large production of fermentation alcohol for industrial use.

The direct ethylene hydration process was first commercialized in 1947 in the U.S.A. and is gradually replacing the indirect process as new capacity is required and older plants are phased out. Though there has been a large amount of research effort on alternative catalysts and processes, none of these has been commercialized, and the present direct hydration process does not appear likely to be superseded in the immediate future.

The main use of ethanol as a chemical intermediate was formerly for oxidation to acetaldehyde and acetic acid, but this has declined as alternative routes to these chemicals have been developed. However, the alternative market as a solvent has increased to compensate for this decline. It is contained in lacquers, adhesives, insecticides, cosmetics and antiseptics, to mention only a few of the many uses in this area. It is the largest-volume oxygenated solvent in use. World capacity (excluding U.S.S.R.) was greater than 1×10^6 tons p.a. in 1968.

The indirect hydration of propylene was first commercialized in the U.S.A. in the early 1920's. It was one of the first commercially produced chemicals to be derived from petroleum. The direct hydration process has been slower to supersede the indirect process than for ethanol. In contrast to ethanol, catalysts other than supported phosphoric acid have been commercialized, notably reduced tungsten oxide, W_2O_5. Judged by the patent literature, the latter had the disadvantage of requiring higher pressures and higher water:propylene ratios, though, being a solid, it had the advantage of being able to withstand liquid water. Most recent direct hydration capacity has been based on the phosphoric acid catalyst.

The major use for propan-2-ol is for dehydrogenation to acetone, though this has declined because of the wide availability of acetone as a co-product of phenol manufacture (see Chapter 7, p. 181). Other uses are as a solvent in the cosmetic and surface-coating industries, and as an anti-freeze. World capacity in 1968 was about 1.5×10^6 tons p.a.

A small amount of butan-2-ol is made by indirect hydration, and is used mainly for the production of ethyl methyl ketone; a small amount of propan-1-ol is formed as a minor product from propylene hydration. Higher alcohols are not made commercially by hydration. The interest is mainly in linear primary alcohols for plasticizer manufacture, and these cannot be made by hydration.

CHAPTER 7

The Choice of a Synthetic Route

The chosen route to a particular product may not be the simplest chemical process. A number of additional factors such as the cost and availability of starting materials, commercial value of by-products, energy requirements, etc., help to determine the choice. These factors may be influenced by the location of the factory and the nature of the market in which the product is to be sold. In the present Chapter we consider five important chemicals, show how new processes have superseded old ones and discuss the reasons that led to the adoption of particular routes.

Vinyl Chloride

(A. H. Jubb, ICI Corporate Laboratory, Runcorn)

Vinyl chloride is almost entirely used in the manufacture of plastics, particularly poly(vinyl chloride) (PVC), and thus its consumption is directly connected to that of the latter. PVC is the oldest of the main thermoplastics (see Chapter 8) and, although polymers of vinyl chloride were produced before World War II, the rapid growth in its use has taken place over the last twenty years (Figure 7.1). The U.S.A. plant capacity of 2×10^6 tons p.a. in 1969 is likely to be at least 3×10^6 tons by the end of 1973. Over the years, the general overcapacity in the U.S.A. has led to a decline in the price of vinyl chloride from about 13 U.S. cents/lb in the early 1950's to between 4 and 5 U.S. cents/lb since 1968.

Vinyl chloride was developed as an acetylene derivative and this situation continued through the 1950's, particularly in Europe (Table 7.1):

$$CH{\equiv}CH + HCl \xrightarrow[\substack{\text{on carbon} \\ 160-250°C}]{HgCl_2} CH_2{=}CHCl \qquad \varDelta H = -184 \text{ kJ mol}^{-1}$$

$$80-85\% \text{ yield at high conversion}$$

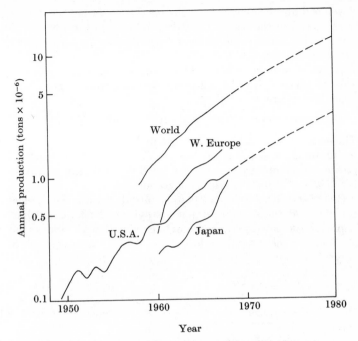

Figure 7.1. World production of polyvinyl chloride (PVC).

This process is well established and operates at moderate acetylene conversion to give above 90% yield of vinyl chloride, or at high conversion with the yield falling to about 85%.

The production of vinyl chloride by alkaline treatment of 1,2-dichloroethane was an established literature reaction. With the growth of the petrochemical industry ethylene became more abundant and this led to research activity aimed at developing ethylene-based vinyl chloride processes. Such routes were

Table 7.1. Vinyl chloride plant capacity, proportion based on acetylene (%)

Country/Area	1955	1960	1965	1966	1967	1968
U.S.A.	65	55	41	40	38	26[a]
W. Europe	100	96	75	73	62	48[b]

[a] U.S. capacity in 1972 = 10% based on acetylene.
[b] Total W. European capacity = 2.0×10^6 tons p.a. (1968).

developed and commercialized in the U.S.A. during the 1950's; exploitation in the rest of the world followed in the 1960's: this different pattern of exploitation was mainly due to higher ethylene prices outside the U.S.A.

Ethylene can be readily chlorinated in the liquid phase or vapour phase to yield 1,2-dichloroethane, the former being generally preferred:

$$CH_2{=}CH_2 + Cl_2(g) \longrightarrow ClCH_2CH_2Cl \qquad \Delta H = -201 \text{ kJ mol}^{-1}$$

The reaction is carried out at 25–50°C in dichloroethane as solvent. Small amounts of metal chlorides are used as catalysts and the pressure is maintained at 0.1–$2.0 \times 10^6 \text{ Nm}^{-2}$. Yields of 95% have been claimed at 98% ethylene conversion.

One of the first developments to be introduced was the thermal cracking of the dichloroethane (ethylene dichloride) which improved the process economics by removing the need for the use of alkali and the consequent loss of one chlorine atom per mole of vinyl chloride.

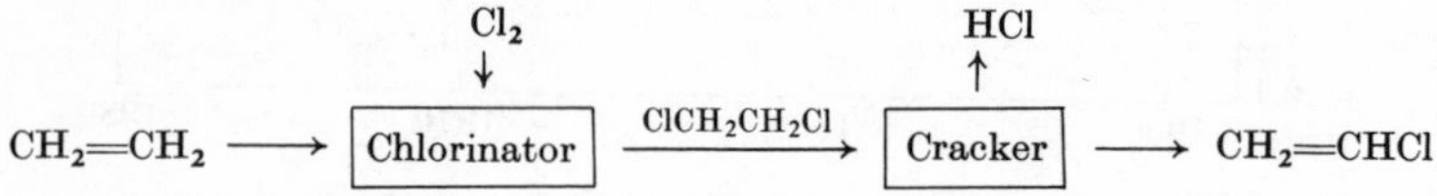

The cracking process is carried out at about 500°C over pumice or kaolin, the dichloroethane being dehydrochlorinated to vinyl chloride in 95% yield; conversion is limited to 50% per pass. This process produces HCl as a by-product, which in many situations could lead to problems of utilization. Three ideas have been developed to overcome this problem. The first links the ethylene-based plant to an acetylene-based plant, the latter utilizing the HCl produced by the former. This technique is the basis of the so-called 'balanced process'.

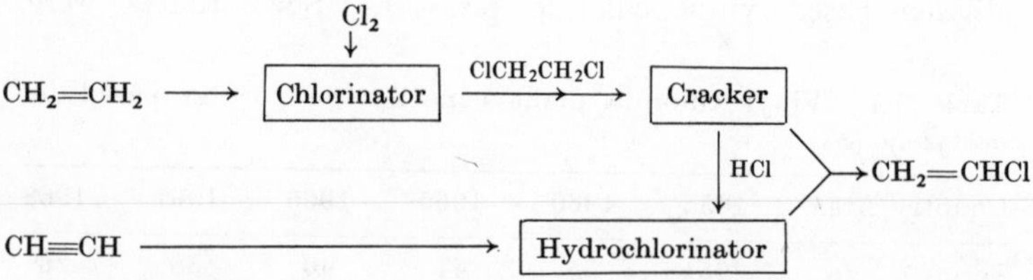

A logical extension of this type of process is the use of a naphtha cracking process to obtain ethylene and acetylene in the correct proportions for vinyl chloride production. In the Societé

Belge de l'Azote process, naphtha is cracked to an equimolar mixture of ethylene and acetylene and, after removal of unwanted constituents, the acetylene in the gas stream is caused to react with the HCl obtained from dichloroethane cracking; the dichloroethane is obtained by reaction of the ethylene in the gas stream with chlorine after removal of vinyl chloride (See Scheme 1). The Kureha process, developed in Japan, is a further variant of this theme.

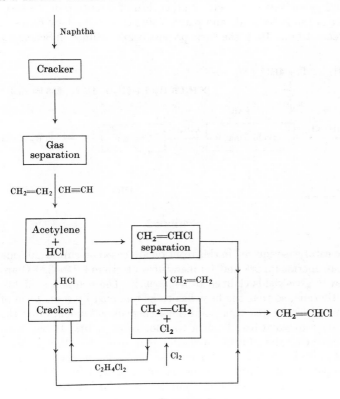

Scheme 1

A second method for overcoming the HCl problem in the basic ethylene process is to convert the HCl into chlorine by either an electrochemical or a modified Deacon process. Many attempts have been made to improve the basic Deacon process:

$$4HCl + O_2 \xrightarrow[\text{400–600°C}]{\text{Cu or Fe salts}} 2Cl_2 + 2H_2O$$

This has resulted in a large patent literature but it has not been commercialized, although Shell have laid claim to a viable process. The presence of a hydrocarbon helps to promote this reaction, leading to the chlorinated hydrocarbon; the Raschig phenol process was the first commercial process to utilize the principle, i.e. oxychlorination.

Oxychlorination provides the third method for utilizing the HCl obtained during chlorination of ethylene. The oxychlorination of ethylene began to receive a great deal of attention in the early part of the last decade and plants using this technology came on stream around 1964, the basic process being outlined in Scheme 2.

$$2CH_2{=}CH_2 + 4HCl + O_2 \longrightarrow$$

$$2CH_2ClCH_2Cl + 2H_2O \quad \varDelta H = -238 \text{ kJ mol}^{-1}$$

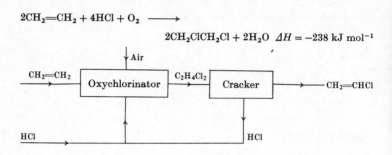

SCHEME 2

The catalysts required in this reaction are based on cupric chloride on an inert support and temperatures of around 240–320°C are used to give yields claimed to be 90–95%. The reaction is highly exothermic, so that the heat has to be removed by some form of cooling. Under reaction conditions there is a tendency for the catalyst to volatilize, leading to poor catalyst life. It has been claimed that the addition of alkali metals overcomes this problem and indeed, whether by this or other undisclosed means, several Companies lay claim to long-life catalyst systems. The basic reaction can be employed in a number of variations depending on the starting materials available to the manufacturer (e.g. ethylene + chlorine, or ethylene + chlorine + HCl, etc.).

One possible extension of these processes would be the one-stage production of vinyl chloride from ethylene, which would exclude the need to isolate the intermediate dichloroethane. Thus, the chlorination and dehydrochlorination steps would be combined in one reactor:

$$2CH_2{=}CH_2 + 2HCl + O_2 \longrightarrow 2CH_2{=}CHCl + 2H_2O$$

This potential process has been subject to much patent activity and discussion in the literature but has not been industrialized, possibly owing to a mismatch between the optimum conditions required for the two reactions which, to date, has led to low yields.

The Lummus Co. have recently announced development of a process (Transcat process) for vinyl chloride based on the single reactor, involving catalytic chlorination of ethane. This is claimed to integrate chlorination, oxychlorination and dehydrochlorination within a single reactor system. No details about the catalyst

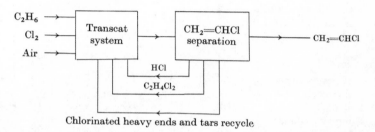

Chlorinated heavy ends and tars recycle

have been published (one patent claims a mixture of copper and potassium chloride), but vinyl chloride yields greater than 80% based on ethane and 95% based on chlorine have been claimed. This process is said to have about a £8.6/ton advantage over ethylene-based processes, largely derived from the use of the less costly ethane (cf. Table 7.2).

Table 7.2. Vinyl chloride cost comparison; cost of reactants in £/ton of vinyl chloride (1972 costs) (Source: The Lummus Co.)

	Process A[a]	Process B[b]	Transcat process
Acetylene at £69/ton	14.6	—	—
Ethylene at £28/ton	6.9	13.8	—
Ethane at £8.6/ton	—	—	5.2
Chlorine at £23.7/ton	14.6	15.5	13.8
Total raw material cost	36.1	29.3	19.0

[a] Combined chlorination of ethylene and hydrochlorination of acetylene.
[b] Combined chlorination and oxychlorination of ethylene.

A broad comparison of the costs of acetylene and ethylene-based processes is given in Table 7.3.

Table 7.3. Cost comparison of various vinyl chloride processes (£/ton of vinyl chloride; 1972 costs scale 10^5 tons p.a.)

	Carbide	Naphtha-based acetylene	Balanced carbide route[a]	Ethylene oxy-chlorination	Mixed gas
Capital investment ($£10^6$)	2.10	6.90	2.76	3.36	5.41
Cost of main raw materials:					
Chlorine at £24.4/ton	15.2	15.7	15.2	15.7	15.7
Carbide at £28/ton	38.2	—	18.1	—	—
Naphtha at £10.1/ton	—	9.1	—	—	9.1
Ethylene at £35.5/ton	—	—	8.1	17.0	—
Total	53.4	24.8	41.4	32.7	24.8
Utilities and auxiliary chemicals	3.2	7.6	4.4	3.8	7.0
Labour	1.9	1.4	1.5	0.8	1.4
Running costs:					
Maintenance	0.9	2.8	1.1	1.4	2.2
Tax and insurance	0.4	1.4	0.6	0.6	1.1
Other indirect costs	0.4	1.4	0.6	0.6	1.1
Depreciation charges	2.4	8.0	3.1	3.9	6.2
Return on capital	2.2	7.1	2.8	3.4	5.6
Total	6.3	20.7	8.2	9.9	16.2
Plant cost, including return on capital	64.8	54.5	55.5	47.2	49.4

[a] Acetylene based on carbide route.

In addition to these variants, there have also been limited patent claims to the use of ammonium chloride oxychlorination of ethylene and one specific literature claim to the oxidation of ethyl chloride over a chromium oxide catalyst supported on alumina. However, there is, as yet, no indication of viable commercial processes evolving from those studies.

Phenol

(M. H. DILKE, Research and Development Department, BP Chemicals International Limited, Epsom)

Although small amounts are still obtained from coal tar, a very high proportion of the present world production of phenol (ca.

3.3×10^6 tons p.a.) is made synthetically. Production of synthetic phenol is particularly interesting because there are probably more processes in use than for any other heavy organic chemical. Six, and possibly seven, different routes are used and will be outlined below. Some are usually known by the name of the starting material, others by that of an intermediate or a process stage; yet others are accorded the name of the Company associated with their development. The abbreviated name used here is given in parentheses after the process heading.

Alkali Fusion of Sodium Benzenesulphonate (Sulphonation Process)

This was the first commercial synthesis of phenol, a plant having been built in Germany as early as 1890. It is a four-stage process, as indicated by the following equations:

$$C_6H_6 + H_2SO_4 \longrightarrow C_6H_5SO_3H + H_2O$$
$$2C_6H_5SO_3H + Na_2SO_3 \longrightarrow 2C_6H_5SO_3Na + SO_2 + H_2O$$
$$C_6H_5SO_3Na + 2NaOH \longrightarrow C_6H_5ONa + Na_2SO_3 + H_2O$$
$$2C_6H_5ONa + SO_2 + H_2O \longrightarrow 2C_6H_5OH + Na_2SO_3$$

The first stage does not go to completion unless the water is removed; one of the methods used to achieve this is separation as its azeotrope with benzene. Some sodium sulphate is always formed in stage two from the excess of sulphuric acid. The sodium benzene-sulphonate is fused with caustic soda at about 350°C to produce sodium phenoxide, from which phenol is recovered by using the sulphur dioxide obtained in the second stage. The phenol is produced as an aqueous solution from which it is recovered by distillation.

Alkaline Hydrolysis of Chlorobenzene (Dow Process)

This process was developed by the Dow Chemical Company and involves two stages:

$$C_6H_6 + Cl_2 \longrightarrow C_6H_5Cl + HCl$$
$$C_6H_5Cl + NaOH \longrightarrow C_6H_5OH + NaCl$$

Chlorination is carried out at about 40°C, with a ferric salt as catalyst to minimize the formation of di- and tri-chlorobenzene. The monochlorobenzene is separated by distillation and hydrolysed at about 350°C and ca. 3×10^7 Nm^{-2} to give a lower aqueous layer containing sodium phenoxide and sodium chloride, and an upper layer of diphenyl ether which is recycled. The sodium phenoxide is acidified with HCl from the first stage to liberate the phenol.

Vapour-phase Hydrolysis of Chlorobenzene (Raschig–Hooker Process)

This is the only vapour-phase process for the manufacture of phenol and proceeds as follows:

$$C_6H_6 + HCl + \tfrac{1}{2}O_2 \longrightarrow C_6H_5Cl + H_2O$$
$$C_6H_5Cl + H_2O \longrightarrow C_6H_5OH + HCl$$

the overall reaction being:

$$C_6H_6 + \tfrac{1}{2}O_2 \longrightarrow C_6H_5OH$$

The process was developed by the firm of Raschig in Germany about 1930 on the basis of known reactions and has been improved considerably in recent years by the Hooker Chemical Company in the U.S.A.

A mixture of benzene, hydrogen chloride vapour, air and steam is passed over the catalyst (e.g. Al_2O_3–$CuCl_2$–$FeCl_3$) at about 230°C. Approximately 98% of the acid and 10% of the benzene is converted at 90% selectivity into chlorobenzene: the remainder of the benzene consumed goes to dichlorobenzenes or is burned. In the original Raschig process the chlorobenzene was separated from the products and hydrolysed in the vapour-phase over a calcium phosphate catalyst at 420°C. The conversion per pass was 10–15%, about 90–95% thereof to phenol, the remainder to benzene. In the Hooker process, a modified catalyst is used which also converts dichlorobenzenes into phenol, thus making the separation of chlorobenzene and dichlorobenzenes unnecessary. A single series of four columns is used to separate phenol, aqueous hydrochloric acid, benzene, chlorobenzene, dichlorobenzenes and tars, all except the phenol and tars being recycled to the appropriate part of the process.

Decomposition of Cumene Hydroperoxide (Cumene Process)

$$C_6H_5CH(CH_3)_2 + O_2 \longrightarrow C_6H_5C(CH_3)_2OOH \longrightarrow$$
$$C_6H_5OH + (CH_3)_2CO$$

This process was developed independently by the Distillers Company Ltd. in the U.K. and Hercules Powder Company in the U.S.A. and is described in detail in Chapter 5 (p. 134).

Oxidation of Toluene (Toluene Process)

The basic reactions are:

$$2C_6H_5CH_3 + 3O_2 \longrightarrow 2C_6H_5COOH + 2H_2O$$

$$2C_6H_5COOH + \tfrac{1}{2}O_2 \xrightarrow{\text{Cu salts}} C_6H_5COOC_6H_4COOH\text{-}o \longrightarrow$$

$$C_6H_5OH + C_6H_5COOH + CO_2$$

the overall reaction being:

$$C_6H_5CH_3 + 2O_2 \longrightarrow C_6H_5OH + CO_2 + H_2O$$

Liquid-phase oxidation of toluene to benzoic acid on a commercial scale was carried out in Germany during World War II. Toluene was oxidized with air at 140°C and about 0.3×10^6 Nm^{-2} in the presence of cobalt salts. The crude product, containing about 50% of benzoic acid, was treated to recover the acid and unconverted toluene, and any benzyl alcohol or benzaldehyde was recycled to the oxidation.

Copper-catalyzed oxidation of benzoic acid to phenol, discovered independently by the Dow Chemical Company and the California Research Corporation, proceeds through benzoylsalicylic acid which is hydrolysed and decarboxylated to phenol and benzoic acid. Benzoic acid, containing cupric benzoate as catalyst and magnesium benzoate as promoter, is heated to 230–240°C at 35–70×10^3 Nm^{-2} with steam and air. Phenol and benzoic acid are vapourized continuously and pass to a distillation column where phenol and water are separated overhead and benzoic acid returns from the base to the oxidizer. A purge of liquid phase from the oxidizer is removed, treated to remove tars and recycled.

Dehydrogenation of Cyclohexanol/Cyclohexanone (Cyclohexane Process)

This process was developed by the Scientific Design Company, after which Company it is sometimes named:

$$C_6H_6 + 3H_2 \longrightarrow C_6H_{12}$$

$$C_6H_{12} + O_2 \longrightarrow C_6H_{10}O + H_2O$$

$$C_6H_{12} + \tfrac{1}{2}O_2 \longrightarrow C_6H_{11}OH$$

$$C_6H_{10}O \longrightarrow C_6H_5OH + 2H_2$$

$$C_6H_{11}OH \longrightarrow C_6H_5OH + 3H_2$$

Hydrogenation of benzene to cyclohexane is a well-established process, and oxidation of cyclohexane to a mixture of cyclohexanone and cyclohexanol is used in one route to caprolactam (see p. 191). The Scientific Design Company has developed these reactions to the stage of a commercial phenol process. In particular, they have developed an improved catalyst for the final dehydrogenation.

Direct Oxidation of Benzene

Although many attempts have been made to oxidize benzene directly to phenol, it has been found difficult to limit the production of biphenyl and related compounds and to reduce burning. Numerous initiators or promoters, including γ-radiation, have been tested in attempts to overcome these difficulties. It is improbable that any successful commercial process has yet evolved, although it is obviously a very desirable goal. In recent years, efforts have been directed towards oxidizing benzene with air or oxygen, usually under pressure, in the presence of acetic acid and a catalyst such as a promoted palladium salt, to obtain a mixture of phenol and phenyl acetate; both liquid- and vapour-phase processes have been examined and although some measure of success has been achieved no commercial process has yet been found.

Comparison of Processes for the Manufacture of Phenol

Although the use of the sulphonation and the Dow process is declining, six different processes are operated by many different Companies in various parts of the world. This indicates that the decision as to which is the 'best process' depends very much on local situations. Some of the factors that have to be considered when deciding between the various processes are raw-material cost, process efficiency, capital cost, by-product production and purity of product.

Most of the processes start directly or indirectly from benzene which is thus a common factor; the toluene required for the toluene process is, however, a cheaper raw material. Two of the processes require the costly chlorine. If cyclohexanol/cyclohexanone are regarded as starting materials in the cyclohexane process their cost may be reduced if they are simultaneously produced for the manufacture of caprolactam. Overall yields of 80–90% of phenol from benzene have been quoted for the processes based on benzene (other than direct oxidation), while the overall yield from toluene in the toluene process is probably only 70–75%. Additionally, chlorination processes require special materials of construction to avoid corrosion problems which lead to increased capital charges.

The only process in which by-product plays an important part is the cumene process where 0.6 kg. of acetone is produced per kg. of phenol; hence the selling price of acetone is vital to the overall economics of this process. However, by-product disposal can also play an important part in the economics of other processes. The disposal of sodium sulphite, possibly to the paper pulp industry, is important in the operation of the sulphonation process. Whereas polychlorobenzenes are by-products in the Raschig version of the vapour-phase hydrolysis of chlorobenzene, they are hydrolysed in the Hooker version and no longer constitute a disposal problem. Diphenyl ether is a by-product from the alkaline hydrolysis process, and benzoic acid from the toluene process—they can be either sold or recycled. If cyclohexane is regarded as the starting material in the cyclohexane process, hydrogen can be considered as a by-product.

Modern phenol processes (e.g. cumene) produce phenol of extremely high purity, such as is necessary for caprolactam production for example.

The first commercial plant to operate the cumene process was constructed in 1953 and others followed in quick succession. Since that time economic advantages have been claimed for the Raschig–Hooker, toluene and cyclohexane routes, but further plants operating the cumene process continue to be built and now account for about 80% of the world production of synthetic phenol.

The major outlet of phenol is still the manufacture of phenolic resins, and a recent survey of the Western European markets gave the following breakdown for its utilization: phenolic resins 38%, caprolactam 25%, adipic acid 18%, diphenylolpropane 13%, alkylphenols 4%, others 2%.

Acrylonitrile

(A. H. JUBB, ICI Corporate Laboratory, Runcorn)

Acrylonitrile was first manufactured commercially in Germany just before World War II and, during the war period, demands for Buna rubber were supplied from plants at Ludwigshaven and Leverkusen (total capacity about 10^3 tons p.a.). During the war the U.S.A. also commenced to produce nitrile rubber. The introduction of acrylic fibres into the textile market during the early 1950's resulted in a significant demand for acrylonitrile. Thus the world plant capacity for the latter increased from 2×10^4 tons p.a.

in 1950 to 1.5×10^6 tons in 1970 (estimated at 2.3×10^6 tons in 1972).

The original German process was based on acetylene and HCN, whilst the first American plant, built by American Cyanamid during 1940 in New Jersey, was based on ethylene oxide and HCN. The latter process was also used by Union Carbide, but all other plants built throughout the world during the 1950's were based on acetylene. In 1960 Sohio brought on stream a plant for ammoxidation of propylene at Lima, Ohio. Subsequently, the bulk of new acrylonitrile capacity throughout the world was based on propylene ammoxidation, mainly Sohio technology; but several other Companies have developed and commercialized similar processes (see Chapter 6). Additionally, many of the acetylene-based plants have been closed down, particularly in Europe.

du Pont later patented a process based on the reaction of propylene with nitric oxide; this can be viewed as a two-stage variant of the Sohio process. There are many references to du Pont building a plant at Beaumont, Texas, in 1962, based on this process, but it is believed that the plant was converted to the standard ammoxidation process, if indeed it was ever based on the nitric oxide technology. Knapsack-Griesheim (a subsidiary of Hoechst) developed a process based on acetaldehyde and HCN and ran a pilot plant on a reported scale of 350 tons p.a.; once again, there are some references to the construction of a full-scale plant, but these have not been confirmed.

Asahi Chemical Industry Co. (Japan) recently reported the development of a process based on ethylene and HCN, whilst ICI and Power Gas Ltd. have claimed the joint development of a process based on the ammoxidation of propane.

Acetylene-based Process

$$CH\equiv CH + HCN \longrightarrow CH_2=CHCN$$

The reaction can be carried out either in the vapour phase at 400–500°C over a metal cyanide catalyst or in the liquid phase with an aqueous catalyst solution of cuprous chloride–HCl (pH 1–3). The former process gives poor yields of acrylonitrile whilst the latter gives an 80% yield based on acetylene. In the liquid-phase process, acetylene and HCN (molar ratio 6–10:1) are passed into the aqueous catalyst solution at 70°C and essentially atmospheric pressure. The gases from the reactor are washed countercurrently with water, and the acrylonitrile is recovered as a 3% aqueous solution. This is then stripped with steam to produce an overhead stream which separates into two phases on cooling. The acrylonitrile layer is purified by distillation, and the lower aqueous layer is passed back to the stripper.

Propylene Ammoxidation

$$2CH_2{=}CHCH_3 + 2NH_3 + 3O_2 \longrightarrow 2CH_2{=}CHCN + 6H_2O$$

The basic details of this process are given in Chapter 6.

Reaction of Ethylene Oxide with HCN

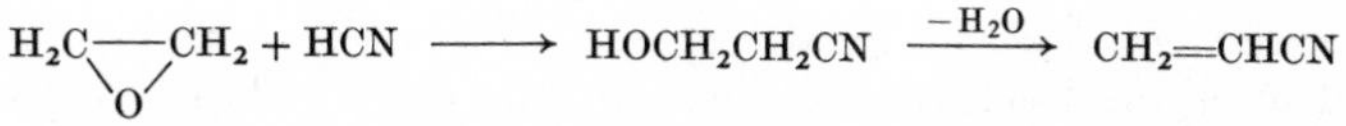

$$H_2C\overset{\diagdown}{\underset{O}{\diagup}}CH_2 + HCN \longrightarrow HOCH_2CH_2CN \xrightarrow{-H_2O} CH_2{=}CHCN$$

Hydrogen cyanide reacts exothermally with ethylene oxide in the presence of a basic catalyst such as diethylamine, to produce ethylene cyanohydrin. The reaction is carried out batchwise or continuously in either ethylene cyanohydrin or water as solvent. It is claimed that the cyanohydrin is produced in 95% yield and, after purification by vacuum-distillation, it is dehydrated in a continuous reactor in the vapour phase at 250–350°C with active alumina as catalyst. Liquid-phse dehydration can also be carried out at 200°C in the presence of sodium formate or calcium oxide catalysts. The yield of acrylonitrile is reported to be 90%.

Reaction of Acetaldehyde with HCN

HCN reacts with acetaldehyde at 10–20°C, to give acetaldehyde cyanohydrin in 90% yield. Dehydration is achieved at 600–700°C with a phosphoric acid catalyst, and yields of about 90% are claimed.

$$CH_3CHO + HCN \longrightarrow CH_3CH(OH)CN \xrightarrow{-H_2O} CH_2{=}CHCN$$

Reaction of Propylene and NO

$$4CH_2{=}CHCH_3 + 6NO \longrightarrow 4CH_2{=}CHCN + N_2 + 6H_2O$$

The only reports of this process are contained in patents; it is carried out at 700°C over a supported silver catalyst.

Ammoxidation of Propane

$$CH_3CH_2CH_3 + NH_3 + 2O_2 \longrightarrow CH_2{=}CHCN + 4H_2O$$

It has been reported that Power Gas Co. and the ICI Corporate Laboratory (Runcorn) have jointly been carrying out development work on this reaction and, although that work was not complete in all aspects, were seeking to license the process. Basically, the reaction is carried out under conditions very similar to those used in the Sohio process but with a different, undisclosed catalyst. The acrylonitrile yield is believed to be about 65% and acetonitrile and HCN by-products are produced at about the same level as in the propylene process. The potential value of this process would be in areas where there is an abundance of cheap natural gas, particularly if propylene were in short supply.

Reaction of Ethylene with HCN

During the 1960's Asahi Chemical Industry Co. (Japan) began patenting various aspects of processes to acrylonitrile based on ethylene and propionitrile. Recently they have disclosed the development of a process based on oxycyanation of ethylene:

$$2CH_2{=}CH_2 + 2HCN + O_2 \longrightarrow 2CH_2{=}CHCN + 2H_2O$$

The reaction is carried out at about 360°C in the presence of HCl over a fixed-bed palladium–caesium–vanadium catalyst. 100% HCN conversion is claimed at low ethylene conversion levels; acrylonitrile yields are reported to be 74% based on ethylene and 88% based on HCN. Acetonitrile, carbon monoxide, carbon dioxide, chloropropionitrile, ethylene dichloride and vinyl chloride are the main by-products.

Economic Considerations for Acrylonitrile

During the late 1950's the unit value of acrylonitrile produced in the U.S.A. was in the range 29–26 cents/lb. based on acetylene costing around 11–13 cents/lb. During the same period propylene costs were around 2 cents/lb. and, although the ammoxidation technology was more complicated and correspondingly more costly than that based on acetylene, its lower raw-material costs more than offset this difference.

This is reflected in the fall in U.S. acrylonitrile unit value from 22 cents/lb. in 1960 to 14 cents/lb. or less in 1962, brought about by the introduction of propylene ammoxidation units. In the event, the current prices to large consumers is probably nearer 12 cents/lb. dictated by the latter technology. The current raw-material costs alone of the acetylene, ethylene oxide and acetaldehyde routes would be in the range 8–11 cents/lb. Thus, none of these processes can compete with propylene ammoxidation although there could be a case for continuing to operate an old acetylene-based plant under certain conditions. The European and the Japanese pictures are very similar to the American except that plant capacities only became significant after 1960 and the impact of the propylene process began to be felt after about 1965. These facts are reflected in the manner in which plant capacities have been based on the different processes over recent years (cf. Table 7.4).

Table 7.4. World acrylonitrile capacity: by percentage per process and the total by weight

	1950	1955	1960	1965	1967	1971
Ethylene oxide–HCN	95	35	12	6	—	—
Acetylene—HCN	5	65	80	40	18	7
Propylene–NH$_3$	—	—	8	45	75	93
Propylene–NO	—	—	—	9[a]	7[a]	—
Acetaldehyde–HCN	—	—	—	—	—	—
Total capacity (10^6 tons p.a.)	0.017	0.90	0.287	0.687	0.964	1.85

[a] There is doubt about this plant being based on propylene–NO.

Any commercialization of the propane ammoxidation route will require a significant price differential ($\not< £10$/ton) between propane and propylene, a situation that is unlikely to occur in the U.S.A. in view of the foreseen shortage of natural gas. It is difficult to assess any future impact that the Asahi process (ethylene–HCN) might make; in the absence of any detailed flowsheet it is not possible to estimate the plant capital costs; at first sight, it would not appear to have any cost advantage over propylene technology even if cheap HCN were available from the latter plants.

Acetone

(B. YEOMANS, Research and Development Department, BP Chemicals International Limited, Hull)

The importance of acetone to world commerce is reflected in the rapid progression of feedstocks used, from natural to refinery products, and by the wide range of manufacturing processes that have been developed since 1900.

Acetone was produced initially by the dry distillation of wood, but this process could not satisfy the demands during World War I when acetone was required for use in dope for airplane wings and in the manufacture of cordite. This led to the development of a route based indirectly on fermentation alcohol, which was oxidized to acetic acid for neutralization with lime to give the calcium salt for pyrolysis.

$$(CH_3COO)_2Ca \xrightarrow{400°C} (CH_3)_2CO + CaCO_3$$

As might be expected, such a process, which yields only one acetone molecule from two of acetic acid, was rapidly superseded; it was replaced by a carbohydrate (e.g. molasses) fermentation process developed by Chaim Weizmann. A feature of this process was the simultaneous production of butan-1-ol in equal weight-yield to acetone. The Weizmann process comprises four steps: culture growth, sterilization of the molasses, fermentation and distillation. A pure stock of bacteria spores is added to sterilized glucose and is allowed to grow at 87°F for a day before transfer to a sterilized solution of molasses, ammonium sulphate and buffers for fermentation; the pH is maintained between 5 and 7 by addition of ammonia, and after ca. 2 days' fermentation the mixture containing ca. 2% of solvents is steam-distilled to give a 50% solvent distillate. Normally the yield of solvents is 28–33 wt.-% of the sugar content of the molasses. This process dominated acetone production until the middle 1920's. In 1920, however, propan-2-ol was produced by the Standard Oil Company of New Jersey through hydration of propylene in what is thought to be the first petrochemical process. Predictably, this development was followed rapidly by new technology for the oxidative conversion of propan-2-ol into acetone and today this is the most important route to acetone. The conversion of propan-2-ol is accomplished either by straight dehydrogenation or by oxidative dehydrogenation (see Chapter 4, p. 93).

Dehydrogenation

Copper or nickel, or their alloys, oxides or salts, may be used as catalyst but the best results are obtained with zinc oxide (e.g. ZnO 7%, Na_2CO_3 2% on pumice). The main competing reaction is dehydration to propylene and di-isopropyl ether and, although it requires a higher temperature for reaction than the copper or nickel catalyst, zinc oxide gives a better balance of yield and selectivity to acetone (90% conversion of propan-2-ol with 98%

selectivity to acetone). The pumice support is only a means for obtaining a large surface area of zinc oxide.

$$(CH_3)_2CHOH \xrightarrow[\text{400–500°C}]{\text{ZnO–pumice}} (CH_3)_2CO + H_2$$

$$\Delta H = +66.4 \text{ kJ mol}^{-1} \text{ at } 327°C$$

The zinc oxide catalyst has a long life (1–1.5 years) but requires treatment with air at ca. 500°C every 10–14 days to remove carbon and other organic deposits.

Oxidative Dehydrogenation

$$(CH_3)_2CHOH + \tfrac{1}{2}O_2 \longrightarrow (CH_3)_2CO + H_2$$
$$\Delta H = -182 \text{ kJ mol}^{-1} \text{ at } 295°C$$

Nickel or platinum, but preferably silver or copper, is used to catalyse the oxidation at temperatures within the range 400–600°C. In contrast to straight dehydrogenation, the reaction is strongly exothermic and the selectivity to acetone (85–90%) is lower.

The Weizmann fermentation process declined further in importance as another catalyst emerged for chemically transforming fermentation alcohol into acetone. The ketonization step was accomplished in the vapour phase at 400–450°C over a Fe_2O_3–ZnO–$CaCO_3$ catalyst (e.g. 2:7:1 by weight). This process was operated mainly until the end of World War II but is currently used in developing countries only (e.g. Pakistan) where there is a local supply of low-cost fermentation alcohol.

The period immediately following World War II saw the development of the cumene–phenol process which yields one mole of acetone as co-product for every mole of phenol (see Chapter 5, p. 134). In essence, here too the acetone is derived from propylene.

$$CH_3CH{=}CH_2 + C_6H_6 \xrightarrow{H^+} C_6H_5CH(CH_3)_2 \xrightarrow[\substack{\text{130°}\\ \text{pH 8.5–10.5}}]{O_2}$$

$$C_6H_5C(CH_3)_2OOH \xrightarrow[\substack{\text{10\% w/w } H_2SO_4\\ \text{ca. 60°}}]{H^+} C_6H_5OH + (CH_3)_2CO$$

This is currently the main competitor of the propan-2-ol route, since more than 70% of the world phenol is now made in this way.

In other recent developments, acetone has been isolated as a valuable co-product of processes designed primarily to yield other oxygenated products, such as the manufacture of acetic acid by oxidation of paraffins. Two such processes have achieved commercial significance: (*a*) the Celanese vapour-phase oxidation of a natural gas fraction comprising mainly propane and butane (operated at Bishop, U.S.A.) which closed down in 1972; and (*b*) the BP liquid-phase oxidation of a light naphtha cut (predominantly C_5–C_7 branched-chain paraffins), which is increasing in importance. In both cases many complex free-radical chain reactions are involved, typified by:

$$RH \longrightarrow R\cdot$$
$$R\cdot + O_2 \longrightarrow RO_2\cdot$$
$$RO_2\cdot + RH \longrightarrow ROOH + R\cdot$$
$$ROOH \longrightarrow Products$$

The competition of propan-2-ol, acrylonitrile and acrylic acid and its esters for propylene as a feedstock is likely to increase the importance of paraffin-oxidation routes to acetone. Like all multiproduct processes, however, expansion is geared to demand for a product range and the extent to which such processes will contribute to demand for acetone will be determined primarily by growth in the demand for other oxygenated chemicals, particularly acids.

On the other hand, the recently developed Wacker–Hoechst process based on the direct palladium-catalyzed oxidation of propylene is likely to become a major route to acetone only where more profitable outlets for propylene are unavailable. The reaction takes place by the following steps at 50–$120°C/5$–10×10^6 Nm^{-2} and pH 2–3:

1. $PdCl_2 \cdot 2H_2O \rightleftharpoons [PdCl_2(OH)H_2O]^- + H^+$

2. $[PdCl_2(OH)H_2O]^- + CH_3CH{=}CH_2 \longrightarrow$
$$[PdCl_2(OH)(CH_3CH{=}CH_2)]^- + H_2O$$

3. $[PdCl_2(OH)(CH_3CH{=}CH_2)]^- + H_2O \longrightarrow$
$$Pd + 2Cl^- + H_2O + (CH_3)_2\overset{+}{C}OH$$

4. $(CH_3)_2\overset{+}{C}OH \longrightarrow (CH_3)_2CO + H^+$

The rate-controlling step is the mass transfer of propylene and oxygen which are fed into a sparged tank reactor. Acetone is obtained in 98% selectivity. The acetone is separated from the reaction mixture as a distillate. The palladium catalyst is kept in solution by redox reaction with cupric chloride and air since the reduced analogue, cuprous chloride, is readily re-oxidized by air.

5. $Pd + 2CuCl_2 \longrightarrow PdCl_2 + 2CuCl$

6. $2CuCl + 2HCl + \frac{1}{2}O_2 \longrightarrow 2CuCl_2 + H_2O$

Uses of Acetone

The demand for acetone in U.S.A. has risen from 1.8×10^5 tons in 1947 to 6.4×10^5 tons in 1968 and was expected to rise to 8.2×10^5 tons by 1972 with a growth rate of 6% p.a. The U.S. production capacity for acetone was 6.9×10^5 tons in 1968 and was expected to rise to 10^6 tons by 1970. There is, therefore, a need to discover new outlets for acetone. The end-uses of acetone are evenly distributed between solvents, derivative solvents, chemical intermediates and methyl methacrylate (see Chapter 8 and Figure 7.2, p. 184).

At the end of World War II acetone was used mainly in solvent applications, but the proportion of sales to this outlet has consistently dwindled, possibly owing to decline in the production of cellulosic plastics. The major derivative solvents produced from acetone include isobutyl methyl ketone, diacetone alcohol, 'hexylene glycol' (2-methylpentane-2,4-diol) and isophorone. Bisphenol A (p,p'-isopropylidenediphenol) is an important chemical derivative of acetone which is used for alkyd surface-coating applications, epoxy resins and polycarbonates. Other derivatives of acetone are used as drugs, vitamins, cosmetics and rubber additives. The major growth area (10% p.a.) is as methyl methacrylate for acrylic plastics in aircraft and automotive glazing and signs, and for construction applications.

Chemical Derivatives of Acetone (cf. Figure 7.3a,b, p. 185)

Diacetone alcohol is produced by the base-catalysed aldolization of acetone under mild conditions (ca. $+5°C$). The reaction is quite selective; a little triacetone alcohol, $HOC(CH_3)_2CH_2$-$COCH_2C(CH_3)_2OH$, is the main by-product. The main outlet for diacetone alcohol is as 'hexylene glycol' which is obtained by hydrogenation over Raney nickel (at $100°C$ and 0.7×10^6 Nm^{-2}).

Diacetone alcohol and triacetone alcohol can be dehydrated to mesityl oxide and phorone, $(CH_3)_2C{=}CHCOCH{=}C(CH_3)_2$, respectively, by mildly acidic reagents (e.g. phosphoric acid), but the commercial outlets for these unsaturated ketones are very small.

Isophorone is produced by the base-catalysed aldolization of acetone under severe conditions (ca. $210°C$ and 3×10^6 Nm^{-2}). The reaction is not selective and, even at the low conversions selected to isophorone ($\leqslant 10\%$), similar amounts of ketonic by-products are also produced. The co-products are reconverted (base-catalysed hydrolysis at ca. $240°C$) to acetone and isophorone in order to obtain good process economics, and ca. 90% selectivity

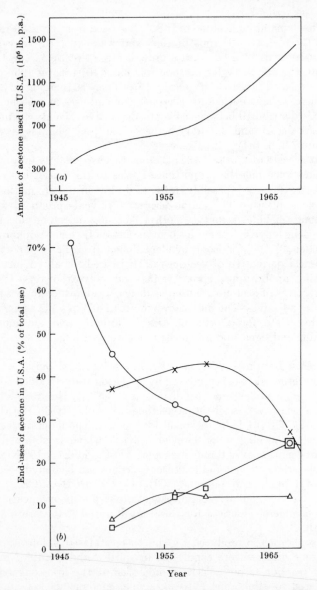

Figure 7.2. Estimated end-use patterns for acetone in U.S.A., 1947 to 1968. (*a*) Growth in acetone usage: 1957–1967, 6% p.a.; 1967–1972, 6% p.a. (estimated). (*b*) End-use key: ○, solvent; ×, derivative solvent; □, methyl methacrylate; △, chemical derivatives (bisphenol A, etc.).

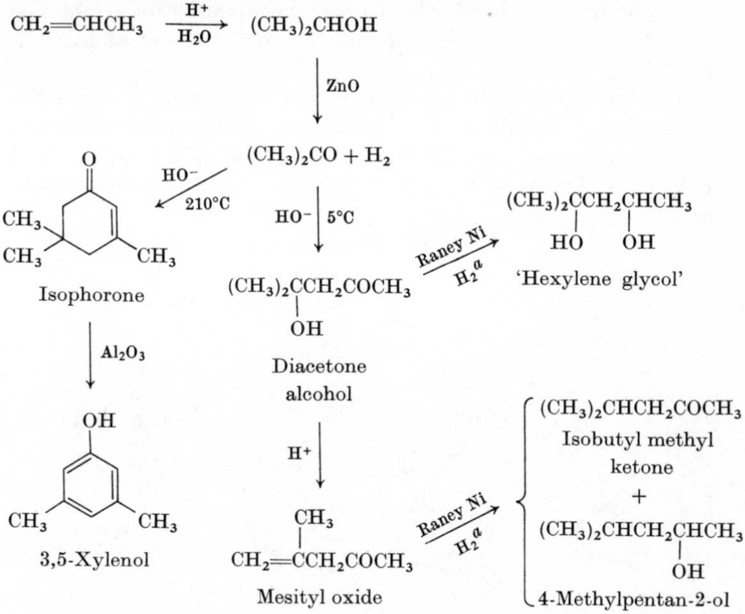
$CH_2{=}CHCH_3 \xrightarrow[H_2O]{H^+} (CH_3)_2CHOH$
ZnO
$(CH_3)_2CO + H_2$
HO^-
210°C
Isophorone
Al_2O_3
3,5-Xylenol
HO^- 5°C
$(CH_3)_2C(OH)CH_2COCH_3$
Diacetone alcohol
Raney Ni
H_2 [a]
$(CH_3)_2C(OH)CH_2CH(OH)CH_3$
'Hexylene glycol'
H^+
$CH_2{=}C(CH_3)CH_2COCH_3$
Mesityl oxide
$(CH_3)_2CHCH_2COCH_3$
Isobutyl methyl ketone
+
$(CH_3)_2CHCH_2CH(OH)CH_3$
4-Methylpentan-2-ol

[a] By-product with acetone.

(*a*)

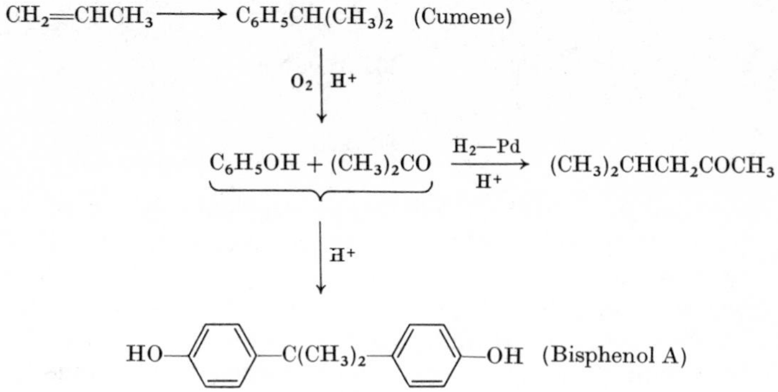
$CH_2{=}CHCH_3 \longrightarrow C_6H_5CH(CH_3)_2$ (Cumene)
O_2 H^+
$C_6H_5OH + (CH_3)_2CO$
H_2—Pd
H^+
$(CH_3)_2CHCH_2COCH_3$
H^+
HO—C(CH₃)₂—OH (Bisphenol A)

(*b*)

Figure 7.3 (*a*). Commercial productions of acetone from propylene via propan-2-ol.

Figure 7.3 (*b*). Commercial production of acetone from propylene via cumene.

to isophorone is claimed. The main chemical outlet for isophorone is as 3,5-xylenol which is obtained by pyrolysis over alumina at ca. 500°C.

3,5-Xylenol is chlorinated for use in a bactericide. Veba has produced the diacids, diols and diamines (see Scheme 3) from isophorone as precursors for the manufacture of polyamides and polyesters, but these processes have not become commercially established.

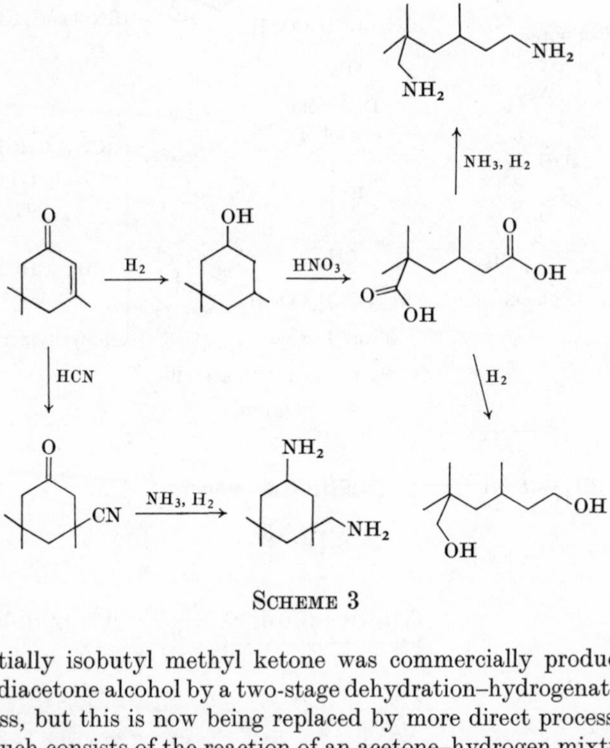

SCHEME 3

Initially isobutyl methyl ketone was commercially produced from diacetone alcohol by a two-stage dehydration–hydrogenation process, but this is now being replaced by more direct processes. One such consists of the reaction of an acetone–hydrogen mixture at 100°C and 7×10^6 Nm^{-2} over a cation resin impregnated with palladium. A 30% conversion to isobutyl methyl ketone is obtained together with a little di-isobutyl ketone. Another variant consists of the reaction of a 2 : 1 mixture of propan-2-ol and acetone over a didymium oxide–copper chromite catalyst at 250°C and 5×10^6 Nm^{-2}. A 30% conversion to isobutyl methyl ketone calculated on propan-2-ol is obtained together with a little more di-isobutyl ketone.

Methyl methacrylate (see Chapter 8, p. 229) is produced by a two-stage process. Preparation of acetone cyanohydrin, $(CH_3)_2C(OH)CN$, by base-catalysed addition of hydrogen cyanide to acetone under mild conditions is followed by hydrolysis to methacrylamide hydrogen sulphate and then by esterification with methanol.

$$(CH_3)_2C(OH)CN \xrightarrow{\text{Aq. 85\% w/w } H_2SO_4/MeOH}$$

$$CH_2{=}C(CH_3)COOCH_3 + NH_4HSO_4$$

Bisphenol A is produced by the acid-catalysed alkylation of phenol with acetone under mild conditions (ca. 50°C). Strong acid, usually dry hydrogen chloride, is used as catalyst together with methanethiol or thioglycollic acid as an accelerator:

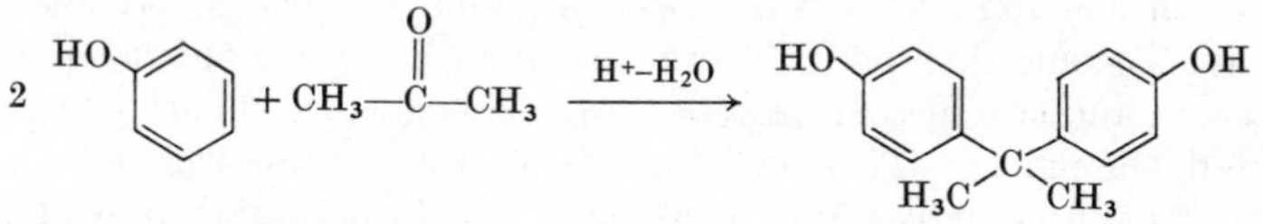

Other carbonyl-containing compounds such as acetaldehyde or laevulic acid may also be used to afford analogues of bisphenol A, but the best yields are obtained with acetone.

Caprolactam*

(A. H. Jubb, ICI Corporate Laboratory, Runcorn)

ϵ-Caprolactam was first synthesized and found to give a polymeric product in the late nineteenth century. Carothers is reported to have investigated fibres produced from caprolactam polymers during 1930 and Schlack's researches led to the first commercial production of poly-(ϵ-caproamide), commonly known as nylon 6, in Germany in the early 1940's. Virtually all caprolactam output goes into the production of nylon 6 which, together with its older partner nylon 6,6, represent the major polyamides used in the production of synthetic fibres (see Chapter 9). In 1969 polyamides accounted for 41% of the 4.4×10^6 tons world production of synthetic fibres and, although their consumption will continue to grow in absolute terms, they are expected, in due course, to take second place to polyester fibres. Nylon 6 has about a 40% share of

* Systematic name: 6-hexanolactam.

world polyamide markets but distribution by geographic area is not uniform. Thus, for largely historical reasons, nylon 6 accounts for about 20% of the U.K. market (50% overall in EEC) and 25% of the U.S. market but 75% of the Japanese market.

In addition to its main use in fibre applications, a limited amount of nylon 6 is used in plastics. Caprolactam is also an intermediate in the production of L-lysine (see Chapter 15).

World caprolactam capacity was about 1.4×10^6 tons in 1969, 17% of which was in the U.S.A. (demand growth rate 17% p.a. during the period 1962–1968). Japan has played a dominant rôle in the growth of caprolactam markets and the 1972 demand has been estimated at about 0.45×10^6 tons.

Processes

Phenol was originally the major feedstock used in caprolactam production but in recent years cyclohexane-based processes have become dominant. There are seven basic commercial processes, and two further processes are reported to be at early stages of development. Hydroxylamine is used as an intermediate in some of these processes and there are several potential routes for making it. Thus, overall, there is a relatively complicated set of possible reaction pathways (Figure 7.4*a* and *b*; see pp. 189, 190). No process is outstandingly superior: the final choice of process generally depends on a variety of local factors.

Phenol Route

The Allied Chemical Corporation operates a 130×10^3 p.a. caprolactam plant at Hopewell, Va., U.S.A., using a variant of the German war-time process. Phenol is catalytically hydrogenated to cyclohexanone; the unconverted phenol (recycled) and small amounts of co-product cyclohexanol are separated by distillation. The ketone is treated with ammonia and hydroxylamine sulphate solution, to yield the oxime. This then undergoes a Beckmann rearrangement (Part 2, p. 166) in the presence of 20% oleum. Both stages are claimed to operate in about 95% yield. A crude product stream is removed continuously, neutralized with ammonia, and solvent-extracted. After removal of the solvent, the caprolactam is purified by recrystallization.

Hydroxylamine is produced from ammonium carbonate via the nitrite; this is the Raschig process.

This multistage route has an important bearing on the overall cost. The commercial viability of the process is also critically dependent on obtaining credit, through fertilizer applications, for

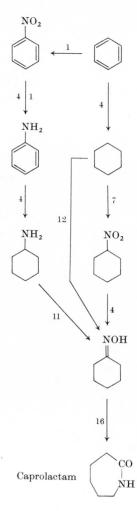

Figure 7.4 (a). Some routes to caprolactam.

Reagents: 1, HNO_3–H_2SO_4. 2, Hydrodealkylation. 3, O_2. 4, H_2. 5, *Via* cumene. 6, O_2–Cu salt (Dow process). 7, HNO_3. 8, Ketene. 9, HO^-. 10, $-H_2$. 11, H_2O_2. 12, $NOCl/h\nu$. 13, H_2NOH. 14, CH_3COOOH. 15, NH_3. 16, H_2SO_4. 17, $LiCl$. 18, Cyclization. 19, $(NO)HSO_4$.

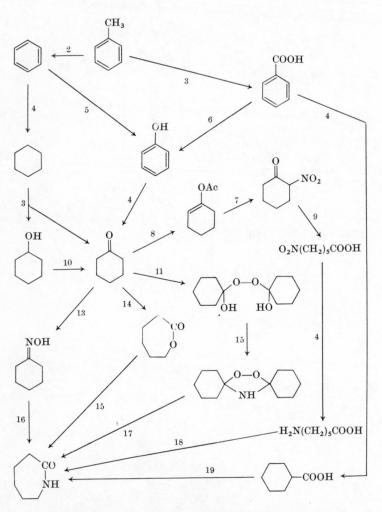

Figure 7.4 (b). Some further routes to caprolactam.
For reagents see Fig 7.4(a)

$$(NH_4)_2CO_3 \xrightarrow[NO_2]{NO} 2NH_4NO_2 \xrightarrow[SO_2]{NH_3 + H_2O}$$

$$2HON(SO_3NH_4)_2 \xrightarrow{H_2O} (NH_2OH)_2 \cdot H_2SO_4$$

| Hydroxylamine | Hydroxylamine |
| disulphonate | sulphate |

the ammonium sulphate produced in both the caprolactam (1.8 tons per ton of lactam) and the oxime (2.7 tons per ton of lactam) stages.

Dutch State Mines has developed a similar process based on phenol and has constructed plant with a capacity of 100×10^3 tons p.a. at Geleen. It has also improved the oximation process by the introduction of the 'hyamophosphate' route to hydroxylamine, thus reducing the ammonium sulphate producton to 1.8 tons per ton of lactam. The main feature of the latter process is the production of hydroxylamine by hydrogenation of nitrate ions over a noble-metal catalyst. An even more recent development is a bisulphate lactam process which yields no by-product ammonium sulphate. Neutralization after the Beckmann rearrangement is controlled to give ammonium bisulphate which is then pyrolysed to yield sulphur dioxide which is in turn converted into oleum for recycle.

A further variant of the phenol route was developed in 1951 by Emser Werke AG. In this, the hydroxylamine is produced by reduction of NO, which also lowers the co-production of ammonium sulphate by 50% compared with the Raschig process.

$$2NO + 3H_2 + H_2SO_4 \longrightarrow (NH_2OH)_2 \cdot H_2SO_4$$

Cyclohexane Air-oxidation Routes

Liquid-phase air-oxidation of cyclohexane in the presence of small amounts of cobalt salts at 145–170°C and 0.8–$1.2 \times 10^6\ Nm^{-2}$ yields cyclohexanol, cyclohexanone and a variety of acid and ester by-products. The cyclohexane conversion is kept to a low level (<10%) to minimize formation of these by-products. Several Companies operate processes that cover this stage. Variations include a process developed by Stamicarbon (4–6% conversion); one by Scientific Design utilizes boric acid (to enhance the yield of cyclohexanol and cyclohexanone and the cyclohexanol: cyclohexanone ratio) at around 10% cyclohexane conversion; others use more traditional technology.

Unconverted cyclohexane is separated from the crude product by distillation, and esters are saponified with aqueous alkali. The cyclohexanol and cyclohexanone are separated by distillation and the former is converted into the latter by catalytic dehydrogenation. The ketone is then converted into caprolactam as described above.

Toluene Route

At Torviscosa in Italy, Snia Viscosa operates a process based on toluene. First the toluene is catalytically oxidized at about 10^6 Nm^{-2} and 150–170°C to benzoic acid, which is then hydrogenated in the liquid phase over a noble-metal catalyst. The resulting cyclohexanecarboxylic acid rearranges to caprolactam with evolution of CO_2 when treated with nitrosylsulphuric acid ($HNOSO_4$). About 50–60% of the organic acid is converted in each pass and unchanged material is recovered for recycling. The mechanism is in dispute.

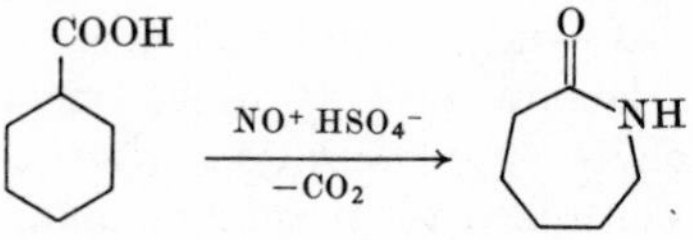

The nitrosylsulphuric acid is obtained from ammonia, oxygen and sulphuric acid.

$$2NH_3 + 3O_2 \longrightarrow N_2O_3 + 3H_2O$$
$$N_2O_3 + 2H_2SO_4 \longrightarrow 2HNOSO_4 + H_2O$$

The overall process leads to the co-production of about 4.2 tons of ammonium sulphate per ton of caprolactam, but it is reported that ANIC and Snia Viscosa are jointly planning to build a plant to produce 80×10^3 tons p.a. based on an improved version of this technology giving only 2 tons of sulphate per ton of lactam.

Photonitrosation of Cyclohexane

In 1951, Toray Industries (Japan) began research on a process based on the light-initiated reaction between cyclohexane and nitrosyl chloride. It now has a plant capacity of 90×10^3 tons p.a. and this is being extended to 140×10^3 tons p.a.

The continuous process yields caprolactam in over 80% yield, based on cyclohexane, and ammonium sulphate production is only 2.2 tons per ton of caprolactam. The heat of rearrangement of the oxime dihydrochloride salt is 107 kJ mol^{-1}, compared with

187 kJ mol⁻¹ for the pure oxime, so that temperature control is easier. One of Toray's main research achievements was the development of the high-pressure mercury photochemical lamp having 60 kW capacity and giving 400 g of caprolactam per kWh. Since wavelengths below 3650 Å enhance tar formation, absorbers are added to the lamp cooling water, and filters are used to minimize the effect. Lamp efficiencies fall and the optimum replacement time depends on economic factors, including lamp cost, rate of production and cost of electricity.

The nitrosyl chloride required for the process is obtained from nitrosylsulphuric acid (see above).

$$HNOSO_4 + HCl \longrightarrow NOCl + H_2SO_4$$

Caprolactone* Process

In 1966, Union Carbide commissioned a 20×10^3 ton p.a. caprolactam plant at Taft, La, U.S.A., based on the amination of caprolactone produced by oxidation of cyclohexanone with peracetic acid. This technology integrates well with the peracetic acid unit which is also on the site. There is no by-product ammonium sulphate, but co-production of acetic acid (1.1 tons per ton of lactam) will restrict the utility of the process to those areas requiring acetic acid (unless the peracetic acid were produced from acetic acid/H_2O_2, in which case the acid could be recycled).

The peracetic acid is usually obtained by the following reaction:

$$2CH_3CHO \xrightarrow{\text{Oxidn.}} \underset{\substack{\text{Acetaldehyde} \\ \text{monoperacetate}^\dagger}}{CH_3\overset{\overset{\displaystyle O}{\|}}{C}O\overset{\overset{\displaystyle CH_3}{|}}{C}HOH} \longrightarrow \underset{\substack{\text{Peracetic} \\ \text{acid}}}{CH_3COOOH} + \underset{\text{(recycled)}}{CH_3CHO}$$

Nitrocyclohexanone Route

Techni-Chem has evolved a new process for the production of caprolactam which does not produce any by-product. The process has been run continuously on a pilot plant at Wallingford, Conn., U.S.A., but not, so far, on the full scale.

The first, key step in the process is the selective mononitration of cyclohexenyl acetate in acetic anhydride solvent. Acetic acid formed in the reaction is used to produce ketene, which is then

---

* Systematic name: 6-hexanolactone.

† Systematic name: 1-hydroxyethyl peracetate.

treated with cyclohexanone to form the cyclohexenyl acetate. The nitrocyclohexanone is ring-opened by aqueous base, and the resulting solution hydrogenated to ϵ-aminocaproic acid (6-amino-hexanoic acid). The latter is thermally cyclized to caprolactam, which is extracted from the aqueous solution by organic solvents and purified by recrystallization.

Nitrocyclohexane Route

du Pont is reported to have a 25×10^3 ton p.a. caprolactam plant operating on the nitrocyclohexane route. The crucial step is believed to be the reduction of this intermediate to the oxime, and a large number of patents have been filed covering a variety of reducing agents, including zinc–chromium/H_2, ferrous salts and sulphites. A specific du Pont patent claims 80–85% oxime yield with co-production of 8–10% of cyclohexylamine. The oxime is rearranged in the normal manner, and again leads to formation of by-product ammonium sulphate. Alternatively, it is claimed that caprolactam can be obtained directly from nitrocyclohexane by reduction and dehydration.

1,1′-Dihydroxydicyclohexyl Peroxide Route

Recently, BP Chemicals (U.K.) Ltd. has filed several patent applications covering the reaction of cyclohexanone with hydrogen peroxide in a suitable solvent. The intermediate peroxide is treated with ammonia to yield 1,1′-peroxydicyclohexylamine which is then thermally decomposed in molten caprolactam in the presence of a lithium halide catalyst. Cyclohexanone, the co-product, is recovered by distillation and recycled.

The process is in a relatively early stage of development and it is difficult to estimate its commercial potential. Once again, no ammonium sulphate is produced in the process, but one potential disadvantage appears to be cleavage of the relatively expensive peroxyamine into equimolar proportions of product and cyclo-hexanone, which then has to be recycled.

Cyclohexylamine Route

Aniline can be hydrogenated to cyclohexylamine in the presence of a nickel–cobalt catalyst, and the reaction of this amine with an excess of hydrogen peroxide in the presence of a molybdate salt gives cyclohexanone oxime in a reported 95% yield. It is claimed that this process has been commercialized in the Soviet Union.

Vickers-Zimmer has developed a process in which cyclohexyl-amine is hydrolysed to cyclohexanol at elevated temperatures.

Table 7.5. Operating costs for the major caprolactam processes (1970 costs)

Raw materials	Phenol route (Allied Chem. Corp.)		Dutch State Mines route		SNIA process		Toray process	
	Unit consumption (ton)	£/ton caprolactam	Unit consumption (ton)	£/ton caprolactam	Unit consumption (ton)	£/ton caprolactam	Unit consumption (ton)	£/ton caprolactam
Phenol (£65/ton)	0.89	57.9						
Toluene (£21/ton)					1.11	23.3		
Cyclohexane (£37/ton)			1.06	39.2			0.93	34.4
Other chemicals		46.9		51.1		60.7		33.2
Total raw materials cost		104.8		90.3		84.0		67.6
By-product credits:								
Ammonium sulphate	4.6	−26.5	4.5	−26.0	4.1	−23.7	2.29	−13.2
Chlorocyclohexane							0.064	−2.5
Net raw materials cost		78.3		64.3		60.3		51.9
Utilities		12.3		13.3		14.2		21.7
Labour and supervision		3.3		1.4		1.9		2.3
Maintenance		8.1		5.6		4.6		10.0
Overheads		11.4		6.9		6.5		12.5
Operating costs (£/ton of caprolactam)		113.4		91.5		87.5		98.4

The alcohol is then dehydrogenated to the ketone which is converted into caprolactam by standard routes.

Other Processes

There are several potential routes to caprolactam that involve intermediate production of adiponitrile, but none of them is commercially viable.

Conclusions

Table 7.5 gives indicative operating costs for the four main processes and Table 7.6 indicates the reported capital investments for the four processes. It can be clearly seen how the economics is influenced by credits for ammonium sulphate. Where the latter are available the SNIA process appears to be marginally the cheapest route. The Toyo Rayon process assumes a more dominant position where there is a problem with ammonium sulphate disposal.

Table 7.6. Capital investment for the four main caprolactam processes (based on *European Chemical News*)

Process	Plant capacity $\times 10^{-3}$ tons p.a.	Capital cost[a] ($£10^6$)
Allied Chem. Corp.	25	6.7
Dutch State Mines	40	6.1
Snia Viscosa	40	6.2
Toray Industries	50	8.8

[a] 1969 prices.

In recent years, caprolactam prices have fallen to £170–200 per ton. This has been due to a combination of falling production costs and effects of overcapacity in a period of relatively high rate of growth in consumption.

CHAPTER 8

Plastics and Elastomers

(M. E. B. JONES, ICI Corporate Laboratory, Runcorn)

Introduction

Macromolecules are conveniently divided for study into the naturally occurring and the synthetic. Although the former are of immense importance, forming the foundation of both animal and vegetable life, this Chapter will be primarily concerned with the chemistry of the principal synthetic polymers; these constitute the major proportion of all plastic materials in everyday use and, in addition, form an ever increasing proportion of all elastomeric materials.

The modern plastics industry dates from the discovery of cellulose nitrate (Part 4, p. 93) by Alexander Parkes in 1862. Within three years Parkes was displaying a collection of articles made from his new material and prophesying its possible utility in the manufacture of knife-handles, combs, shoe soles, tape, pipes, battery cells and waterproof fabrics. But, despite its 100 years of commercial life, cellulose nitrate has never become a large-tonnage plastic.

In contrast, the phenol–formaldehyde resins, invented during the early part of the twentieth century by Leo Baekland, have become one of the major plastic materials. Baekland recognized that earlier investigations of the phenol–formaldehyde reaction had failed to produce a homogeneous product. By controlling the proportions of these two materials and by recognizing the catalytic function of acids and bases he produced a material which moulded readily under the action of heat and pressure. So successful was his discovery that when he died in 1944 the annual world production of phenolic resins was approaching 2×10^5 tons.

The commercial development of polystyrene and polyvinyl chloride and of co-polymers such as those based on vinyl chloride–vinyl acetate occurred during the late 1920's and early '30's, but up to that time the chemistry of polymerization reactions and indeed the intrinsic nature of the polymeric materials themselves had received little attention. It was not until the investigations by Staudinger that the macromolecular hypothesis was accepted. During the 1920's he proposed long-chain formulae for polystyrene, polyoxymethylene and rubber and received support for his hypothesis from the X-ray investigations by Major and Mark of cellulose and of other crystalline polymers.

In 1929, W. H. Carothers began his work which culminated in the development of nylon and neoprene. During these studies he synthesized polymeric molecules by well-known organic reactions and by doing so emphasized the relationship between synthetic and natural polymers.

World War II gave a major incentive for the development of many new synthetic polymers, and the ten years from the mid

Table 8.1. Plastics production (tons $\times 10^{-3}$) in 1971

	U.S.A.	U.K.	Japan
Thermoplastics			
Low-density polyethylene	2010	305	951
High-density polyethylene	848	63	394
Polypropylene	560	78	627
Polystyrene and co-polymers	1710	168	695
Polyvinyl chloride and co-polymers	1482	325	1031
Other vinyls	290	50	110
Acrylics	90	15	60
Cellulosics	75	10	10
Polyamides	50	15	22
Sub-totals	7115	1029	3890
Thermosetting			
Phenolic	476	65	218
Urea and melamine	300	170	647
Alkyd	272	63	102
Epoxy	71	14	19
Polyester	360	38	128
Polyurethanes	410	48	94
Sub-totals	1889	398	1208
Miscellaneous (all types)	420	50	89
Total	9424	1477	5187

World plastics production 32.1×10^6 tons.

1930's to the mid 1940's produced several major materials including polyethylene, polytetrafluoroethylene, poly(ethylene terephthalate), the silicones and new synthetic rubbers.

Since the war an expanding research effort has provided an impressive array of novel polymeric structures and polymerization procedures, perhaps the most outstanding developments being the discovery by Ziegler of specialized catalysts for low-pressure polymerization of ethylene and Natta's development of catalysts for the stereospecific polymerization of α-olefins. Thus the polymer chemist now has a better understanding of polymerization reactions and, with an ability to correlate polymer structure and properties, can now tailor-make polymers to suit many special needs. The impressive results are illustrated in Table 8.1 in terms of tons for the year 1971.

Definitions and Classification

Definitions

A polymer is commonly understood to be a molecule of high molecular weight that is built up from many, much smaller chemical units. To a first approximation, most polymers can be described in terms of their structural units. For some polymers, which are built up from bifunctional (or bivalent) units, the repetition of units is essentially linear, that is the units are joined together in a chain formation such as

$$—M—M—M—M—M—M— \quad \text{or} \quad —(—M—)_n—$$

to form a linear polymer. In such cases M is termed the repeat unit and n is the degree of polymerization. Together, these define the molecular weight of the polymer. Most of the useful synthetic linear polymers have molecular weights between 1.0×10^4 and 1.0×10^7. Such linear chains must also have terminal groups on each end of the chain and, depending on the chemistry of the polymerization reaction, these may be chemically similar to the repeat unit or quite different. Poly(ethylene terephthalate) and polystyrene are typical examples of linear (or thermoplastic) polymers, and in the polyester the repeat unit contains both terephthalic acid and ethylene glycol residues:

$$-\left[-CO-\bigcirc-COOCH_2CH_2O-\right]_n-$$

while in the case of polystyrene the structural unit and the monomer (styrene) from which the polymer is formed are the same, only the arrangement of bonds having changed during the polymerization.

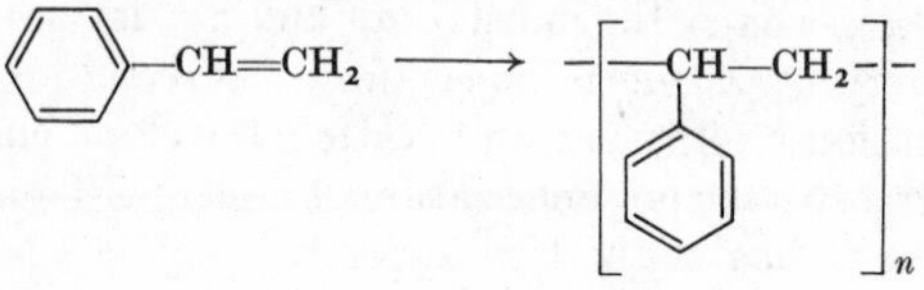

However, linear polymers are not the only form which synthetic polymers may take, for in cases where the structural units are tri- or poly-functional there the polymer chains become branched or interconnected to form three-dimensional networks. Glycerol is an example of a polyfunctional unit and introducing a small proportion into a poly(ethylene terephthalate) preparation leads to a three-dimensional or cross-linked product, which may be represented schematically as:

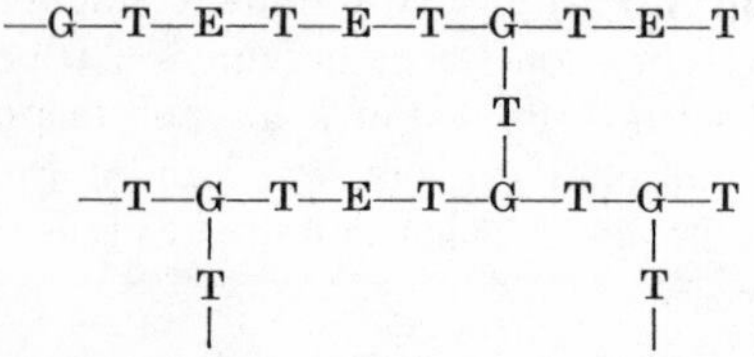

where T = terephthalic acid residue,
E = ethylene glycol residue
and G = glycerol residue.

In such cases the concentration of cross-links between chains is related to the concentration of trifunctional groups (G) in the polymer composition. The phenol–formaldehyde resins (Part 1, p. 171; also this chapter, p. 235) are an example of a highly cross-linked polymer in which the formaldehyde is difunctional and the phenol is di- or tri-functional. Such cross-linked (or thermo-setting) polymers are not conveniently described in terms of their repeat units and are usually represented as shown above.

Classification

Carothers distinguished two types of polymerization process, called addition and condensation polymerization. Condensation polymers were defined as those in which the repeat unit lacked certain atoms that were originally present in the starting materials from which the polymer was formed. An example of a condensation polymer is poly(tetramethylene adipate):

$$n\text{HO(CH}_2)_4\text{OH} + n\text{HOOC(CH}_2)_4\text{COOH} \longrightarrow$$
$$\text{H}-[-\text{O}-(\text{CH}_2)_4\text{OOC(CH}_2)_4\text{CO}-]_n-\text{OH} + (2n-1)\text{H}_2\text{O}$$

Here condensation takes place between glycol and diacid with the elimination of one water molecule for each ester link formed.

Addition polymerizations were recognized as being chain reactions, involving a chain carrier that could be either an ion or a radical (Part 1, p. 65). The polymer was formed from a bi- or poly-functional starting material (or monomer) by addition of one monomer to another without the loss of any atom. Polymers formed from vinyl monomers are the most common examples of this type of polymer. A typical example of an addition polymerization is the benzoyl peroxide-initiated polymerization of methyl methacrylate; the peroxide initiator may be decomposed thermally to provide the initial free radical; this reacts with a methacrylate molecule to open the double bond, leaving an intermediate with an unpaired electron which then adds to successive monomer unit, thus giving the growing chain; growth continues very rapidly until termination occurs, typically by two growing radicals reacting with one another to form the polymer molecule. The sequence of events is therefore as in the following scheme.

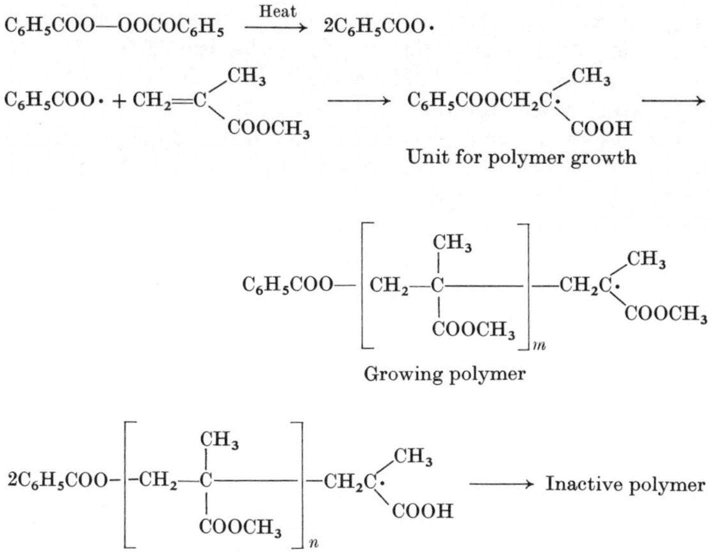

$$\text{C}_6\text{H}_5\text{COO}-\text{OOCOC}_6\text{H}_5 \xrightarrow{\text{Heat}} 2\text{C}_6\text{H}_5\text{COO} \cdot$$

Unit for polymer growth

Growing polymer

There are difficulties associated with Carothers' original classification, in that some polymers formed by condensation processes can also be produced by alternative procedures that do not involve any compositional change. For example, nylon 6 can be prepared either from ϵ-aminocaproic acid (6-aminohexanoic acid), a process involving loss of water, or from caprolactam (6-hexanolactam), without any overall compositional changes:

$$\longrightarrow \quad -[-NH(CH_2)_5CO-]_n- \quad \xleftarrow[-H_2O]{} \quad NH_2(CH_2)_5COOH$$

Similarly, polyureas may be considered as being formed by the addition of a diamine to a di-isocyanate (Part 2, p. 281) or by the condensation of a diamine with diphenyl carbonate:

$$H_2N-R-NH_2 + OCN-R-NCO$$

$$\longrightarrow -[-HN-R-NHCO-]_n-$$

$$H_2N-R-NH_2 + C_6H_5OCOOC_6H_5 \quad \xrightarrow{-2C_6H_5OH}$$

Therefore Carothers' original distinction has been modified and the concepts of step-reaction polymerization and chain-reaction polymerization were introduced. In this scheme all polymerizations are classified according to the mechanism of the reaction without regard to the loss or otherwise of reaction products. The formation, for example, of polyurethanes from mixtures of a diol and a di-isocyanate, is classified as step-reaction polymerization along with the classical polycondensation involving the loss of a small molecule. This mechanistic classification of polymers gives rise to the following distinguishing features of the two types of polymerization: In *step-polymerization* the starting material disappears early in the reaction and the molecular weight rises progressively during the polymerization; long reaction times are generally required to achieve high molecular weights and at any point in the reaction all the molecular species are present in a calculable distribution. In contrast, during *chain polymerization* the monomer concentration decreases progressively throughout the reaction and high polymer is present almost at once. Its molecular weight remains relatively constant during the reaction, and long reaction times increase the yield of polymer but not the molecular weight.

Relation between Chemical Structure and Polymer Properties

Among the most important properties of both thermoplastic and thermosetting synthetic polymers are their softening behaviour, strength, rigidity, toughness and stability towards radiation or oxidative and chemical attack. The flow, or melt, properties of thermoplastics are also important with respect to their conversion into fabricated forms.

It is possible to relate the performance of all the synthetic polymers to their chemical structure and among the major factors determining the physical properties are the type and the distribution of the chemical bonds in the polymer molecule, its symmetry and conformation and its molecular weight.

Bonding in Polymer Chains

The polymers described in this Chapter consist of structural units held together by covalent bonding. Although this bonding holds the individual atoms in place in the polymer chain it does not satisfactorily account for all the observed properties of plastics and rubbers and, to do this, other bonding forces must be taken into account. These are generally intermolecular forces and though much weaker than the primary covalent bonds they nevertheless play an important role in determining the properties of the material.

Undoubtedly the strongest of these secondary forces is the hydrogen bond and, where it exists, e.g. in the polyamides and polyureas, it plays a major role in determining the properties of the polymer. For example, the high melting points of the nylons are due almost entirely to the presence of this type of bonding. Dipole and related induction forces must also be taken into account in any attempt at structure–property correlation and such dipole–dipole interactions are present in, for example, polyesters. Finally, dispersion forces are present in all molecules, and in non-polar materials such as the hydrocarbon polymers they account for essentially all the intermolecular forces.

Conformations of Polymer Chains

In the first studies of this subject, by Staudinger, it was shown that the phenyl groups in polystyrene were attached to the backbone of the polymer in a regular fashion to alternate carbon atoms, i.e. in a head-to-tail arrangement (see p. 204). Other possible arrangements are head-to-head and random, also illustrated.

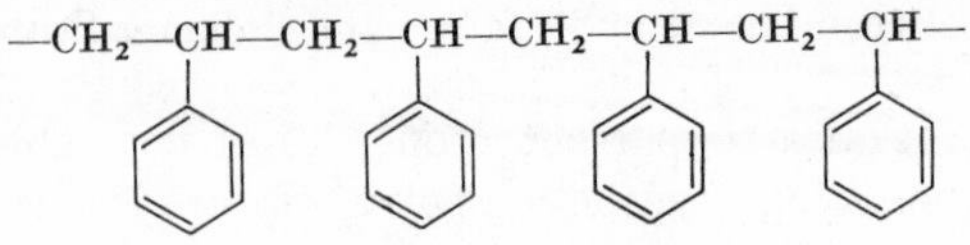

Head-to-tail

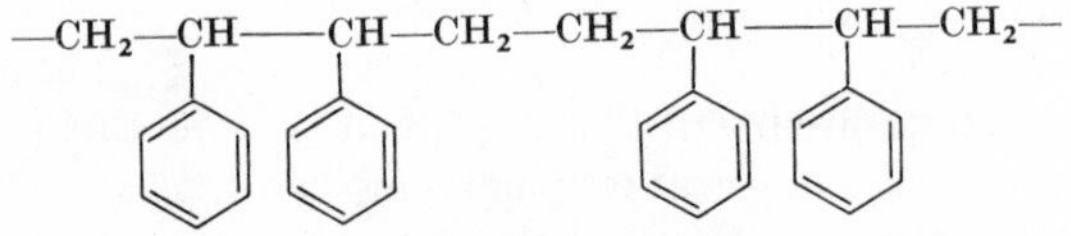

Head-to-head

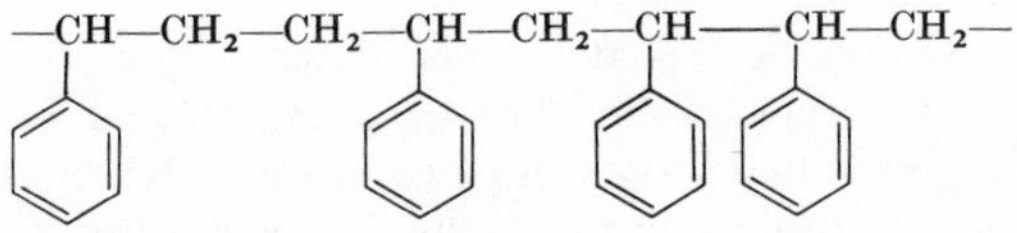

Random

These possibilities arise because in principle the addition of the initiating radical or ion to a monomer of the vinyl type, $CH_2{=}CHR$, could occur at either carbon atom of the double bond, depending on the stability of the resulting product. The majority of vinyl polymers have a predominantly head-to-tail arrangement.

A second variation, which is most commonly found in polyethylene, is branching. Here, the atom at the site of the branch becomes trifunctional and leads to a molecule with a feather-like structure due to "short-chain branching".

Finally, there is the conformation to be taken into account when considering the structures of polymers made from monomers of the type $CH_2{=}CHR$. Staudinger recognized that in polymers derived from these monomers the carbon atoms with the R-substituent could be asymmetric with either an *R*- or an *S*-configuration. Since this theoretical deduction such stereospecific syntheses have been achieved, notably in the case of the stereoregular polymers prepared from α-olefins. The nomenclature used to describe polymers of this type is due to Natta and provides names for the three possible arrangements. In these, the substituents are on the same side of the polymer chain (isotactic),

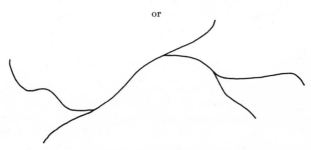

Short-chain branching

or

Long-chain branching

alternately above and below the main chain (syndiotactic), or disposed in a random fashion (atactic).

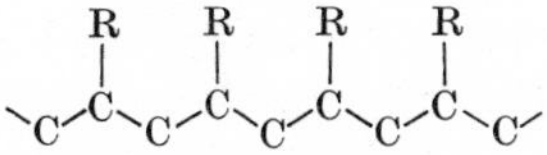

Isotactic

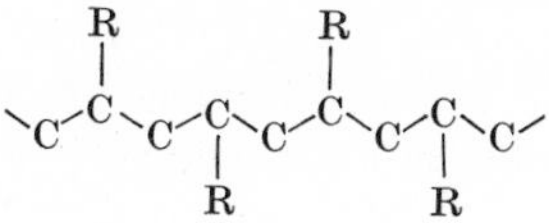

Syndiotactic

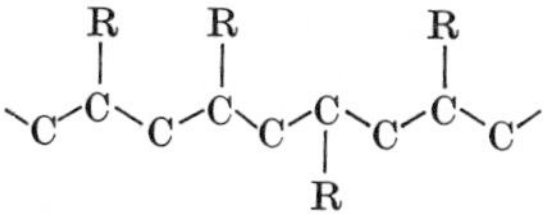

Atactic (random)

Polymers produced from 1,2-disubstituted olefins can also exist in several stereoregular forms, while polymers of 1,3-dienes that retain one residual double bond per repeat unit can also contain sequences with differing conformations, but it is outside the scope of this Chapter to consider these more complex possibilities.

Crystallinity

Although many synthetic polymers give discrete X-ray diffraction patterns, these X-ray studies show that no polymer is 100% crystalline and that crystalline (ordered) and amorphous (disordered) regions occur together in the same 'semicrystalline' polymer. Evidence of crystallinity is found most frequently in symmetrical polymer structures. Irregularities such as branch points, co-monomer units, cross-links, atactic structures, etc., usually result in reduced crystallinity. Increasing the level of such irregularities can give completely amorphous materials. As a result totally atactic polymers, highly cross-linked resins and co-polymers with a high level of co-units are generally completely amorphous.

Well-crystalline polymers generally have relatively sharp melting points (Tm's) which can be determined by a variety of optical and thermal techniques. As a result of the presence of significant levels of crystallinity, fabrication procedures which involve melt processing, e.g. extrusion or injection moulding, must be conducted at temperatures above the melting point.

Glass Transition Temperature

There are many examples of non-crystalline (amorphous) polymers and, although these materials do not exhibit true melting, at sufficiently low temperatures they all become stiff and rigid. The limiting temperature varies widely, depending upon the polymer. It can be as low as $-70°C$ for natural rubber or as high as $+100°C$ for poly(methyl methacrylate). This temperature at which amorphous polymer changes from a flexible, rubbery state to a rigid, glassy condition is called the glass–rubber transition temperature (Tg). In molecular terms this process can be envisaged as the temperature at which relatively large segments of the main polymer chain becomes mobile. Below the Tg this large-scale motion is restricted and the amorphous material is stiff and rigid; above the Tg it is rubbery owing to mobility of long segments of the polymer chain.

Since the Tg is a characteristic of the amorphous region and as no crystalline organic polymer is ever 100% crystalline, 'crystalline' polymers also exhibit a Tg in addition to a melting transition. When a well-crystalline polymer is heated, its passage through the Tg does not necessarily produce an obvious softening, since the now rubbery amorphous regions are still reinforced by the crystalline segments; only when Tm is reached does the polymer soften in the conventional sense.

Mechanism of Chain Polymerization

Radical Polymerization

Free-radical chain reactions have been discussed in some detail previously (Part 1, pp. 11–12, 59, 65; Part 2, pp. 126–129; Part 3, pp. 311–317). The free-radical polymerization of ethylene (cf. Part 1, p. 66) can be described in the following reaction sequence:

Initiation: $R \cdot + C_2H_4 \longrightarrow RC_2H_4 \cdot$

Propagation: $RC_2H_4 \cdot + C_2H_4 \longrightarrow RCH_2CH_2CH_2CH_2 \cdot$

Transfer: (a) intramolecular, $R(CH_2)_4CH_2 \cdot \longrightarrow$
$$R\dot{C}H(CH_2)_3CH_3$$

(b) intermolecular, $RCH_2 \cdot + R'(CH_2)_nR'' \longrightarrow$
$$RCH_3 + R'\dot{C}H(CH_2)_{n-1}R''$$

(c) to tert. H, $RCH_2 \cdot + (CH_3)_3CH \longrightarrow$
$$RCH_3 + (CH_3)_3C \cdot$$

Termination: (a) by disproportionation
$$RCH_2CH_2 \cdot + R'CH_2CH_2 \cdot \longrightarrow$$
$$RCH{=}CH_2 + R'CH_2CH_3$$

(b) by combination
$$RCH_2 \cdot + R'CH_2 \cdot \longrightarrow RCH_2CH_2R$$

The conditions under which the polymerization is carried out can influence the relative rates of these various reactions, and so the structure of the polymeric product can be determined by control of reaction pressure, temperature, amount of solvent, etc. For example, the rate of transfer increases faster with temperature than does the rate of propagation; so, as reaction temperature is raised, the molecular weight of the product falls.

The nature of the initiation step can be all important. Common initiators include dibenzoyl peroxide (Part 2, p. 128), di-*tert*-butyl peroxide (Part 2, p. 125) and azodi-isobutyronitrile (Part 2, p. 173). These are all molecules with comparatively weak bonds which can be broken thermally to yield free radicals ($R \cdot$ in the above sequence).

Chain Polymerization by Ionic Mechanisms

Anionic or cationic initiation can be used to produce polymers from certain substituted ethylenes. Typical anionic catalysts include the strong bases such as sodium naphthalene or sodium amide, while typical cationic polymerization catalysts are the Lewis acids such as aluminium trichloride and boron trifluoride. Ionic polymerizations are typically very fast and are usually carried out at low temperatures in a low-boiling diluent which assists in dissipating the heat of reaction.

It is possible to relate the polarity of a monomer double bond to its susceptibility to polymerization by cationic, anionic or free-radical processes. For example, a monomer containing an electron-rich double bond, as a result of having an electron-donating substituent, will tend to form a more stable carbonium ion and hence be more susceptible to cationic polymerization. Typical monomers that polymerize cationically are vinyl ethers, isobutene and dihydrofuran. On the other hand, monomers with electron-withdrawing groups attached to the double bond tend to polymerize anionically and olefins that are strongly activated, such as vinylidene dicyanide, will be catalyzed by weak bases although most anionically polymerizable monomers need strong bases as catalysts. Between these two extremes, however, there are many substituted ethylenes that have at most only weakly electron-withdrawing substituents and polymerize by free-radical mechanisms.

Anionic Polymerization. Many of these reactions proceed best at low temperatures but a limited number are better carried out at room temperature or slightly above. Some of them have some degree of stereospecificity associated with them, particularly where heterogeneous catalysts are employed, and this is associated not only with simple monosubstituted olefins but also with the 1,4- and 1,2-polymerization of butadienes. In a conventional anionic polymerization, for example, styrene in liquid ammonia catalysed by potassium amide, it is found that (*a*) the rate increases as the reaction temperature is reduced, (*b*) molecular weight is independent of amide ion concentration, (*c*) molecular weight, at constant conversion, is dependent on potassium ion concentration, (*d*) molecular weight depends upon initial monomer concentration, (*e*) molecular weight increases with decreasing temperature, (*f*) the polymer contains one nitrogen atom per mole, (*g*) the polymer contains no residual double bond and (*h*) no potassium amide is consumed. The experimental observations are explained by the following mechanism:

Initiation: $K^+NH_2^- + C_6H_5CH{=}CH_2 \longrightarrow$
$$NH_2CH_2\bar{C}HC_6H_5 + K^+$$

Propagation: $NH_2CH_2\bar{C}HC_6H_5 + CH_2{=}CHC_6H_5 \longrightarrow$
$$NH_2CH_2CH(C_6H_5)CH_2\bar{C}HC_6H_5$$

Termination: $NH_2CH_2CH(C_6H_5)CH_2\bar{C}HC_6H_5 + NH_3 \longrightarrow$
$$NH_2{-}[{-}CH_2CH(C_6H_5){-}]_{n+1}CH_2CH_2C_6H_5 + NH_2^-$$

Cationic Polymerization. Polymerization takes place in the presence of carbonium ions, so with Lewis-acid catalysts a co-catalyst is necessary. This effects the polymerization, and propagation consists of addition of monomer to carbonium ion. Nevertheless, in some cases a co-catalyst is not required since some acids can generate their own cation and the polymerization steps are then analogous to a free-radical mechanism.

The polymerization of isobutene catalysed by boron trifluoride in the presence of a trace of water is a typical example:

Initiation: $BF_3 + H_2O \longrightarrow {}^-BF_3OH\ H^+$

$\qquad\qquad {}^-BF_3OH\ H^+ + CH_2{=}C(CH_3)_2 \longrightarrow (CH_3)_3C^+\ {}^-BF_3OH$

Propagation: $(CH_3)_3C^+\ \bar{B}F_3OH + CH_2{=}C(CH_3)_2 \longrightarrow$

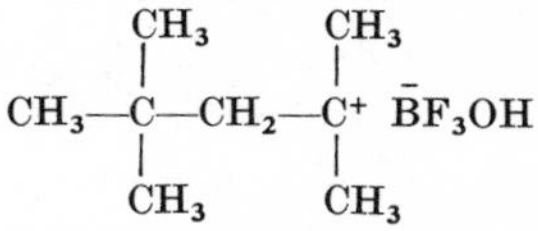

$$CH_3-\underset{\underset{CH_3}{|}}{\overset{\overset{CH_3}{|}}{C}}-CH_2-\underset{\underset{CH_3}{|}}{\overset{\overset{CH_3}{|}}{C^+}}\ \bar{B}F_3OH$$

Termination: several mechanisms, e.g.

(i) kinetic

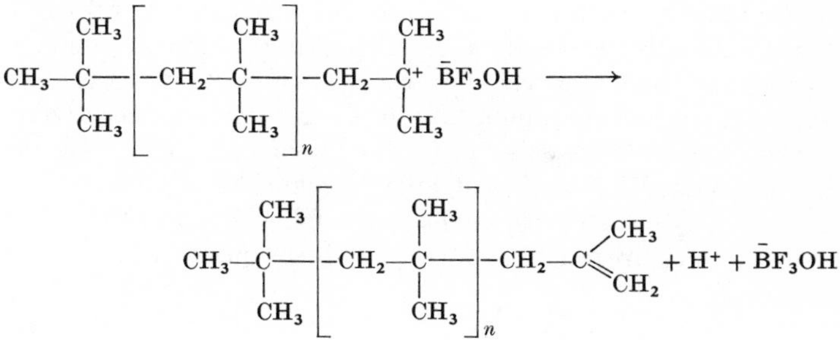

$$CH_3-\underset{\underset{CH_3}{|}}{\overset{\overset{CH_3}{|}}{C}}{\left[-CH_2-\underset{\underset{CH_3}{|}}{\overset{\overset{CH_3}{|}}{C}}-\right]}_n-CH_2-\underset{\underset{CH_3}{|}}{\overset{\overset{CH_3}{|}}{C^+}}\ \bar{B}F_3OH \longrightarrow$$

$$CH_3-\underset{\underset{CH_3}{|}}{\overset{\overset{CH_3}{|}}{C}}{\left[-CH_2-\underset{\underset{CH_3}{|}}{\overset{\overset{CH_3}{|}}{C}}-\right]}_n-CH_2-C{\overset{CH_3}{\underset{CH_2}{<}}}\ + H^+ + \bar{B}F_3OH$$

(ii) combination

$$CH_3-\underset{\underset{CH_3}{|}}{\overset{\overset{CH_3}{|}}{C}}{\left[-CH_2-\underset{\underset{CH_3}{|}}{\overset{\overset{CH_3}{|}}{C}}-\right]}_n\underset{\underset{CH_3}{|}}{\overset{\overset{CH_3}{|}}{C^+}}\ \bar{B}F_3OH \longrightarrow$$

$$CH_3-\underset{\underset{CH_3}{|}}{\overset{\overset{CH_3}{|}}{C}}{\left[-CH_2-\underset{\underset{CH_3}{|}}{\overset{\overset{CH_3}{|}}{C}}-\right]}_n-CH_2-\underset{\underset{CH_3}{|}}{\overset{\overset{CH_3}{|}}{C}}-OH + BF_3$$

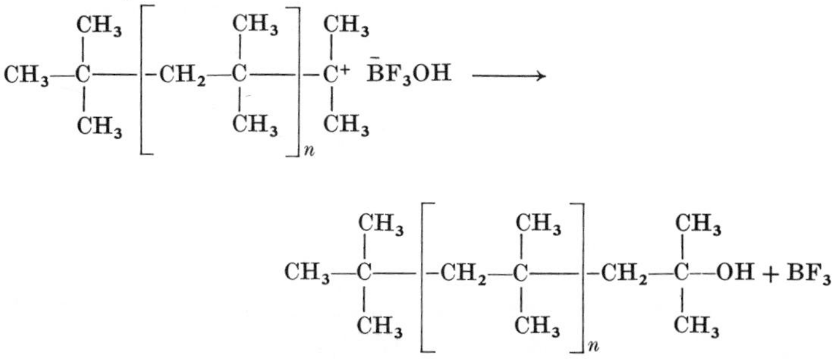

(iii) **transfer to monomer**

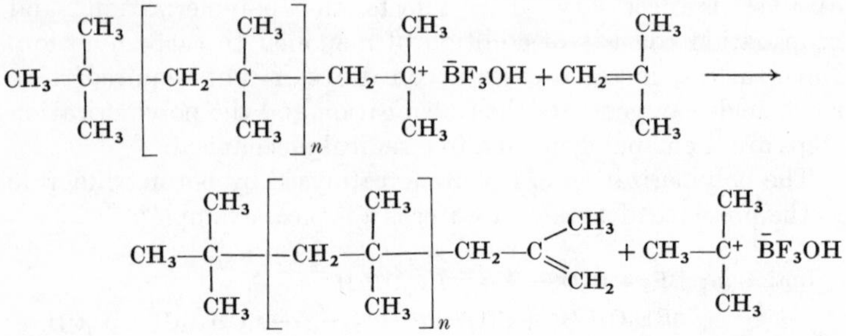

The actual termination step(s) will therefore depend upon the particular combination of monomer, catalyst and polymerization conditions.

Co-ordinated Catalysts

Co-ordinated catalysts are conveniently divided into two classes: (i) those prepared from a metal alkyl or hydride, i.e. a reducing agent, in combination with a reducible transition-metal halide (Ziegler–Natta catalysts) and (ii) a reduced metal oxide which is usually carried on a support such as alumina. Common examples of these two types are triethylaluminium with titanium tetrachloride, and chromium oxide on silica–alumina, respectively.

These catalysts are effective for the polymerization of α-olefins and conjugated dienes. The detailed mechanism of the action of these catalysts is still disputed but the following equations seem to describe the reactions occurring with a typical Ziegler–Natta catalyst:

$$TiCl_4 + AlR_3 \longrightarrow TiCl_3R + AlR_2Cl$$
$$TiCl_3R \longrightarrow TiCl_3 + R\cdot$$
$$TiCl_3 + AlR_3 \longrightarrow TiCl_2R + AlR_2Cl$$

or

$$TiCl_4 + 2AlR_3 \longrightarrow TiCl_2R_2 + 2AlR_2Cl$$
$$TiCl_2R_2 \longrightarrow TiCl_2R + R\cdot$$
$$TiCl_2R \longrightarrow TiCl_2 + R\cdot$$
$$TiCl_2R + AlR_3 \longrightarrow TiClR_2 + AlR_2Cl$$
$$TiCl_2 + AlR_3 \longrightarrow TiClR + AlR_2Cl$$

Any proposed mechanism for these catalysts must take into account the stereoselective nature of the products, i.e. the formation of isotactic poly-α-olefins and *cis*-1,4-, *trans*-1,4- or 1,2-iso-tactic polybutadienes.

The rate of polymerization with such catalysts depends on the concentration of titanium, the olefin, the nature of the alkylating agent and the ratio of alkylating agent to titanium tetrachloride. The true catalyst is almost certainly a complex co-ordination compound containing titanium, aluminium, halogen and alkyl groups, in which the catalyst surface serves as a mould, and whichever configuration is taken by the first monomer influences the addition of the next monomer molecule.

Polymerization Processes

It is convenient to divide polymerization processes into those involving the pure monomer, or a homogeneous solution of the pure monomer, from those that employ a heterogeneous dispersion of the monomer suspended in a solvent, usually water. We can divide the first two processes into *bulk polymerization* and *solution polymerization* and the heterogeneous processes into *suspension polymerization* and *emulsion polymerization*. We shall illustrate each by reference to actual industrial systems.

Bulk Polymerization

The radical polymerization of vinyl chloride is sometimes carried out by a bulk process. The process is a heterogeneous reaction since polyvinyl chloride (PVC) is insoluble in vinyl chloride monomer. The attractions of a bulk process are that emulsifiers, protective colloids, etc. are not required. A typical bulk process uses azodi-isobutyronitrile as initiator and is carried out in two stages with a final temperature of 60°C. The first stage produces conversion of ca. 10% of monomer and during the second stage the mixing speed is reduced and conversions of up to 70% are achieved. The change in agitation is claimed to be responsible for the favourable particle size distribution which is obtained. The bulk polymerization of styrene is homogeneous since polystyrene is soluble in monomer. These reactions were developed mainly in Germany and they are the preferred techniques for producing polymer of high optical clarity and superior electrical properties. Severe chemical engineering problems are presented since the heat of reaction, which is considerable (63 kJ mol^{-1}), must be dissipated and the reaction involves the conversion of a reactive, mobile liquid into a highly viscous polymeric material. In a typical continuous bulk process pure styrene monomer is pre-polymerized to about 30% conversion by thermally initiated polymerization. The syrup of polymer in monomer is run into a second reaction vessel where it encounters progressively higher temperatures up to a maximum of 200°C. This ensures substantially complete polymerization and the molten polymer from this second reaction is then cut into strips which are freed from volatile matter, chipped and stored. The high final temperatures are necessary so that molten polymer can be pumped around the plant, but at these temperatures there is a tendency to produce species of low molecular weight either by re-equilibration of high polymer or directly from monomer.

Bulk polymerization of methyl methacrylate (p. 230) is employed in the production of acrylic sheet, which is prepared by pouring a syrup of polymer in monomer into a casting cell consisting of two polished glass plates fitted around three edges with a gasket. The syrup is prepared by treating pure methyl methacrylate at about 95°C with a peroxide, e.g. benzoyl peroxide, and, when the required degree of conversion has been achieved, cooling to minimize further polymerization. This preparation of pre-polymer syrup can be difficult to control and an alternative is to dissolve the required quantity of polymer in monomer. The casting cell is held together with clips which can contract to take up any shrinkage during further polymerization. The filled cells pass through ovens where polymerization is continued, initially at 40°C and finally at 95°C for up to an hour to complete the polymerization. After a total reaction time of about 15 hours the cell is cooled and the sheet of polymer is removed.

Solution Polymerization

Addition polymerizations are usually exothermic and may proceed uncontrollably unless the heat is removed. We have seen above that dissipation of heat is important in the bulk polymerization of styrene. Solution polymerizations overcome the problems of heat dissipation and the handling of viscous melts which are associated with the bulk process. Solvent and monomer are mixed in a continuous fashion before passing into a multi-zone reactor where the solution is first heated to initiate polymerization and then in subsequent zones cooling is applied to moderate the rate of reaction. In the final phase the solution is again heated to complete polymerization and to assist in the removal of solvent in the final 'devolatilizing' vessel. Polystyrene homopolymer produced by all these procedures is available in a number of grades.

For example, if improved strength and toughness are major requirements, then polymers of higher molecular weight than those of the general-purpose grade are available. Such a modification might be required without the loss of clarity associated with the rubber-modified forms described below. Higher-softening grades are also produced by reducing the content of volatile materials below that associated with the general-purpose grades. Internal lubricants such as liquid paraffin may also be added to improve the melt-flow characteristics of certain high-molecular-weight polymers, but this usually results in a lowering of the softening point and these products are generally used only in complex moulding operations.

Suspension Polymerization

Suspension polymerization procedures are employed to make most of the PVC currently produced in the U.S.A. These employ a water phase which contains suspended vinyl chloride monomer. The quality of the polymer produced is dependent upon the choice of suspending agents since this controls the particle size, shape and porosity. The nature of the particles in turn influences bulk density, handling properties, absorption of plasticizer and processing characteristics. Typical suspending agents include poly-(vinyl alcohol), water-soluble cellulose derivatives and gelatine; these are employed at levels of between 0.05% to 0.5% by weight. Emulsifiers such as sulphonated oils and esters are sometimes included in a polymerization mixture in amounts of between 0.07% to 0.03% by weight and are reported to assist in the control of the surface of the suspended particle. The radical initiators used for suspension polymerization must be soluble in monomer (e.g. dodecyl peroxide) and are used at polymerization temperatures of 50–60°C under pressures of around 0.7×10^6 Nm^{-2}. Typical polymerization times under these conditions are from 12 to 15 hours. Reducing the temperature increases molecular weight and at the same time reduces the amount of branching in the polymer. A variety of mechanical factors such as the stirring speed during polymerization and the design and location of baffles in the polymerization vessel can affect particle size and hence the nature of the product. When polymerization is complete the polymerization vessel is discharged, residual monomer stripped and the suspension centrifuged to remove the water. The particles are washed with water before drying and bagging.

Suspension polymerization is probably the most widely used procedure for styrene and in a batch process it may be carried out by suspending styrene in water. The polymerization is initiated by a monomer-soluble peroxide such as benzoyl peroxide. Since removal of heat from a low-viscosity system is relatively straightforward this technique does not have the particular problems associated with the bulk processes. The suspension of particles is stabilized by adding suspending agents such as poly(vinyl alcohol), and further particle control can be exercized by agitation rates, vessel design, etc. On completion of the polymerization any unconverted monomer and other volatile substances are removed by steam-distillation. The disadvantages of this technique are poor utilization of reaction vessel space since the greater part is occupied by water, and the form of the product which frequently needs an extrusion and chipping operation to convert it into a form suitable for subsequent fabrication.

Suspension polymerization of methyl methacrylate was developed for the preparation of polymer suitable for melt fabrication since the product of bulk polymerization is of very high molecular weight and unsuitable for this application. In common with other suspension polymerizations there are difficulties in controlling particle size and particle agglomeration, and careful reactor design and choice of suspending agents, protective colloid, etc. are necessary to achieve adequate control. Typical suspending agents are aluminium oxide, talc and magnesium carbonate, while the sodium salt of poly(methacrylic acid) can be used as a protective colloid. The suspension polymerization technique is also used to prepare co-polymers containing minor amounts of acrylate co-monomer such as ethyl acrylate, and such co-polymers offer advantages of improved thermal stability.

Emulsion Polymerization

Emulsion polymerization differs from suspension polymerization in that a soap is added and micelles are formed (see Chapter 12). The monomer is either dispersed into droplets which are stabilized by an adsorbed layer of soap molecules, or it is solubilized in the soap micelle. Emulsion polymerization techniques are extensively used in Europe for vinyl chloride (p. 224). As in suspension polymerization, water is the continuous phase and it acts as heat-transfer medium. Oxygen inhibits the polymerization and care must be taken to remove all traces from the polymerization vessel and from the water. Initiators may be either water- or monomer-soluble but water-soluble initiators such as potassium persulphate are the most common. 'Redox' initiators are also used, particularly for low-temperature polymerizations, where the initiator is activated by a reducing agent such as sodium or hydrogen sulphite. In these emulsion systems, sulphonate esters are commonly used as emulsifying agents in combination with water-soluble polymers such as poly(vinyl alcohol); the latter function as protective colloids to reduce agglomeration. The choice of emulsifier and protective colloid is crucial since they may well remain in the final product.

Compounding Additives

Because of the limited thermal stability of polymer and its tendency to adhere to metal during fabrication it is usual to compound (i.e. mix) PVC with a number of additives. By these compounding procedures it is possible to produce a wide range of useful materials. In addition to polymer a PVC compound may therefore contain one or more of the following additives: plasticizer, stabilizer, pigment, ultraviolet stabilizer, filler, lubricant and extender. It is beyond the scope of this Chapter to try to cover all these modifications in detail and only the more important compounding ingredients will be mentioned.

Stabilizers. The most obvious indication of thermal degradation of PVC is a progressive change in colour. From an initial water-white it will change at the normal processing temperature of 150–200°C, through pale yellow, orange and brown to black, and therefore a stabilizer is a material which retards this colour formation. Other tests for stabilizers, based, for example, on rates of dehydrochlorination, are not as reliable as the colour-formation test. The most important class of stabilizer comprises lead compounds; these react with any hydrochloric acid evolved during decomposition to form lead chloride. Basic lead carbonate is probably the most widely used but it has the disadvantage of liberating carbon dioxide which under certain conditions can lead to foamed products. For this reason lead sulphide is being used more frequently, especially in rigid PVC compounds. Other lead compounds are used in more specific applications, for example dibasic lead phthalate is a stabilizer used in heat-resistant insulation applications and for 'high-fi' gramophone records. Although lead compounds are the most widely used stabilizers, other metal salts have found increasing application, particularly zinc and cadmium stearate, octanoate and palmitate. Another group of stabilizers comprises organotin compounds, which found use initially because they gave water-white formulations; however, the original tin compounds such as dibutyltin dioctanoate conferred limited heat stability and are being replaced by alternatives such as dibutyltin maleate.

Plasticizers. The consumption of PVC plasticizers in Great Britain exceeds that of many plastics and only polyethylene, polystyrene and PVC itself are produced in greater amounts. Basically a plasticizer is a non-volatile solvent for PVC that can be readily mixed with polymer at ca. 150°C to increase flexibility and to improve toughness. Because of the relatively large size of the plasticizer molecule they have very low rates of diffusion through the polymer at ambient temperatures. Phthalates are the most important type of plasticizer and these are commonly prepared from phthalic acid or anhydride and an alcohol with about 8 carbon atoms. For economic reasons di(isooctyl) phthalate and the esters of the C_7–C_9 'oxo' alcohols are the most commonly used. Use of phosphate plasticizers has not grown as rapidly as that of phthalates, but because of its better price structure the consumption of trixylyl phosphate has increased at a greater rate than that of tritolyl phosphate; both give compounds of improved fire resistance compared with equivalent formulations of phthalates and their use also enhances the solvent resistance of the product. However, they are both more toxic than the phthalates and their compounds are embrittled more severely at low temperatures. For applications that demand good resistance to low temperatures, the aliphatic esters such as dibutyl sebacate or dioctyl azelate are commonly employed.

Extenders. Unlike the plasticizers, extenders are not compatible with PVC and compatability is achieved by addition of a plasticizer. Extenders are normally cheaper than plasticizer and can be used to replace up to about one-third of it without significantly altering the properties of the compound.

Lubricants. These are used to minimize adhesion of polymer to the metal surfaces of processing machinery such as mills, colanders and extruders. They are normally incompatible with polymer, so that during a processing operation a film of lubricant is formed between the polymer and the processing surfaces. Stearic acid, calcium stearate and normal lead stearate are examples of lubricants. The formulation of unplasticized PVC, in particular, demands a careful choice of lubricants and commercial grades may contain a mixture of two or more lubricants.

THE MAIN PLASTICS

Polyethylene, $-[-CH_2CH_2-]-$

Polyethylene was discovered in 1932 during research by ICI investigators into reactions under high pressure. A mixture of ethylene and benzaldehyde compressed to 1.4×10^8 Nm^{-2} at 170°C gave a white solid that was recognized as polyethylene. This product was essentially what is now known as low-density polyethylene, its relatively low density and melting point as compared to polymethylene being due to the presence of chain branching. The production of polyethylene by the high-pressure process expanded during the war and in 1945 it had reached 15×10^3 tons p.a. Post-war expansion of production was very rapid, stimulated by advances in production procedures and fabrication techniques. However, the major development of the post-war period was Ziegler's discovery of high-density polyethylene in 1953, when it was found that the product of reaction of titanium tetrachloride with aluminium alkyls was capable of polymerizing ethylene at ambient temperatures and atmospheric pressure. The high-molecular-weight product was more crystalline than low-density polyethylene owing to its more linear structure and this resulted in a higher-melting and stronger polymer. At about the same time as Ziegler's discovery Phillips Petroleum Co. developed a medium-pressure (4×10^6 Nm^{-2}) process for high-density polyethylene; the catalyst was essentially CrO_3 supported on a SiO_2–Al_2O_3 carrier. The polymer produced by this process is generally more linear than the Ziegler product.

Low-density (High-pressure) Polyethylene

Under rather severe conditions of pressure and temperature ethylene will give polymer of high molecular weight. The polymerization proceeds by a free-radical mechanism in the bulk (i.e. liquid ethylene), in the gas phase, in emulsion or in solution. If the conditions are chosen correctly, polymerization can be initiated by many free-radical producing agents, e.g. peroxides, hydroperoxides, persulphates, azo-compounds or oxygen itself. The choice of initiator determines the temperature and pressure range within which the reaction operates (see Table 8.2 for examples).

In spite of the fact that the original ICI, oxygen-initiated, high-pressure process was the first to be developed, the mechanism is still not completely understood although it is clear that it differs from similar processes employing alternative radical initiators. It

Table 8.2.   Conditions for formation of low-density polyethylene

Initiator	Pressure (10^6 Nm^{-2})	Temperature (°C)
α,α'-Azobisisobutyronitrile	700	55
Benzoyl peroxide	75	75
tert-Butyl hydroperoxide	75	100
Oxygen	100–150	100–190

seems likely that a peroxide is formed *in situ*, since the reaction exhibits an induction period during which oxygen is consumed. Subsequently two types of polymerization have been observed, an initial fast polymerization which is sometimes followed by a slower polymerization. The former reaction occurs only under quite specific conditions of oxygen concentration, temperature and pressure. The slower polymerization will continue at much lower pressure but only after the first reaction has commenced. The characteristics of the oxygen-initiated polymerization have led to its being compared to a 'degenerate explosion' and the formation of an intermediate polyperoxide has been postulated that subsequently undergoes homolysis. The observed induction period is attributed, in part, to formation of this intermediate and such a mechanism fits relatively well the observed kinetics of the system.

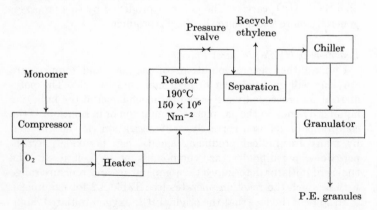

Figure 8.1   Flow diagram of ICI bulk high-pressure process for polyethylene

Polymerization initiated by alternative, free-radical sources, e.g. di-*tert*-butyl peroxide, has different characteristics to the oxygen-initiated polymerization and the optimum conditions depend on the nature of the initiator. For example, using benzoyl peroxide as initiator, it has been shown that the use of higher pressures tends to suppress chain branching and produces polymers with melting points of 128–133°C.

A flow diagram of a typical bulk high-pressure process is given in Figure 8.1.

For all these processes the monomer (ethylene) must be available in a state of high purity and for oxygen-initiated polymerizations the oxygen content must be accurately known and should be less than 0.002% by weight. When other initiators are employed it should be less than 0.005% by weight. The gaseous monomer is compressed before passing to the reactor where it is heated and where the pressure is adjusted. Alternatively, the ethylene may be liquefied at more modest pressure but lower temperatures before being fed into the reaction vessel. The reaction vessel is designed so that it permits accurate pressure control and gives rapid heat dissipation and an even flow of reactants and products. Inadequate control of the reaction conditions can result in the formation of cross-linked polymers and even to explosive decomposition. Tubular reactors have advantages over the more conventional, stirred autoclaves in terms of more effective control and in such tubular reactors the temperature along the reaction zone normally varies, increasing to a maximum before falling again towards the end of the reactor. For oxygen-initiated reactions, optimum conditions are approximately 150×10^6 Nm^{-2} at 180–200°C, and under these conditions the polymer obtained has a density of 0.92 g (cc)$^{-1}$. Ethylene at this pressure is an incompressible fluid and the reaction can be considered as solution polymerization in liquid monomer. When other initiators are employed they are usually in solution and in certain cases diluents may also be used, their main function being to assist in heat dissipation and the removal of polymer from the reaction zone; some diluents may also serve as chain-transfer agents, although specific transfer agents are frequently added to effect further control of molecular weights and degree of branching.

Polymer recovery from high-pressure polymerizations is relatively straightforward, no purification is required and unconverted ethylene is recovered by depressurizing; the molten polymer is then extruded, chipped when cold and bagged.

High-density (Low-pressure) Polyethylene

The description of this type of polyethylene is conveniently subdivided on the basis of the type of catalyst used for the polymerization reaction; these are the Ziegler catalysts and the metal oxide catalysts.

Ziegler Catalysts. These are complex materials, generally produced in either a soluble or an insoluble form. The insoluble (or precipitated) catalysts are the more commonly employed and are formed by reaction of a solution of a transition-metal compound, typically $TiCl_4$ or $VOCl_3$ with an organometallic compound such as triethylaluminium, chloro(diethyl)aluminium, butyl-lithium or diethylzinc. The reaction of $TiCl_4$ with chloro(diethyl)aluminium gives initially an ethyl radical and a precipitate of metastable $TiCl_3$, the radical disproportionating to ethane and ethylene:

$$TiCl_4 + (C_2H_5)_2AlCl \longrightarrow C_2H_5AlCl_2 + C_2H_5TiCl_3$$
$$C_2H_5TiCl_3 \longrightarrow C_2H_5\cdot + TiCl_3$$
$$2C_2H_5\cdot \longrightarrow C_2H_6 + C_2H_4$$

An excess of metal alkyl is necessary after this reduction is complete if a useful catalyst is to be produced. The active catalyst can be regarded as a reduced, solid titanium chloride with alkyl-aluminium compound(s) chemisorbed on the surface. The valency of the transition metal in these catalysts depends on the molar ratio of the two components and the reducing strength of the organometallic component. The molar ratio of the two components affects the polymerization rate and the molecular weight of the polymer, but useful catalysts are normally produced with aluminium:titanium ratios between 1:1 and 2:1. These catalysts are sensitive to the method of preparation, and their activity can be affected by the order of addition of the components, the concentration and temperature of the solutions used, and the nature and purity of the diluents used for the polymerization. Aliphatic hydrocarbons are the diluents most commonly used, since polar liquids such as ethers, ketones, esters, etc., generally inhibit polymerization, as do traces of oxygen, carbon dioxide or water.

The polymerization is normally carried out at temperatures well below the melting point of the polymer, under which conditions the polyethylene is virtually insoluble. The polymerization occurs in a slurry or suspension and, on its completion, the product is removed as a slurry of polymer containing the catalyst. All the catalyst handling is carried out under an atmosphere of inert gas such as nitrogen, but an active catalyst can be prepared in a number of ways, e.g. the components may be fed continuously in the correct proportions into the reaction zone, they may be pre-reacted and the isolated product fed to the reactor, or they may be pre-reacted and the active compound added as a slurry in the diluent. Polymerization temperatures are typically between $50°$ and $70°C$ at a pressure of 2×10^5 Nm^{-2}. Typical reaction times for batch reactions are 1–4 hours and times for continuous processes are similar. The molecular weight of the product can be controlled in several ways, for example, by the formulation and concentration of the catalyst, the temperature of the reaction or addition of hydrogen. The polymerizations are normally terminated by addition of alcohols, which solubilize the catalyst and assist subsequent purification of the polymer; the product is then simply isolated by filtration or centrifuging and may be further washed to reduce the content of catalyst residues.

Metal Oxide Catalysts. These catalysts, often referred to as "Phillips' catalysts", consist of chromium oxide supported on silica or a silica–alumina carrier. They are prepared by impregnating the carrier with an aqueous solution of CrO_3 or a chromium salt such as the nitrate or chloride. The normal concentration of chromium is 2–3% by weight of the support and the catalyst is activated by heating at $400–800°C$ in dry air. Promoters such as BaO, CuO or Fe_2O_3 may also be added before heating. It has been found that the proportion of hexavalent chromium and the molecular weight of the polymer decrease as the temperature of the catalyst activation step increases.

The polymerization is a solid–liquid suspension process in which polymerization occurs at the surface of the catalyst suspended in a hydrocarbon diluent such as 'iso-octane' or cyclohexane. Unlike the Ziegler polymerization, which may be either a batch or a continuous process, Phillips' polymerizations are all run continuously, at temperatures near or above the melting point of the polymer $(125–160°C)$ and pressures of about 4×10^6 Nm^{-2}. Catalyst concentrations of 0.5% based on diluent are typical and this material is normally dispersed evenly throughout the polymerization zone of the reactor. Pure monomer is introduced into the reactor at its base, polymer solution containing some suspended catalyst is removed at the top, and fresh catalyst is regularly added to maintain the catalyst concentration in the reactor. The resulting polymer solution is freed from unconverted ethylene, filtered to remove suspended catalyst, and cooled to precipitate the polymer which is then separated by centrifuging.

Polypropylene, $-[-CH(CH_3)CH_2-]_n-$

Before the mid-1950's all attempts to produce polyolefins other than those from ethylene and isobutene had given only low-molecular-weight materials of little commercial value. This is because radical transfer (involving hydrogen abstraction) competes with the radical-addition reactions involved. But in 1954 G. Natta, following the work of Ziegler, found that certain bi-metallic, Ziegler-type catalysts could produce high-molecular-weight, crystalline polymers from propylene or from many other α-olefins, such as 4-methylpent-1-ene and but-1-ene. By subtle variations in the catalyst composition and form, Natta found that different types of high-molecular-weight polypropylene could be obtained and that these showed markedly different properties. Further study showed these differing properties were due to the stereoregularity discussed earlier in this Chapter. The isotactic form of high molecular weight was seen to resemble high-density polyethylene in many respects, while the atactic form was amorphous and of low strength. The presence of the methyl group attached to alternate carbon atoms in the chain can affect polymer properties in a number of ways. For example, in comparison to polyethylene, it stiffens the polymer chain and increases the crystalline melting point. But it can also interfere with symmetry and lead to a fall in the overall level of crystallinity, and the most significant effect is production of polymers of differing tacticities. The isotactic configuration is the most regular, as the methyl substituents are all arranged on the same side of the chain back-bone. These molecules cannot now crystallize in the same fashion as polyethylene (planar zigzag) since the steric hindrance of the methyl groups prohibits this, and so isotactic polypropylene crystallizes in a helical form in which there are three monomer units per turn of the helix. Commercial polymers are up to 95% isotactic, the remainder being atactic (random). In these products this small percentage of atactic material may be present as discrete polymer molecules of high molecular weight or as blocks of varying length in an otherwise isotactic chain. The general effects on polymer properties of variations in the degree of tacticity are now known: for example, atactic polymer is an amorphous, rubbery material, soluble in many organic solvents and of little commercial value, and on the other hand isotactic polymer is a stiff, highly crystalline material that dissolves only at temperatures above the polymer melting point. There are several similarities between preparation of polypropylene by Ziegler–Natta catalysts and preparation of high-density polyethylene by Ziegler catalysis. The monomer must be carefully purified and dried, and care must be taken to ensure the purity and dryness of the other components of the reaction mixture. A typical catalyst might be prepared from titanium trichloride and chloro(diethyl)aluminium. The two components are allowed to react in an inert hydrocarbon diluent under a nitrogen atmosphere, to give a slurry of active catalyst containing about 10% of solids. In a batch process this slurry is charged to an autoclave, along with monomer and further hydrocarbon diluent. The reaction is run under modest pressure at about 60°C for up to 8 hours. Since the reaction temperature is well below its crystalline melting point (170°C) the polymer is insoluble in the diluent and is formed as a suspension. When reaction is complete the vessel contains mainly isotactic polymer with a little atactic polymer plus diluent, catalyst and unconverted propylene. Residual monomer is flashed off before the slurry is transferred to a centrifuge where most of the diluent is removed along with most of the atactic polymer, which is soluble in the hydrocarbon diluent. The insoluble isotactic polymer is then treated in the centrifuge with methanol containing a little hydrochloric acid, which decomposes and solubilizes the catalyst. After further methanol washings the polymer is dried, blended with appropriate additives such as antioxidants, extruded and cut into pellets.

Isotactic polypropylene of high molecular weight has found widespread applications, major ones being as an injection-moulding material, in the preparation of packaging films and in the spinning of fibres.

Poly(vinyl chloride), $-[-CHClCH_2-]_n-$

In terms of annual production, poly(vinyl chloride) (PVC) is one of the two major plastic materials, the other being polyethylene. The commercial success of PVC has largely been due to the development of effective stabilizers, as the unmodified homopolymer has limited thermal stability and cannot be readily processed in the melt. An alternative approach to improving processability has been to lower the fabrication temperature by co-polymerization, e.g. with vinyl acetate, but this has had more limited success. The first stabilizer discovered was tritolyl phosphate and the blending of this and related non-volatile liquids with polymer to give stabilized, plasticized PVC is now one of the major activities of the plastics industry.

PVC was first produced in commercial quantities in Germany and the U.S.A. during the years before World War II. Production in Great Britain began during the war, when PVC was employed as, amongst other things, a rubber substitute for cable insulation. In the immediate post-war period developments were concerned largely with plasticizers, although some progress was made in rigid, unplasticized grades. Co-polymers were also developed for such applications as floor coverings and gramophone records. The major development of recent years has been concerned with the polymerization process itself, and the processing characteristics of the polymer have been greatly influenced by control of the size, size distribution, porosity, etc. of the polymer particles produced.

A comparison of the properties of polyethylene and PVC in relation to their structure is useful. In the case of PVC, the pendant chlorine atoms give increased inter-chain interaction, resulting in a harder, stiffer polymer with a much higher glass-transition temperature. Because of the chlorine, PVC is also much more polar than polyethylene and it has a higher dielectric constant. X-ray studies show that only a small proportion (5%) of the polymer is crystalline, and commercial PVC has essentially an atactic structure although there is some evidence of short syndiotactic sequences. There is also evidence, based on reduction experiments, that commercial PVC contains a significant though variable amount of branching.

Since it is known that structurally similar chlorinated hydrocarbons of low molecular weight are thermally stable, the pronounced instability of PVC is somewhat surprising and it has been suggested that it is due to structural defects arising from branching or tacticity changes, etc. Commercially PVC is prepared by a free-radical polymerization of vinyl chloride monomer. These polymerizations are initiated by peroxides, azo-compounds, persulphates, etc. and are normally operated as suspension-emulsion or bulk-polymerization processes.

Poly(tetrafluoroethylene), $-[-CF_2CF_2-]_n-$

Polymerization of tetrafluoroethylene was first described in 1941 and semi-commercial production began shortly afterwards. The monomer, $CF_2{=}CF_2$, polymerizes by free-radical initiation in the presence of a small amount of oxygen. Pressures of about

7×10^6 Nm^{-2} are needed with temperatures of 50–200°C. The product is commonly produced as aqueous dispersions containing up to 60% of solids, but granulated or powdered products are available for extrusion or moulding applications.

Although the polymer is linear and must be classified as thermoplastic, it does not behave in a conventional manner since it does not become fluid even at temperature as high as 450°C. A gel is formed at about 350°C; at this temperature there is sufficient molecular mobility for the particles of polymer to adhere together and most forming processes are based on this property. Forming methods are therefore comparable to those employed in powder metallurgy and ceramic technology.

The main application for the dispersions is the preparation of surface coatings which, like the mouldings, need to be heated above 300°C to produce a smooth surface. Such coatings have excellent chemical resistance and are also widely used for the preparation of non-stick surfaces.

Polystyrene, $-[-CH(C_6H_5)CH_2-]_n-$

Polystyrene homopolymer is the third most important thermoplastic material produced in Great Britain. Like polyethylene and poly(vinyl chloride) its production was stimulated largely as a result of World War II, after which there was a considerable surplus of production capacity for both styrene monomer and polymer. Because of this surplus and the considerable knowledge of production techniques which had been built up, it was found possible to produce polystyrene as a cheap, general-purpose thermoplastic. Since this polymer has an attractive combination of such properties as transparency, rigidity, easy processability and low cost, the consumption of polystyrene has risen spectacularly since the war.

Commercial-grade polymers are normally produced by radical polymerization of styrene monomer:

$$C_6H_5CH{=}CH_2 \xrightarrow{\;R\cdot\;} -[-CH(C_6H_5)CH_2-]_n-$$

As produced, this is an atactic polymer because of the random spatial arrangement of the benzene rings and it is therefore essentially an amorphous, transparent material. The effect of the bulky pendant rings is to stiffen the chain in comparison to polyethylene, and this together with relatively strong intermolecular forces results in a high glass-transition temperature (95°C) and a material that is hard and stiff at room temperature. The attractive combination of properties have led to wide use of polystyrene as an injection moulding and vacuum-forming polymer. In addition, its low thermal conductivity combined with a simple process for conversion into a foam provides a material that is widely used for thermal insulation. The principal limitations of polystyrene are its brittleness and limited mechanical stability at elevated temperatures. Because of the former, many manufacturers have made modifications to their materials in attempts to improve their toughness (see below).

The polymerization of styrene is a chain reaction which may be initiated ionically, by an appropriate free-radical source or thermally. The procedures employed commercially involve bulk, suspension, solution and emulsion techniques. The last of these is used mainly for polystyrene latex production, since the emulsifiers used tend to affect adversely the electrical and optical properties of the polymer.

Rubber Modification. For many potential applications polystyrene is considered to be too brittle. Its toughness can be improved by co-polymerization, for example with butadiene, but such modifications generally result in a marked loss of both stiffness and softening point before adequate toughness can be achieved. The technique most commonly employed is therefore to blend polystyrene with a minor amount of a rubber, such as polybutadiene or a styrene–butadiene rubber (typically containing about 30% of styrene).

The polystyrene and rubber may be blended in a number of ways, the most common procedure being to dissolve the rubber in styrene monomer and then polymerize the whole in the normal fashion. As a result, these blends now contain not just polystyrene and rubber but also significant proportions of graft co-polymer in which short polystyrene side chains are attached to the rubber molecule. This leads to a blend with impact behaviour noticeably improved over that of simple polystyrene–rubber blends. To obtain the best performance from such modifications careful control over resin–rubber compatibility is necessary. Complete compatibility gives little improvement in toughness, while complete incompatibility leads to poor resin–rubber adhesion and poor impact performance. By the grafting operation it is possible to control compatibility and achieve the optimum balance of properties.

Styrene Co-polymers. Commercial production of styrene–acrylonitrile co-polymers (SAN) has been under way for several years. These materials show significant gains in stiffness, softening point and solvent resistance over homopolystyrene. Though they are somewhat tougher than polystyrene itself but are still too brittle for some more stringent applications such as metal replacement, a range of rubber-modified SAN polymers, known as ABS, was introduced. These are either blends of SAN polymers with a butadiene–acrylonitrile rubber or inter-polymers of polybutadiene with acrylonitrile and styrene. The former type is normally prepared by lightly cross-linking the rubber to reduce compatibility, while the latter modification is effected by adding acrylonitrile and styrene to a rubber latex, adding an initiator such as persulphate and polymerizing to give a blend which is a mixture of SAN, polybutadiene and a polybutadiene–SAN graft co-polymer.

It can be seen that by alteration of composition and reaction conditions a wide range of ABS-type polymers can be produced. Although more costly than polystyrene, these ABS polymers have an attractive combination of toughness, surface finish and hardness and have found many applications in domestic appliances, telephone receivers, suitcases, motor-car panels, etc.

Polyoxymethylene (Polyacetal), $-[-CH_2O-]_n-$

Although homopolymers of formaldehyde have been known for over one hundred years and in spite of formaldehyde having been available as a cheap, bulk chemical for over forty years, commercial production of a polyacetal plastic did not begin until 1959, when the du Pont Co. began production in the U.S.A. Three years later a second producer began to sell an acetal co-polymer formed from trioxan and a cyclic ether. Since that time other Companies have begun manufacture of these two basic types of polyacetal.

The primary properties of the polyacetals, namely high modulus, strength, hardness, adequate toughness and relatively high melting point (190°C) place them in the group of synthetic polymers known as engineering thermoplastics, a term which is generally applied to the stronger plastics such as polycarbonates, nylons and polysulphones. It implies the general characteristics mentioned above together with such properties as low creep under load and high mechanical strength which are retained over a wide temperature range.

The existence of two types of acetal polymers has been mentioned; the differences arise from alternative synthetic approaches to the problem of polymer stability. Formaldehyde itself polymerizes readily by a chain reaction in aqueous and non-aqueous systems. The reaction is rapid and exothermic. It is initiated by various ions to give structures that are generally accepted to be polyoxymethylenes:

$$nCH_2O \longrightarrow HOCH_2[OCH_2]_{n-2}OCH_2OH$$

This polymerization is reversible, an effect which is readily understood in terms of the hemiacetal structure of the end groups. Molecular weights greater than 30,000 are needed for a mechanically useful polymer, but the thermal instability is such that depolymerization occurs at the melting point and melt processing is therefore impossible. Selective reaction of the end groups by either etherification or esterification improves stability and these 'end-capping' reactions form an important part of polyacetal technology. Formaldehyde may also be co-polymerized with, for example, styrene or butadiene; this interrupts the sequence of carbon–oxygen links and also improves the stability since it prevents the stepwise unzipping of polyformaldehyde chains. Although co-polymers of formaldehyde have not achieved commercial importance, those of trioxan, which utilize the same stabilizing effect, do have commercial value.

Trioxan polymerizes readily by a cationically initiated ring-opening polymerization to give linear polyoxymethylene. Homopolymer produced from pure trioxan with, for example, boron trifluoride–ether as catalyst, is identical with other polymers prepared from formaldehyde as regards its chemical, thermal and mechanical properties. It too requires stabilization, and co-polymerization offers a convenient method. But as many of the attractive characteristics of polyoxymethylenes are a direct result of the high level of crystallinity (70–75%) it is essential that the major proportion of this crystallinity be retained in the final fabricated form. Therefore care must be taken in the selection of the type and amount of the co-monomer so that it has the minimum effect on crystallinity.

As with the majority of commercial polymerization procedures, care must also be taken to ensure that the starting materials are of the highest purity. The large-scale preparation of polyacetal from formaldehyde involves the addition of pure, anhydrous formaldehyde to an inert diluent at sub-zero temperatures. Diluents such as propane, cyclohexane and aromatic hydrocarbons are described in the patent literature, and addition of dispersion aids such as poly(ethylene oxide) is said to help production of a finely powdered product, which in turn assists subsequent end-group stabilization reactions. Catalysts such as metal carbonyls, phosphines, stibines and amines initiate polymerization and the molecular weight of the product can be controlled by addition of chain-terminating and transfer agents such as water, ethanol or formic acid. To reduce the concentration of hemiacetal groups the resulting polymer powder is treated with acetic anhydride to acetylate the terminal hydroxyl group. The powdered, end-capped polyacetal is then washed, dried and stored.

Processes employing diluents such as methylene chloride and cyclohexane are employed for the co-polymerization of trioxan with cyclic ethers such as 1,3-dioxolane. Monomers, chain-transfer agents and initiators are dissolved in the diluent, the polymer being precipitated as a fine powder. Useful initiators are boron trifluoride–ether and various other Friedel–Crafts catalysts. Stabilization of the polymer is effected by mild basic hydrolysis of the powdered co-polymer; this releases formaldehyde from the chain ends until a carbon–carbon link derived from the co-monomer is reached. These terminal groups stabilize the co-polymer during subsequent thermal processing operations.

At the moment the United States accounts for most of the world production of acetal resins. The major uses are in the automobile industry and, in general, metal replacement parts. Other principal users of acetals are the communications industry, plumbing and the domestic-appliance markets.

Poly(methyl methacrylate),
$$-\left[-CH_2-\underset{\underset{COOCH_3}{|}}{\overset{\overset{CH_3}{|}}{C}}-\right]_n-$$

Although the utility of this polymer was recognized during the 1920's it was not until a commercially viable route to methyl methacrylate monomer had been developed that the polymer became an economic proposition. Poly(methyl methacrylate) sheets became prominent during World War II as a result of their use for aircraft windows, and the number of applications has multiplied in recent years, particularly with the introduction of grades of polymer designed for use in injection moulding and extrusion processes. These applications make use of this polymer's unusual combination of high transparency, high surface hardness and excellent ultraviolet stability. Commercial polymer is prepared by radical-initiated chain polymerization of the monomer:

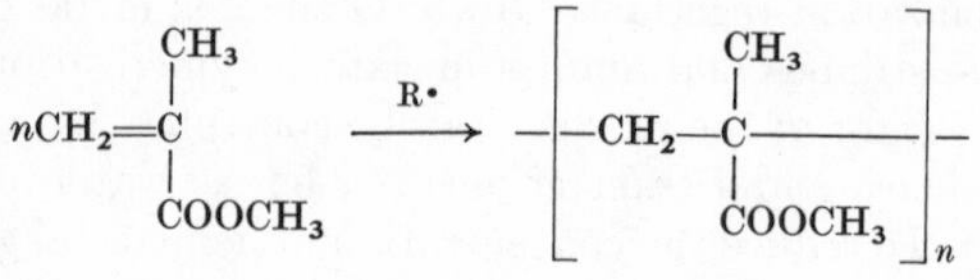

As a result of the random spatial arrangement of the pendant methyl and ester groups attached to alternate carbon atoms of the chain, this is an atactic polymer which is amorphous and therefore transparent. Because of these two substituents, chain flexibility is somewhat restricted and this together with the polarity of the ester group results in polymer that is relatively hard and stiff and has a Tg of 110°C. Poly(methyl methacrylate) is somewhat tougher than general-purpose polystyrene but is still relatively brittle in comparison to other thermoplastics such as polyethylene and PVC; it is the optical properties that are important since the polymer absorbs very little light and has transmission values of up to 90% or more.

Methyl methacrylate polymerizes readily even during storage under ambient conditions. Inhibitors such as hydroquinone are therefore normally present and must be removed by distillation or washing before polymerization is attempted. Free-radical initiators such as peroxides or azo-compounds are normally employed, but oxygen must be excluded so as to minimize side reactions. In commercial practice there are two polymerization techniques, a suspension polymerization giving polymer suitable for injection moulding and extrusion, and a bulk process producing polymer in the form of sheet or rod.

Polyesters

These polymers are conveniently divided into the cross-linked (or thermosetting) and the linear (or thermoplastic) variety. Polyesters became significant in commercial terms during the early 1900's when cross-linked alkyd resins began to be used as surface coatings. Linear polyesters were first studied by Carothers during the 1930's, but they were not exploited until the next decade brought the discovery of poly(ethylene terephthalate) which was developed as a fibre-forming polymer. The cross-linked polyester laminating resins, used in large quantities today for the preparation of glass-reinforced structures, were first made available during 1946 and these materials currently constitute the major proportion of cross-linked polyester production.

Cross-linked Polyesters

Before being cross-linked, i.e. in the uncured state, these materials are generally either viscous liquids or low-softening solids. They are prepared by stepwise polymerization of a glycol, such as propylene glycol, $HOCH(CH_3)CH_2OH$, with a mixture of saturated and unsaturated dicarboxylic acids, for example phthalic and maleic acid, the latter providing reactive sites for the subsequent cross-linking of the polyester. Before use, these unsaturated polymers are mixed with a liquid monomer such as styrene or methyl methacrylate. This has the effect of lowering the viscosity, which improves the application characteristics of the material and provides the means whereby cross-linking can be achieved. Before applying these resins to a reinforcement, typically glass fibre, a free-radical initiator is added. Cross-linking (or cure) then occurs by chain reaction of the styrene and the residual unsaturation in the polyester chains. Glass fibre is the most common form of reinforcement since it generally provides the strongest product for a given cost; it is available in the form of cloth, chopped strand mat (consisting of short fibres, approximately 5 cm. long, bonded together) or in pre-formed shapes made by depositing short fibres on to a mould and spraying with a binder. To provide the adhesion between fibre and resin which is essential for adequate mechanical properties in the final composite, it is usual to provide the fibre with a chemical finish that promotes resin–glass bonding. The most important of these finishes are based on silanes, e.g. trichloro(vinyl)silane, which reacts with the hydroxyl groups on the glass surface to give a vinyl group chemically bonded to the glass, as in the illustration.

During the cross-linking of the polyester resin it is believed that these vinyl groups become incorporated into the resin by co-polymerization to give a chemical bond between resin and glass.

As already mentioned, propylene glycol is most commonly used in the preparation of these polyesters. Ethylene glycol gives rise to products of inferior compatibility with styrene owing to their inherently higher symmetry, while the other commercially available glycols, such as diethylene and dipropylene glycol, give cured products with lower softening points and greater water uptakes. Most general-purpose resins contain either maleic anhydride and/or fumaric acid as the unsaturated component in the polyester chain, while the preferred saturated acid is phthalic acid, which is normally used as the anhydride. The amount of maleic or fumaric acid in the polyester-forming mixture is normally adjusted to provide the required level of cross-linking sites since many of the final properties are dependent upon the cross-linking density in the cured product. Phthalic is preferred over other possible saturated acids because of its low price and because it provides a relatively stiff link in the chain and so contributes to the rigidity of the cured resin. These polyesters are conventionally synthesized by heating the mixture of diacids and/or anhydrides with glycol in the required proportions under a blanket of nitrogen for up to 10 hours at temperatures between 130° and 200°C. A slight excess of glycol is normally employed and high-boiling hydrocarbons can be used to assist in removal of the water of reaction by azeotropic distillation. The course of the reaction can be followed by analysis of the residual carboxyl content in the polymerization mixture, and when the desired degree of reaction has been achieved the mixture is cooled and diluted with the appropriate amount of styrene (or methyl methacrylate). Such mixtures may also contain small amounts of inhibitors such as hydroquinone to prevent premature gelation and improve storage life. Before application of this product to the reinforcement a peroxide is added, normally 0.5–2% of the polyester weight, and curing is then carried out. For smaller articles, where control of the exothermic reaction is not difficult this cure may be at elevated temperatures, e.g. 100°C, under moderate pressure; but, for large structures or where application of heat is difficult, a room-temperature cure is employed. Benzoyl peroxide is commonly used for elevated-temperature cures while at ambient temperatures ethyl methyl ketone peroxide or cyclohexanone peroxide in admixture with an accelerator such as cobalt naphthenate is usual.

The largest application for glass-reinforced polyesters (g.r.p.) is in the building industry where sheeting used for roofing and insulation purposes accounts for about 30% of resin production.

G.r.p.'s are also widely used in boat-building, chemical plants, sports equipment and the transport industry.

Linear Polyesters

The best-known member of this class of material is poly(ethylene terephthalate) (Part 1, p. 170) which is widely used in the form of fibre and oriented film. The chemistry of this polymer is discussed in detail in Chapter 9 and it is sufficient to note here that this polyester has only recently been available in a form suitable for processing as a conventional thermoplastic moulding material; to date, however, its plastic applications are rather limited.

Polycarbonates, $-[-OCOORO-]_n-$

The most important thermoplastic polyester used in non-fibre applications is polycarbonate. The first of this class became available commercially in 1958 and it is essentially the polyester derived from 'carbonic acid', H_2CO_3. Linear polycarbonates can be prepared from the reaction of an aliphatic diol or a bisphenol with a carbonic acid derivative. In practice, only the aromatic polycarbonates derived from bisphenols are of commercial importance and, in particular, that derived from p,p'-isopropylidenediphenol (bisphenol A), which is the only polymer to be produced in significant quantities. The production of this polymer has risen sharply in recent years and, stimulated by price reductions, world capacity for polycarbonate now exceeds 150×10^3 tons p.a.

All commercial grades of polycarbonate are prepared from bisphenol A, which is the product of the acid-catalysed reaction of phenol and acetone (see p. 187). Since carbonic acid does not exist in the free state, polyester-forming derivatives such as phosgene or diphenyl carbonate are employed. There are two step-polymerization processes that are employed in polycarbonate manufacture. These are a high-temperature ester-interchange reaction and a low-temperature solution reaction of phosgene and bisphenol A in the presence of an acid acceptor.

The ester-interchange reaction is carried out by heating bisphenol A and diphenyl carbonate together in the melt at 200°C and under reduced pressure until about 85% of the theoretical quantity of phenol has been removed (see p. 234).

The temperature is then raised progressively to nearly 300°C and the vacuum to 130 Nm^{-2} or less. Under these conditions the molecular weight rises steadily as the remaining phenol is removed

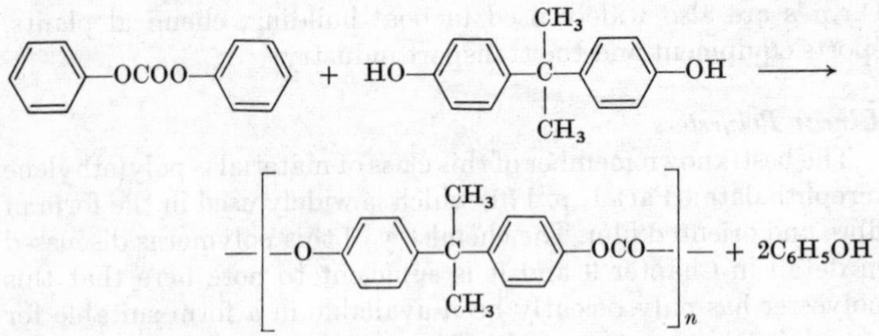

$$\text{C}_6\text{H}_5\text{—OCOO—C}_6\text{H}_5 \;+\; \text{HO—C}_6\text{H}_4\text{—C(CH}_3)_2\text{—C}_6\text{H}_4\text{—OH} \longrightarrow$$

$$\left[\text{—O—C}_6\text{H}_4\text{—C(CH}_3)_2\text{—C}_6\text{H}_4\text{—OCO—} \right]_n \;+\; 2\text{C}_6\text{H}_5\text{OH}$$

and when the required viscosity has been reached the polymer is extruded from the reaction vessel, cooled, chipped and stored.

A variation on this procedure is to use up to a two-fold excess of diphenyl carbonate in the initial mixture. Under these conditions the free phenolic groups are consumed more rapidly under the initial reaction conditions at 200°C and this effectively protects these relatively sensitive groups during the latter stages of the reaction at the higher temperature. Under these conditions, with an excess of diphenyl carbonate, the initial product is the bis-(phenyl carbonate) of bisphenol A, i.e. *p,p′*-isopropylidenedi-(phenyl phenyl carbonate):

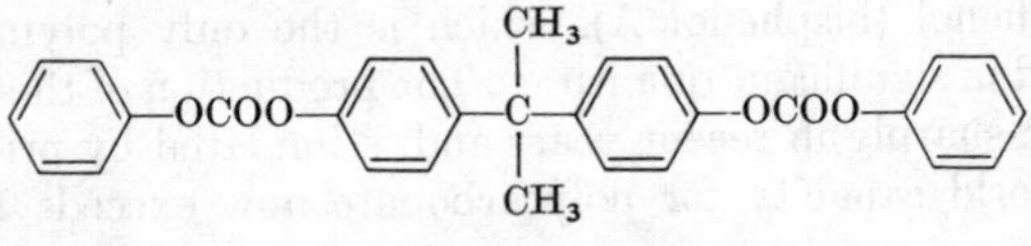

$$\text{C}_6\text{H}_5\text{—OCOO—C}_6\text{H}_4\text{—C(CH}_3)_2\text{—C}_6\text{H}_4\text{—OCOO—C}_6\text{H}_5$$

and polymerization then proceeds by removal of diphenyl carbonate.

The alternative commercial synthesis of polycarbonate is carried out at ambient temperatures and involves the reaction of phosgene with bisphenol A either in an aqueous emulsion or in an organic solvent. In the former case an aqueous solution of the diphenoxide is emulsified with a water-immiscible organic liquid such as methylene chloride, and phosgene is passed into the stirred mixture. Catalysts such as quaternary ammonium salts are required and the polymer formed is retained in solution in the organic phase which is separated and washed before removal of solvent to isolate the product. In the alternative procedure the bisphenol A is dissolved in a solvent such as pyridine which also acts as acid-acceptor for the hydrochloric acid also produced by the reaction. Phosgene is bubbled into the solution at 30°C and within a few minutes pyridinium chloride is precipitated. As the polymerization proceeds the viscosity of the solution increases and when the required molecular weight has been reached the polymer is isolated by addition of a second organic liquid such as methanol, which dissolves the pyridinium salt and precipitates the polymer.

The polycarbonate prepared from bisphenol A by these processes has a particularly attractive combination of properties. It is strong, rigid and transparent, the last of these properties being due to its essentially amorphous character; and because of its high Tg (140°C) it retains a high level of mechanical properties at temperatures above 100°C. In addition, it is one of the toughest thermoplastics and it is this combination of properties which has led to its application in uses as diverse as domestic mouldings, babies' bottles, housings for electrical apparatus and metal replacement in engineering uses.

Phenol–Formaldehyde Resins

These are highly cross-linked materials obtained by step-polymerization between phenol and formaldehyde (Part 1, p. 171). They were the earliest of the truly synthetic polymers and have been employed in a wide range of applications since the beginning of this century. The annual production of these thermosetting materials has risen steadily over the last fifty years, but in recent years the growth rate has not matched those of the major thermoplastics such as polyethylene and polystyrene.

Although these resins have been in use for many years their detailed structure remains obscure, largely owing to the difficulties associated with the insoluble, infusible nature of the cured products. However, it is known that the various structures that can be formed depend very much on the ratio of phenol to formaldehyde. For practical utility it is essential that the initial product of the phenol–formaldehyde reaction be a fusible material of low molecular weight. This low-molecular-weight material is subsequently transformed into the final, shaped, cross-linked product by heat under pressure. The initial, fusible product of the reaction can be of two types, called novolak and resol resins, respectively.

Novolaks

Novolaks are prepared under acid catalysis by reaction of formaldehyde with a slight excess of phenol, typically in a molar ratio 0.8:1.0. Under these conditions there is a relatively slow formation of *o*- and *p*-(hydroxymethyl)phenol.* These then react rapidly with further phenol to form methylenediphenols, the ratio of the three possible isomers being determined by the pH of the reaction (see annexed scheme).

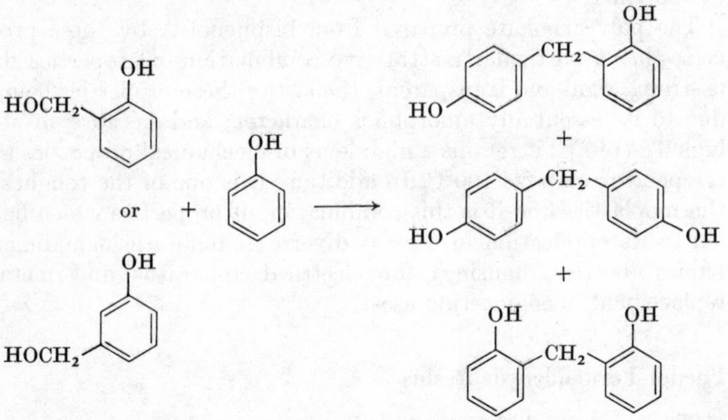

Subsequent reaction of these bisphenols with more formaldehyde is relatively slow and results in the production of polynuclear phenols but, because there is an overall excess of phenol in the reaction, the molecular weight of the polyphenols is limited and they contain an average five aromatic rings, e.g. (**1**).

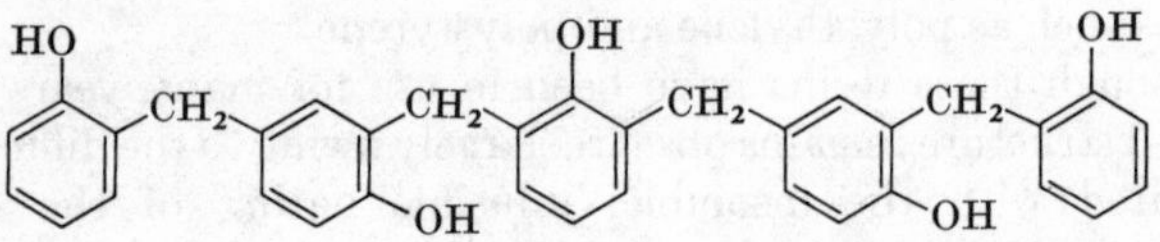

(**1**)

When the novolak preparation is complete the products contain no reactive hydroxymethyl groups and so no further reaction can occur. Therefore, in order to cross-link these resins it is necessary to add a source of formaldehyde before the heating/fabrication process. Typical formaldehyde sources that can be employed are paraformaldehyde and hexamethylenetetramine (**2**) (Part 1, p. 73). Heating such mixtures produces further methylene bridges and formation of an infusible, insoluble, cross-linked product.

---

* CH$_2$OH groups are commonly called 'methylol' rather than 'hydroxymethyl' in the plastics industry. The latter is now the systematically correct name.

(2)

Resoles

Resole, in contrast to novolak, resins are obtained under basic conditions from phenol–formaldehyde mixtures that contain an excess of formaldehyde. In this case, as a result of the different catalyst, the formation of (hydroxymethyl)phenols is rapid and the subsequent condensation to di- and poly-phenolic structures is slow. This difference in the relative rates of these reactions results in the formation of bis- and tris-(hydroxymethyl)phenols as well as the monoalcohol. In addition, the polynuclear products obtained under these conditions are generally of somewhat lower molecular weight than those obtained in the preparation of novolak resins, and a typical liquid resole may only contain an average of two aromatic rings. Since these resins now contain reactive hydroxymethyl groups, further heating results in cross-linking due to reaction of these groups with unsubstituted activated positions on the aromatic rings; here addition of a formaldehyde source is unnecessary. Both novolak and resole resins are soluble and they soften and flow at relatively low temperatures. They are converted into hard, insoluble cross-linked products at elevated temperatures, in the presence of a formaldehyde source if this is appropriate. Investigations of the mechanism of hardening indicate that at temperatures below ca. 150°C cross-linking occurs by condensation of hydroxymethyl groups on adjacent benzene rings and by condensation of these groups with the unsubstituted activated positions of adjacent rings (see scheme on p. 238).

It is thought that other cross-linking reactions can occur above 150°C; for example, formation of quinone methides (Part 2, p. 247) by condensation between phenolic hydroxyl and ether links, which results in the typical dark colour of phenolic mouldings at these higher temperatures.

The nature of the catalyst employed during the cross-linking reaction is important; for example, resole can be cured at room temperature by the use of strong acids, but as the pH of the mixture rises the rate of cure falls, passing through a minimum at pH 7 and rising again under alkaline conditions.

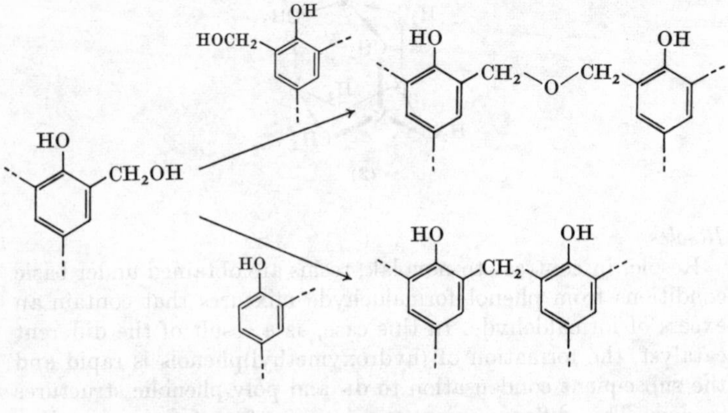

Phenol–formaldehyde resins are used in a variety of moulding applications, e.g. in the electrical industry where they are used for insulation and as casings for electrical apparatus. In the automobile industry they are used for the manufacture of distributor caps and fuse-boxes, and elsewhere when electrical insulation combined with good heat-resistance is required. They are used also in the preparation of laminated materials where the reinforcing materials are, for example, paper or fabric. In recent years there has been an increasing interest in foam prepared from these resins because of the superior flame-retardant behaviour over polystyrene-based foams.

Urea–Formaldehyde Resins

Like the phenolic resins, urea–formaldehyde resins are crosslinked, insoluble and infusible in the final cured form and, as for the phenolics, a product of low molecular weight is produced initially and is subsequently cross-linked during application or fabrication. The commercial production of urea–formaldehyde resins began during the 1920's and the scale of production has since risen steadily, products being developed for use as adhesives, textile auxiliaries, moulding powders, surface coatings and insulating foams. The mechanisms of the resinification reaction between urea and formaldehyde is still not fully understood and, as with the phenol–formaldehyde resins, this is the direct result of the intractability of the final product.

The initial reaction of urea with formaldehyde is carried out under neutral or very slightly alkaline conditions, i.e. at pH 7–8.

This leads to the formation of mono- and di-(hydroxymethyl)ureas, i.e.

$$2H_2NCONH_2 + 3CH_2O \longrightarrow$$
$$H_2NCONHCH_2OH + HOCH_2NHCONHCH_2OH$$

If the ratio of formaldehyde to urea is high enough to produce a significant proportion of di(hydroxymethyl)urea, then the product can be heated under mildly acidic conditions to give a cured product. The subsequent reactions are unknown but several possibilities have been proposed: for example, condensation of adjacent hydroxymethyl groups to form ether bridges, hydroxymethyl–imino condensation, methyleneurea formation and production of cyclic structures.

The preparation of a urea–formaldehyde moulding powder may be taken as an example of resin production. These moulding powders are widely used because of their low cost, good electrical properties, stiffness and wide colour range. The first stage of the preparation is to adjust the pH of the urea–formaldehyde mixture to 7–8 by addition of base, usually sodium hydroxide. This mixture is stirred at 40°C for up to one hour, the pH not being allowed to fall throughout this time. The resin may then be concentrated by removal of some of the water or dried to give a powder. In common with the phenol–formaldehyde resins it is the usual practice to incorporate a filler, such as wood pulp or wood flour, before fabrication. A variety of pigments may also be incorporated at this stage if a coloured product is desired. Finally a catalyst is needed if adequate rates of cure are to be achieved: these catalysts are generally acids or compounds capable of generating acid at elevated temperatures; examples of such 'latent' catalysts are trimethyl phosphate, ammonium sulphamate and ethylene sulphite. Since the storage stability of these resins is limited, a stabilizer such as hexamethylenetetramine may also be incorporated if desired.

Polyurethanes

Polyurethanes are produced by the reaction of di- or tri-isocyanates with glycols or triols, for example, 'toluene-2,4-di-isocyanate' (4-methyl-*m*-phenylene di-isocyanate) with tetramethylene glycol, as shown on p. 240.

Other isocyanates commonly employed commercially are p,p'-methylenedi(phenyl isocyanate), $CH_2(C_6H_4NCO)_2$, and hexamethylene di-isocyanate, $OCN(CH_2)_6NCO$.

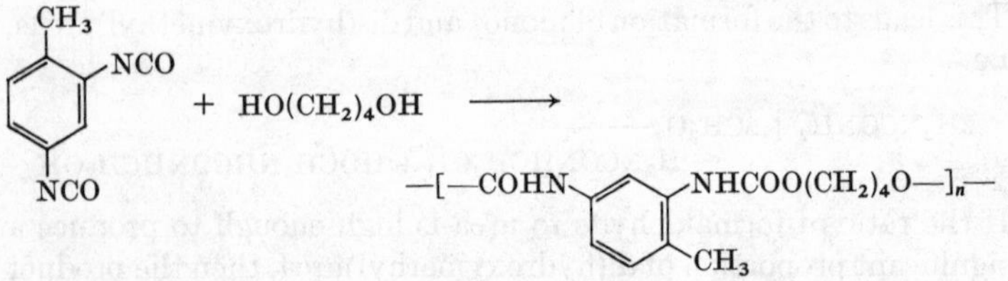

$$—[—COHNNHCOO(CH_2)_4O—]_n—$$

The combination of di-isocyanate and diol to form a polyurethane normally leads to linear polymers, but cross-linked
polyurethanes can be obtained if a polyisocyanate, e.g. the
tri-isocyanate (3) or a triol, e.g. glycerol $HOCH_2CH(OH)CH_2OH$,

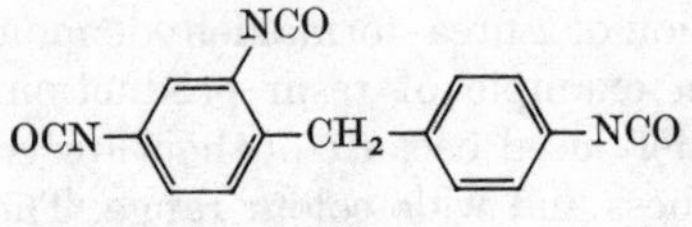

(3)

is used to replace all or part of the difunctional reactants in such
polymer-forming reactions.

The linear polymers find application as elastomeric yarns and
in these cases the diol may be a polyether having terminal
hydroxyl groups such as poly(tetramethylene glycol),

$$HO-[—(CH_2)_4—O—]_n—H$$

The major application for cross-linked polyurethanes is in the
manufacture of foams. The polyols used for this application are
commonly hydroxyl-terminated aliphatic polyesters, manufactured from an excess of glycol with an aliphatic dicarboxylic acid.
Poly(propylene adipate) (4) is an example of such a polyester.

$$HOCH(CH_3)CH_2O—[—CO(CH_2)_4COOCH_2CH(CH_3)O—]_n—H$$
(4)

The terminal hydroxyl group can be treated with a polyisocyanate
to give cross-linked products, and if a little water is added during
this reaction some urethane groups are converted into reactive
amine groups with simultaneous evolution of carbon dioxide.
Under the proper condition this gas can be used to expand the
polymerizing mixture to give a foam. The properties of these
polymers can be varied almost without limit by varying the ratio
of polyester or polyether to the polyisocyanate or by changing the
characters of the polyhydric compound. Similarly the addition of

varying amounts of water followed by further isocyanate reaction with the amine groups so formed produces polyurethane–polyureas with modified properties.

Elastomers

Natural Rubber

For very many years natural rubber was the only known material of high elasticity and it has been the subject of intensive study in both the vulcanized and the unvulcanized state. A brief account of this work on natural rubber will assist in the understanding of the differences between natural and the various synthetic materials (see Part 4, p. 221, for biosynthesis).

As collected from the tree, rubber latex is a dispersion of rubber particles in an essentially aqueous medium; it shows the general properties of any emulsion and, for example, can be coagulated. An average latex yields up to 40% by weight of dried extract, of which approximately 90% is rubber hydrocarbon. The composition of this rubber hydrocarbon corresponds to the molecular formula $(C_5H_8)_n$, a fact noted by Faraday as early as 1826. It was soon discovered that this hydrocarbon was unsaturated, since many substances would combine with it to give products whose subsequent analysis indicated one double bond per C_5H_8 group. The destructive distillation of rubber was shown to yield some isoprene, C_5H_8, and, as the polymerization of isoprene gave a substance with rubber-like properties, it was a reasonable conclusion that natural rubber was built up by addition of isoprene units. Ozonolysis showed that these units were arranged in a regular sequence, although it was not until much later that the influence of stereoisomerism was understood. The existence of this isomerism explains the difference in properties between rubber, poly-(*cis*-1,4-isoprene), on the one hand, and gutta-percha, poly-(*trans*-1,4-isoprene), on the other.

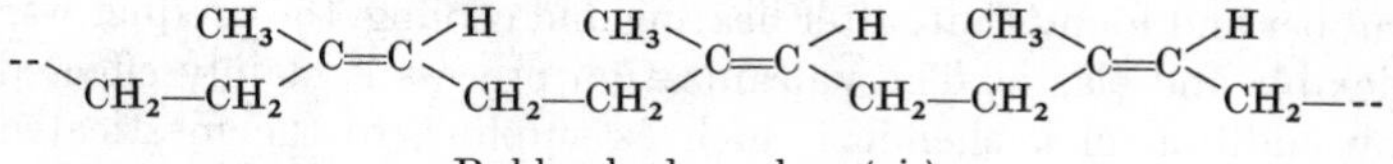

Rubber hydrocarbon (*cis*)

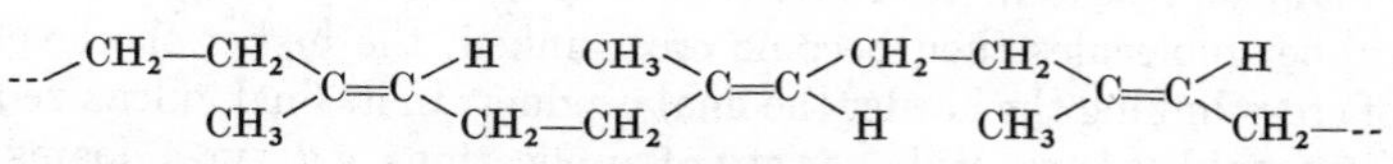

Gutta-percha (*trans*)

In contrast to natural rubber, gutta-percha is virtually inelastic and therefore of much less general utility. Although vulcanizable, like rubber, its major applications are dependent upon its thermoplastic character and its excellent chemical resistance. It found early application in the insulation of submarine cables and more recently as the covering material for golf-balls.

Because of the unsaturated nature of the rubber molecule its chemistry is theoretically that of an ethylenic material. However, an important additional aspect of this chemistry is that there are other reactive sites in the macromolecular chain, namely the carbon atoms α to the double bonds. The hydrogen atoms at these sites are readily removed, giving rise to highly reactive free radicals.

Although the rubber molecule undergoes the characteristic addition reactions of unsaturated molecules these are not always simple reactions, being frequently complicated by side reactions involving substitution and/or cyclization. For example, the reaction of the rubber molecule with oxygen is most important and is usually characterized by scission of the polymer chain which leads to plasticization and the phenomenon known as 'ageing'. These effects can be seen even when very small amounts of oxygen are absorbed; for instance, 1% by weight of adsorbed oxygen can result in complete degradation and production of a material that is of little utility.

Much work has been done on the chemistry of the vulcanization of the rubber hydrocarbon molecule and sufficient evidence has been collected to show that the process involves the formation of intermolecular bonds (cross-linking).

The process of vulcanization is the transformation of rubber from an essentially plastic condition into an essentially elastic state and the chemical and physical changes produced by this process have made possible the enormous expansion in the applications of rubber. Vulcanization was discovered in 1840 by Charles Goodyear who incorporated sulphur into masticated rubber and found that, after heating and cooling, the product was flexible and elastic. The vulcanization process is mainly effected by addition of a chemical, such as sulphur, to the masticated rubber and, when this mixture is homogeneous, it is placed into a mould and heated, to above 110°C in the case of sulphur; the rubber molecules then become cross-linked, the higher the level of cross-linking the harder the final product. In its final vulcanized form, rubber has a wide variety of applications, e.g. tyres, foams, footwear, cable insulation, etc.

Vulcanized rubber, therefore, consists of a three-dimensional network in which the total number of cross-links must be limited to retain elasticity and flexibility. If the number of cross-links is progressively increased more rigid structures are obtained until ultimately hard, brittle products, e.g. ebonite, are the result.

The raw rubber, before vulcanization, is elastic, soluble and sensitive to quite modest temperatures. It is therefore common practice first to reduce the molecular weight to improve processability and subsequently, by light cross-linking, to regenerate the rubbery state and raise the level of elastic properties. The reduction in molecular weight is generally achieved mechanically, for example, by shearing the raw-rubber in a twin-roll mill; this degraded material may then be more readily mixed with the various additives, antioxidants, fillers, pigments, etc., and the resulting mixture is then cross-linked or vulcanized. Sulphur is the most widely used vulcanizing agent; it is commonly employed in conjunction with an accelerator at 150–200°C.

Unfilled, vulcanized natural rubber has high tensile strength in the stretched condition. This strength is the direct result of crystallization of the polymer in the stretched condition, but when abrasion- and tear-resistance are required properties then finely divided fillers such as carbon black are added. Similarly, oxidation-resistance is improved by the addition of antioxidants, but it has been found that for many applications an improvement in cost-effectiveness can be obtained by using one of the many synthetic alternatives that are now available.

Synthetic Rubbers

Together these form one of the most important groups of commercial, organic polymers and, although the bulk of synthetic rubbers are based on dienes or olefins, there are many other speciality materials, including polyurethanes, fluorinated rubbers, polyacrylates, polysulphides and silicones. Although natural rubber has a higher world consumption than any single synthetic rubber, the growth rate of the natural material is much lower than, for example, that of styrene–butadiene rubber (SBR).

Styrene–Butadiene Rubber (SBR). This is currently the major general-purpose synthetic rubber. It came into wide use during World War II in response to a shortage of natural rubber. The co-polymer normally contains about 75% of butadiene and 25% of styrene, and of the butadiene units approximately 80% are *trans*-1,4- with the remainder 1,2-units. The co-polymer is conventionally prepared in an aqueous emulsion system with, for example, persulphate as initiator and a mercaptan (thiol) as chain-transfer agent. The product is obtained in the form of a latex, similar to natural rubber, and subsequent handling and fabrication operations are similar in that some mastication is required before vulcanizing under conditions similar to those employed for natural rubber.

Synthetic Poly-(cis-1,4-*isoprene*). This has been available since the mid-1950's and although it is similar in most respects to the natural polymer it differs somewhat in that the *cis*-content is generally 90–95%. The commercial success of this material has been somewhat limited by its inferior performance in comparison with the natural product.

Synthetic Poly-(trans-1,4-*isoprene*). This isomer is also produced commercially and, in contrast to the *cis*-polymer, is frequently used in preference to the natural equivalent (gutta-percha) because of its lower cost and more consistent behaviour.

Ethylene–Propylene rubbers. These are the most recent development in synthetic rubbers and they are the direct result of developments in the organometallic Ziegler-type catalysts. A finely divided catalyst, prepared typically by reaction of vanadium oxychloride with a trialkylaluminium, is most commonly used. In an inert hydrocarbon diluent, these catalysts give co-polymers with typical ethylene unit contents of 70–80%. Since the reactivity of ethylene in the co-polymerization is so much greater than that of propylene the feedstocks normally contain a relatively large excess of propylene.

In order to provide suitably reactive sites for subsequent vulcanization it is common practice to incorporate a small proportion of a third monomer into the feedstock. This is typically a non-conjugated diene, for example, dicyclopentadiene, in which the relative reactivity of the two unsaturated sites is such that only one takes part in the polymerization, leaving the second available as a site for subsequent cross-linking.

Polyisobutene. Polyisobutene was first prepared by the low-temperature (−80°C) solution-polymerization of isobutene initiated by boron trifluoride. The high-molecular-weight material had some rubber-like properties in addition to outstanding chemical and oxidative stability. But it was not until methods of cross-linking this material had been found that any commercial interest developed. When small amounts of isoprene or other conjugated diene were incorporated into the polymer chain vulcanization became possible and the current grades of butyl rubber were developed. The most important polymers incorporate 1–5% of diene; these co-polymers are generally prepared by feeding in a mixture of monomers of the required composition in a diluent such as methyl chloride and initiating the cationic polymerization with, for example, aluminium trichloride. The molecular weight of the polymer varies inversely with the amount of co-monomer and, to obtain a uniform product, conversion is normally restricted to 60%. The product, in the form of a slurry, is then vigorously stirred with water to deactivate the catalyst and remove water-soluble contaminants.

Bibliography

J. K. Stille, *Introduction to Polymer Chemistry*, Wiley, New York, 1962.

J. A. Brydson, *Plastics Materials*, Butterworths, London, 1969.

F. Billmeyer, *Textbook of Polymer Science*, Interscience, New York, 1962.

W. R. Sorenson and T. W. Campbell, *Preparative Methods of Polymer Chemistry*, Interscience, New York, 1968.

P. W. Morgan, 'Condensation Polymers: By Interfacial and Solution Methods', *Polymer Rev.*, 1956, **10**.

R. A. V. Raff and J. B. Allison, *High Polymers*, Vol. XI, *Polyethylene*, Interscience, New York, 1956.

R. H. Boundy and R. F. Boyer, *Styrene, Its Polymers, Copolymers and Derivatives*, Hafner, New York, 1952.

M. Kaufman, *The History of PVC*, Maclaren, London, 1969.

T. O. J. Kusser, *Polypropylene*, Reinhold, New York, 1960.

H. V. Boeing, *Unsaturated Polyesters*, Elsevier, Amsterdam, 1967.

C. P. Vale and W. G. K. Taylor, *Aminoplastics*, Iliffe, London, 1964.

A. Whitehouse, E. Pritchett and G. Barnett, *Phenolic Resins*, Iliffe, London, 1967.

CHAPTER 9

Fibres

(J. R. HOLKER, Shirley Institute, Manchester)

Introduction

Fibres played an important part in the development of polymer chemistry (see Chapter 8), for the early work of Staudinger (1920) on the structure of cellulose and rubber led to the concept of these substances as long-chain molecular entities (Part 4, p. 83) rather than 'colloidal' aggregates of small molecules. This work, later supported by the X-ray diffraction studies of Meyer and Mark (1927) on cellulose, signalled the beginning of a real understanding of macromolecular structure that was soon followed by the first rational approaches to the synthesis of fibre-forming polymers by Carothers and his colleagues. Man-made fibres had been known, of course, for much longer. Regenerated cellulose, as viscose rayon and cuprammonium rayon (Part 4, p. 93), was produced around the turn of the century, whilst the first claims for a wholly synthetic polymer that might yield a fibre were made in 1913 in respect of poly(vinyl chloride), although the patent was not then exploited. The first truly synthetic fibre to appear commercially was nylon 6,6 (Part 1, p. 172) in 1938, followed in rapid succession by nylon 6, Pe-Ce (PVC, after-chlorinated), 'Vinyon', a vinyl chloride–vinyl acetate co-polymer (1939), 'Saran' (vinyl chloride–vinylidene chloride, 1940), fibres based on polyacrylonitrile (1945), and lastly 'Terylene' (polyethylene terephthalate) (Part 1, p. 170) in 1949. Since that period no major new fibre has emerged, but the production and properties of existing fibres have been improved. Many specialized fibres of relatively less importance have been and are being developed, indicating scope for research in this field.

Despite the phenomenal growth of the man-made fibre industry, cotton still accounts for over 50% of the total world production (Table 9.1); its use is exceeded by that of regenerated and synthetic fibres only in some highly developed countries.

Table 9.1. Production of fibres (1972 and 1970) (lb. $\times 10^{-6}$)

| | 1972 | | 1970 | |
	World	U.K.	World	U.K.
Cotton	28,286	337	24,700	436
Wool	3272	423	3515	430
Rayon	8159	453	7570	465
Polyamide	5336	313	4130	315
Polyester	5523	269	3580	175
Acrylic	2794	242	2210	250
Other	—	—	950	—

The three main types of purely synthetic fibre account for over 90% of the market for synthetics, with little possibility of any change in the foreseeable future except that polyester will displace nylon as the most important fibre. Any new fibres developed are likely to be for specific end-uses rather than for the mass market.

For the organic chemist, interest in the man-made fibres centres on the development of new and improved syntheses of fibre-forming polymers and intermediates. Little attention is given to chemical treatment of the fibres themselves, many of which are indeed fairly inert. In contrast, the (bio)chemistry of natural fibre formation is relatively neglected, compared with the extensive effort that has gone into structural studies and into methods of improving their properties by chemical modification. This difference in emphasis will be clear from the necessarily condensed account of fibre chemistry which follows and which is prefaced by some general comments on the requirements for fibre-forming polymers. The fibres have been divided conventionally into three main groups—natural, regenerated and synthetic.

A basic requirement for fibre formation is that the extended polymer molecules must be at least about 1000 Å (100 nm) long, i.e. the minimum molecular weight is around 10×10^3. This value can, of course, be exceeded; the molecular weight of undegraded cotton cellulose, for example, may be as high as 5×10^5. With synthetic fibres, the molecular weight is limited insofar as the polymer must have a melt or solution viscosity suited to the spinning process. Most melt-spun fibres have molecular weights in the region $10-20 \times 10^3$; solution-spun fibres may have higher values. Textile fibres exhibit some degree of crystallinity and/or molecular orientation along the fibre axis. These inherent properties of natural fibres are imparted to regenerated and synthetic fibres in the spinning, drawing and heat-setting operations. Control of these parameters effectively determines the physical, and to some extent the chemical, properties of the final product. Realization of this structural order requires, in turn, some degree of molecular order or symmetry to facilitate the close packing necessary for the creation of strong interchain forces, whether these arise through hydrogen-bonding, dipolar association or van der Waals' attraction. A *too* perfectly crystalline fibre, however, although it may have high strength, would be too rigid and inextensible in processing and in use; since chemical reactivity resides almost exclusively in the disordered regions of fibres, it would also be extremely difficult to dye. The strongest synthetic fibres are probably no more than 50–60% crystalline; acrylic fibres are unusual in that they show little evidence of true crystallinity but have a high degree of two-dimensional order. Interchain forces in the disordered regions of fibres may be augmented by covalent or ionic cross-links, as in the case of wool and resin-treated cotton.

The effect of structure on the thermal behaviour of fibres is important. Whereas the natural fibres, because of their highly polar character, tend to be degraded without melting, most of the synthetic fibres are thermoplastic. Some are sufficiently stable above their melting point to enable them to be spun directly from the molten polymer; nylons 6 and 6,6, poly(ethylene terephthalate) and polypropylene are in this class. Fibres that are not thermally stable, notably the acrylics, cellulose acetates, poly(vinyl alcohol) and poly(vinyl chloride), are obtained rather more laboriously by dissolution of the polymer in a solvent and extrusion of this solution into hot air, to evaporate the solvent, or into a non-solvent coagulating bath. Where possible, the melt-spinning route is clearly preferable. Fibres of low melting point are at an obvious disadvantage for many uses; fabrics and furnishings containing them are readily damaged, e.g. by overhot irons, cinders and cigarette stubs. Dimensional stability at elevated temperatures (100°C or even 150°C) is also desirable, since this effectively governs the severity of conditions under which fabrics may be processed and laundered or dry-cleaned. Dyeability is a most desirable fibre property. Whilst natural fibres possess good accessibility to aqueous dyestuff solutions and ample dye-acceptor sites, dyeing of the more hydrophobic synthetic fibres has necessitated development of new dyes and techniques and modification of the polymers by incorporation of co-monomers deliberately to break up structural regularity, thus increasing accessibility, and to provide dye-acceptor sites (see Chapter 10). Almost all fibres are colourless, so that they can be coloured to order; exceptions are some heat-resistant speciality fibres based on condensed ring systems. Fibres may be delustred by addition of inorganic pigment, usually a small amount of titanium dioxide. Photo-initiated oxidation (photo-tendering) of fibres causes discolouration and a lowering of the molecular weight with a consequent deterioration in mechanical properties. Both natural and synthetic fibres are affected. Amongst the latter, the acrylics are most resistant, the nylons and poly-propylene less so. Photo-oxidation is a radical reaction, which may

Table 9.2. Amino-acid composition[a] of wool, silk fibroin and collagen (cf. text on p. 250)

Amino-acid	Wool	Fibroin	Collagen
Glycine	5.6	33.3	24.5
Alanine	3.9	26.1	9.8
Valine	5.7	2.6	1.7
Leucine	8.4	0.7	3.2
Isoleucine	3.6	0.9	1.3
Phenylalanine	4.0	1.2	2.2
Proline	5.9	0.5	11.3
Serine	9.3	13.2	3.2
Threonine	6.3	1.3	1.7
Hydroxyproline	—	—	9.1
Tyrosine	5.4	11.5	0.5
Aspartic acid[b]	6.9	2.4	5.5
Glutamic acid[b]	14.9	2.1	13.7
Arginine	9.2	1.0	8.9
Lysine	3.3	0.9	3.2
δ-Hydroxylysine	—	—	1.0
Histidine	1.0	0.4	0.6
Tryptophan	0.7	0.5	?
Cystine	11.2	Traces	
Cysteine	0.3	Traces	
Methionine	0.4	0.1	—

[a] Wool in g. amino-acid/100 g; fibroin and collagen in g. amino-acid *residues*/100 g.

[b] Partly as the amides, asparagine and glutamine.

be accelerated by traces of some metals, such as copper and iron, and by certain dyestuffs, which are converted into highly active species under the influence of light.

Natural Fibres, I. Animal Fibres

The three main groups in this Section are animal hairs, silk and leather. They are all proteins (Part 4, pp. 305–341), belonging respectively to the keratin, fibroin and collagen groups, and they differ markedly in chemical and physical structure. Wool is, of course, the most important hair fibre; its structure shows quite significant chemical differences between breeds of sheep and even between fibres from a single fleece. Variations are also found in silks from different sources, although commercial interest is largely centred on one type, and in collagens performing different functions (i.e. whether from skin, muscle or bone, etc.). Wool and silk are usable once the major non-fibrous impurities (wool-wax and sericin, respectively) have been removed; collagens, however, need considerable after-treatment to convert them into leather, principally tanning to give them adequate stability. Some typical amino-acid compositions are given in Table 9.2 on p. 249 (see Part 4, pp. 308–309 for amino-acid structures).

Wool

The morphological and chemical structure of wool is highly complex and still incompletely understood. Wool fibres are not homogeneous, but they consist mainly (ca. 90%) of spindle-shaped cortical cells (sometimes enclosing medullary cells) sheathed in a cuticle, whose scale-like structure is responsible for the characteristic felting properties of the fibres. Differential staining reveals a bi-component structure, the fibres being composed of two hemi-

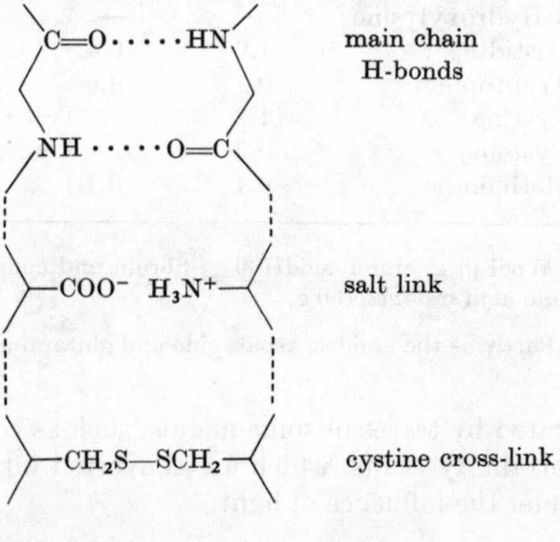

C=O·····HN   main chain H-bonds

NH·····O=C

—COO⁻ H₃N⁺—   salt link

—CH₂S—SCH₂—   cystine cross-link

cylinders, the so-called ortho- and para-cortex, twisted together. Cortical cells in the two components differ slightly in structure.

Considering the wool fibre as a whole, its unique mechanical properties are the result of a variety of inter- and intra-molecular interactions, ranging from van der Waals' forces, through the hydrogen-bonding of main-chain amide groups and polar side chains, and the ionic bonds ('salt links') between acidic and basic side chains, to the covalent cystine cross-links. The wet strength of a wool fibre is lower at pH 1 than in distilled water, presumably through rupture of the salt links by back-titration (neutralization)

Wool can be separated chemically into three major components which are themselves heterogeneous, viz. a fibrous oriented protein (α-keratose), a cuticular protein (β-keratose) and a non-fibrous matrix (γ-keratose) in which the α-keratose is dispersed. Because it is highly cross-linked, wool is insoluble in all non-degrading solvents, and an essential step in separating the components is the fairly specific oxidation of cystine cross-links with, e.g. peracetic acid, to cysteic acid side chains as formulated.

$$
\begin{array}{ccc}
| & | & | \\
CO & NH & CO \\
| & | & | \\
CHCH_2SSCH_2CH & \longrightarrow & CHCH_2SO_3H \\
| & | & | \\
NH & CO & NH \\
| & | & |
\end{array}
$$

The oxidized α- and γ-keratose are reprecipitated by dilute acid or strong electrolytes. Alternatively, the cystine links may be reduced to cysteine to facilitate solubilization. α-Keratose is characterized by a high content of amino-acids with small side chains, which favours molecular close packing (cf. silk fibroin) and is largely responsible for the X-ray diffraction pattern of wool. γ-Keratose, on the other hand, is an amorphous material containing most of the original cystine, upon which much of the technological chemistry of wool depends. When a wool fibre is stretched by over 20%, the X-ray pattern undergoes a radical change, complete at about 50% stretch; this change is associated with a progressive shift in the conformation of the crystalline regions, the α-keratin $\rightarrow$ β-keratin transformation, from a helical to a fully extended zigzag structure of the type found in silk fibroin.

The highly heterogeneous nature of wool makes a complete structural determination by conventional techniques difficult; the principal accessible N-terminal amino-acids, found by dinitrophenylation, (cf. Part 4, p. 320) are serine, threonine, glycine, alanine and valine.

The rich variety of reactive side chains makes wool keratin amenable to modification of its properties by chemical methods; they are also the seat of many inherent weaknesses, notably sensitivity to thermal discolouration and photo-initiated oxidation and to attack by even mild alkalis. Much of this chemistry centres on the cystine cross-links. Treatment of wool with aqueous alkali, or even with boiling water, converts bound cystine into lanthionine, probably through the sequence:

$$>CHCH_2SSCH_2CH< \xrightarrow[\text{(CN}^-)]{\text{HO}^-} \quad >C{=}CH_2 + [S] + HSCH_2CH< \longrightarrow$$

$$>CHCH_2SCH_2CH< + Na_2S(NaCNS)$$

Lysinoalanine cross-links may also be formed:

$$>C{=}CH_2 + H_2N(CH_2)_4R \longrightarrow >CHCH_2NH(CH_2)_4R$$

Reducing agents, for example an excess of a thiol, convert cystine into cysteine; the reaction is readily reversed by mild oxidizing agents or air. More stable cross-links are formed by treatment of reduced wool with dihalides:

$$RCH_2SSCH_2R' \longrightarrow$$
$$RCH_2SH + HSCH_2R' \xrightarrow[\text{alkali}]{Br(CH_2)_nBr} RCH_2S(CH_2)_nSCH_2R'$$

This cystine–cysteine interconversion is the basis of current processes to impart a 'permanent set' to wool, i.e. to fix a fabric or garment in a configuration that resists creasing or retains deliberately inserted pleats in wear and during laundering or cleaning. Briefly, the steps involved are: (i) fission of the restraining cystine cross-links, facilitating (ii) simultaneous molecular rearrangement under the influence of hydrogen-bond breaking agents (moist steam, aqueous urea) to more stable structures ($\alpha \to \beta$-keratin), and (iii) reformation of cystine and other, more stable cross-links.

Small amounts of reducing agent suffice for reaction (i), since once initiated it is propagated by an 'unzipping' mechanism (thiol–disulphide interchange) as illustrated diagrammatically.

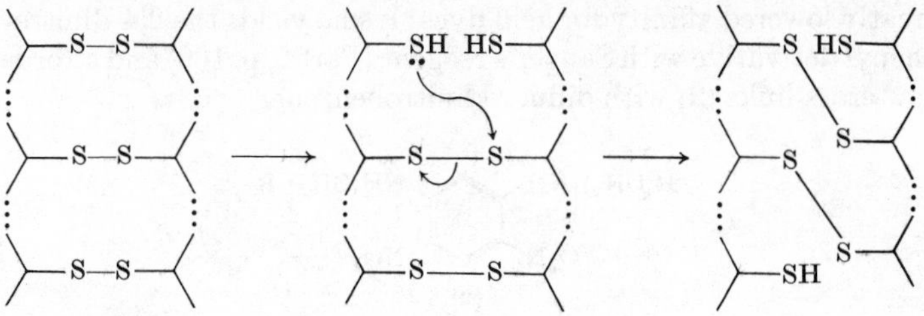

Typical reducing agents include ammonium thioglycollate (which may cause problems of residual odour and of staining through reaction with traces of iron), ethanolamine sulphite, and sodium hydrogen sulphite at pH 4–6. The last reagent yields initially a thiolsulphonate (Bunte salt):

$$RCH_2SSCH_2R' \longrightarrow RCH_2SH + NaOSO_2SCH_2R'$$

Reductive fission and subsequent re-formation of cystine cross-links by oxidation with, e.g., bromates is also the basis of chemical methods for permanent waving of hair (see Part 4, p. 316).

Another industrially important reaction of wool is that with aqueous chlorine, hypochlorites or organic compounds that release chlorine at a controlled rate under the correct conditions of pH and temperature, e.g. dichloroisocyanuric acid (**1**). This reaction is

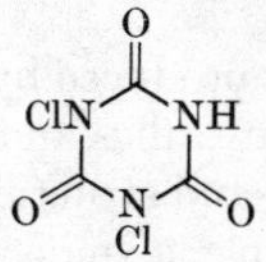

(**1**)

imperfectly understood; it involves extensive attack on cystine and tyrosine. By limiting reaction to the fibre surface, the scales are weakened or destroyed. The propensity of fibres to felt together under compression, and hence cause shrinkage of a fabric through the ratchet-like function of these scales, is thus reduced. Pre-treatment with chlorine also serves to 'activate' wool-fibre surfaces and ensure better keying of the polymer coatings now used to reduce felting shrinkage.

Because the amino-acids with reactive side chains occur mainly in accessible regions of the wool fibres, they can often be modified in high yield. Apart from cystine, which has already been discussed, the basic side chains can, e.g., be acetylated to give fibres with a

greatly lowered affinity for acid dyes; lysine yields the 2,4-dinitro-
phenyl derivative with Sanger's reagent (Part 2, p. 160) and affords
new cross-links (2) with difluorodinitrobenzene.

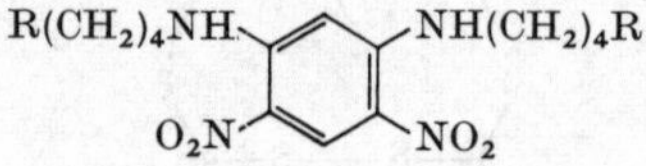

$$R(CH_2)_4NH\overbrace{}^{} NH(CH_2)_4R$$

$$O_2N \qquad\qquad NO_2$$

(2)

Cross-linking with diepoxides (i), epichlorohydrin (ii), and glycol
dimesyl esters (iii) also probably involves amino-groups, whilst

$$\xrightarrow{\text{i}} \quad R'NHCH_2CH(OH)RCH(OH)CH_2NHR'$$

$$2R'NH_2 \xrightarrow{\text{ii}} \quad R'NHCH_2CH(OH)CH_2NHR'$$

$$\xrightarrow{\text{iii}} \quad R'NH(CH_2)_nNHR'$$

different classes of reactive dyes (see Chapter 10) depend on amino-
and hydroxyl groups for fixation.

Silk

Silks are fibrous proteins produced by spiders and a great variety
of insects, in particular butterflies and moths. They differ widely
in amino-acid composition and structure, each being tailored for a
specific purpose; for example, a spider will produce different silks
for its life-line and web. The commonest function of insect silk is as
a cocoon to protect the pupa, and it is the cocoon spun by the larva
(silkworm) of the moth *Bombyx mori* that provides most of the
silk of commerce; lesser amounts are obtained from the wild Indian
and Chinese Tussah silk moths.

The threads of *B. mori* silk are actually twin cores of a highly
oriented fibrous protein, fibroin, cemented together with non-
fibrous material, sericin, which is readily removed with boiling
soap solution (degumming). Fibroin and sericin differ in composition (Table 9.3). Fibroin is characterized by a high content of the
small amino-acids, glycine, alanine and serine, in that order, which
together account for 86% of the total amino-acid nitrogen. Sericin
is rich in serine, glycine and aspartic acid; some of the aspartic and
glutamic acid is present as asparagine and glutamine, which yield

Table 9.3. Amino-acid composition of fibroin and sericin (per 1000 amino-acid residues)

	Fibroin	Sericin		Fibroin	Sericin
Glycine	445	147	Arginine	5	36
Alanine	293	43	Histidine	2	12
Valine	22	36	Tyrosine	52	26
Leucine	5	14	Phenylalanine	6	3
Isoleucine	7	7	Proline	3	7
Serine	121	373	Tryptophan	2	—
Threonine	9	87	Methionine	1	—
Aspartic acid	13	148	Cystine (half)	2	5
Glutamic acid	10	34	(NH$_3$	—	86)
Lysine	3	24			

ammonia on hydrolysis. Metastable aqueous solutions of fibroin are obtained by dialysis of much-diluted solutions of the protein in cupriethylenediamine solutions or, better, aqueous lithium thiocyanate; they are readily denatured to a fibrous gel. Digestion of these solutions with chymotrypsin, which breaks peptides specifically at the carboxyl group of aromatic amino-acids, i.e. at tyrosine residues in the case of fibroin, results in the separation of a granular crystalline polypeptide, $Gly_{29}Ala_{20}Ser_9Tyr$, whose X-ray powder diagram corresponds to that of intact fibroin. This peptide is therefore believed to constitute the crystalline regions (some 60%) of silk fibroin. Its structure (**3**) was largely determined by classical partial hydrolysis, dinitrophenylation and chromatographic separation of DNP-peptides (Part 4, p.320); the *C*-terminal sequence has recently been resolved by mass spectrometry of the completely *N*-methylated peptide as SerGlyAlaGlyAlaGlyTyr.

GlyAlaGlyAlaGly[SerGly(AlaGly)$_n$]$_{12}$SerGlyAlaGlyAlaGlyTyr

(**3**) n is usually 2

The isolable hexapeptide SerGlyAlaGlyAlaGly thus constitutes much of the molecule. The position of serine residues is determined by subjecting the peptide to an *N*- to *O*-acyl rearrangement before dinitrophenylation and partial hydrolysis as shown on p. 256.

The soluble peptides from chymotryptic digests contain a high proportion (ca. 40%) of two octapeptides, $Gly(AlaGly)_3Tyr$ and $Gly(Gly_3Ala_2Val)Tyr$, and a tetrapeptide, GlyAlaGlyTyr. These plus the crystalline polypeptide account for about 75% of the amino-acid residues in fibroin; the remaining 25% is made up

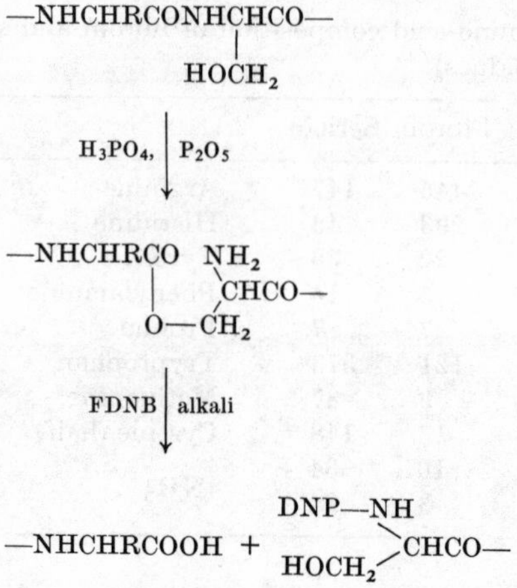

$$—NHCHRCONHCHCO—$$
$$\mathrm{HOCH_2}$$

$$\mathrm{H_3PO_4,} \quad \mathrm{P_2O_5}$$

$$—NHCHRCO \quad NH_2$$
$$\searrow CHCO—$$
$$O—CH_2$$

FDNB $\Big|$ alkali

$$DNP—NH$$
$$—NHCHRCOOH + \qquad \searrow CHCO—$$
$$HOCH_2$$

largely of amino-acids with polar or bulky side chains. A fibroin molecule is therefore envisaged as a three-segment chain, comprising (*a*) a highly ordered segment of 84 amino-acids (overall length ca. 30 nm), largely containing the repeat sequence SerGlyAlaGlyAlaGlyAlaGly, (*b*) a less ordered segment, due to the presence of valine and tyrosine with their rather large side-chains (this yields the octa- and tetra-peptide sequences), and (*c*) a disordered segment containing polar and bulky amino-acids.

The whole chain may contain some 2500 amino-acid residues, giving a molecular weight around 200×10^3. So compact is the segment (*a*) (mainly glycine and alanine) that the extended molecules (β-fibroin) are able to pack closely together in a 'pleated-sheet' structure (Figure 9.1) of antiparallel chains with a high degree of inter-chain —CO---HN— hydrogen-bonding. These sheets, in turn, are stacked together into ribbon-like microfibrils, with a cross-section of about 6 nm in the plane of the sheets by 2 nm in the side-chain direction.

Mechanisms leading to formation of such a highly ordered structure from an initially viscous, non-fibrous secretion, and the molecular conformations of the protein before spinning, are still undetermined.

As a textile fibre, silk is now an expensive luxury. It possesses considerable disadvantages compared with other fibres, in respect of wear resistance, liability to creasing, and high sensitivity to damage by photo-initiated oxidation. Many attempts have been made to produce silk-like polymers of the α-amino-acids, but with little success, although other fibrous polymers are now available

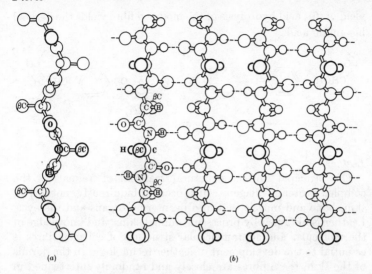

Figure 9.1. The pleated-sheet structure of fibroin: (*a*) edge view; (*b*) view perpendicular to sheet. [Reproduced, by permission, from S. W. Fox and J. F. Foster, *Introduction to Protein Chemistry*, Wiley, New York, 1957.]

that are claimed to have some silk-like properties (see Qiana, Chinon). Strong intermolecular hydrogen-bonding and a high crystallinity confer on silk its high strength (both wet and dry), whilst the essential extensibility is ensured by the greater molecular mobility created by the bulkier amino-acids in disordered regions. Silk, like other protein fibres, has a high moisture regain (about 10%), the absorption occurring mainly at peptide links and other polar sites in the disordered regions.

Apart from reacting with amino end-groups, 1-fluoro-2,4-di-nitrobenzene converts combined tyrosine and lysine of silk into *O*- and *N*-DNP derivatives (**4, 5**). Similarly, reactive difluoro-

compounds may yield the corresponding bis-derivatives; for example, 3,3′-dinitro-4,4′-difluorodiphenyl sulphone affords a high

yield of (**6**); acid hydrolysis of the modified fibre yields the derived
bis-amino-acid.

(**6**)

Leather

The principal fibrous structural material of animals is the
complex protein, collagen, which occurs ubiquitously, notably in
skin, bone and muscle, but also in internal organs and the eyes.
Leather is obtained by processing animal skins, and knowledge of
the molecular and supramolecular structure of collagen fibres is
essential to the development of leather technology. In the dermis
of the skin, these fibres are closely and randomly intertwined in
a dense three-dimensional network resembling a felt; hence leather
has been referred to as 'Nature's non-woven textile'.

The various collagens differ in composition; a typical amino-
acid analysis is given in Table 9.2 (p. 249). A significant common
feature is the high content of proline and hydroxyproline which pro-
foundly influences the molecular structure. Classical acid hydrolysis, allied to proteolysis with trypsin, chymotrypsin and other
peptidases, particularly the highly specific collagenase from *Clostridium histolyticum* which splits only sequences GlyProXGlyProX
at the X–Gly link, suggests that the primary main-chain amino-
acid sequences in collagen are arranged in a block co-polymer
type of structure, with alternating polar and non-polar sequences. The structure is unique in that glycine almost exclusively occupies every third position in the chain. The non-polar
segments are rich in peptides of the type $(GlyProX)_n$. Many synthetic polytripeptides of this and similar structure, e.g. $(GlyXPro)_n$,
have been prepared and their X-ray crystallographic behaviour
has been examined. The resemblance between collagen and the
peptide $(GlyProPro)_n$ is particularly close. These non-polar
sequences, which constitute about 35% of the collagen molecule,
are largely responsible for its ordered (X-ray) structural component and determine its thermal stability. The latter property (effect
of heat) refers to the susceptibility of the fibre to contract sharply
to about one-third of its length when heated to a temperature (T_s)
that is characteristic for the particular type of collagen and is

directly related (about 27° higher) to the denaturation temperature of the protein in solution. Broadly speaking, T_s increases as the combined proline plus 4-hydroxyproline content increases. These two cyclic amino-acids account for some 25% of the total amino-acid content of mammalian skin collagen. Their influence is due to the rigidity they impose on the protein molecule through rotational restriction at the N—C-1 bond of the pyrrolidine ring and the C-1—C═O bond, and to the change in direction of the chains at each heterocyclic centre (7).*

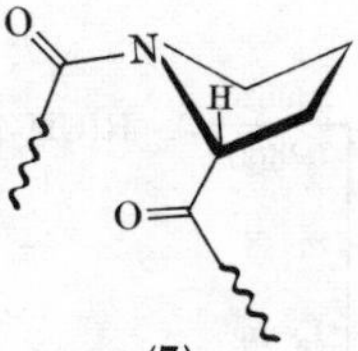

(7)

X-ray diffraction data indicate a structure in which each polypeptide chain has a helical conformation, with three amino-acid residues per turn; the collagen molecule contains three of these so-called α-chains, wound in a 'super-helix' and held together by inter-chain hydrogen bonds. Two of the chains are identical and each has a molecular weight around 100×10^3. The compact glycine residues at every third position in each chain are in turn accommodated in the 'core' of the composite helix, with the other more bulky amino-acids on the periphery.

Sequences containing the polar amino-acids are rich in aspartic acid, glutamic acid, lysine and arginine. Some of the glutamic acid appears to be bound through the γ-carboxyl groups, since succinaldehydic acid (8) has been isolated by the reactions shown. Native collagen contains no detectable N-terminal group.

$$-NHCHCH_2CH_2CONH- \xrightarrow[\text{(2) NH}_2\text{OH}]{\text{(1) MeOH}} -NHCHCH_2CH_2CONH- \xrightarrow{\text{FDNB}}$$
$$\quad | \qquad\qquad\qquad\qquad\qquad\qquad\qquad | $$
$$\quad COOH \qquad\qquad\qquad\qquad\qquad\quad CONHOH$$

$$-NHCHCH_2CH_2CONH- \xrightarrow[\text{(Lössen)}]{\text{NaOH}} -NHCHCH_2CH_2CONH- \xrightarrow{\text{H}_2\text{O}}$$
$$\quad | \qquad\qquad\qquad\qquad\qquad\qquad\qquad\quad | $$
$$\quad CONHO—DNP \qquad\qquad\qquad\qquad\quad NH_2$$

$$-NH_2 + OHCCH_2CH_2COOH + H_2N-$$

(8)

* In this Chapter, wavy lines denote chains of indefinite length.

Non-helical fragments, known as telopeptides, can be detected in collagen; some of these are terminal appendages to the main α-chains, whilst others may be attached at branching points along their length. These are the peptides removed by treatment of *native* collagen with proteolytic enzymes; they are also the seat of a few thermally and chemically unstable cross-links (1–2 per 1000 amino-acid residues). The exact nature of the cross-links is unknown, but enzymic oxidative deamination of the telopeptide lysine and hydroxylysine side-chains is apparently implicated in their formation:

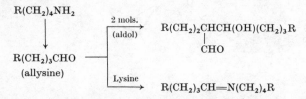

Progressive oxidation and cross-linking gradually lower the solubility of collagen. Tanning, an essential step in the conversion of collagen into leather, is effected by the introduction of stable cross-links, the simplest of which result from the action of form-aldehyde on, probably, lysine, glutamine and asparagine side-chains, e.g.:

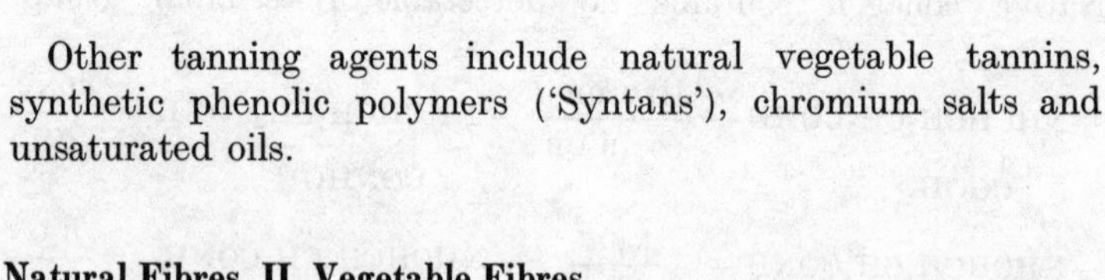

Other tanning agents include natural vegetable tannins, synthetic phenolic polymers ('Syntans'), chromium salts and unsaturated oils.

Natural Fibres, II. Vegetable Fibres

The vegetable fibres are derived from the stems (flax, jute, hemp), leaves (sisal), or seed hairs (kapok, cotton) of plants, and differ considerably in morphology. The basic fibrous constituent of vegetable matter is cellulose, a linear polysaccharide composed solely of 1,4-linked β-D-glucopyranoside units (**8A**), (Part 4, p. 83).

(8A)

Huge amounts of cellulose fibres are also obtained from wood pulp, but they are too short for direct conversion into textiles. They are used in the production of paper and of regenerated cellulose (rayon) fibres (see p. 272). The cellulose is always accompanied by non-fibrous polymeric materials—pectins, hemicelluloses and lignin—in varying amounts. Only cotton, as the most important textile fibre, will be considered here.

Cotton

Obtained from the seed hairs of *Gossypium* spp., cotton is a pure form of cellulose. The fibres are up to 2 inches long, rather flat, and convoluted. The cellulose is laid down in a complex structure as layers of fibrils spiralling around the fibre axis and enclosing a hollow central canal or lumen. Most of the small amount of non-cellulosic matter is found in an enveloping sheath or primary wall.

The X-ray diffraction pattern of cotton is typical of a partly crystalline polymer. It indicates a unit cell containing four cellobiose units, with adjacent polymer chains running antiparallel, and a degree of crystallinity of about 70%, a value higher than obtained by measurements of chemical accessibility (see below). The fine structure of the elementary fibrils into which the individual molecular chains are aggregated is still debatable. A long-held view—the fringe micellar theory—is that within each fibril individual molecules pass through both crystalline and non-crystalline (amorphous) regions, only the latter being accessible to reagents. Current opinion, however, is that each fibril is itself completely crystalline and that the accessible regions are represented by molecules exposed at dislocations in fibrillar surfaces.

Accessibility is an important feature of fibre structure and several methods have been developed for its measurement in cotton, each of which relates, of course, only to the conditions under which the measurement was made. Perhaps the most elegant and important, since it is a vapour-phase reaction and does not involve a solvent that may swell the fibre and hence have a disordering effect, is the use of deuteriation coupled with infrared spectroscopy. This method was first applied to films of regenerated and bacterial cellulose. When specially prepared films of cotton are exposed to D_2O vapour, an exchange reaction occurs with all accessible OH groups: thus the broad 'disordered' O—H stretching band

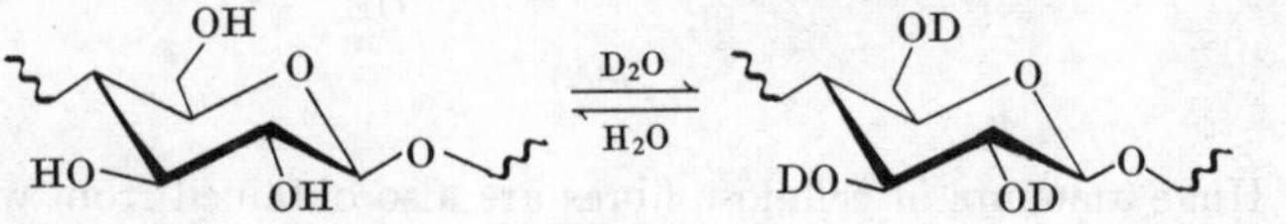

disappears and a new O—D band is found at 2530 cm^{-1}; the well-defined O—H band system at 3350 cm^{-1} of the 'ordered' OH groups remains unchanged. From the intensities of these bands relative to that of the unchanged C—H band at ca. 2900 cm^{-1}, the proportion of OH groups in ordered (crystalline) and disordered regions can be calculated. For purified native cotton, some 58% occur in the crystalline region. A similar value is obtained from moisture-regain measurements. Thus all but the most hydrophobic fibres, such as polypropylene, contain sorbed water in an amount that, generally speaking, increases with increasing humidity of the atmosphere and decreases with a rise in temperature. These moisture-regain values at equilibrium under average conditions, say 65% relative humidity and 20°C, range from 15% for highly hydrophilic fibres down to less than 1% for a polyester. Cotton has a moisture regain of 6–7%. Accessibility is derived from the ratio of the number of moles of water absorbed per 1,4-anhydroglucose unit to the amount that would be taken up by a fully accessible cotton, which is 1.53 moles.

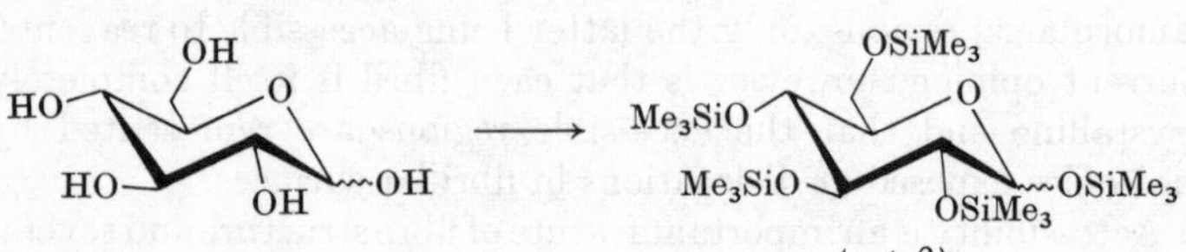

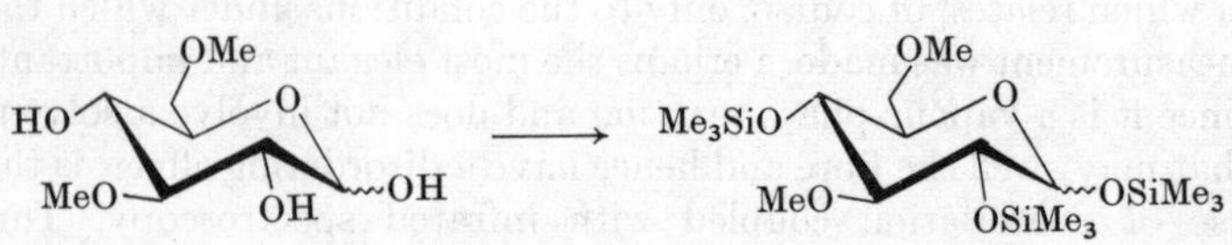

An elegant refinement of the chemical measurement of accessibility, which provides quantitative support for the concept of a totally crystalline structure for cotton, uses a modification of classical methylation techniques (Part 4, pp. 49 and 86). Repeated treatment of cotton, soaked in 2M sodium hydroxide solution, with dimethyl sulphate in dimethyl sulphoxide results in a continuous, but progressively slower, increase in the degree of methylation until a point is reached at which further methylation is clearly disturbing the originally inaccessible regions. At this point, the distribution of methoxyl groups can be determined by total acid hydrolysis of the fibre to a mixture of glucose and methylglucoses, which is then trimethylsilylated and separated quantitatively into its individual components by gas–liquid chromatography, as formulated at the foot of p. 262.

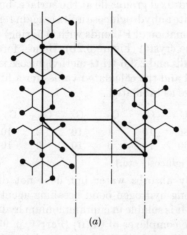

(a)

Figure 9.2(a). Basic structural unit of cellulose. [Reproduced, by permission, from R. Jeffries, J. G. Roberts and R. N. Robinson, *Textile Res. J. Polymer Sci.*, **17**, 168 (1955).] (See text on p. 264.)

(b)

Figure 9.2(b). Conformation of cellulose, showing hydrogen bonding between the OH group on C-3 and the ring-oxygen atom on C-5. (See text on p. 264.)

The distribution so obtained corresponds closely to a value calculated for a structure in which the crystalline elementary fibril consists of a bundle of 80 cellulose chains in a block 10×8. Forty-eight of these chains lie within the crystal and are therefore inaccessible, thus yielding only glucose on hydrolysis. A molecular model shows that anhydroglucose units in the broad pair of faces have their OH groups on C-6 or C-2 and C-3 projecting alternately outwards and thus accessible (Figure 9.2a, p. 263); the OH group on C-3, however, is strongly hydrogen-bonded to the neighbouring C-5 (ring) oxygen and of low reactivity. In this face, then, methylation can only yield 2- or 6-O-methylglucose. In the narrow pair of faces, all the hydroxyl groups lie at the surface, but the OH group on C-3 of alternate anhydroglucose units is again rendered inaccessible through formation of H-bonds with C-5 (ring) oxygen that are directed into the crystal (Figure 9.2b). Hence, formation of equal amounts of 2,6-di- and 2,3,6-tri-O-methylglucose is predicted. The results obtained and the calculated values for a fibrillar bundle of 10×8 chains are as tabulated.

	G	G_2	G_3	G_6	$G_{2,3}$	$G_{2,6}$	$G_{3,6}$	$G_{2,3,6}$
Found	56	10	1	10	2	10	1	9
Calc.	60	10	0	10	0	10	0	10

($G_2 = 2$-O-methylglucose, etc.)

Cotton readily absorbs water but does not dissolve even in solutions of strong hydrogen-bond breaking agents such as LiBr, $ZnCl_2$ and urea; it is soluble in cuprammonium hydroxide, aqueous ethylenediamine complexes of Cu(II) (Part 4, p. 93) ('cuene') and Cd(II) ('cadoxene'), and similar reagents. Apart from the loss of some reducing end-groups, through conversion by a rather complex route into acidic end-groups (Part 4, p. 42), cotton is chemically stable towards aqueous alkali. However, sodium hydroxide solution (5M or above) produces a change in morphology (to a more cylindrical, non-convoluted fibre almost free of lumen) and in crystal structure (the cellulose I $\rightarrow$ cellulose II transformation). This process, known as mercerization, is of immense practical importance, since it can improve the strength, lustre and dyeability of cotton. Similar changes, but with conversion not into cellulose II but into a third structure, are effected by brief treatment with anhydrous liquid ammonia, a powerful swelling agent for cotton ('Prograde' process).

Hydrolysis of cotton with hot, dilute mineral acid ultimately yields D-glucose; the reaction involves random glycosidic bond scission and various oligomers, such as cellotetraose, cellotriose, and cellobiose can be detected at intermediate stages (Part 4, p. 69). An initially rapid breakdown is followed by a slower reaction in which the loss in weight is a linear function of time; the cellulose attacked in the initial reaction has been equated to the accessible portion of the fibres. Good yields of cellobiose octa-acetate are

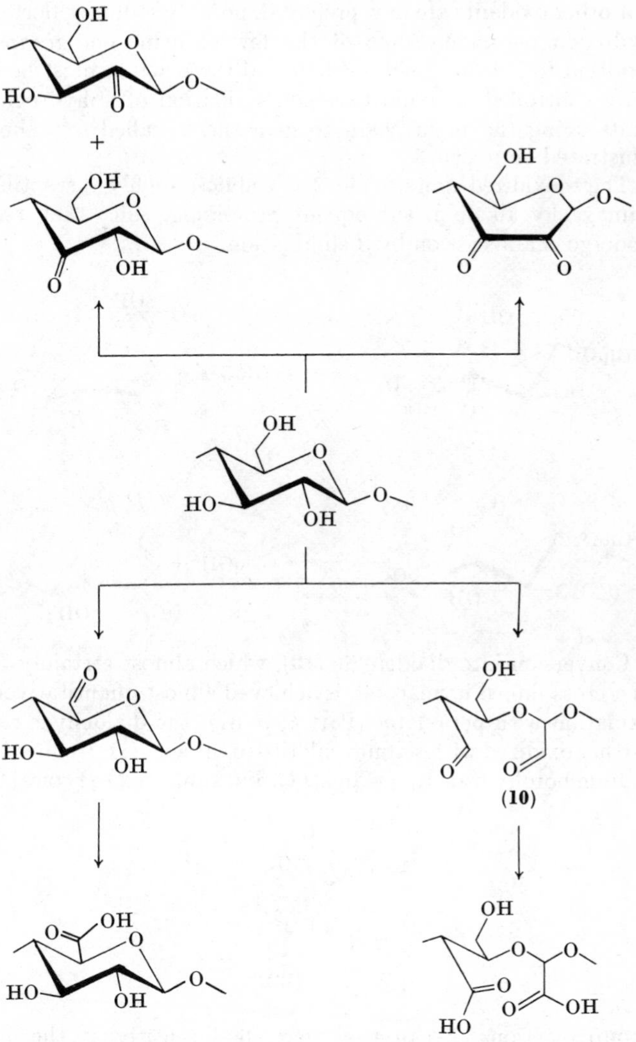

obtained by acid-catalysed acetolysis of cotton with acetic anhydride.

Bleaching, to remove coloured impurities, is an essential stage in the purification of cotton, after treatment with dilute alkali (scouring) to remove hemicellulose, pectin, waxes and residual seed-cases. Classically, sodium hypochlorite ('chemic') was used, but other oxidants are now preferred, notably sodium chlorite and hydrogen peroxide. Some of the few reducing end-groups are oxidized to gluconic acid residues and the reaction must be carefully controlled to avoid excessive oxidation of anhydroglucose units along the main chain to give the so-called oxycelluloses illustrated on p. 265.

These oxidized units render the cellulose highly susceptible to damage by alkali in subsequent processing, since they readily undergo chain-scission by β-elimination:

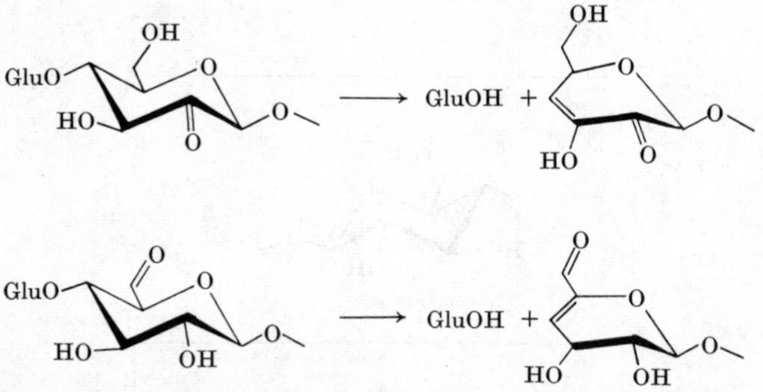

Conversion into dialdehyde (**10**), which almost certainly exists as a cross-linked hemiacetal, is achieved almost quantitatively by oxidation with periodates (Part 4, p. 87); the dialdehyde can be further oxidized with sodium chlorite to the acid (**11**), reduced with sodium borohydride to the diol (**12**), and subjected to typical C=O

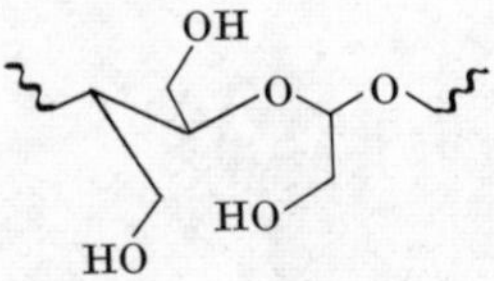

(12)

group reactions. Because of their alkali-sensitivity, the dialdehydes and their derivatives have found no practical application.

Nitrogen dioxide oxidizes cotton with some specificity to the uronic acid (**9**); gauze fabric so modified has use as a haemostatic wound dressing.

Cotton can be modified by a variety of typical substitution and addition reactions, e.g. etherification, esterification, hydroxyalkylation with epoxides, and Michael-type additions; these are normally effected in aqueous systems. Because cotton is impervious to most organic solvents, reactions carried out in them are largely restricted to the fibre surface, and special techniques are necessary to achieve deeper penetration; similar arguments apply to rayon, wool and other hydrophilic fibres. If the reagent is not too readily hydrolysed, the fibres may simply be swollen with an aqueous phase; otherwise, the water must then be removed by a solvent-exchange process (e.g. water → DMF, water → ethanol → benzene). In a few cases, the reagent may be applied in the vapour phase. It must be remembered, too, that fibres behave as semipermeable membranes and undergo only surface reactions when treated with polymeric reagents.

Simple ethers of cotton obtained by treatment with alkyl halides and alkali include —OMe, —OEt, —OCH$_2$COOH, —OCH$_2$CONH$_2$, —OCH$_2$P(O)(OH)$_2$, —OCH$_2$Ph, —OCHPh$_2$, —OCH$_2$CH$_2$NH$_2$ and OCH$_2$CH$_2$NEt$_2$; the last may be obtained also by reaction with the salt (**13**), formed *in situ*. Generally speaking, the mechanical properties are adversely affected even at a low degree of substitution (D.S.).

$$\underset{}{\overset{+}{ClCH_2CH_2NHEt_2}}Cl^- \xrightarrow{HO^-} H_2C\!\!-\!\!CH_2 \xrightarrow{Cell-OH} CellOCH_2CH_2NEt_2$$
$$\underset{Et\ \ Et}{\overset{\diagdown+\diagup}{N}}\ Cl^-$$

(**13**)

Treatment of cotton with mixtures of acetic anhydride, acetic acid and perchloric acid yields fibrous acetates with a wide range of D.S.; acetylation has also been achieved with acetyl chloride–pyridine and with ketene. Higher acyl esters, mesylates, tosylates and so on can all be prepared. Many carboxylic acids react directly with cotton in the presence of trifluoroacetic anhydride as an 'impeller', e.g. acetic, benzoic, acrylic and higher fatty acids. Michael addition of acrylamide in the presence of alkali introduces the —OCH$_2$CH$_2$CONH$_2$ group, with some carboxyethylation through concomitant hydrolysis.

Cyanoethylation has been widely investigated:

$$CellOH + CH_2:CHCN \xrightarrow{OH} CellOCH_2CH_2CN$$

Reaction with isocyanates in dry pyridine yields carbamates and acid-catalysed addition of vinyl ethers gives acetals:

$$CellOH + RNCO \longrightarrow CellOCONHR$$
$$CellOH + CH_2{=}CHOR \longrightarrow CellOCHMeOR$$

Displacement reactions are less common. Tosylated cotton reacts with ammonia, with the introduction of some amine groups; a recent example is the preparation of a putative chlorodeoxy-derivative of low D.S. by treatment with thionyl chloride–dimethyl-formamide, a Vilsmeier-type reagent (Part 3, p. 100).

$$SOCl_2 + Me_2NCHO \longrightarrow$$
$$[Me_2N{:}CHCl]^+ \ Cl^- \xrightarrow{CellOH} CellCl + Me_2NCHO + HCl$$

By and large, these reactions do not produce any improvement in the mechanical properties of cotton and most are uneconomical to operate commercially. They may bring about modification of the dyeing properties, confer greater immunity to microbiological attack, as, e.g., in the case of controlled (surface) cyanoethylation or acetylation, or improve the heat-resistance, and so on. Other examples are discussed below.

Cross-linking reactions of cotton cellulose are of great importance since they form the basis of processes for imparting ease-of-care properties, i.e. recovery from creasing, retention of pressed-in creases (durable press) and smooth-drying after laundering. The simplest cross-link is that given by formaldehyde, as in $CellOCH_2OCell$, but the reaction is notoriously difficult to control and losses in strength and abrasion resistance, which invariably accompany cross-linking, may become unacceptably high. Of many variants on the process, the use of gaseous formaldehyde and sulphur dioxide with fabrics having closely controlled moisture content (ca. 6%) offers promise; the strong acid catalyst required is believed to be produced *in situ* and destroyed on drying the fabric:

$$CH_2O + SO_2 + H_2O \rightleftharpoons HOCH_2SO_3H$$

Good crease-recovery in both wet and dry states is obtained with less than 1% 'add-on' of formaldehyde.

At present, however, most easy-care treatments rely on the use of bis-(*N*-hydroxymethyl)ureas and similar compounds, notably (**14**) to (**20**). These compounds are readily obtained from the parent

$$HOCH_2NHCONHCH_2OH$$

(**14**)

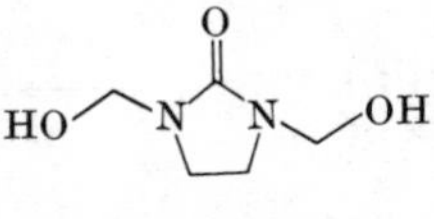

(**15**) 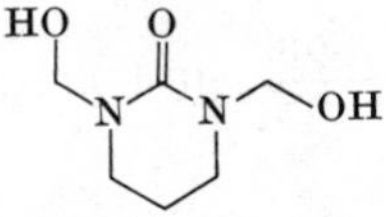　　　　　　(**16**) 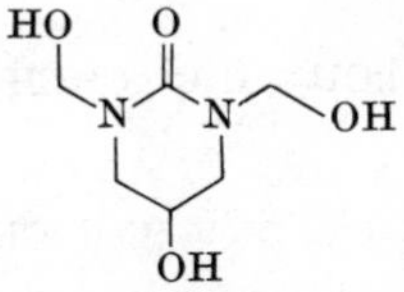　　　　　　(**17**)

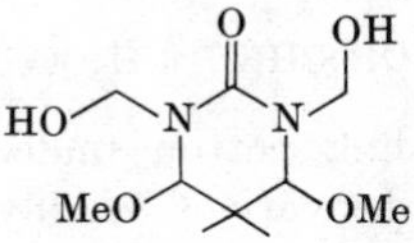

(**18**) 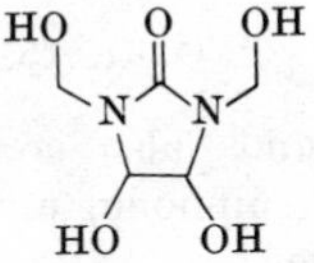　　　　　　(**19**) 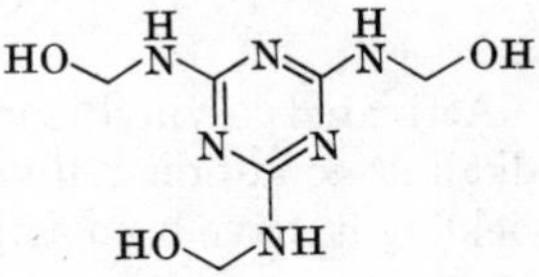　　　　　　(**20**)

urea (or melamine for **20**) and formaldehyde under slightly alkaline conditions; the hydroxylated urea for (**19**) is produced from urea and glyoxal. They are applied to cotton along with a Lewis acid, e.g. $Zn(NO_3)_2$ or $MgCl_2$, and are then heated typically at ca. 150°C for 3–4 min., to effect cross-linking, as shown for (**15**):

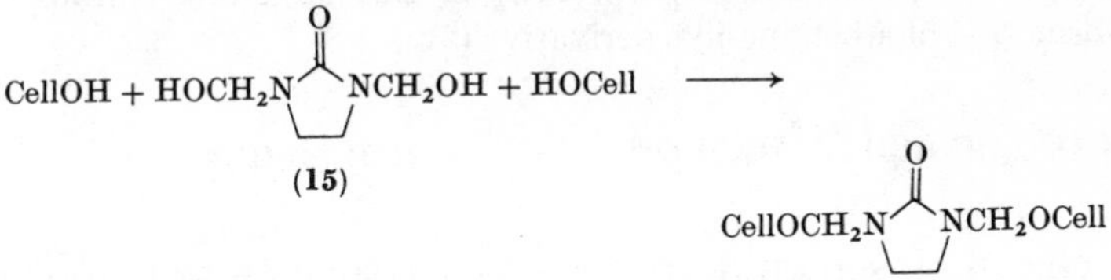

Some polymerization of the urea derivative may also occur. The amount of reagent required is about 5%.

A disadvantage of the urea derivative (**14**), and of the other reagents if they undergo partial hydrolysis instead of reacting completely with the cotton, is their susceptibility to hypochlorites which may be used as bleaching agents in laundering. They yield *N*-chloro-derivatives which decompose when heated (e.g. ironed), with liberation of HCl, causing discolouration and tendering (hydrolysis) of the fabric.

Cotton can be cross-linked with epichlorohydrin or diepoxides and aqueous alkali, or with diepoxides and a Lewis acid catalyst, typically BF_3 generated *in situ* by thermal decomposition of $Zn(BF_4)_2$.

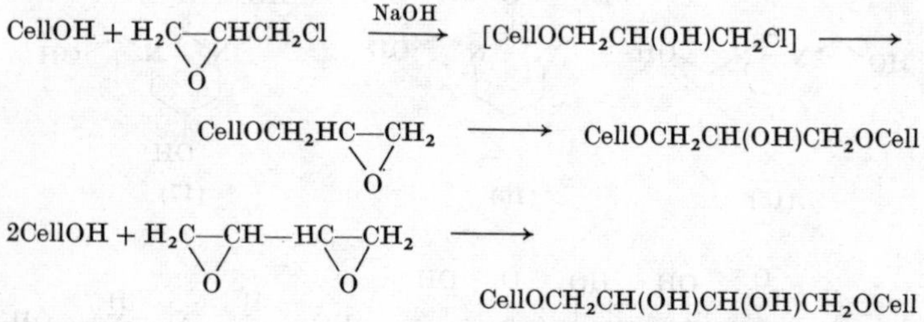

$$CellOH + H_2C{-}CHCH_2Cl \xrightarrow{NaOH} [CellOCH_2CH(OH)CH_2Cl] \longrightarrow$$
$$\overset{\diagdown\diagup}{O}$$

$$CellOCH_2HC{-}CH_2 \longrightarrow CellOCH_2CH(OH)CH_2OCell$$
$$\overset{\diagdown\diagup}{O}$$

$$2CellOH + H_2C{-}CH{-}HC{-}CH_2 \longrightarrow$$
$$\overset{\diagdown\diagup}{O}\qquad\overset{\diagdown\diagup}{O}$$
$$CellOCH_2CH(OH)CH(OH)CH_2OCell$$

Activated divinyl compounds also cross-link cotton under alkaline conditions; divinyl sulphone, and derivatives readily yielding it, have been employed:

$$2CellOH + CH_2{=}CHSO_2CH{=}CH_2 \xrightarrow{HO^-}$$
$$CellOCH_2CH_2SO_2CH_2CH_2OCell$$

The compound (**21**), obtained from thiodiglycol, $S(CH_2CH_2OH)_2$, reacts very rapidly with cotton, through transient formation of the highly activated unstable trivinylsulphonium ion ('Sulfix A' reagent); an after-treatment with Na_2SO_3 eliminates the undesirable ion-exchange properties of the cross-link through formation of a betaine-like derivative (**22**).

$$(NaOSO_2OCH_2CH_2)_2\overset{+}{S}CH_2CH_2OSO_3^- \xrightarrow{HO} [(CH_2{=}CH)_3S^+\,HO^-] \longrightarrow$$
$$\textbf{(21)}$$

$$CellOCH_2CH_2\overset{+}{S}CH_2CH_2OCell^+\,HO^- \xrightarrow{Na_2SO_3} CellOCH_2CH_2\overset{+}{S}CH_2CH_2OCell$$
$$\underset{CH{=}CH_2}{|} \qquad\qquad\qquad \underset{CH_2CH_2SO_3^-}{|}$$
$$\textbf{(22)}$$

Cotton cross-linked in a swollen state, i.e. whilst still wet, has good recovery from creasing when wet, giving good smooth-drying properties, but poor recovery from dry-creasing; a second treatment with one of the di-(*N*-hydroxymethyl)urea types of compound is necessary to confer dry crease-recovery.

Launderable flame-proofed cotton is obtained by treatment with the phosphonium salt (23) (THPC), and the corresponding 'free base' (THPOH) obtained by careful reaction with one equivalent of NaOH. THPOH in solution is probably largely the phosphine (24):

$$PH_3 + 4CH_2O + HCl \longrightarrow \overset{+}{P}(CH_2OH)_4Cl^- \xrightarrow[HO^-]{} P(CH_2OH)_3$$

$$\text{(23)} \qquad\qquad\qquad \text{(24)}$$

The chloride is applied from aqueous solution and the fabric is then treated with gaseous ammonia ('Proban' process), whereby cross-linked polymers of unknown constitution are deposited within the fibres; they probably contain structures of the type (25). Alter-

$$\begin{array}{c} {>}NCH_2 \\ {>}NCH_2 \end{array}\!\!\!{>}PCH_2\overset{|}{N}CH_2P\!\!\!\begin{array}{c} {<}CH_2N{<} \\ {<}CH_2N{<} \end{array}$$

$$\text{(25)}$$

natively, the P compounds may be applied together with a melamine–formaldehyde precondensate, dried, and cured at high temperature (>140°C).

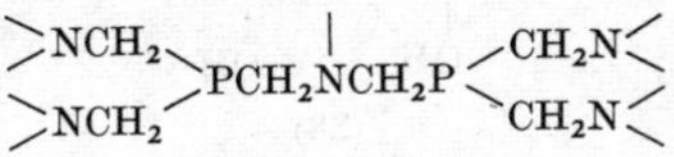

The phosphoramide (26) cross-links cotton and also imparts flame-retardancy but presents health hazards, whilst a recent process ('Pyrovatex') depends for fixation on acid-catalysed condensation with compounds such as (27) under curing conditions typical of those for the urea derivative mentioned above. The

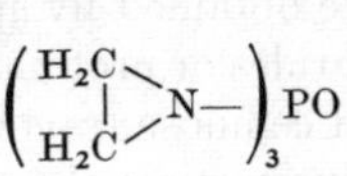

$$\left(\begin{array}{c} H_2C \\ | \\ H_2C \end{array}\!\!\!{>}N{-} \right)_{\!\!3}\!PO \qquad (MeO)_2P(O)CH_2CH_2CONHCH_2OH$$

$$\text{(26)} \qquad\qquad\qquad \text{(27)}$$

parent amide of (27) is readily available from an Arbuzov reaction (Part 3, p. 241) between 3-chloropropionamide and a trialkyl phosphite or addition of a dialkyl phosphite to acrylamide.

$$(MeO)_3P + ClCH_2CH_2CONH_2$$

$$\text{or} \qquad\qquad\qquad \longrightarrow (MeO)_2P(O)CH_2CH_2CONH_2$$

$$(MeO)_2PH + CH_2{=}CHCONH_2$$

Some 3–5% of phosphorus is necessary to impart reasonable flame-retardancy. Many nitrogen-containing compounds, notably melamine derivatives, have a synergic action when used in combination with phosphorus compounds.

Flame retardants may actually lower the temperature at which cellulose undergoes pyrolysis, and they drastically alter the course of the reactions; a much higher proportion of water and carbonaceous matter may be formed, at the expense of the more highly flammable volatile matter and tar, of which laevoglucosan, 1,6-anhydro-β-D-glucopyranose (**28**) is the major constituent. This is the so-called 'dehydration' mechanism held to account for the action of many phosphorus and boron derivatives. Volatile phosphorus compounds produced may also act in the vapour phase as

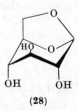

(**28**)

radical-inhibitors of flame-propagation; organohalogen (except fluorine) compounds, which are effective flame-retardants for many polymers, may also function in this way, yielding halogen radicals as the active species. Their activity can be greatly enhanced by antimony trioxide, which apparently yields the relatively volatile antimony (oxy)halides.

Regenerated Fibres

This class comprises those fibres that are obtained by spinning solutions of naturally occurring polysaccharides or proteins into a suitable coagulating medium. Regenerated cellulose (rayon) is by far the most important member; of lesser interest are the alginate fibres from seaweed and fibres derived from vegetable protein such as casein.

Rayon

Two methods of manufacture have been developed in both of which highly purified cellulose from wood pulp is used. Bemberg (or cuprammonium) rayon is obtained by spinning a solution of cellulose in cuprammonium hydroxide into an acid coagulating bath. Viscose rayon, produced in vastly greater quantity, is obtained by conversion of cellulose into a soluble partial xanthate

$$CellOH + CS_2 + NaOH \longrightarrow CellOCSSNa$$

by reaction with CS_2 and sodium hydroxide solution and subsequent regeneration of cellulose by spinning this solution into an acidic coagulant. In detail, the process is highly complex, but the main steps may be summarized as follows (cf. Part 4, p. 93).

1. The fibrous cellulose is impregnated with concentrated NaOH solution and allowed to age or 'ripen' in air. Residual hemicelluloses are removed and the degree of polymerization of the fibres falls as the result of oxidation and chain-scission. Addition of small amounts of oxidant, e.g. H_2O_2, accelerates this degradation. Chain-scission, the result of β-elimination reactions outlined in the Section on oxidized cotton (cf. Part 4, p. 42), is followed by rearrangement of some liberated reducing end-groups to stable acidic end-groups, as formulated.

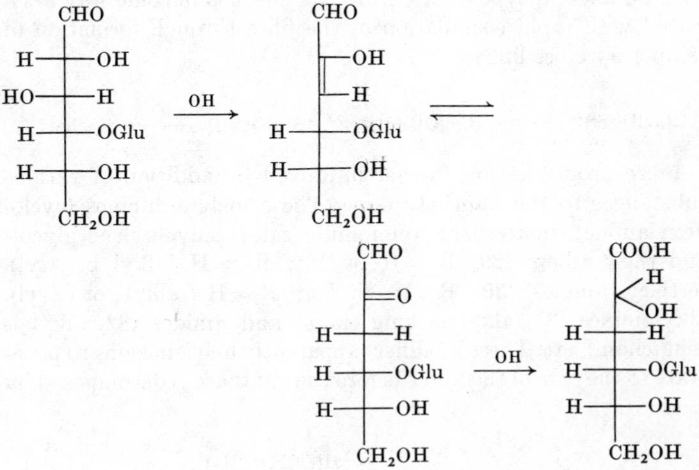

2. When the extent of degradation of the cellulose has reached the required value, the shredded, alkali-impregnated fibres are mixed with CS_2, in excess of the required degree of substitution, which is in the region of 0.5–0.6, to allow for side-reactions. The latter include formation of Na_2CO_3, Na_2CS_3 and Na_2S. Cellulose xanthate is formed as an orange granular solid.

3. The xanthate is dissolved in dilute NaOH solution and allowed to 'ripen' further; some hydrolysis occurs, the degree of substitution falling to about 0.35–0.4, and the solution becomes less viscous. In addition, there is some redistribution of xanthate groups, involving (i) migration from C-2 where preferential substitution occurs initially, and from C-3 to C-6, and (ii) substitution

on glucose residues that were not accessible to reagents in the heterogeneous xanthation stage.

4. The viscose 'dope' is spun into an acid-salt coagulating bath, to yield filaments that are washed and oriented by drawing (stretching).

A simple coagulating bath would contain only H_2SO_4–Na_2SO_4, the actual concentrations depending upon the conditions of spinning and fibre properties required; in modern rayon production $ZnSO_4$ (up to 6%) is also commonly added. Standard rayon fibres have a so-called 'skin-core' structure, with the cellulose molecules in the peripheral (skin) regions more highly oriented. Progressively increasing the amount of $ZnSO_4$ leads ultimately to strong fibres with an all-skin type of structure; its effect is in some way associated with rapid coagulation of the fibre through formation of temporary cross-links:

$$2CellOCSS^- \longrightarrow [CellOCSS^-\ Zn^{2+}\ {}^-SCSOCell] \xrightarrow{H^+} 2CellOH$$

Fibre properties are further improved by addition of various substances to the xanthate dope. These include amines (cyclohexylamine), quaternary ammonium salts, polyalkylene glycols and their ethers (**29**, R = H or Me; R' = H, alkyl or aryl), tertiary amines (**30**, R = H or Me; R' = H, alkyl or aryl), alkylamines (**31**, alkyl is long-chain) and amides (**32**, alkyl is long-chain), etc. These modifiers apparently help zinc ions to penetrate to the core of the fibres before the xanthate is decomposed by acid.

(**29**)　　$H(OCHRCH_2)_nOR'$　　$\begin{matrix} H(OCHRCH_2)_n\!\diagdown \\ H(OCHRCH_2)_m\!\!-\!\!N \\ H(OCHRCH_2)_l\!\diagup \end{matrix}$　　(**30**)

(**31**)　　$AlkN\!\!\begin{matrix} \diagup(CH_2CH_2O)_nH \\ \diagdown(CH_2CH_2O)_mH \end{matrix}$　　$AlkCON\!\!\begin{matrix} \diagup(CH_2CH_2O)_nH \\ \diagdown(CH_2CH_2O)_mH \end{matrix}$　　(**32**)

Variations in the basic process, such as degree of degradation, degree of polymerization, degree of xanthation, use of additives, coagulant composition, etc., coupled with changes in the spinning and drawing conditions, permit production of a range of fibres with widely differing properties. Notable are high-tenacity tyre-cord rayons (a major outlet) and the high wet-modulus (HWM) rayons that more closely resemble cotton in physical properties and in having a fibrillar structure. Polynosic fibres are a type of HWM fibre, which, unlike standard rayon, will withstand the swelling action of concentrated (>5M) sodium hydroxide and can, therefore, be used in blends with cotton that are to be mercerized.

Viscose rayon has the cellulose II crystal structure (cf. mercerized cotton). Its molecular weight is lower than that of cotton (ca. 250 units for standard fibres, ca. 500 units for a polynosic fibre) whilst X-ray and infrared–deuteriation measurements indicate that its structure is considerably less ordered; this is reflected in a higher moisture regain and greater chemical reactivity. The reactions of viscose rayon parallel those of cotton; it may be dyed with the same types of dye and can be given improved minimum-care properties by treatment with bis(hydroxymethyl)urea derivatives. The fibres can also be modified by use of spinning dope additives. For example, the addition of tris(dibromopropyl) phosphate, $(CH_2BrCHBrCH_2O)_3P(O)$, or of an alkoxyphosphazene (**33**)

$$(AlkO)_2P \underset{N \underset{P}{\searrow} N}{\overset{N \searrow N}{\parallel}} P(OAlk)_2$$
$$(OAlk)_2$$

(**33**)

imparts flame-retardancy. Fibres that are dyeable with acid (wool) dyes and also accept direct dyes more readily are obtained by addition of certain basic substances to the dope; a great many basic compounds and polymers examined in this context failed because they drastically lowered the photostability of dyestuffs on the fibres.

Cellulose Acetate

The two fibres in this, the second most important, group of regenerated fibres, are the so-called diacetate or secondary acetate, with a practical degree of substitution (D.S.) around 2.4, and the triacetate (D.S. usually 2.9–2.95). Acetylation of high-purity wood pulp or cotton linters, pre-swollen in hot acetic acid to improve accessibility, is effected batchwise with acetic anhydride–acetic acid–sulphuric acid (or other strong acid catalyst); some acetolysis occurs concomitantly. The product dissolves as the reaction nears completion and is isolated by precipitation with dilute acetic acid. Acetylation can also be effected in methylene chloride, another solvent for the triacetate. A heterogeneous process is also available, in which the diluent (e.g. benzene or CCl_4) is a non-solvent and perchloric acid is the preferred catalyst; the triacetate is isolated simply by filtration. Use of ketene as the esterifying agent has also been extensively studied.

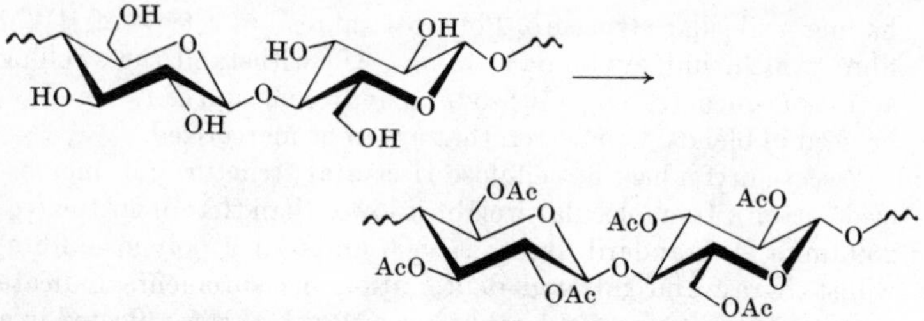

Because acetylation is heterogeneous in the earlier stages, attempts to produce a diacetate by direct acetylation give a product containing fractions with a wide distribution of D.S. around the desired mean. This causes difficulties at the spinning stage and yields inferior fibres. The diacetate is therefore obtained from the triacetate by controlled acid-catalysed hydrolysis, simply by adding water and sulphuric acid to the acetylation mixture and warming at 45–50°C until the product is completely soluble in acetone, the spinning solvent. The treatment also serves to remove sulphate ester groups, which if left in the fibre would slowly undergo hydrolysis.

$$\text{CellOSO}_3\text{H} + \text{H}_2\text{O} \rightleftharpoons \text{CellOH} + \text{H}_2\text{SO}_4$$

The liberated sulphuric acid would then catalyze main-chain hydrolysis and some acetolysis, with a consequent loss in fibre strength, and would cause discolouration during heat processing. This problem does not arise with a perchloric acid catalyst.

Both the diacetate (m.p. ca. 230–250°C) and triacetate (m.p. >300°C) are too unstable to be melt-spun; they are usually dry-spun into a hot-air chamber from, respectively, concentrated (20–25%) solutions in acetone containing a few percent of water as a viscosity-depressant or CH_2Cl_2–MeOH (or EtOH) (90:10). Triacetate can also be wet-spun directly from the acetylation mixture into aqueous acetic acid.

Fully and partially substituted cellulose acetates are obtained by heterogeneous acetylation of cotton with acetic anhydride and phosphoric acid or of regenerated cellulose (viscose rayon) with gaseous acetic anhydride, but these processes are not exploited commercially.

Cellulose acetate fibres have only low to medium tenacity. In the triacetate this may be related to the separation of molecular chains imposed by the bulky acetyl side-groups and hence to weak inter-chain forces, despite the regularity of structure that is reflected in its well-defined X-ray diffraction pattern. With secondary acetate fibres, irregularity of substitution further disrupts interchain forces and adversely affects their ability to orient, although this is partly offset by hydrogen-bonding through free OH groups in adjacent chains. Cellulose diacetate, through its more open structure and free OH groups, is more hydrophilic than the triacetate, with a moisture regain of 7% compared with 3–4%, and can be more readily dyed at any given temperature; however, the greater thermal stability of the triacetate enables it to be dyed at higher temperatures. Problems of static electrification, soil retention and wettability (e.g. with dye liquors) can be minimized by controlled saponification of triacetate fibre surfaces with dilute sodium hydroxide ('S' finish); there is little or no main-chain degradation.

Synthetic Fibres, I. Polyamides

Polyamides belong to one of two main classes of polymer, the AB type (34) that may be regarded as polymers of ω-amino-acids, and the AABB type (35) obtained from the polycondensation of diamines and dicarboxylic acids. Co-polymers containing both types of structure can be prepared.

$$\sim\!\!NH(CH_2)_nCO\!\!\sim \qquad \sim\!\!NH(CH_2)_nNHCO(CH_2)_mCO\!\!\sim$$

(34) Nylon $n + 1$ (35) Nylon $n, m + 2$

Of the many polymers examined—aliphatic, alicyclic and aromatic—only nylon 6 (34, $n = 5$) and nylon 6.6 (35, $n = 6, m = 4$) are produced in quantity, the latter preponderating slightly. Their acceptance over other nylons lies in the relative ease of processing them into fibres with a good overall balance of mechanical and chemical properties, coupled with the availability of economical methods for the synthesis of the intermediates on a large scale from mainly petrochemical sources.

Nylon 6

Nylon 6 (36) is prepared by thermal polymerization of capro-lactam (6-hexanolactam) (Part 2, p. 229) in an inert atmosphere at temperatures up to 270°C. An initiator is needed, usually water, 6-aminohexanoic acid or nylon 6.6 salt (see below) (1–2%), and evidence points to an addition mechanism.

$$n \; \underset{NH}{\overset{O}{\bigcirc}} \; + H_2O \longrightarrow HO—[—CO(CH_2)_5NH—]_{n-1}—CO(CH_2)_5NH_2$$

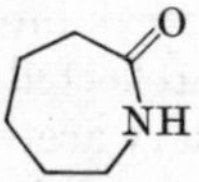

$$(36)$$

Small amounts of monocarboxylic acid, e.g. acetic acid, may be added as chain-regulator, to control the molecular weight of the polymer by blocking amine end-groups. Some 10% of monomer present at equilibrium must be extracted with water before the fibre is melt-spun. Small amounts of cyclic oligomers (**37**, $n = 1, 3, 4$ up to 8) are also formed. Further reconversion into monomer

$$\begin{array}{c} NH(CH_2)_5CO \\ | \qquad\qquad | \qquad (37) \\ [CO(CH_2)_5NH]_n \end{array}$$

occurs to a limited extent during spinning.

The various routes to caprolactam and their relative importance are discussed in Chapter 7.

Nylon 6.6

Nylon 6.6 (**38**) is obtained by thermal polycondensation of hexamethylenediamine with adipic acid (Part 1, p. 172). The preparation of these intermediates has been discussed in previous Chapters.

To produce polymers of high molecular weight, the reactants must be in exactly stoichiometric proportions and this is achieved by first mixing them in methanolic solution, when the pure 1:1 salt ('nylon salt') separates out. A concentrated solution of this salt is heated in an inert atmosphere to about 270°C under pressure; steam is then bled off and the condensation completed under a vacuum.

$$HOOC(CH_2)_4COOH + H_2N(CH_2)_6NH_2 \longrightarrow$$

$$\sim\!\!\sim\!CONH(CH_2)_6NHCO(CH_2)_4\!\sim\!\!\sim$$

$$(38)$$

An excess of either reagent would cause premature termination of the polymerization; molecular weight is better controlled by addition of a small amount of acetic acid to the system, so that the polymer has some —NHAc end-groups. Control of molecular weight is necessary to give a polymer with the optimal viscosity for successful melt-spinning consistent with good mechanical properties of the fibre; typical values are $10-15 \times 10^3$ for both nylon 6 and nylon 6.6.

Drawn fibres of nylon (both types) appear from X-ray data to have a layered sheet structure in the crystalline regions, with extensive intermolecular hydrogen-bonding (Figure 9.3). However,

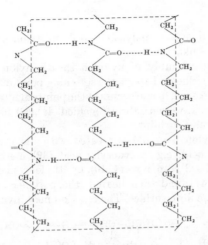

Figure 9.3. Hydrogen bonding in nylon 6. [Reproduced, by permission, from D. R. Holmes, C. W. Bunn and D. J. Smith, *J. Polymer Sci.*, **17**, 168 (1955).]

infrared spectra indicate that the nylons, like cellulose, contain a considerable proportion of NH groups that are not H-bonded and these groups readily undergo H—D interchange with D_2O vapour.

Amine end-groups facilitate the uptake of acid dyes and reactive dyes by nylon. They are estimated in solution by titration with HCl in phenol; accessible end-groups in the solid fibre can be determined by conversion into the 2,4-dinitrophenyl derivative. Acid end-groups can be titrated with $KOH-HOCH_2CH_2OH$ in hot benzyl alcohol.

The difference in melting temperature (Tm) between nylon 6 (215–220°C) and nylon 6.6 (265–270°C) is exploited in bicomponent fibres made by spinning both polymers simultaneously through the same orifice under conditions such that the molten polymer streams do not mix. The resulting fibre can be given a side-by-side (i) or sheath-core (ii) structure; a third class of bicomponent fibre (iii) consists of fine fibrils of one polymer dispersed in a matrix of a

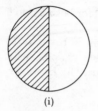
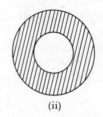

(i)   (ii)   (iii)

second polymer, e.g. polyester fibrils in nylon 6 ('Source') or in nylon 6.6 ('Enkatron').

When an assembly of nylon 6 sheath–nylon 6.6 core fibres ('Heterofil') is heated above the softening point of nylon 6 and subjected to pressure, fusion occurs at the points of intersection to yield a type of non-woven fabric ('melded' fabric) useful in carpets, upholstery and curtains.

When nylon 6 or 6.6 is treated with an aqueous polymeric carboxylic acid, e.g. polyacrylic acid (**39**) methyl vinyl ether–maleic acid 1:1 co-polymers (**40**), or ethylene–maleic acid 1:1 co-polymers (**41**), and then heated, the polymer coating is firmly bound to the fibre surface. The reaction may involve (*a*) acylation

$$\sim CH_2-CH\sim \qquad \sim CH_2CH-CH-CH\sim \qquad \sim CH_2-CH_2CH-CH\sim$$
$$\quad\ \ \ | \qquad\qquad\qquad\ \ \ |\quad\ \ \ |\quad\ \ \ | \qquad\qquad\qquad\qquad |\quad\ \ \ |$$
$$\quad\ COOH \qquad\qquad OMe\ COOH\ COOH \qquad\qquad HOOC\quad COOH$$

(**39**)   (**40**)   (**41**)

of end-groups, (*b*) transamidation or (*c*) imide formation. The

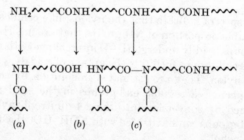

(*a*)   (*b*)   (*c*)

process can be used to stiffen nylon and to impart better soil-release properties and may occur accidentally through overdrying of nylon yarns sized with polyacrylic acid before weaving.

Other AB-type Nylons

The range of aliphatic nylons containing the repeat unit $-CO(CH_2)_nNH-$ ($n = 3\text{--}12$) are all known, but none has reached production on a scale approaching that of nylon 6 ($n = 5$). They show a typical alternating effect in Tm (see Chapter 8, p. 206) with an overall downward trend as the aliphatic chain lengthens.

n	2	3	4	5	6	7	8	9	10	11	12
Tm(°C)	330	265	260	230	223	233	200	209	188	190	179

The homologues below nylon 6 are of great interest on account of their hydrophilicity; nylon 3 has a moisture regain of about 10% at 21°C and 65% relative humidity.

Of these polymers, nylon 11 ('Rilsan'), nylon 7 ('Enant', U.S.S.R.) and recently nylon 4 ('Tajmin') have been produced as fibres; nylon 12, although not used as a primary fibre, is used in hot-melt adhesive threads. Nylon 11 and nylon 12 are also used as engineering plastics.

Nylon 3. Anionic polymerization of 3-propiolactam yields a fibre-forming polymer, but unfortunately the monomer is difficult to prepare. However, substituted β-lactams can readily be obtained from the appropriate olefin (ethylene is too inert) and yield poly-β-amides with anionic catalysts:

$$SO_3 + ClCN \longrightarrow ClSO_2NCO$$

$$ClSO_2NCO + R^1R^2C{=}CR^3R^4 \longrightarrow \begin{array}{c} R^1R^2C{-}CR^3R^4 \\ | \quad | \\ ClSO_2N{-}CO \end{array} \xrightarrow{H_2O}$$

$$\begin{array}{c} R^1R^2C{-}CR^3R^4 \\ | \quad | \\ HN{-}CO \end{array} \longrightarrow \sim\!NH{-}CR^1R^2{-}CR^3R^4{-}CO\!\sim$$

1,2-Dimethylpropiolactam $\overline{\text{NHCHMeCHMeCO}}$ exists in racemic *cis-* and *trans-*forms, which give polymers with repeat units in the *erythro-* and *threo-*configuration, respectively; the *erythro-*form has the greater thermal stability. Fibres are obtained from these polymers by wet-spinning from, e.g. methanol–calcium thiocyanate. Poly-3-aminobutyric acid, $\sim$NHCHMeCH$_2$CO$\sim$, yields fibres with many similarities to natural silk. Acrylamide can be polymerized to nylon 3 by a Michael-type self-addition.

Nylon 4. Pyrrolidone (4-butyrolactam) is obtained from acetylene (see Chapter 4, p. 77). Anionic polymerization with CO_2 as initiator yields a polymer of high molecular weight that can be melt-spun, despite the fact that the lactam–polymer equilibrium completely favours the monomer at 280°C.

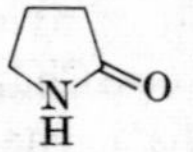

$$\text{(pyrrolidone structure)} \longrightarrow -CO(CH_2)_3NH-$$

Nylon 7. One route to this fibre is based on 6-hexanolactone:

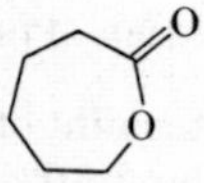

$$\text{(caprolactone structure)} \longrightarrow Cl(CH_2)_5COOH \xrightarrow[\text{EtOH}]{\text{KCN}} CN(CH_2)_5COOEt \xrightarrow{H_2}$$

$$H_2N(CH_2)_6COOEt \xrightarrow{H_2O} H_2N(CH_2)_6COOH \xrightarrow{\Delta} \text{nylon 7}$$

Another, developed in Russia and Israel, starts from the telomerization of ethylene with CCl_4:

$$CCl_4 + nCH_2{=}CH_2 \xrightarrow{R_2O_2} Cl(CH_2CH_2)_nCCl_3 \xrightarrow[(n=3)]{H_2SO_4,\ H_2O}$$
$$\text{(42)}$$

$$Cl(CH_2)_6COOH \xrightarrow[100°]{NH_3} H_2N(CH_2)_6COOH \xrightarrow{\Delta} \text{nylon 7}$$

Nylon 9. The telomerization route for nylon 7, with the appropriate homologue (**42**, $n = 4$), has been used to give nylon 9, whilst a more recent process is based on the production of 9-amino-nonanoic acid from soyabean oil.

Nylon 11. Nylon 11 is obtained by polymerization of 11-aminoundecanoic acid which is produced from castor oil (see Chapter 2).

Nylon 12. Trimerization of butadiene affords cyclododeca-1,5,9-triene, which can be converted into dodecanolactam, the nylon 12 precursor, by several standard routes (see Chapter 4).

Other AABB-type Aliphatic Nylons

Only nylon 6.10 and to a lesser extent nylon 10.10 appear to have been examined on a commercial scale, the former chiefly as

$$CH_3(CH_2)_5CH(OH)CH_2CH{=}CH(CH_2)_7COOH$$
$$\downarrow \text{NaOH} < 200°$$
$$CH_3(CH_2)_5COCH_3 + HOCH_2(CH_2)_8COOH$$
$$\downarrow \begin{array}{c} \text{NaOH, H}_2\text{O} \\ 240° \end{array}$$
$$CH_3(CH_2)_5CH(OH)CH_3 + HOOC(CH_2)_8COOH$$

coarse filaments for the bristle industry. The C_{10} diacid, sebacic acid, is obtained by high-temperature alkali-scission of ricinoleic acid (from castor oil). Polycondensation with hexamethylene-diamine or decamethylenediamine yields the nylon in the usual way.

Alicyclic Polyamides (Qiana)

A novel group of polyamide fibres is obtained by the conden-sation polymerization of di-(4-aminocyclohexyl)methane [4,4'-methylenedi(cyclohexylamine)] with C_8, C_{10} and C_{12} dibasic fatty acids or their chlorides. The amine is obtained by hydrogenation of the condensation product of aniline and formaldehyde, the reaction being controlled to give a product containing at least 50% of the most nearly planar *trans-trans*-isomer (**43**).

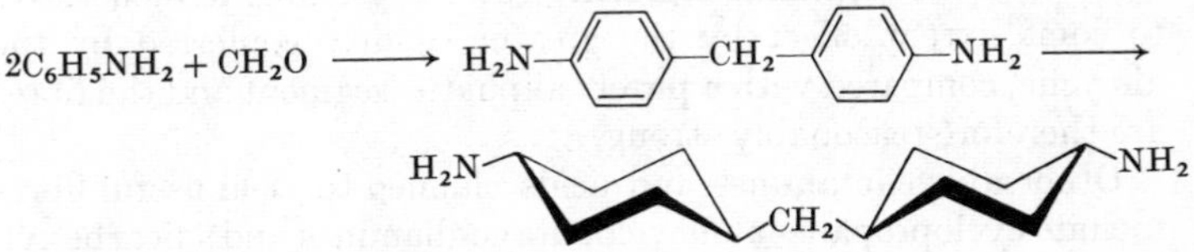

$$2C_6H_5NH_2 + CH_2O \longrightarrow H_2N\text{—}\bigcirc\text{—}CH_2\text{—}\bigcirc\text{—}NH_2 \longrightarrow$$

(**43**)

Steric control is essential to the production of a spinnable material, since the ratio of *trans-trans : trans-cis : cis-cis*-isomers governs to some extent the physical properties, in particular Tm (melting temperature) and Tg (glass–rubber transition tempera-ture) of the polymer (cf. Chapter 8, p. 206).

The melt-spun fibre from (**43**) and probably the C_{12} diacid can be made into fabric with silk-like aesthetics (Qiana). This fibre has about the same tenacity as nylon 6.6 but better recovery from small strains and slightly lower moisture regain. A feature of this group of fibres is the high Tg value compared with those of nylon 6 and 6.6, which is attained without an accompanying rise in Tm to

Table 9.4. Comparison of some properties of Qiana fibres and those of nylon 6.6

	Tm (°C)	Tg (°C) Dry	Tg (°C) Wet	Moisture regain (%)
Qiana fibre (C_8 acid)	240	140	—	3.2
Qiana fibre (C_{12} acid)	205	135	100	3.0
Nylon 6.6	265	90	<20	4.3

a value unacceptably high for processing of the polymer. Some quoted figures are shown in Table 9.4.

An important consequence of the high Tg value is that Qiana fibres remain in the glassy state during fabric laundering. It is the sudden cooling of fibres from above to below the Tg (e.g. by cold rinsing), before creases have had the opportunity to smooth out, that results in these creases being set into the fabric until it is again heated above the Tg. Qiana fabrics are therefore free from this limitation and have excellent non-iron properties; nylon fabrics will shed creases even if suddenly cooled, provided they are kept damp.

The disparity in length of the acid and amine segments in these alicyclic nylons precludes the high degree of interchain hydrogen-bonding found in nylons 6 and 6.6, and close-packing is inhibited by the bulk of the amine segment. These weakening influences are to some extent offset by the greater rigidity conferred by the alicyclic, compared with a purely aliphatic, segment and the fibres are therefore reasonably strong.

Other alicyclic amines and acids claimed to yield useful fibres include cyclopropane- and cyclobutane-diamines and -dicarboxylic acids, cyclobutylenedi(methylamines), *trans*-1,4-cyclohexylene-diamine and -di(methylamines) and 4,4'-methylenedi(cyclohexane-carboxylic acid).

Synthetic Fibres, II. Polyesters

'Terylene'

Of the many polyesters that have been examined, poly(ethylene terephthalate) ('Terylene') still virtually monopolizes this class, for both practical and economic reasons. Terephthalic acid is derived from *p*-xylene by catalytic oxidation directly (Part 1, p. 170) or as the half-ester in two stages via *p*-toluic acid; it is also obtainable by carboxylation of potassium benzoate or isomerization of potassium phthalate (see p. 128). Production of fibre-grade polymer by direct polyesterification with ethylene glycol is possible, but requires terephthalic acid of very high purity. More usually, the acid is converted first into the easily purified dimethyl ester (**44**), which undergoes ester interchange with an excess of ethylene glycol to yield an intermediate mixture of glycol terephthalate and oligomers. This is further condensed at elevated temperature under reduced pressure with elimination of glycol, to a high polymer suitable for melt-spinning; antimony oxide is used as a catalyst. A cyclic trimer is also produced in minor amounts.

$$MeOCO-\langle\bigcirc\rangle-COOMe \longrightarrow$$

(44)

$$HOCH_2CH_2OCO-\langle\bigcirc\rangle-COOCH_2CH_2OH + oligomers \xrightarrow[Sb_2O_3]{\Delta}$$

$$HOCH_2CH_2OCO-\langle\bigcirc\rangle-COOCH_2CH_2\!\sim\!\!OCO-\langle\bigcirc\rangle-COOCH_2CH_2OH$$

Success is highly dependent on careful purification of materials and exact process control. At the spinning stage, water must be rigorously excluded, otherwise rapid hydrolysis of the ester occurs at the high temperatures necessary (270–280°C), with a consequent lowering of molecular weight. The filaments are hot drawn (>80°C) to induce orientation.

X-ray diffraction data indicate that molecules in the crystalline regions of oriented polyethylene terephthalate fibres are almost fully extended, co-planarity of the aromatic rings (Figure 9.4, p. 286) permitting a high degree of close packing. A consequence of this rather rigid structure is that polyester fibres are stiffer than other common synthetic fibres. The orientation and absence of polar groups result in a very hydrophobic fibre with negligible moisture regain. The almost complete resistance to swelling in water presented a serious obstacle to the development of polyester fibres, since it was necessary to find new high-temperature dyeing methods and dye carriers (mainly aromatic compounds such as biphenyl, diphenyl ether, and methyl salicylate that dissolve dyes and are absorbed by the polyester) in order to obtain satisfactory rates of dyeing and depths of shade.

Poly(ethylene terephthalate) is susceptible to hydrolysis by strong alkali, with progressive erosion of the fibre surfaces, and to aminolysis; hydrazine hydrate rapidly destroys the fibres. The

$$\sim\!\!OCH_2CH_2OCO-\langle\bigcirc\rangle-CO-O- \xrightarrow{N_2H_4}$$

$$HOCH_2CH_2OH + H_2NNHCO-\langle\bigcirc\rangle-CONHNH_2$$

terephthalate is otherwise chemically inert. Soiling of polyester fabrics is a serious problem, ascribable to the ease with which they acquire a static charge, which then attracts oppositely charged soil particles, and to their strong affinity for oils, fats and greases,

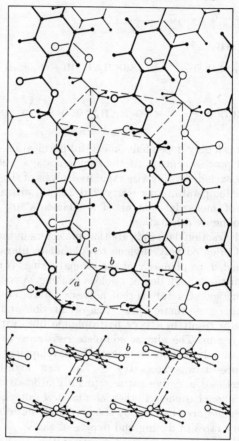

Figure 9.4. Crystal structure of polyethylene terephthalate. [Reproduced, by permission, from H. F. Mark, S. M. Atlas and E. M. Cernia, *Man-made Fibres, Science and Technology*, Vol. III, Interscience, New York, 1968.]

which retain soil tenaciously. This disadvantage can be minimized by treatment with a small quantity of non-fibrous polymer (Permalose TG), which resembles the polyester in structure but has a proportion of the glycol groups replaced by polyoxyethylene

$$\sim OCH_2CH_2OCO-\!\!\!\bigcirc\!\!\!-COO(CH_2CH_2O)_n-CO-\!\!\!\bigcirc\!\!\!-CO-O\sim$$

(45)

groups (cf. **45**; *n* is fairly small). The polymer is integrated into the surfaces of the fibres when they are heated briefly at 180–200°.

Many co-polymers, retaining ethylene terephthalate as the major component, have been made in efforts to create fibres that are more accessible (i.e. less highly ordered in structure), more hydrophilic, more readily dyed, or dyeable with dyes other than the disperse types normally used. Both glycol and acid units have been replaced. Substitutes for the latter include isophthalate (**46**), *p*-hydroxybenzoate (**47**) and, for basic dye-affinity, sulphoisophthalate residues (**48**).

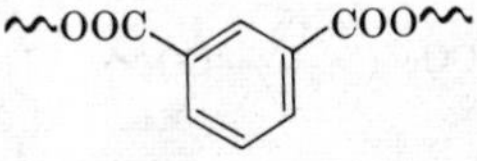

| (46) | (47) | (48) |

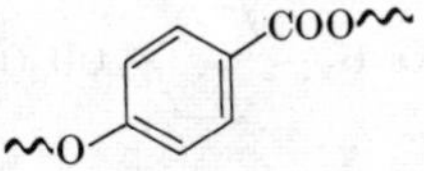

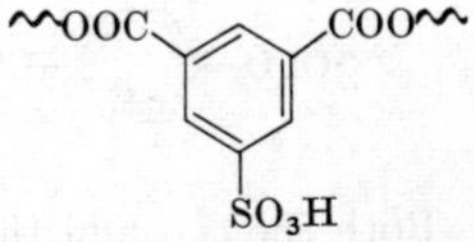

Polyester fibres burn readily, but because they are thermoplastic and thus tend to drip away from a burning fabric, they are not classed as highly flammable. However, in combination with cellulosic fibres, which provide a carbonaceous skeleton, the polyester is prevented from dripping and becomes highly hazardous. The detailed mechanism of the initial stages of pyrolysis is imperfectly understood, although formation of vinyl ester groups occurs as illustrated.

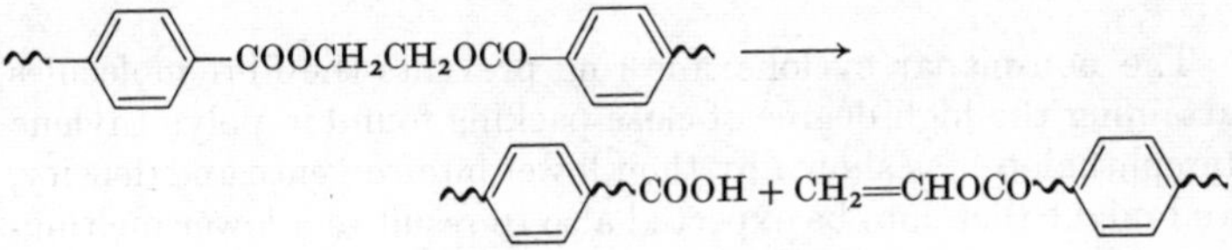

Formation of benzene and toluene under anaerobic conditions at ca. 600°C serves as a means of identifying polyester fibres by pyrolysis–GLC techniques.

Other Polyester Fibres

Of the many polyesters examined as competitors to poly(ethylene terephthalate), perhaps the best-known is the terephthalate of 1,4 - cyclohexanedimethanol (1,4 - bishydroxymethylcyclohexane) ('Kodel'). Selective reduction of dimethyl terephthalate yields the alicyclic diester (**49**), which is then further hydrogenated over copper chromite to the diol (**50**). Transesterification of dimethyl terephthalate with this diol yields the polymer, which can be melt-spun, drawn, and heat-set in the usual way.

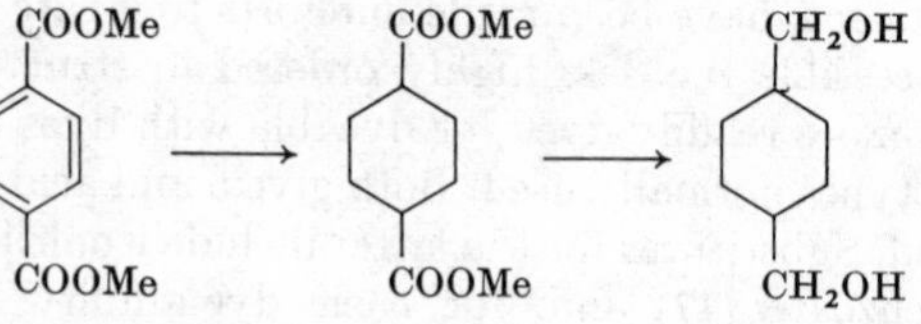

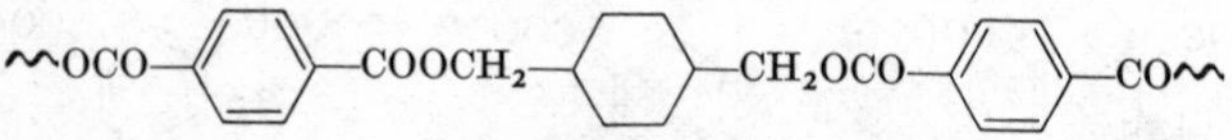

Both the *cis*- and the *trans*-isomer of the diol (50) are known, and polymers made from mixtures of these show an increase in melting point and glass-transition temperature with increase in *trans*-diol content. In practice, the diol is usually about 70% *trans*-isomer, which probably has the diequatorial conformation (51); however, rapid cooling of the molten polymer as it emerges from the spinneret may produce metastable conformers.

The non-planar cyclohexane ring prevents the fibre molecules attaining the high degree of close-packing found in poly(ethylene terephthalate), as shown by their lower birefringence and density, and might therefore be expected also to result in a lower melting-point. In fact, this effect is more than offset by the increased rigidity imposed by the diol ring system and both Tm and Tg of the fibres (70% *trans*-diol) are higher than the values for poly(ethylene terephthalate). The greater resistance to hydrolysis than that of poly(ethylene terephthalate) fibres is ascribed to steric hindrance in the diol segment. Susceptibility to thermal and actinic oxidation, however, is greater.

A polyester that has aroused recent interest is the fibre ('A-Tell') derived from *p*-hydroxybenzoic acid; it is said to have silk-like properties. Additionally, the polyester derived from butane-1 4-diol and terephthalic acid has been developed in the U.S.A. Two aliphatic polyesters of interest are polypivalolactone (52) and

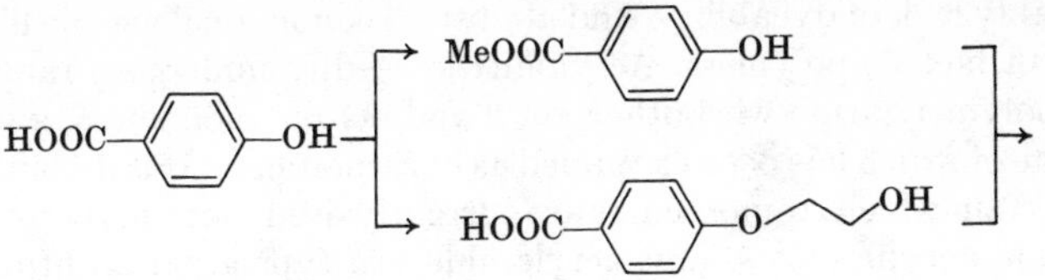

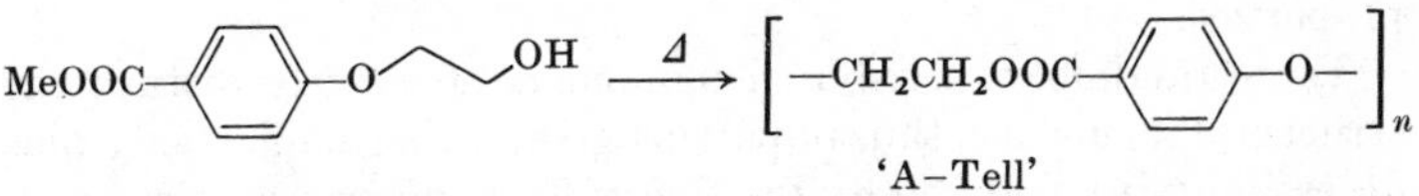

'A–Tell'

polyglycollic acid (53). Fibres of (52) have a melting point of about 240°C and reasonable stability to hydrolysis is conferred by steric hindrance.

$\sim\!\!\sim\!\!OCH_2CMe_2CO\!\!\sim\!\!\sim$ $\sim\!\!\sim\!\!OCH_2CO\!\!\sim\!\!\sim$

        (52)   (53)

Fibres of (53) have no textile value, but their low resistance to hydrolysis is turned to advantage in fibres for surgical sutures ('Dexon'), which are slowly hydrolysed and dissolve in body fluids without toxic or other undesirable side effects.

Synthetic Fibres, III. Acrylic Fibres

By definition, these fibres contain at least 85% of acrylonitrile; a separate group, containing 35–85% of acrylonitrile, are classed as 'modacrylics' (see below). Because of their wool-like aesthetics when suitably textured, acrylic fibres have their main outlets in knitwear rather than in woven goods, and in carpets and upholstery. The various routes to acrylonitrile are discussed in Chapter 7 (p. 175).

Radical polymerization of acrylonitrile occurs readily in aqueous suspension with standard redox catalysts, the polymer separating as a powder of molecular weight 75–150×10^3. The polymer is then dry-spun from solution in dimethylformamide into hot air or wet-spun from dimethylformamide, dimethylacetamide or aqueous sodium thiocyanate into a suitable aqueous coagulating bath. In some processes, solution polymerization, in solvents of low transfer constant to give a sufficiently high molecular weight (e.g. in aqueous NaSCN), is now used to yield the spinning dope directly. Fibres of acrylonitrile homopolymer present practical problems, notably lack of dyeability, and almost all commercial acrylic fibres are in fact co-polymers. Acrylonitrile readily undergoes random co-polymerization with other vinyl and acrylic monomers, a wide range of which has been examined as fibre modifiers. It is difficult to ascertain which co-monomers are currently used, but vinyl acetate, simple acrylic esters, and acrylamide are typical; if addition is restricted to below 15%, the fibres retain acceptable mechanical properties.

These randomly distributed monomers effectively disturb the structural regularity, thus imparting greater thermoplasticity plus accessibility to, and affinity for, dyestuffs. Some preparations may deliberately include sufficient (less than 5%) of a third co-monomer with a high affinity for specific types of dye, e.g. acidic monomers for basic dyes and vice-versa. A large number of these co-monomers have been suggested; itaconic acid, isobutene-1-sulphonic acid and sulphoalkylacrylamides for basic dyes, and vinylpyridines for acid dyes, are examples. Partial hydrolysis to introduce COOH groups, and reaction with hydroxylamine, have also been used to provide acceptor sites for, respectively, basic and acid dyes.

$$\text{NC}\!\!-\!\!\underset{\text{NC}\!\!-\!\!}{\Big]}\!\!-\!\!\text{COOH} \xleftarrow{\text{H}_3\text{O}^+} \text{NC}\!\!-\!\!\underset{\text{NC}\!\!-\!\!}{\Big]}\!\!-\!\!\text{CN} \xrightarrow{\text{H}_2\text{NOH}} \text{NC}\!\!-\!\!\underset{\text{NC}\!\!-\!\!}{\Big]}\!\!-\!\!\text{C}\!\!\underset{\text{NHOH}}{\overset{\text{NH}}{\diagup}}$$

Despite having a predominantly atactic structure, polyacrylonitrile forms quite strong fibres. X-ray diffraction data indicate a good degree of lateral order, but there is no evidence of true crystallinity; this is not surprising since the molecules lack the tacticity necessary for regular axial alignment. Intermolecular bonding can be ascribed to strong dipolar interaction between C=N groups on neighbouring chains, the energy of which may equal or exceed that of H-bonds in cellulose and polyamide fibres. In consequence, polyacrylonitrile fibres are soluble only in highly polar solvents. Attempts to produce stereoregular polyacrylonitrile fibres have failed; anionic polymerization yields a coloured product.

Crimping of acrylic fibres, so that they more closely resemble wool, is widely practised and modern techniques make use of bicomponent fibres. For example, a side-by-side bicomponent fibre is spun from two acrylic polymers containing different amounts of co-monomer such as vinyl acetate, and is then heated. The two components relax (shrink) to different extents and the net result is that the fibre curls. The crimp is permanent at temperatures up to those used in the relaxation treatment.

Modacrylic Fibres. These co-polymer fibres, by definition containing 35–85% of acrylonitrile, were originally developed to improve the properties of PVC fibres (e.g. to raise the softening point) and facilitate spinning from common solvents; they now find use as flame-retardant fibres, alone or in blends with others. They are basically random co-polymers of vinyl chloride or vinylidene dichloride with acrylonitrile, obtained by conventional radical polymerization; examples are 'Dynel', 'Teklan', 'Verel' and 'Kanekalon'. They are all solution-spun. Because of their irregular structure and low acrylonitrile content, interchain bonding is lower in these fibres than in the true acrylics. Consequently the fibres are usually weaker, dissolve in less polar spinning solvents, and soften at lower temperatures than the true acrylics, whilst retaining their chemical resistance and resistance to photo-degradation. A modacrylic fibre more closely resembling the true acrylics is now available ('Acrilan Modacrylic'). Many other halogen-containing co-monomers have been suggested for production of modacrylic fibres, e.g. $CH_2=CHBr$, $CH_2=C(CH_2Cl)CN$ and $CH_2=CHCOOCH_2C(CH_2Br)_3$, but these are expensive and have not found commercial application.

Synthetic Fibres, IV. Polyolefins

Polyethylene

Polymerization of ethylene at high pressure and temperature with a peroxide or other radical initiator yields a low-melting (110°C), low-density (0.92) product, in which a significant amount of chain-branching is evident and interferes with close packing of the polymer molecules. A more regular, linear polymer of molecular weight ca. 20×10^3 is obtained by polymerization on a supported heavy-metal oxide or Ziegler-type catalyst at low pressure. This polymer can be melt-spun at 300°C to yield, after stretching, rather weak fibres with a melting point around 130°C and density 0.95. Their waxy feel, relatively low melting point, zero moisture regain, and ease of soiling render polyethylene fibres unsuitable for use in apparel, except protective clothing, although they may be used in shrinkage yarns and as fusible (heat-bonding) components of non-woven fabrics.

Polypropylene

In contrast with polyethylene, polypropylene fibres have attained considerable importance. They are derived from the largely isotactic polymer, which is produced by low-pressure, low-tem-

perature polymerization over a Ziegler-type catalyst in an inert
hydrocarbon solvent; atactic polypropylene has no fibre-forming
characteristics, and the syndiotactic form is not made commercially. The polymer, of melting temperature 165°C and molec-

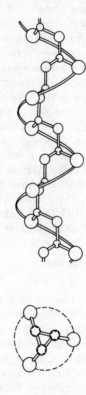

Figure 9.5. Helical structure of isotactic polypropylene. [Reproduced, by
permission, from G. Natta and P. Corradini, *J. Polymer Sci.*, **39**, 36 (1959).]

ular weight up to 4×10^5 is filtered off from the reaction mixture,
freed from catalyst, compounded with antioxidant and
pigment (if required), melt-spun and drawn. A degree of tacticity
around 90% is essential, and the oriented fibres may have a high
degree of crystallinity (50–60%). To minimize steric interaction
between the methyl groups, the essentially linear molecules
take up a helical configuration, with three monomer units per
turn and with the backbone C—C bonds in alternate *trans-* and
gauche-conformations (Figure 9.5).

A large quantity of polypropylene is converted into fibres by film-fibrillation instead of melt-spinning. In one method, an extruded film of polymer is slit into narrow tapes (of value *per se* for making twines and coarse fabrics), which are oriented by hot-stretching. Each tape is then scored by passage over sets of closely spaced pins mounted on a roller. This mechanical action causes the tape to fibrillate longitudinally into a network of fibres, which are then twisted into yarns.

Polypropylene has found uses in cordage, carpet backing, sacks, rugs and blankets, but its low softening point, hydrophobic character and resistance to dyeing have prevented its widespread use in apparel. Great efforts have been made to improve the dyeability of polypropylene, since mass pigmentation of fibres before spinning cannot provide the range of shades and brightness necessary for successful marketing. The absence of dye-receptor sites in polypropylene cannot be remedied by co-polymerization, as in the case of acrylic fibres, because polar monomers would deactivate the catalyst system, whilst chemical inertness precludes modification of the fibres after spinning. One solution to the problem has been the addition of nickel salts to the polymer before spinning; these act as complexing agents for suitable disperse dyes. Again, a second non-fibrous polymer may be dispersed in the polypropylene before spinning, e.g. polyvinylpyridine, which has an affinity for acid dyes. Whilst this type of treatment provides dye sites, the problem of diffusion of dye into the fibre remains and many fibres so modified exhibit ring-dyeing, i.e. only the peripheral regions are coloured. This problem is said to be obviated in an acid-dyeable fibre ('Herculon Type 404') in which dye-receptivity is imparted by a dispersion of microfilaments of a highly polar, hydrophilic, poorly crystalline polymer of undisclosed constitution. The microfilaments provide a pathway for aqueous dye liquors to penetrate throughout the fibre. Polypropylene dyed in this way may be suitable for carpet pile yarns or apparel fabrics.

Other Polyolefins

Of the higher polyolefins, the isotactic polymer from 4-methyl-pent-1-ene itself obtained by anionic dimerization of propylene, yields extremely light fibres (*d* 0.83) with tenacity and extensibility comparable to those of polypropylene. These fibres are, however, much higher melting (Tm 240°C) and far less liable to shrink on laundering or dry-cleaning. Polystyrene and styrene–acrylonitrile co-polymers are produced as rather thick extruded filaments for brush 'bristles'.

Elastomeric Fibres

Natural Rubber

Fibres are prepared from rubber, *cis*-polyisoprene (**54**) of molecular weight ca. $50–100 \times 10^3$, by compounding the latex with a vulcanizing agent such as sulphur, a dithiocarbamate (**55**) or a thiuram (**56**) and extruding it into a coagulating bath of dilute

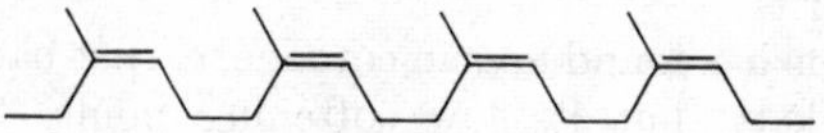

(**54**)

$$(Me_2NCSS)_2Zn \qquad Me_2NCSS\text{---}SCSNMe_2$$
(**55**)   (**56**)

acetic acid. The fibres are then heated until the requisite degree of vulcanization, i.e. cross-linking, has occurred. In the relaxed state, the molecules in rubber are highly coiled and disordered and the high extensibility of the fibres is due to the axial orientation of the molecules under stress; stretched rubber gives an X-ray diffraction diagram typical of a crystalline polymer. Cross-links act as the 'memory' of the fibre, restoring the molecules to their original coiled configuration on release of stress. If too many cross-links are created, the over-vulcanized fibre will be brittle and inelastic.

Polyurethanes

Synthetic elastomeric fibres are still almost exclusively polyurethanes. The collective term 'Spandex' is given (U.S. Federal Trade Commission) to fibres containing at least 85% of a segmented polyurethane, i.e. a block copolymer of 'soft' and 'hard' segments. The 'soft' segments are of two basic types, either polyethers or polyesters of molecular weight up to 4×10^3; the 'hard' segments are produced by reaction of these 'soft segments' with a di-isocyanate. Typical polyethers are polyethylene glycol (**57**), polypropylene glycol (**58**), or block co-polymers of these (**59**), or polytetramethylene glycol (**60**) obtained by polymerization of tetrahydrofuran.

(**57**) $HOCH_2CH_2(OCH_2CH_2)_mOCH_2CH_2OH$

(**58**) $HOCH_2CHMe(OCH_2CHMe)_mOCH_2CHMeOH$

(**59**) $HOCH_2CH_2(OCH_2CH_2)_m(OCH_2CHMe)_nOCH_2CHMeOH$

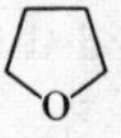

$$\xrightarrow[\text{or } BF_3]{HSO_3F} \quad HO(CH_2)_4O(CH_2)_4O\!\sim\!\sim(CH_2)_4OH$$
(**60**)

Polyesters are commonly obtained by condensation of adipic acid with a slight excess of glycol, so that the chains are terminated by OH groups (**61**); primary end-groups are preferred because of their greater reactivity.

$$HOCH_2CH_2O[CO(CH_2)_4COOCH_2CH_2O]_mCO(CH_2)_4COOCH_2CH_2OH$$
$$(\textbf{61})$$

The co-reactant di-isocyanate is usually aromatic, e.g. a mixture of 2,4- and 2,6-tolylene di-isocyanate (80:20 or 65:35); these compounds are prepared by the action of phosgene on the appropriate amine (**62**). Alternative simpler routes to the isocyanates are

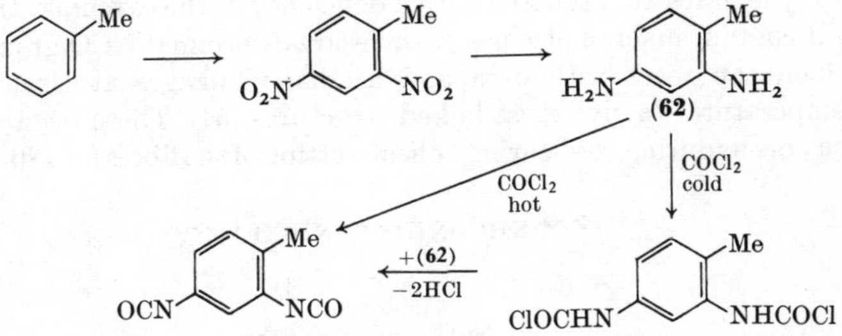

keenly sought; attempts to prepare them directly from the nitro-compound and CO have had limited success.

In a second reaction, the HO-terminal polyether or polyester, HO—X—OH, and an excess of di-isocyanate, OCN—Y—NCO, give an isocyanate-ended polymer:

$$OCN—Y—NHCO(O—X—OCONH—Y—NHCOO—X)_nOCONH—Y—NCO$$

This prepolymer is further extended by reaction with an aliphatic diamine, hydrazine or diacylhydrazine to yield the final product; a small amount of secondary amine may be added as a chain regulator to limit the molecular weight.

$$R_2NH + OCN\sim\sim NCO + H_2NNH_2 + OCN\sim\sim NCO + HNR_2 \longrightarrow$$

$$R_2NCONH(\sim\sim NHCONHNHCONH)_n\sim\sim NHCONR_2$$

Melt-spinning of polyurethanes is precluded by their thermal instability which results *inter alia* in re-formation of di-isocyanates. Fibres are therefore obtained in one of three ways, the simplest being to wet-spin a solution of the polymer in dimethylformamide into a coagulating bath or to dry-spin it into a hot atmosphere. In the other method, known as 'reaction spinning' the final chain-extension is carried out during the spinning operation; the prepolymer is extruded from spinnerets through aqueous hydrazine or diamine, whereupon the very rapid reaction produces a tough skin of polymer on the filaments; the filaments are wound up and drawn into finer fibres; further treatment with diamine, or even hot water, then completes the curing operation in the core of the fibres. These polymers (63) are linear and soluble in polar solvents.

$$\text{\textasciitilde\textasciitilde NCO} + H_2O + \text{OCN\textasciitilde\textasciitilde} \longrightarrow \text{\textasciitilde\textasciitilde NHCONH\textasciitilde\textasciitilde} + CO_2$$
(63)

If the fibres are prepared with a deficiency of the extender, they will contain macromolecules with residual terminal NCO groups, which can react with urea and urethane linkages at elevated temperatures to give cross-linked structures (64). These reactions may occur during core-curing or heat-setting of the fibres (at 180°C).

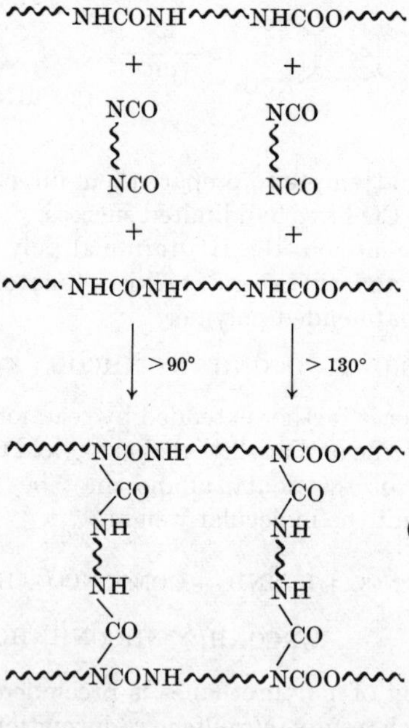

Cross-linking is also achieved if the original linear polyether or polyester is replaced by a slightly branched one (65), e.g. by incorporating small amounts of a triol such as $CH_3CH_2C(CH_2OH)_3$ into the structure:

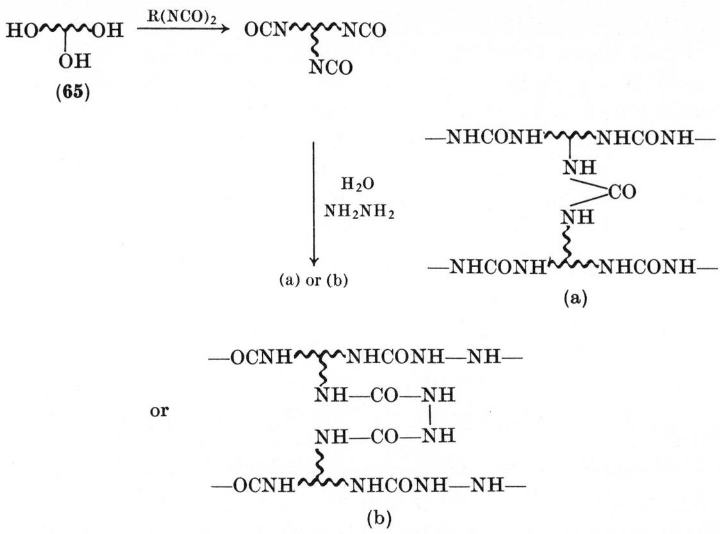

$$\text{HO}\sim\!\!\sim\!\!\text{OH} \xrightarrow{\text{R(NCO)}_2} \text{OCN}\sim\!\!\sim\!\!\text{NCO}$$
$$\underset{\displaystyle\text{(65)}}{\overset{\displaystyle\text{OH}}{|}} \qquad \overset{\displaystyle\text{NCO}}{|}$$

$$\xrightarrow[\text{NH}_2\text{NH}_2]{\text{H}_2\text{O}} \quad (a)\text{ or }(b)$$

—NHCONH$\sim\!\!\sim$NHCONH—
　　　　　NH
　　　　　$>$CO
　　　　　NH
—NHCONH$\sim\!\!\sim$NHCONH—
(a)

or

—OCNH$\sim\!\!\sim$NHCONH—NH—
　　　　NH—CO—NH
　　　　　　｜
　　　　NH—CO—NH
—OCNH$\sim\!\!\sim$NHCONH—NH—
(b)

Cross-links can only be introduced during or after spinning the fibres; if introduced prematurely, they make the polymer infusible and insoluble in all non-degrading solvents, and hence unspinnable.

Polyurethane fibres, such as 'Spanzelle' and 'Lycra', are characterized by two second-order transition temperatures. The lower one ($<0°$) is associated with the 'soft' polyether or polyester segments, the higher one ($>100°$) with the 'hard' aromatic urethane segments. Their X-ray diffraction patterns show no evidence of crystallinity below extensions of ca. 400%; in the relaxed state, therefore, the soft segments must be highly disordered and folded, as are the molecules of natural rubber. The ready extensibility of the molecules under stress is curbed by the interaction of the 'hard' segments, as these are drawn closer together, on account of their bulk and strong interchain hydrogen-bonding forces. Because of these forces, polyurethane fibres retain an unusually high modulus above their first Tg. Long-chain cross-links will also restrict extensibility and assist elastic recovery.

Compared with natural rubber, the polyurethanes have some disadvantages. Their low strength in hot water limits laundering conditions and they are more susceptible to hydrolysis, particularly the polyester types; alkaline agents and exposure to light cause yellowing and the fibres are vulnerable to bleaching agents; hydrazine-extended polymers are particularly susceptible to hypochlorites.

Anidex-type Elastomerics

These fibres, based on acrylic co-polymers, are exemplified by 'Anim/8'. Typically, two co-polymers are prepared, one (**66A**) containing 70% of acrylonitrile, 20% of ethyl acrylate and 10% of *N*-(hydroxymethyl)acrylamide, the other (**66B**) containing the same monomers in the proportion 20:70:10. A solution of the two polymers in dimethylformamide is then dry-spun into hot air, whereupon random cross-linking between A and B chains occurs.

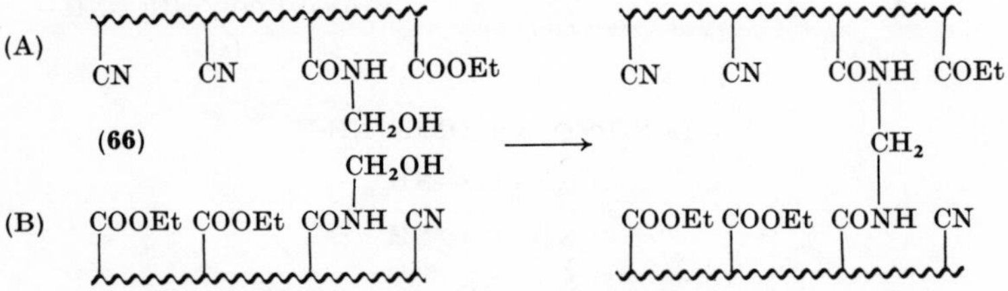
(A)
CN
CN
CONH
COOEt
CH2OH
(66)
CH2OH
(B)
COOEt COOEt CONH CN
CN CN CONH COEt
CH2
COOEt COOEt CONH CN

This gives a highly elastic fibre, insoluble in all solvents, with good resistance to photo-oxidation (hence no yellowing) and hydrolysis. The unique properties are attributed to a balance between the potential fibre-forming properties of the 'hard' segment A, rich in acrylonitrile and the 'soft', rubbery component B.

Poly(vinyl alcohol) (PVA)

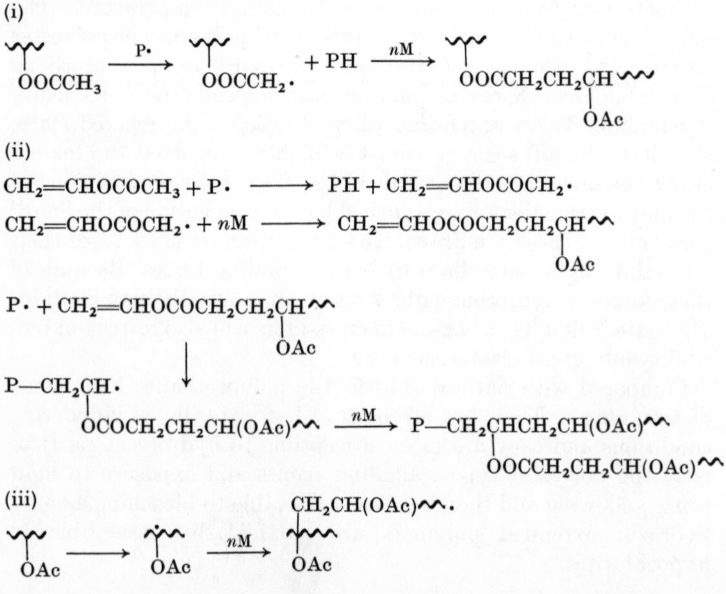
(i)
OOCCH3
P·
OOCCH2·
+ PH
nM
OOCCH2CH2CH
OAc
(ii)
CH2=CHOCOCH3 + P· → PH + CH2=CHOCOCH2·
CH2=CHOCOCH2· + nM → CH2=CHOCOCH2CH2CH
OAc
P· + CH2=CHOCOCH2CH2CH
OAc
P—CH2CH·
OCOCH2CH2CH(OAc)
nM
P—CH2CHCH2CH(OAc)
OOCCH2CH2CH(OAc)
CH2CH(OAc)
(iii)
OAc
OAc
nM
OAc

Production of fibrous poly(vinyl alcohol) is largely peculiar to Japan, where some 150 tons per day are said to be processed. The starting monomer, vinyl acetate, is readily available from ethylene or acetylene. Radical polymerization in methanol or other solvent of low transfer constant yields a vinyl polymer, whose linear structure may be interrupted by branching through radical-transfer from growing polymer chains (P·) to acetyl groups in either the polymer (i) or the monomer (M) (ii) or through generation of radical sites on the polymer backbone (iii).

Catalytic deacetylation or acid hydrolysis of the polymer in methanol then yields poly(vinyl alcohol) as an insoluble powder.

$$\sim\!\!\text{CH}_2\text{CH(OAc)}\!\!\sim \xrightarrow{\text{NaOH–MeOH}} \sim\!\!\text{CH}_2\text{CH(OH)}\!\!\sim + \text{MeOAc}$$

Ester-type branches are also removed, with a consequent reduction in molecular weight and production of some carboxyl end-groups.

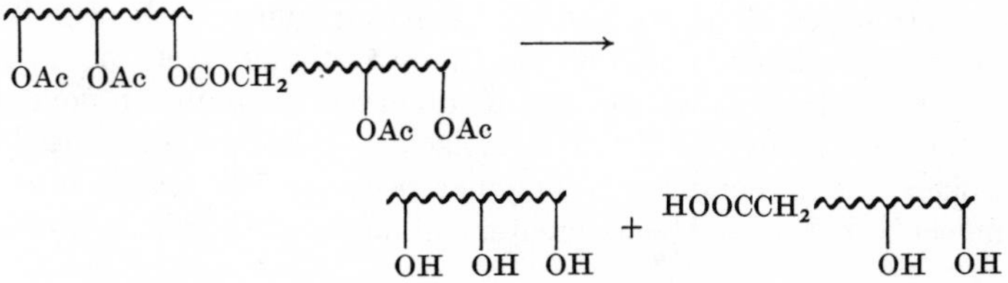

Lowering of the molecular weight of poly(vinyl alcohol) may occur on periodate oxidation; it is ascribed to the presence of structure (**67**) arising either by head-to-head polymerization (*a*) or possibly mutual termination (*b*).

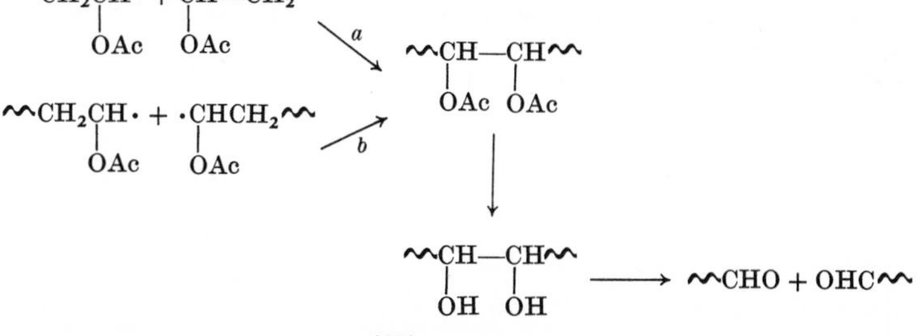

(**67**)

The essentially linear atactic polymer so obtained yields fibres on extrusion of its aqueous solution into a coagulating bath, usually sodium sulphate solution. These fibres are stretched, dried and heated, whereby they are rendered insoluble in water, presumably through the imposition of a high degree of hydrogen-bonding, as in cellulose. They still shrink readily in hot water, however, and a further treatment with formaldehyde, to yield cyclic formal groups and cross-links, is necessary for complete stabilization.

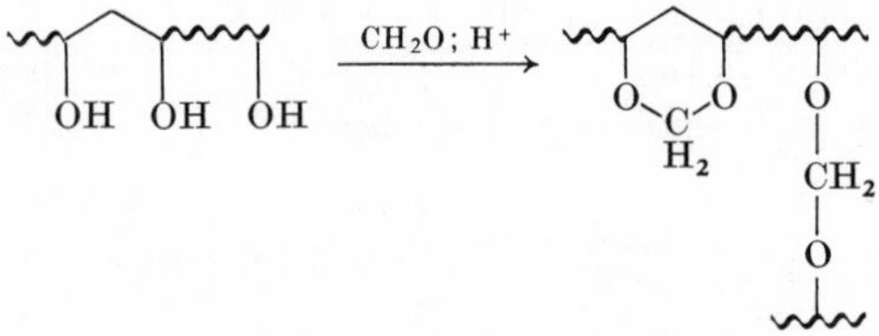

This cross-linking blocks some hydrophilic hydroxyl groups, and occurs exclusively in the disordered regions of the fibre, which are therefore appreciably modified. The effect of polymer tacticity on hydrogen-bonding in poly(vinyl alcohol) is reflected in the sensitivity to water. Thus, the undrawn atactic polymer with poor H-bonding is soluble in cold water; the isotactic polymer, in which intramolecular bonding (68) can predominate, is soluble in boiling water, the relatively few intermolecular H-bonds being easily broken; but the syndiotactic polymer, because of the mainly intermolecular nature and hence greater stability of the bonds, is soluble

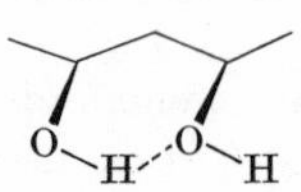

(68)

only at 160°C under pressure. Unfortunately, the stereoregular polymers can only be produced by special methods, such as the cationic polymerization of trimethylsilyl vinyl ether (which yields a markedly syndiotactic structure, as illustrated), followed by hydrolysis.

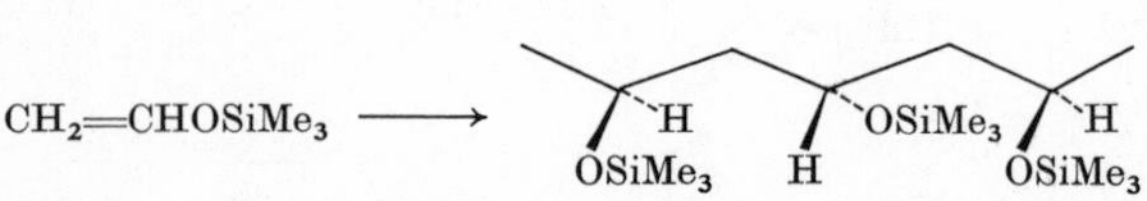

The loosely ordered structure of PVA fibres, and consequently the accessibility of the OH groups, results in a high moisture regain (ca. 6% at 65% relative humidity and 20°C) and ready dyeability with cotton dyes. Affinity for acid dyes is conferred by treatment with aminobenzaldehyde to give cyclic (69) and open acetals (70).

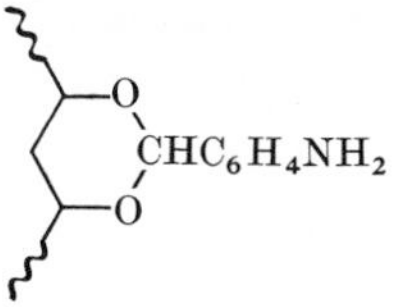

$$\text{CHC}_6\text{H}_4\text{NH}_2$$

(69)

$$\text{O—CH—O}$$
$$\text{C}_6\text{H}_4\text{NH}_2$$

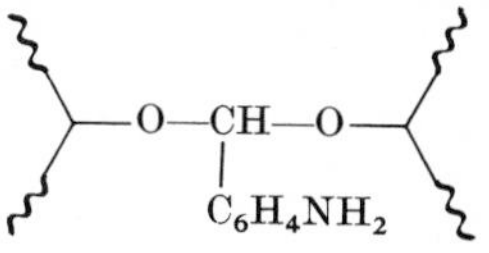

(70)

Treatment with benzaldehyde has been used to improve stability to hot water and impart greater elastic recovery.

Poly(vinyl chloride) (PVC)

Emulsion polymerization of vinyl chloride with a radical initiator readily affords a predominantly atactic polymer containing a few syndiotactic sequences. This polymer is too unstable thermally to be melt-spun, but can be dry-spun satisfactorily from solutions in binary solvent mixtures of CS_2 with, usually, acetone, or from tetrahydrofuran. Before these solvents were discovered, PVC was further chlorinated to give an acetone-soluble polymer (chlorine content ca. 65%); this polymer (71) would have the extra chlorine distributed randomly in the chain and would differ in structure from a vinyl chloride–vinylidene dichloride co-polymer (72). Fibres of low strength ('Vinyon',

$$\sim\sim\text{CH}_2-\text{CHCl}-\text{CHCl}-\text{CHCl}-\text{CH}_2-\text{CCl}_2-\text{CHCl}-\text{CHCl}-\text{CH}_2\sim\sim$$

'Piviacid' (71)

$$\sim\sim\text{CH}_2-\text{CCl}_2-\text{CH}_2-\text{CHCl}-\text{CH}_2-\text{CHCl}-\text{CH}_2-\text{CHCl}-\text{CH}_2-\text{CCl}_2\sim\sim$$

(72)

'Fibre MP') spinnable from acetone can also be obtained by co-polymerization of vinyl chloride (86–90%) with vinyl acetate (14–10%).

The low softening point of the atactic PVC fibres is a considerable obstacle to their use since they are liable to shrink when heated, laundered or dry-cleaned. The recent development of a more crystalline, ordered polymer with a higher softening point should increase the usefulness of this potentially cheap material. Polymerization of the liquid monomer at low temperature (−30°C) yields a more completely syndiotactic polymer with a low incidence of chain-branching. Use of boron alkyls and metal alkyls as initiators poses practical difficulties, and one current process utilizes a radical system comprising cumyl hydroperoxide, sulphur dioxide and sodium methoxide $(1:1.5:1.6)$. The postulated mechanism of initiation is as illustrated. In the absence of alkoxide,

$$NaOMe + SO_2 \longrightarrow NaOSOOMe \longrightarrow Na^+ + MeOSO_2^-$$
$$PhCMe_2\text{—}OOH + MeOSO_2^- \longrightarrow MeOH + PhCMe_2\text{—}OOSO_2^-$$

yields are poor and SO_2 is incorporated into the polymer. With

$$PhCMe_2\text{—}OOSO_2^- \longrightarrow PhCMe_2O\cdot + SO_3^-$$
$$PhCMe_2O\cdot + M \longrightarrow PhCMe_2OM\cdot \xrightarrow{\;nM\;} \text{Polymer}$$
$$(M = \text{monomer})$$

0.1% of initiator, 20% conversion is achieved in a few hours; the process can be operated continuously. Pure solid polymer is isolated by centrifugation and unconverted monomer is recovered and recycled. Fibres (e.g. 'Leavil') are obtained by spinning a solution in hot cyclohexanone into a coagulating bath containing cyclohexanone, water and ethanol (or acetone), followed by conventional drawing and heat-setting. Their X-ray crystallinity is about 35–40%.

PVC fibres are stable towards most acids (except concentrated HNO_3), alkalis, oxidants and reducing agents. They yield HCl by photo-initiated, autocatalytic decomposition or on pyrolysis, and darken, presumably through development of conjugated polyene sequences.

Unlike PVC coatings, PVC fibres do not contain added plasticizers. Use is made of their inherent non-flammability in flame-retardant fabrics. Their tendency to shrink when heated has led to uses as retractile fibres in making fur fabrics with different depths of pile and for producing embossed effects when used in conjunction with other fibres. 'Leavil'-type fibres withstand temperatures up to 130°C and can therefore be dyed with disperse dyes at 110°C under pressure without a carrier. They can be laundered and dry-cleaned without difficulty.

Another approach to chlorofibres that are more easily dyed, hydrophilic, resistant to thermal shrinkage and to swelling by dry-cleaning solvents, yet still remain non-flammable, has been to spin a $1:1$ emulsion of PVC in poly(vinyl alcohol) solution into a coagulating bath. The PVA is then stabilized (i.e. made water-resistant) in the usual way, to give a fibre ('Cordelan') which is a matrix of PVC particles dispersed in PVA.

Poly(vinylidene chloride)

Chlorination of vinyl chloride yields 1,1,2-trichloroethane, which is readily dehydrochlorinated by hot aqueous alkali or thermally (400°C) in the vapour phase to 1,1-dichloroethylene (vinylidene dichloride), b.p. 32°C.

$$CH_2{=}CHCl \longrightarrow CH_2ClCHCl_2 \longrightarrow CH_2{=}CCl_2$$

In the absence of an inhibitor, the monomer readily yields a highly unstable peroxide. Conventional radical polymerization in solution or in aqueous emulsion yields a linear homopolymer that crystallizes readily. X-ray data indicate that steric interference between the pairs of geminal Cl atoms forces the molecules into a helical configuration (Figure 9.6) in the crystalline polymer.

The softening point (>180°C), which lies close to the decomposition temperature (210°C), and the insolubility in all suitable solvents, render the homopolymer unsuitable for melt- or solution-spinning. The commercial fibre ('Saran', chlorine content ca. 70%)

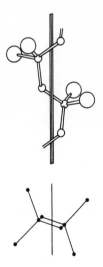

Figure 9.6. Helical structure of crystalline poly(vinylidene chloride). [Reproduced, by permission, from V. M. Coiro, P. De Santis, A. M. Liquori, and A. Ripamonte, *J. Polymer Sci.*, Part B, **4**, 821 (1966).]

is therefore made from a random co-polymer with some 15% of vinyl chloride that can be melt-spun satisfactorily. It is chemically inert, has good light stability, and is self-extinguishing when ignited in air.

High-temperature Fibres

It is possible to design fibres primarily to withstand hostile thermal environments, either for long periods at moderate temperatures or for short periods at very high temperatures (600°C or above). Typical uses are in protective clothing for foundry workers, chemical plant operatives, personnel handling highly flammable loads, and aircrew. 'Nomex', 'Durette' and 'Kermel' are examples of this group of fibres, most of which are polymers or co-polymers of *meta-* or *para-*disubstituted aromatic acids and amines, e.g. (73–76). They are usually produced from the acid chlorides because

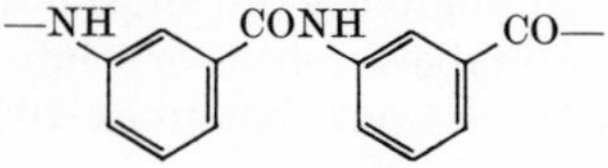

(73)

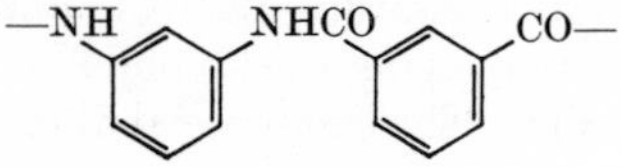

(74)

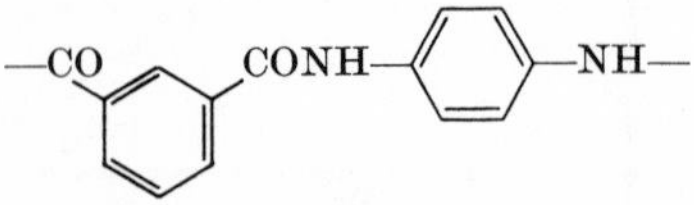

(75)

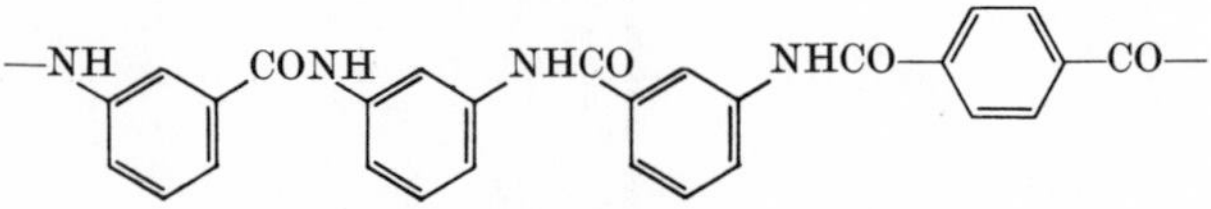

(76)

of practical difficulties in effecting direct melt-condensation of the free acids with appropriate diamines; an extensive study of co-polymers has been directed to the production of materials that have adequate solubility in spinning solvents and are capable of being stretch-oriented under practicable conditions whilst retaining a high polymer melt temperature (>400°C). Homopolymers, particularly of the all-*para* type, may be too intractable for commercial processing.

Aromatic polyimides falling into this class can be obtained from, for example, pyromellitic anhydride. Fibres are solution-spun from the intermediate amic acid (**77**), which is then dehydrated thermally.

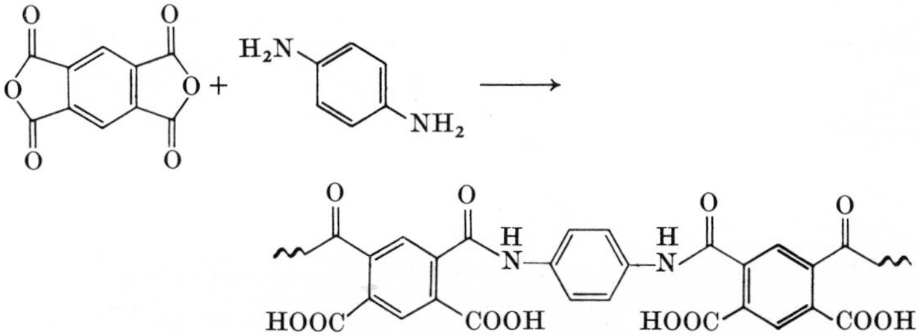

(**77**)

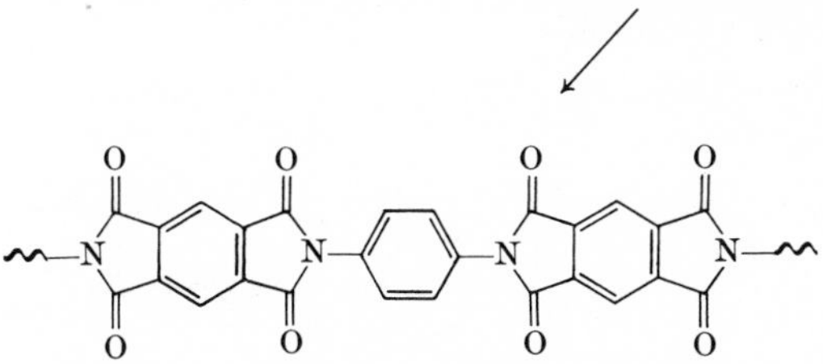

'Kermel' fibres are amide-imides (cf. **78**) prepared from trimellitic acid anhydride and an aromatic diamine.

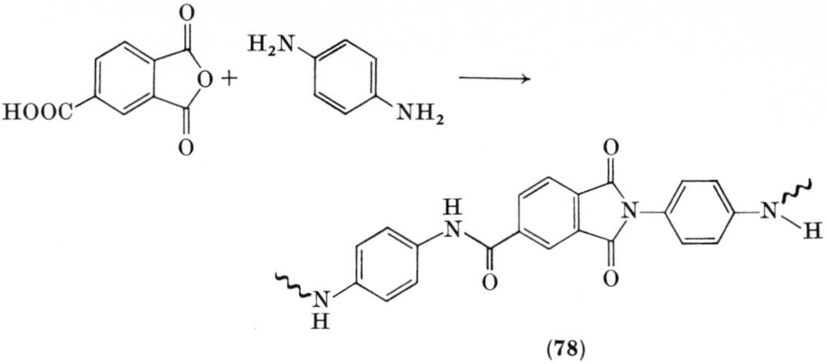

(**78**)

Polybenzimidazole-type fibres (PBI) (cf. **79**), derived from aromatic tetra-amines, have also been prepared.

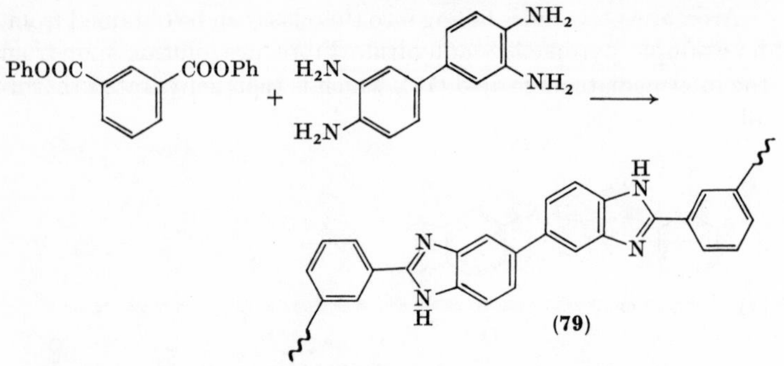

(79)

A more sophisticated fibre (BBB) from naphthalene-1,4,5,8-tetracarboxylic acid and biphenyl-3,3′,4,4′-tetra-amine, has structure (80). It is wet-spun from concentrated H_2SO_4 into dilute (70%) acid and is said to retain 60% of its cold tenacity after being heated at 600°C for a minute.

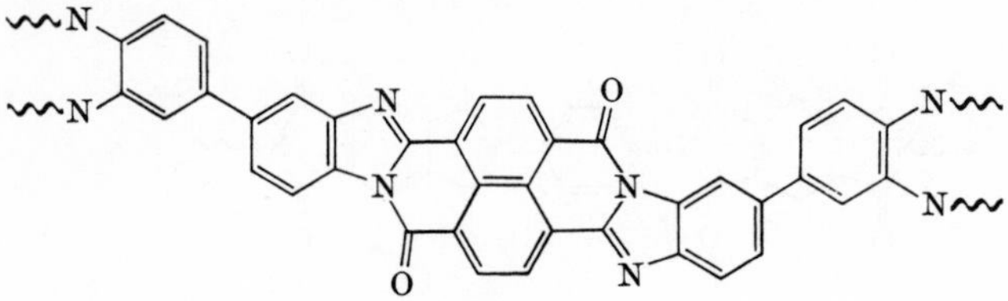

(80)

A group of fibres containing the oxadiazole and thiadiazole systems are derived from hydrazides, e.g. (81).

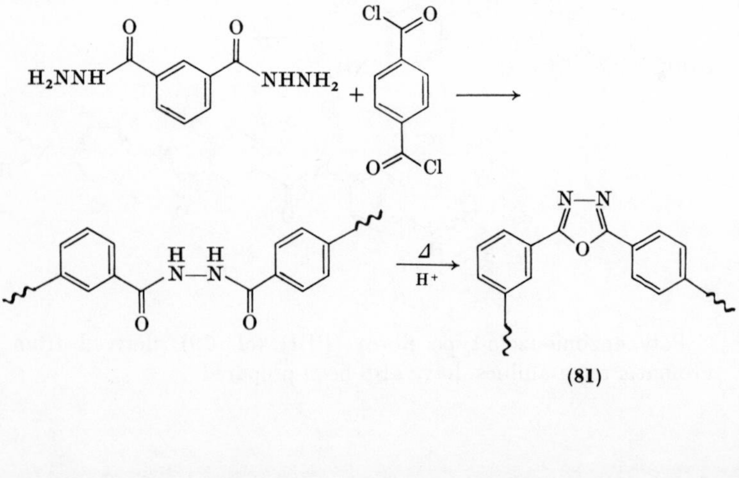

(81)

Related to the last group are fibres derived from terephthaloyl chloride and oxamidrazone by interfacial polymerization. The intermediate polymer (**82**) is treated with a chelating metal to give a co-ordinately cross-linked, non-flammable fibre, 'Enkatherm' (**83**). This fibre is said to be carbonized, yet retain its physical structure at 1500°C.

$$H_2NN{=}C(NH_2)C(NH_2){=}NNH_2 + p\text{-}ClCOC_6H_4COCl \longrightarrow$$

$$-NHN{=}C(NH_2)C(NH_2){=}NNHCOC_6H_4CO-$$

(**82**)

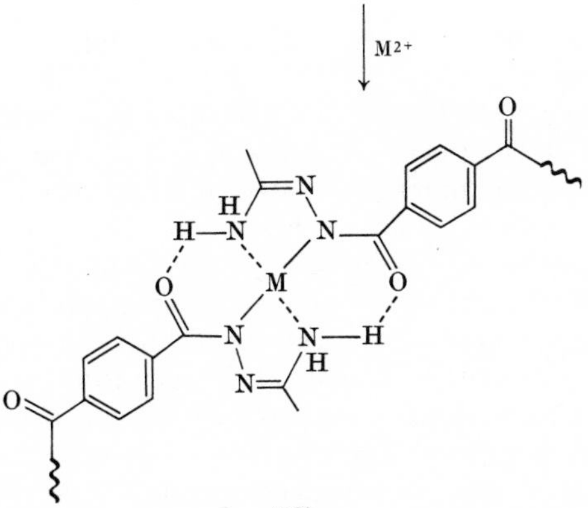

(**83**)

Another novel thermostable fibre, which is also an excellent electrical insulator, can be prepared from cyclohexanone. Controlled condensation yields 2,6-dicyclohexylidenecyclohexanone + 2,6-dicyclohexenylcyclohexanone, which are dehydrogenated to 2,6-diphenylphenol. Oxidative coupling of this phenol (cf. 2,6-dimethylphenol) yields the polyphenylene oxide (**84**), which can be conventionally dry-spun from solvent and drawn at high temperature into the highly crystalline fibre, 'Tenax', with Tm 480°C and Tg 235°C.

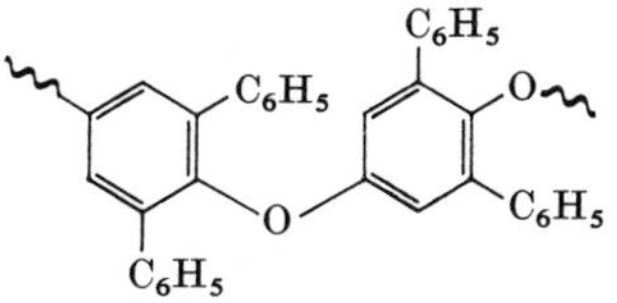

(**84**)

'Kynol' is a golden-brown heat-resistant fibre believed to be obtained basically from a phenolic (phenol–formaldehyde) resin; the polymer is cross-linked as it is extruded. Unlike most fibres, 'Kynol' is amorphous. It has very low strength and cannot be used as a primary textile fibre, but can be made into felts and waddings for lining conventional fabrics. When heated, it decomposes to give mainly CO_2 and water, leaving a carbonaceous fibre skeleton which then has excellent heat-resistance.

Carbon Fibres

The carbon fibres used in structural fibre-resin composites are characterized by high tensile strength and stiffness. They are mainly derived from a special grade of acrylic fibre by a carefully programmed thermal treatment in three stages of increasing severity. First, the acrylic fibres are heated in air at 200–300°C, whilst held under tension to maintain a high degree of orientation. The oxidized fibres are then carbonized in an inert atmosphere by increasing the temperature to 1500°C and are finally graphitized at temperatures reaching 2500–3000°C. The chemistry of this transformation is complex and little understood. Oxygen is introduced into the polymer in the first stage and the product becomes resistant to thermal chain-scission. Several structures have been suggested for this intermediate material, mostly incorporating a fused-ring or 'ladder' system, with a progressive increase in conjugation and hence deepening of colour as the reaction progresses. They include the units (**85–87**).

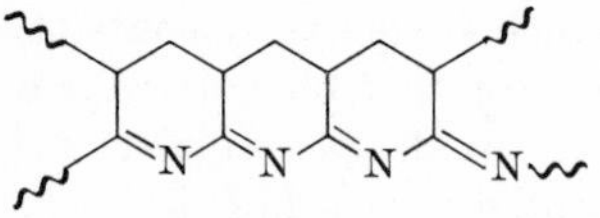

(**85**) 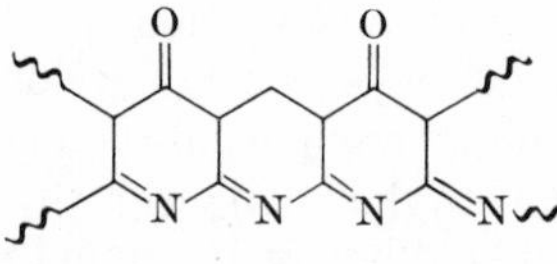   (**86**)

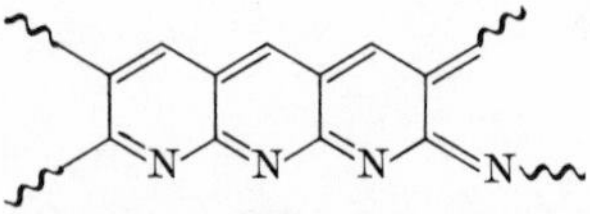

(**87**)

Structures (**85**) and (**86**) are not planar but must assume a random arrangement of half-chair conformations as the atactic $C\equiv N$ groups come into juxtaposition for cyclization. The stabilizing effect of a ladder structure is considered vital to preserve orientation during eventual carbonization. Extensive cross-linking of the polymer chains at the oxidation stage is discounted on the grounds that the fibres are still largely soluble in certain solvents and that their elastic modulus is substantially unaltered.

Carbonization is accompanied by elimination of water, ammonia and hydrogen cyanide, and the polymer becomes insoluble in all solvents, thus heightening the difficulties of structural determination. Annellation probably occurs at this stage, yielding strong low-modulus fibres in which the carbon skeleton has assumed some degree of two-dimensional order. Some nitrogen (ca. 7%) and hydrogen remain even at 1000°C but they are slowly eliminated at higher temperatures. Finally further rearrangement of the carbonized polymer yields crystallites exhibiting three-dimensional order similar to that in graphite. Carbon fibres with properties approaching those of the acrylic-based products can also be made by the controlled pyrolysis of cellulose fibres. Better bonds with the matrix resin in composites can be obtained if the carbon fibre surfaces are 'activated' chemically, e.g. by controlled oxidation with air or concentrated nitric acid to introduce carbonyl or carboxyl groups.

Graft Co-polymers of Fibres

When treated with certain oxidizing agents or exposed to high-energy radiation (γ-rays, accelerated electrons), fibres yield macroradicals capable of initiating graft polymerization of vinyl and acrylic monomers. Typical oxidizing agents are Fenton's reagent ($Fe^{2+}-H_2O_2$; active species $HO\cdot$), persulphates, and certain metallic salts, notably those of Co(III), Mn(IV) and Ce(IV). Natural and regenerated fibres and poly(vinyl alcohol) react readily in aqueous systems, in which they are highly swollen and therefore accessible. Acrylic and nylon fibres are less accessible, although accessibility may be improved by use of mixed solvents, and they are less readily oxidized. Polyester and polyolefins are too resistant to oxidation to undergo chemically initiated grafting readily, but they are more amenable to radiation-initiated reaction; the monomer may be in the vapour phase or in solution in a solvent or solvent mixture that has some swelling action, e.g. dichloroacetic acid + water for polyester; some of the radical sites generated by irradiation are long-lived and are capable of initiating polymerization long after the fibre has been removed from the radiation source.

The massive dose rates obtainable with electron accelerators (150 keV–4 MeV) effect polymerization of reactive monomers, e.g.

allyl acrylate, within seconds, compared with hours for γ-ray sources.

It is difficult to elucidate detailed initiation mechanisms for grafting on to natural fibres because of the multiplicity of potential sites. Any or all of the carbon atoms in the anhydroglucose units of cellulose, for example, may be attacked by OH radicals or by the radical-ion ($SO_4^{-}\cdot$) from persulphate.

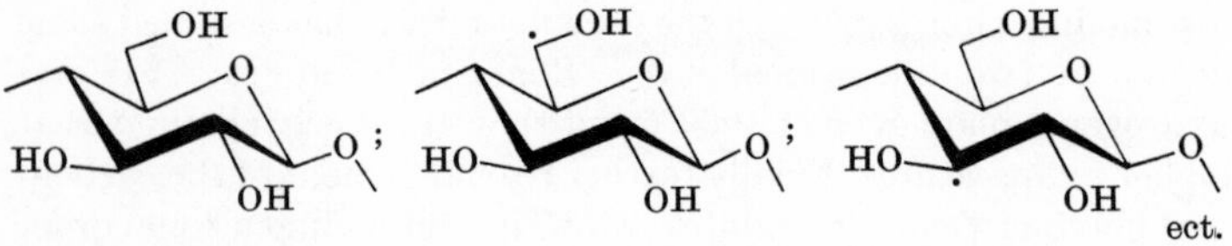

Oxidation by Ce^{4+} salts in acid solution depends not only on the cerium concentration, but also on the acid concentration and on the nature of the anion. The perchlorate is the most powerful oxidant; it slowly oxidizes water to hydrogen peroxide. Pinacol is converted into two mols of acetone by Ce^{4+} salts, but only one mol is isolated if a polymerizable monomer (acrylamide) is added, suggesting by analogy that cellulose may be attacked additionally at the C-2—C-3 bond.

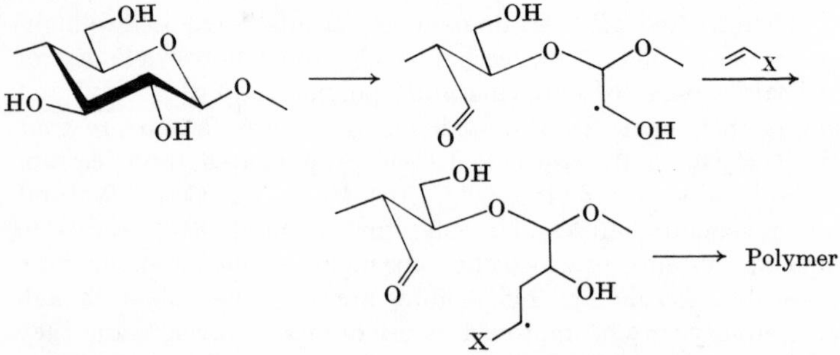

cf. $Me_2C(OH)C(OH)Me_2 \longrightarrow Me_2CO + Me_2\dot{C}(OH)$;

$Me_2\dot{C}(OH) \xrightarrow{\quad} Me_2C(OH)CH_2\dot{C}HX$

In protein fibres, hydroxylated side-groups of combined serine, threonine and tyrosine are all likely initiation sites; cystine cross-links in wool may also be implicated.

Nylon 6 and 6.6 are both oxidized slowly by Ce^{4+} salts, ultimately with extensive chain-scission and formation of acids and carbonyl compounds. In the presence of suitable monomers, notably acrylic acid and acrylamide, graft polymerization occurs; because of the low rates of diffusion of oxidant and monomer, this may be limited to peripheral regions of the fibres. Initiation is believed to take place primarily at the CH_2 adjacent to amide nitrogen:

$$\text{\footnotesize WWCONHCH}_2\text{\footnotesize WW} \longrightarrow \text{\footnotesize WWCONH}\overset{\cdot}{\text{C}}\text{H} \xrightarrow{\overset{X}{\frown}} \underset{\underset{CH_2CHX\cdot}{|}}{\text{\footnotesize WWCONHCH WW}}$$

Polyester and polypropylene fibres are too stable to respond readily to chemical initiation. Radiation initiation on all fibres is largely non-specific; in the presence of air, some primary radicals may yield peroxides, which are also capable of initiating polymerization:

$$\underset{/}{\overset{\backslash}{{}}}\text{CH} \longrightarrow \underset{/}{\overset{\backslash}{{}}}\text{C}\cdot \xrightarrow{O_2} \underset{/}{\overset{\backslash}{{}}}\text{C}-\text{O}-\text{O}\cdot$$

Grafted polymers of, e.g. styrene, acrylonitrile, alkyl acrylates and methacrylates, and vinyl acetate do not have any great beneficial effect on the physical properties of cotton, rayon or wool. However, a graft co-polymer of acrylonitrile on natural protein has been spun commercially by Japanese workers into a fibre ('Chinon') having silk-like properties; an important feature, which may account for the success of these fibres, is that they are stretch-oriented after the grafting reaction.

Grafted polymers with reactive side-groups include those from acrylamide, acrylic and methacrylic acids, vinylpyridines, hydroxyalkyl and (dialkylamino)alkyl acrylates. Treatment of cotton fabrics containing grafted polyacrylamide with formaldehyde and acid gives a cross-linked product with easy-care characteristics; pre-treatment of cotton with N-(hydroxymethyl)acrylamide, followed by radiation-initiated polymerization, gives a similar product (**88**).

$$\underset{\underset{CONHCH_2OCell}{|}}{\overset{\overset{CellOCH_2NHCO}{|}}{\text{\footnotesize wwwwwwwwww}}}$$

(**88**)

Hydrophilic polymers grafted on to nylon and polyester fibres improve their electrical conductivity and soil-releasing properties.

Polymers with OH side-chains may act as receptor sites for reactive dyes, whilst basic polymers and acidic polymers confer affinity for acid and basic dyes, respectively. Nylon containing polyacrylic acid in the calcium salt form is more resistant to 'hole-melting' caused by cigarette ends or cinders.

Unambiguous proof that graft polymerization has occurred, as distinct from formation of occluded homopolymer, is difficult to obtain. Methods used include the following: (1) Fractional precipitation. A nylon–polyacrylic acid graft is partly soluble in 90% formic acid and yields precipitates containing both polymers on progressive dilution with water; only a small amount of ungrafted water-soluble polyacrylic acid can be isolated. (2) Solvent extraction. Acetylation of a rayon–vinylpyridine graft yields a product soluble in CH_2Cl_2 from which only traces of basic polymer can be extracted by dilute acid. (3) Direct extraction. When a cellulose–polyacrylic acid graft is treated with PhNCO to convert Cell-OH quantitatively into Cell-OCONHPh, only a small amount of cellulose tricarbanilate can be subsequently extracted by a solvent in which it is normally soluble. (4) End-group detection. Graft co-polymers in which the fibre moiety can be removed, e.g. by hydrolysis, may leave the grafted vinylic polymer carrying a distinctive end-group:

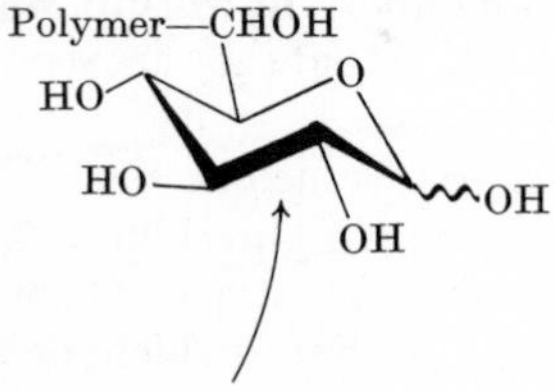

or perhaps $HOOC(CH_2)_4CHNH_2$
                                |
                            Polymer

(from cellulose)   (from nylon 6)

Bibliography

R. W. Moncrieff, *Man-made Fibres*, 5th edn., Heywood, London, 1970.

H. F. Mark, S. M. Atlas and E. Cernia (Eds.), *Man-made Fibres: Science and Technology*, Vols. I–III, Wiley–Interscience, New York, 1968.

S. M. Cockett, *An Introduction to Man-made Fibres*, Pitman, London, 1966.

J. G. Cook, *Handbook of Textile Fibres*, 4th edn., Merrow, Watford, 1968.

I. Goodman, *Synthetic Fibre-forming Polymers*, Roy. Inst. Chem. Lecture Series, 1967, No. 3.

F. Fourné, *Synthetische Fasern*, Wissenschaftliche Verlagsgesellschaft, Stuttgart, 1964.

E. M. Hicks, Jnr. *et al.*, 'The Production of Synthetic Polymer Fibres', *Textile Progress*, 1971, **3**, No. 1.

P. A. Koch, 'Faserstoff-Tabellen', *Textil-Industrie*, 1970, **72** (1), 21, and references therein.

H. Hopff, A. Müller and F. Wenger, *Die Polyamide*, Springer, Berlin, 1954.

I. Goodman and J. A. Rhys, *Polyesters. Vol. I. Saturated Polyesters*, Iliffe, London, 1965.

J. E. McIntyre, *The Chemistry of Fibres*, Arnold, London, 1971.

M. E. Carter, *Essential Fiber Chemistry*, Marcel Dekker, New York, 1971.

K. Götze, *Chemiefasern nach dem Viscoseverfahren*, Springer, Berlin, 1967.

H. Ludewig, *Polyester Fibres: Chemistry and Technology* (trans. B. Buck), Interscience–Wiley, London, 1971.

J. W. S. Hearle and R. H. Peters, *Fibre Structure*, Textile Institute and Butterworths, Manchester–London, 1963.

J. W. S. Hearle and R. Greer, 'Fibre Structure', *Textile Progress*, 1970, **2**, No. 4.

J. E. Bailey and A. J. Clarke, 'Carbon Fibres', *Chemistry in Britain*, 1970, **6**, 484.

R. Meredith, *Elastomeric Fibres*, Merrow, Watford, 1971.

J. M. Buist and H. Gudgeon (Eds.), *Advances in Polyurethane Technology*, McLaren, London, 1968.

P. Wright and A. P. C. Cumming, *Solid Polyurethane Elastomers*, McLaren, London, 1969.

K. Ward, *Chemistry and Chemical Technology of Cotton*, Interscience, New York, 1955.

N. M. Bikales, N. G. Gaylord and H. F. Mark (Eds.), *Encyclopedia of Polymer Science and Technology*, Interscience, New York, 1964–1972.

P. Alexander and R. F. Hudson, *Wool. Its Chemistry and Physics*, 2nd edn., Chapman and Hall, London, 1963.

J. Honeyman (Ed.), *Recent Advances in the Chemistry of Starch and Cellulose*, Heywood, London, 1959.

W. G. Crewther *et al.*, 'The Chemistry of Keratin', *Adv. Protein Chem.*, 1965, **20**, 191.

F. Lucas and K. M. Rudall, *Extracellular Fibrous Proteins: The Silks* in *Comprehensive Biochemistry*, Vol. 26B, p. 475, Elsevier, Amsterdam, 1968.

F. Lucas, J. T. B. Shaw and S. G. Smith, 'The Silk Fibroins', *Adv. Protein Chem.*, 1958, **13**, 108.

W. Traub and K. A. Piez, 'Chemistry and Structure of Collagen', *Adv. Protein Chem.*, 1970, **25**, 243.

N. Ramanathan, *Collagen*, Wiley, New York, 1962.

A. J. Bailey, *The Nature of Collagen*, in *Comprehensive Biochemistry*, Vol. 26 B, p. 297, Elsevier, Amsterdam, 1968.

W. R. Sorenson and T. W. Campbell, *Preparative Methods of Polymer Chemistry*, 2nd edn., Interscience, New York, 1968.

W. J. Roff and J. R. Scott, *Fibres, Films, Plastics and Rubbers*, Butterworths, London, 1971.

CHAPTER 10

Colour Chemistry

(C. V. STEAD, Organics Division, Imperial Chemical Industries Limited, Blackley)

From the very earliest times man has desired to colour the fabrics he possessed and initially he turned to Nature for the materials with which to accomplish this. Certain naturally occurring coloured compounds emerged as useful dyestuffs since they could be applied by one or other of the two dyeing processes that were developed. One of these processes was mordant dyeing. In this the cloth to be dyed was impregnated with a solution of a soluble salt of a metal such as aluminium, iron, chromium or tin, and the insoluble metal hydroxide was precipitated on the fibre. This mordanted fibre was then treated with a solution of a naturally occurring colouring matter which was capable of forming an insoluble chelate compound with the mordant. The colouring matters used came from a variety of sources and differed widely in chemical structure. All were, however, alike in that each contained a chelating system. The most important was alizarin (1), obtained from madder root; this gave red dyeings. Shades varying between red and violet, depending upon the mordant used, resulted from carminic acid (2), the coloured compound present in cochineal extracted from female insects of the species *Coccus cacti*. Logwood, the only natural dye used in any quantity today, is obtained from the heartwood of the tree *Haematoxylon caonpechiancum*; it contains hematoxylein (3) which results from oxidation of hematoxylin (4) present in the tree and gives black dyeings.

The second process was used for dyeing with indigo (5), obtained by oxidation of indoxyl (Part 3, p. 104) which occurs as the glucoside indican (Part 4, p. 116) in the indigo plant. The dyeing process involved reduction of indigo under alkaline conditions to the water-soluble leuco form (6). The cloth was dipped in the solution of this and then exposed to air to allow re-oxidation of the reduced form to the dark-blue insoluble indigo. Indigo is still widely used but the natural product has been almost completely supplanted by synthetic material. Tyrian purple, 6,6′-dibromo-indigo, the highly prized colouring matter obtained from the purple snail *Murex brandaris* is a rare example of a naturally occurring bromo-compound.

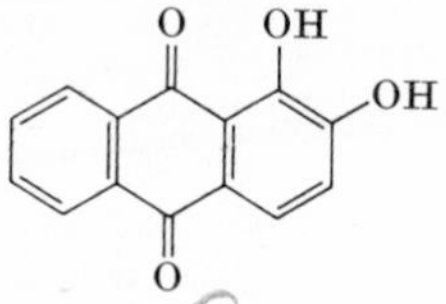

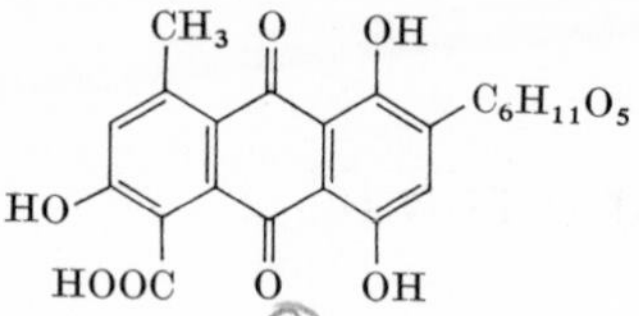

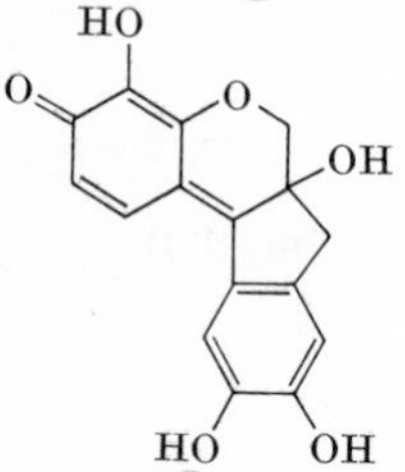

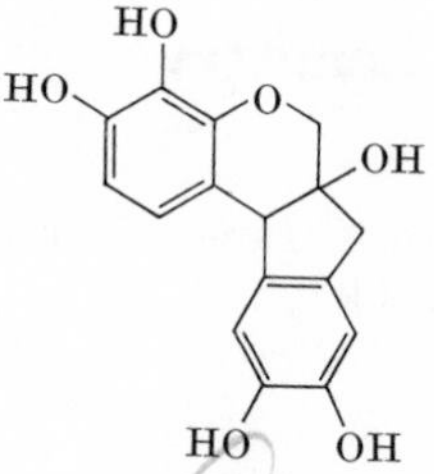

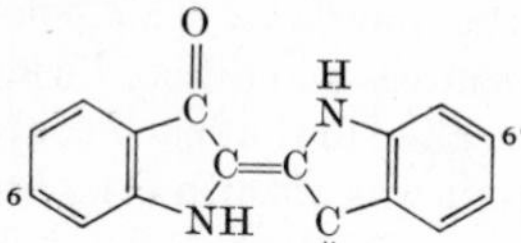

These natural colouring matters presented difficulties in their extraction, which could often be tedious, and in their application which could be lengthy and uncertain in its final outcome. They could not compete with the flood of synthetic dyes which followed the preparation of Mauveine by Perkin in 1856 since the synthetic dyes offered the advantages of a brighter and wider shade range, uniformity of quality, easier application and better fastness properties, i.e. degree of resistance to agencies such as light and washing, at a much lower cost.

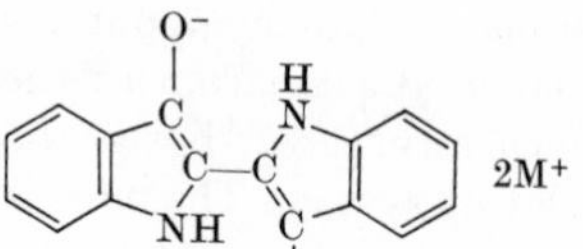

Mauveine, the first commercially successful synthetic dye, was obtained by oxidation of crude aniline by potassium dichromate. The subsequently elucidated constitution (**7**) shows the presence of methyl groups resulting from the considerable proportion of toluidines in the aniline then available. Mauveine gave brilliant purple dyeings on silk, greatly superior to those obtained from natural dyes and its success ensured the founding of the synthetic

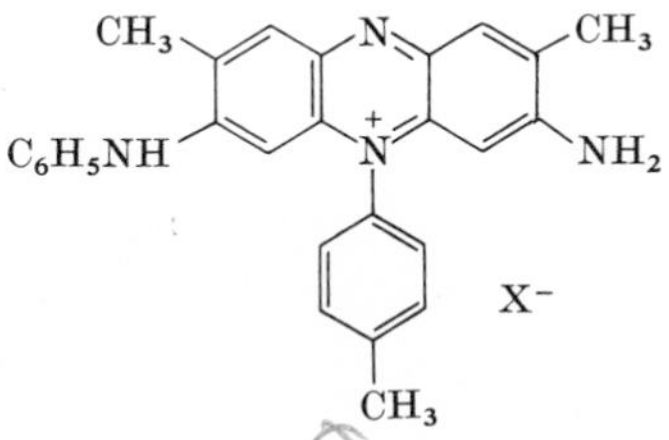

(**7**)

dyestuffs industry. It is a member of the azine class of dyes, a group of cationic dyes noteworthy for their brilliant shades but now of minor interest owing to their relatively poor light fastness.

In addition to their use as dyestuffs, the coloured products of the dyestuffs industry have found a further major outlet as pigments. The difference between these two classes is that a dyestuff is applied to a textile material from an aqueous solution or suspension whereas a pigment is bodily incorporated into a paint or plastic medium. Numerous smaller outlets, such as colour photography and biological stains, exist but attention in this Chapter is focused on the two major uses and on the main chemical classes of coloured compound in use today.

DYESTUFFS

Azo-dyes

Preparation, Structure and Colour

This class of dye is conventionally characterized by the presence of an azo-group —N=N— in the molecule. In certain cases, however, an alternative hydrazone tautomer —NH—N= may be the preferred structural form. The overwhelmingly important method of preparation involves diazotization of an aromatic or heteroaromatic primary amine and coupling of the resulting diazonium salt (**8**) with a suitable coupling component (Part 2, pp. 179–180). Important coupling components are a variety of keto-enolic compounds, in particular acetoacetarylamides and 1-aryl-3-substituted pyrazol-5-ones; phenols and naphthols; anilines, naphthylamines and their *N*-alkylated and *N*-arylated

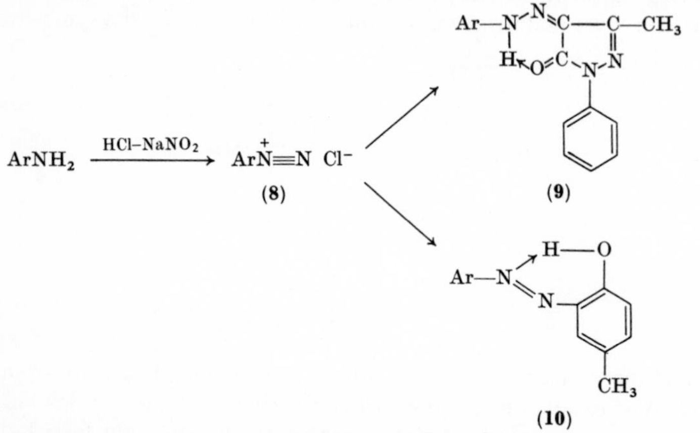

(**9**)

$$ArNH_2 \xrightarrow{\text{HCl–NaNO}_2} Ar\overset{+}{N}{\equiv}N \ \ Cl^-$$

(**8**)

(**10**)

derivatives. Coupling with the hydroxyl-containing components occurs via an ionized form and consequently is commonly carried out in alkaline solution whereas arylamines can be coupled most conveniently in weakly acid medium. Whilst often indiscriminately represented as azo-compounds, the exact structures of the products depend upon the relative energy levels of the possible tautomeric forms. Thus in the case of acetoacetarylamides and pyrazol-5-one coupling components the hydrazone form (e.g. **9**) is favoured whereas simple phenols give the azo-form (**10**). Naphthols usually give tautomeric mixtures.

The ease with which structural variations can be accomplished in this dyestuff class has made the azo-dyes of paramount importance. For their preparation a considerable array of intermediates is required. These are manufactured starting from a relatively restricted range of primary compounds (e.g. benzene, methylbenzenes, phenol, naphthalene) which are available in large quantities originally by the distillation of coal tar. An alternative source, of growing importance, is petroleum; indeed, this is now the major source of monocyclic hydrocarbon primaries in the United States (see Chapter 2). Starting from the ten or twenty important primaries the requisite intermediates are produced by common chemical reactions among which nitration, halogenation, sulphonation, reduction, *N*-alkylation, *N*-acylation and fusion with alkali are particularly important. In the naphthalene series the Bucherer reaction (Part 2, p. 264) is widely exploited.

Of the benzenoid intermediates mentioned above, arylamines most usually arise by nitration followed by reduction. Iron powder in the presence of a catalytic amount of hydrochloric acid is commonly employed as a reducing agent although this method is now being widely replaced by catalytic hydrogenation. Acetoacetarylamides result from reaction of arylamines with diketene, and pyrazol-5-ones are prepared by cyclization of the condensation products of amines with ethyl acetoacetate.

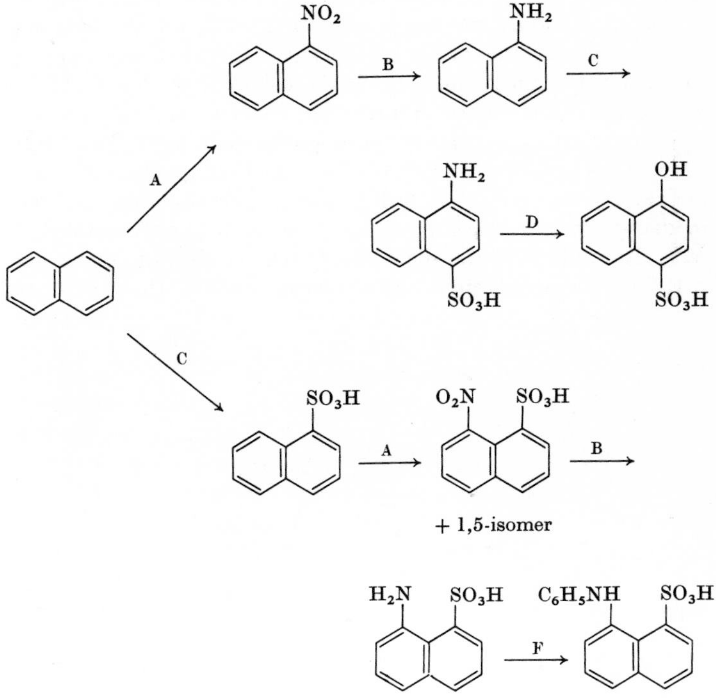

Key: A, Nitration; HNO_3–H_2SO_4 mixture. B, Reduction; Fe–HCl. C, Sulphonation; H_2SO_4 or SO_3/H_2SO_4. D, Bucherer reaction; $NaHSO_3$–NaOH. E, Alkali fusion; NaOH. F, Heat with aniline.

SCHEME 10.1. Preparation of naphthalene intermediates involving a single initial α-substitution (see text on p. 320).

In the naphthalene series the influences directing entry of substituents are capitalized by suitable arrangement of the order of the stages involved so as to produce intermediates of the desired orientation. Substitution occurs most readily at the α-positions and thus nitration and low-temperature sulphonation yield the α-isomers. If, however, the sulphonation (which is reversible) is conducted at higher temperature the introduced sulphonic acid group is relocated to produce the thermodynamically more stable β-isomer. The way in which a strictly limited number of reactions is used to produce important dyestuffs intermediates is illustrated in Schemes 1 and 2.

Monoazo-dyes obtained by a single diazotization and coupling sequence are conventionally designated as of the A→E type, the symbol A standing for the primary amine, E for the coupling component and the arrow signifying 'diazotized and coupled with'. A major factor influencing the colour is the length of the conjugated chain contained in the dye molecule and this depends to a large extent on the coupling component used. Thus with simple benzenoid A components, acetoacetarylamides and pyrazol-5-ones give greenish-yellows, phenols and anilines reddish-yellows, naphthols oranges, and certain aminonaphthols reds. The nature of the diazo-component is also, of course, important. Naphthylamines give deeper shades than anilines;

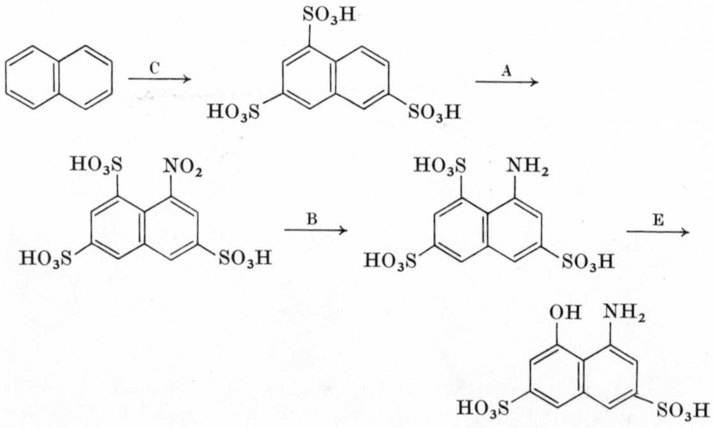

Key as for Scheme 10.1.

SCHEME 10.2. Preparation of 1-amino-8-naphthol-3,6-disulphonic acid.

substituents in the diazo-component can have a profound effect, as will be illustrated below.

In addition to the monoazo-dyes, dyestuffs containing two or more azo-groups within the molecule can be produced by employing more than one sequence of diazotization and coupling stages. Three schemes are available for the preparation of disazo-dyes. Thus an aromatic diamine such as benzidine or 4,4′-diamino-diphenylamine-2-sulphonic acid (D component) can be tetra-azotized, i.e. both amino-groups diazotized, and coupled with two possibly dissimilar equivalents of coupling component. This produces a bisazo-dye of the $E_1 \leftarrow D \rightarrow E_2$ pattern, e.g. (11), advantage being taken of the differing rates of coupling of the two diazonium salt groupings. The same structural type can be arrived at by diazotizing and coupling a nitroarylamine or an acylaminoarylamine and then either reducing the nitro-group with sodium sulphide or hydrolyzing the acylamino-group to yield

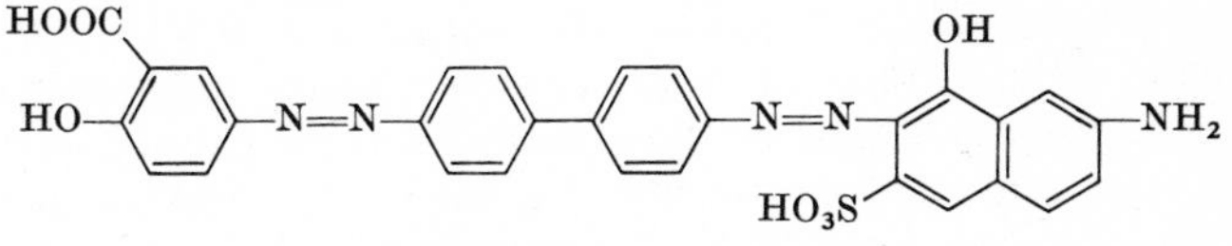

(11) Brown

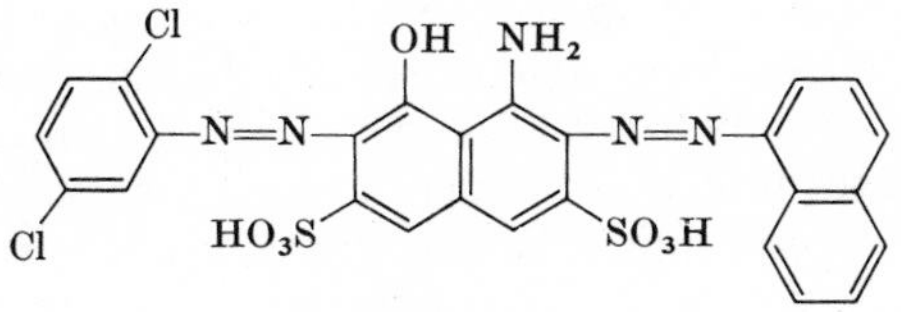

(12) Green

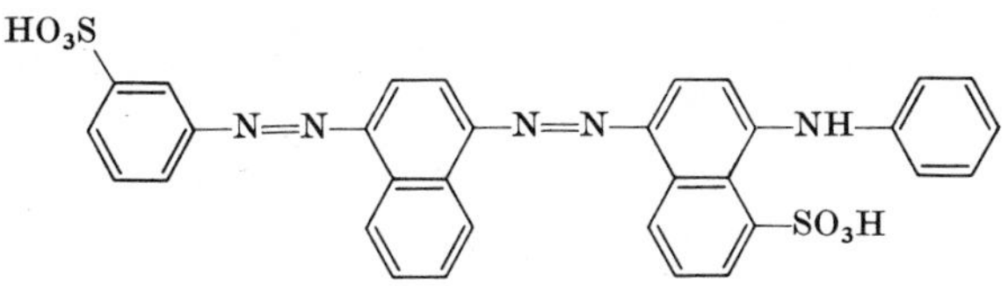

(13) Navy blue

an aminoazo-compound which is then further diazotized and coupled. Similarly, a component capable of coupling twice (Z component), among the most important of which is 1-amino-8-naphthol-3,6-disulphonic acid, may be used to produce $A_1 \rightarrow Z \leftarrow A_2$

disazo-dyes such as (12). The third possibility arises by coupling a diazotized arylamine on to a coupling component containing an amino-group, typically 1-naphthylamine and its 6- and 7-sulphonic acids (in this context termed an M component). Diazotization of the resulting aminoazo-compound and coupling with an E component yields an A→M→E structure (13). The deep shades that can be obtained from these dyes reflect the extensive conjugated systems contained within their molecules. In similar fashion dyes containing three (trisazo), four (tetrakisazo) or even more azo-groups can be prepared; deep shades abound in dyes of this complexity.

A vitally important property of a dyestuff is its light fastness, i.e. resistance to fading on exposure to light. Photochemical degradation of azo-compounds involves scission of the azo-linkage and the rate at which it occurs varies from dye to dye, depending upon the exact constitution. Often substituents that withdraw electrons from the conjugated system, thereby diminishing the availability of electrons on the azo-nitrogen atoms, enhance the light fastness. The presence of bulky substituents located *ortho* to the azo-linkage, thus inhibiting approach to the linkage, is also beneficial. In addition to these measures there exists a further powerful means of protecting the azo-linkage against photo-degradation: it depends upon the facility with which suitably o,o'-disubstituted azo-compounds function as tridentate ligands in the formation of stable complexes with a number of metals, in particular, copper, chromium and cobalt. Such complexes are readily formed by o,o'-dihydroxyazo- or o-carboxy-o'-hydroxyazo-compounds, which themselves conveniently result from employing a 2-aminophenol or anthranilic acid as diazo-component. Less commonly used are analogous azo-compounds in which one or

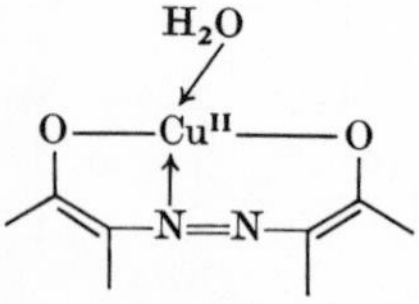

(14) 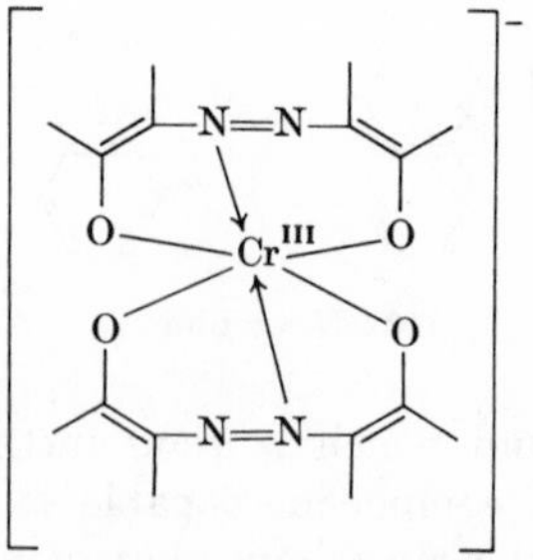   (15) 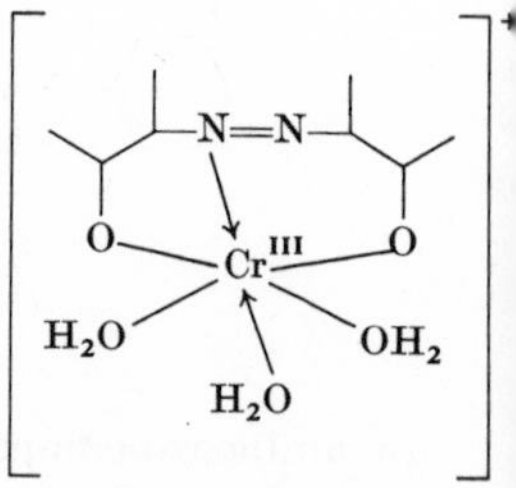   (16)

more of the hydroxyl groups is replaced by an amino-group or groups. Treatment with cupric acetate yields a 1:1 square coplanar complex, in which the situation of the metal atom is as shown in (**14**). Chromium salts at neutral pH values yield a negatively charged, octahedral 1:2 complex (**15**) and at low pH values a positively charged 1:1 complex (**16**). Treatment of the latter complex with a second, dissimilar metallizable azo-dye yields an unsymmetrical, 'mixed' 1:2 complex. The marked preference of cobalt for nitrogen donor ligands results in total expulsion of the co-ordinated water molecules and formation of the symmetrical 1:2 complex even at low pH values. Prior complexing of the cobalt with, e.g., ammonia does, however, enable the reactions to be carried out stepwise. The shades resulting from metal complexes cover the whole spectrum. In general they are duller than those of the unmetallized azo-dyes. Their excellent light fastness has, however, given them a high importance.

The wide variety of readily available intermediates and the excellent preparative methods which proceed in high yield allow ready production of an enormous number of azo-compounds, covering the whole colour range. Their flexibility makes them adaptable to the dyeing of all fibres and for use as pigments. They are the most important group of dyes, accounting for over 50% of all manufactured dyestuffs. It will be convenient to use them to illustrate the facets of dyestuff structure that determine whether a dye is of use on a particular fibre; application as pigments is deferred to a later Section.

Application

When a dyestuff is applied to a fibre an obvious requirement is that the application process should entail as little damage to the fibre as possible. Further, the colouration should be permanent throughout the useful life of the fibre and hence a degree of resistance to washing, as well as to light, is required in a dyestuff. In addition, fastness to certain other agencies is needed, but these are often specific requirements varying from fibre to fibre.

To achieve fastness to washing, some mechanism whereby the dye can be anchored on to the fibre is necessary. In practice five mechanisms are employed. These are formation of a solid solution, formation of a salt linkage, hydrogen-bonding, insolubilization and formation of a covalent bond. The mechanism to be used is determined solely by the chemical and physical properties of the fibre, in particular by the functional groups it contains and the degree to which it is hydrophilic or hydrophobic in character.

Certain synthetic fibres, such as nylon and polyester, are plastic in nature. On heating, they soften and eventually melt. Fibres of this type are readily dyed by *disperse dyes*. The method of application involves heating the fibre in an aqueous dispersion of a water-insoluble dye; during the dyeing process the dye transfers from aqueous dispersion into the fibre wherein it forms a solid solution. To enable the dye to penetrate the fibre a small dye molecule is required. Simple water-insoluble monoazo-dyes are ideal for this application. Apart from yellow dyes, e.g. (**17**), wherein phenolic coupling components are common, these dyes are usually based on N,N-dialkylated aniline coupling components since these allow the ready production of a wide range of shades from small dye molecules. Typical disperse dyes for the relatively hydrophilic nylon are (**18**) and (**19**). The N-alkyl groups on the coupling component are often complex. 2-Hydroxyethyl groups are particularly favoured, these being conveniently introduced by reaction of, e.g., N-ethylaniline with ethylene oxide to give

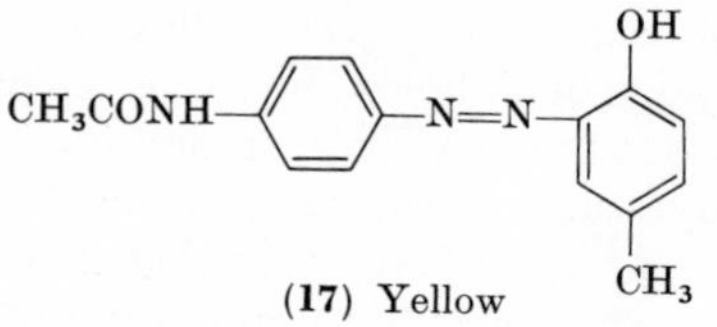

(**17**) Yellow

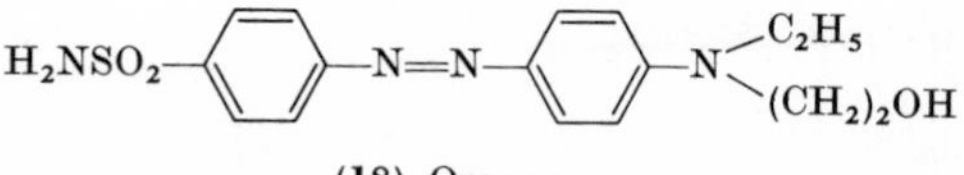

(**18**) Orange

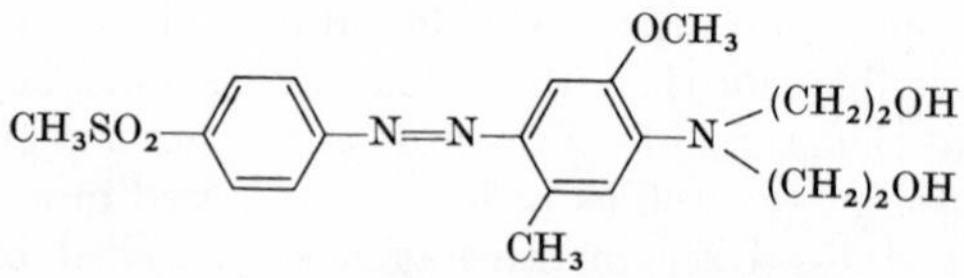

(**19**) Red

N-ethyl-N-(2-hydroxyethyl)aniline used in (**18**). These dyes illustrate the bathochromic shift, i.e. movement of the light absorption to longer wavelength with consequent deepening of colour, caused by incorporating further electron-releasing substituents in the coupling component. Such dyes give very level dyeings on nylon but are often deficient in fastness to washing. They are complemented on nylon by the acid dyes (see below) which give more uneven coverage but have better fastness properties.

Polyester and cellulose triacetate fibres are much more hydrophobic than nylon and penetration of the fibre by the dye is much

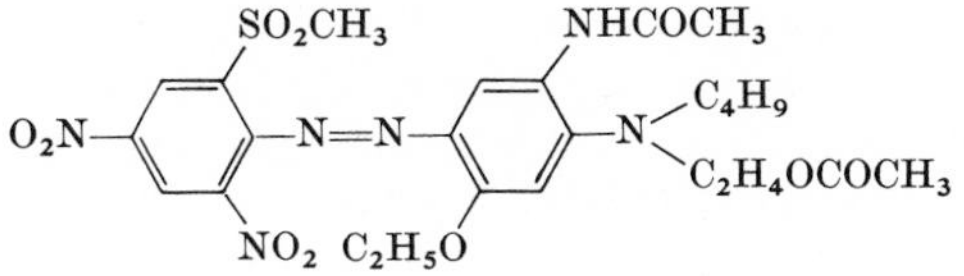

(**20**) Greenish-blue

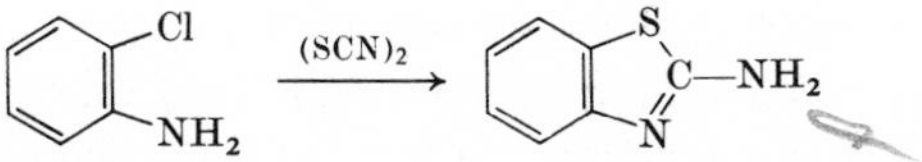

(**21**)

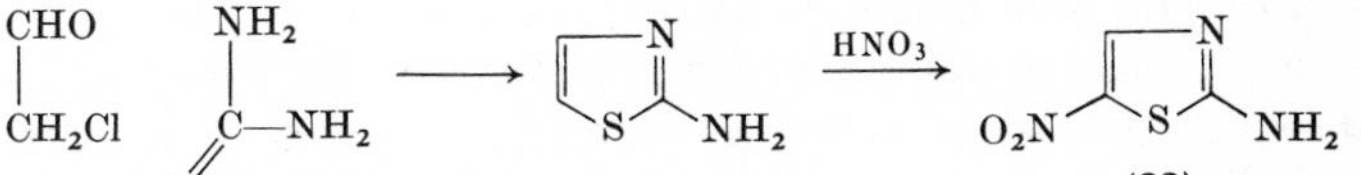

(**22**)

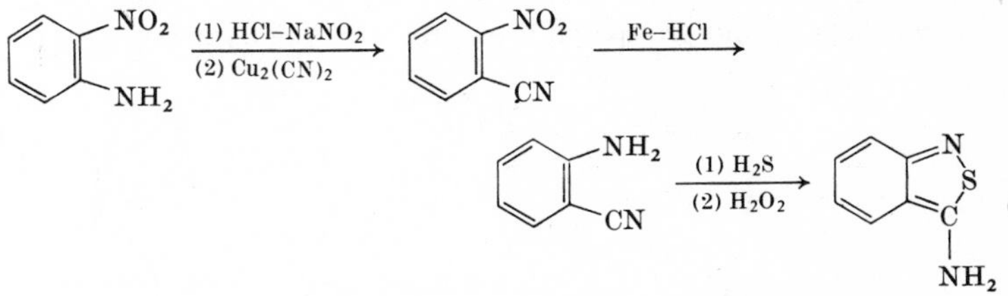

(**23**)

more difficult to achieve. Small molecular size is of even greater importance in dyes for these fibres. Disperse dyes are the only suitable class of dye and their passage from the boiling dye bath into the fibre can be facilitated by the presence of a 'carrier' such as benzyl alcohol in the dye bath. Pressure dyeing at higher temperatures (130°C) removes the need for a carrier, residual traces of which may have a deleterious effect on light fastness. Yellow, orange and red shades are readily obtained from simple substituted anilines→tertiary amines, but it is necessary to use diazo-components containing a number of electron-withdrawing groups in order to achieve blue shades whilst remaining within the constraints on dye size imposed by these fibres. The difficulties in preparing and handling diazo-components such as that in (**20**) have made feeble basic heterocyclic amines such as 2-amino-benzothiazole (**21**), 2-amino-5-nitrothiazole (**22**) and 3-amino-benzisothiazole (**23**) important alternatives. These compounds are often more accessible than the highly substituted anilines, of lower molecular weight and equally effective when used as A components in producing the required bathochromic shift. The alkyl substituents on the tertiary nitrogen atom of the coupling component are often complex, containing groupings such as acetoxy to adjust the overall hydrophilic character of the molecule. The purpose of this is to confer on the dye a resistance to sublimation, which is a necessary feature since dyed polyester fibre articles are often steam-pleated.

The anionic water-soluble mono- and bis-azo-dyes resulting from the use of sulphonated intermediates find use as *acid dyes* for wool and nylon. Both of these fibres contain amino-groups and the mechanism of dye retention depends upon formation of a salt between the fibre amino-group and the dyestuff sulphonic acid group. Typical dyes are (**24**) for wool and (**25**) for nylon. In the latter, prepared by coupling diazotized 4-chloro-2-(trifluoro-methyl)aniline at low pH to 2-amino-8-naphthol-6-sulphonic acid, both nitrogen atoms of the azo-linkage are hydrogen-bonded, one with the amino-group and the other with the hydroxyl group,

(**24**) Greenish-yellow   (**25**) Red

this leading to excellent light fastness. The interesting diazo-component used in this dye is prepared from *m*-chlorotoluene by chlorination to *m*-chlorobenzotrichloride and reaction of this with hydrogen fluoride to yield *m*-chlorobenzotrifluoride which is then nitrated and reduced. Deeper shades such as blue and navy result from the use of bisazo-dyes. On nylon these dyes have excellent fastness to washing but can give unlevel dyeings due to uneven penetration caused by differing degrees of crystallinity of the fibre along its length. In the case of wool, simple dyes of this type need to be forced on to the fibre by application from a strongly acid dye bath; fastness to washing is also limited. These defects are surmounted by incorporation of a long alkyl chain into the molecule as in (**26**). The increase in molecular size together with

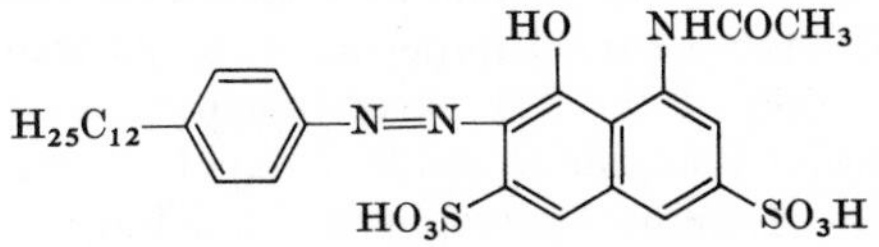

(**26**) Red

the hydrophobic nature of the long alkyl group allows satisfactory dye uptake at near neutral pH values and helps prevent subsequent desorption by water, thus giving increased fastness to washing.

Dyes of this type supply many of the bright shades on wool. The use of wool in men's suiting, however, imposes a requirement for dyes of deep shades, not particularly bright but of excellent light fastness. For this type of outlet the chromium and cobalt complexes of azo-dyes are eminently suited. Initially such complexes were formed *in situ* by dyeing wool with a metallizable dye and then treating it with a suitable metal salt, e.g. a dichromate. This was detrimental to the fibre and a big advance occurred with the introduction of the neutral-dyeing pre-metallized dyes, an example of which is (**27**), prepared by treatment of 2-amino-4-methylsulphonylphenol→2-naphthol with chromium acetate. The first dyes of this type carried a single negative charge due to

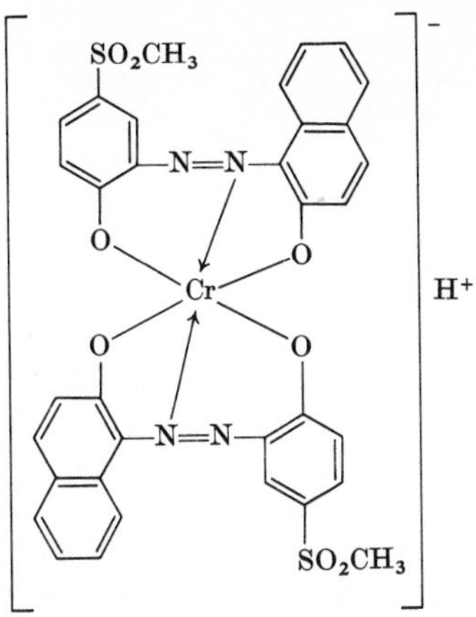

(**27**) Grey

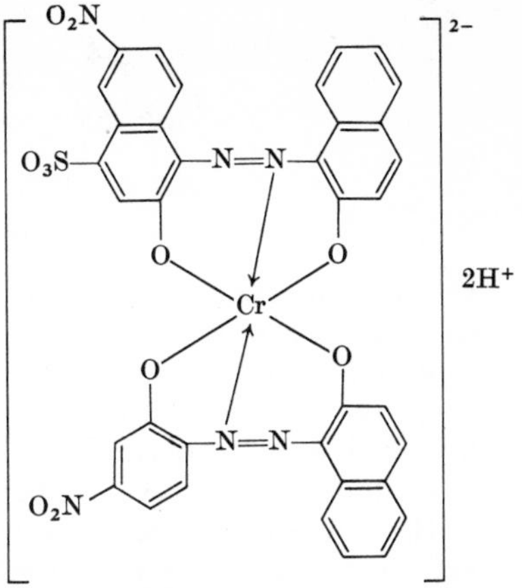

(**28**) Black

the chromium atom. Their moderate ionic character allowed satisfactory uptake from a neutral dye bath and together with their large molecular size conferred excellent fastness to washing. The presence of sulphone or sulphonamide groups was necessary to give a satisfactory level of water-solubility. Recently, mixed complexes such as (28) have assumed importance; preparation of this dye entails formation of the 1:1 chromium complex of 1-amino-6-nitro-2-naphthol-4-sulphonic acid→2-naphthol and subsequent reaction with 2-amino-5-nitrophenol→2-naphthol to yield the unsymmetrical structure shown; in these a further negative charge is contributed by a single sulphonic acid group; this degree of ionic character is not excessive in molecules of this size and satisfactory neutral exhaustion and wet-fastness is still obtained.

The converse of the acid dyes is supplied by the *basic dyes*, cationic compounds solubilized by a quaternary nitrogen atom which are employed in the dyeing of polyacrylonitrile. These dyes form a salt linkage with the carboxylic and sulphonic acid groups introduced into the fibre by the incorporation of suitable monomers at the polymerization stage. Two distinct types of basic-dye structure are exploited. In one the quaternary nitrogen is carried in a pendant side chain (29); in the other it comprises part of a

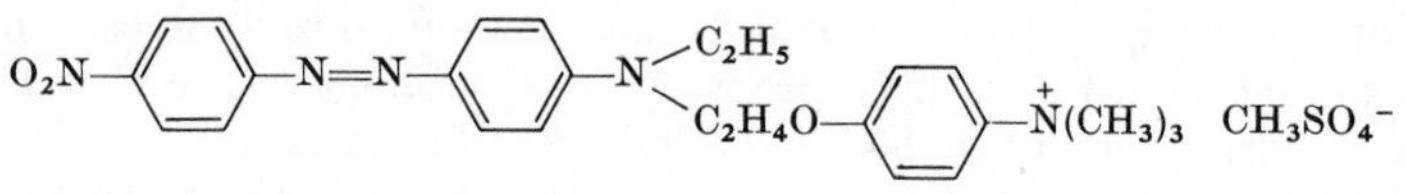

(29) Red

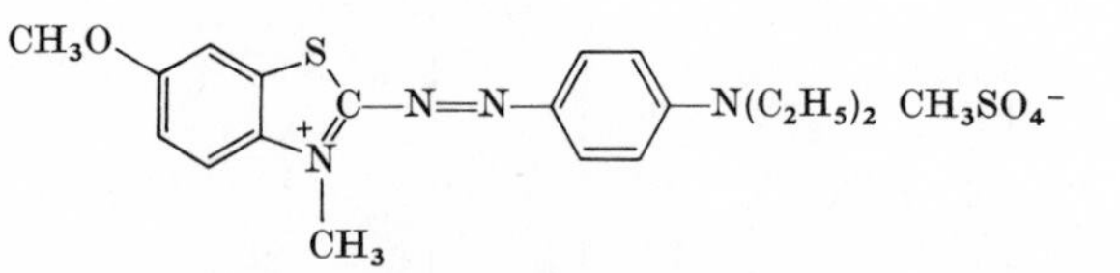

(30) Blue

heterocyclic system within the chromophore (30). Both are conveniently prepared by quaternization of a precursor dye containing a tertiary nitrogen atom. Satisfactory fastness properties ensue from the salt linkage formed, but difficulties are caused by rapid rates of dyeing which cause unlevelness; addition of restraining agents to the dye bath assists in overcoming this.

The two dye-retention mechanisms discussed so far are not applicable to the very hydrophilic cellulosic fibres cotton and viscose since neither are these plastic in nature nor do they contain acidic or basic centres. However, certain water-soluble dyes, e.g. the bisazo-dye Congo Red (**31**) can be directly adsorbed on to

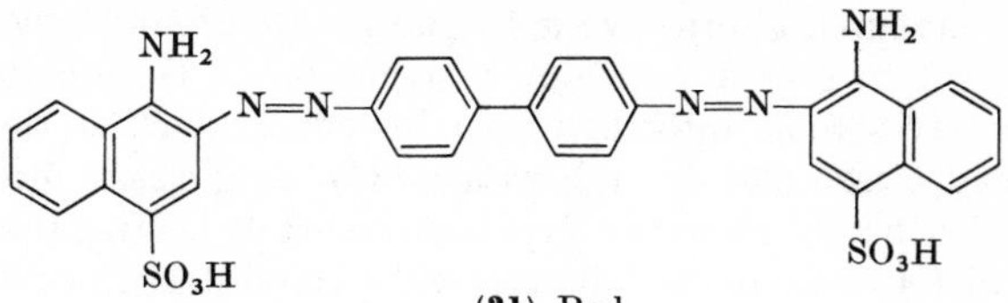

(**31**) Red

cotton from an aqueous dye bath. These are the *direct dyes*, the vast majority of which are azo-compounds. They possess this property of cotton substantivity by virtue of their propensity for hydrogen-bonding either with the fibre or with their own kind, thus building up aggregates within the fibre pores that resist subsequent washing treatments. Structurally they require to be long, planar molecules with the bare minimum of water-solubility required for application from aqueous solution. Various units such as benzidine residues and urea linkages and the azo-group itself assist in conferring cotton substantivity and a

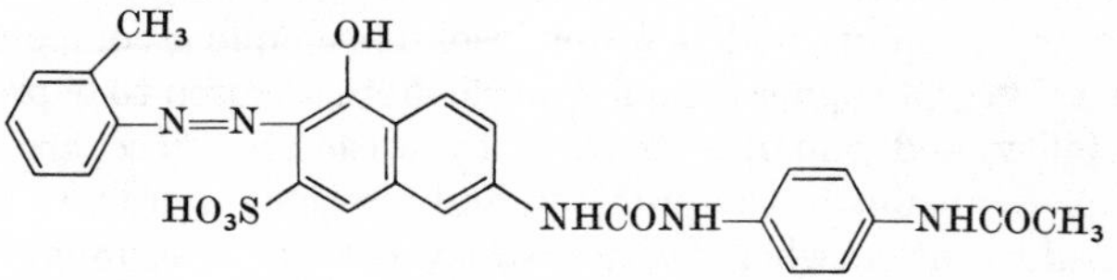

(**32**) Red

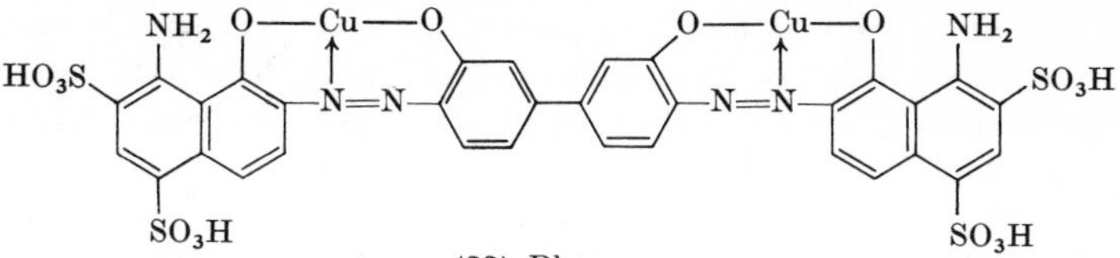

(**33**) Blue

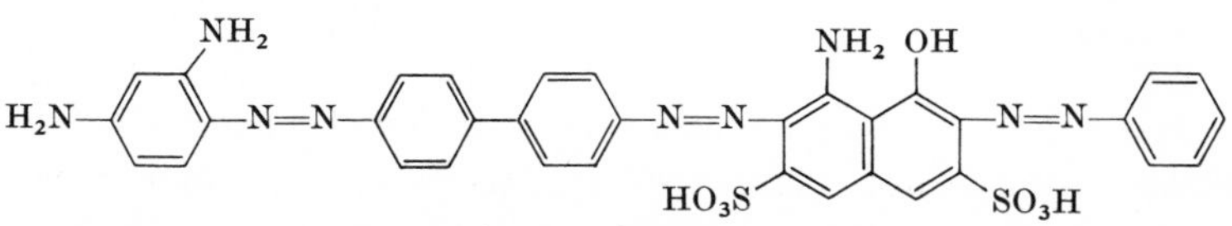

(**34**) Black

multiplicity of such features is required to enhance the effect up to the level of usefulness. In the case of monoazo-dyes it is necessary to incorporate these features into an appendage as in (**32**); a favoured alternative is to phosgenate an aminomonoazo-compound to give a bisazo-urea structure. It is here that the polyazo-dyes come into their own, structures such as the twice coppered bisazo (**33**) and the trisazo (**34**) being classic direct dye types. The first of these two dyes is prepared by heating the bisazo dyestuff 4,4′-diamino-3,3′-dimethoxybiphenyl $\rightrightarrows$ (1-amino-8-naph-thol-2,4-disulphonic acid) with ammoniacal copper sulphate to form the copper complex with concurrent demethylation of the methoxyl groups. The trisazo-dyestuff (**34**) presents a fine example of selective coupling: 4,4′-diaminobiphenyl is tetrazotized and the first, highly reactive diazonium group coupled, at low pH, *ortho* to the amino-group in 1-amino-8-naphthol-3,6-disulphonic acid. Diazotized aniline is next introduced and coupled under alkaline conditions *ortho* to the hydroxyl group, the feebly reactive second diazonium group on the biphenyl unit showing little tendency to react. Finally *m*-phenylenediamine is added, where-upon formation of the third azo-group slowly occurs. Direct dyes are easily applied but the weak reversible forces relied on to hold the dye on to the cloth limit their wet-fastness. After-treatments such as diazotization of an amino-containing direct dye on the fibre and coupling with a water-insoluble coupling component or formation of a copper complex on the fibre are used to depress the solubility and enhance fastness to washing. These are not a complete answer to the problem, and direct dyes are now largely limited to outlets where fastness to severe washing is unimportant, e.g. suit linings. They retain a major outlet in the colouration of paper.

An alternative approach to the dyeing of cotton depends upon forming an insoluble dye on the fibre. This is the dye-retention mechanism relied on in the case of *azoic dyes*. This class utilizes the cotton substantivity shown by arylamides of 2-hydroxy-3-naphthoic acid, the parent compound of which is Naphtol AS (2-hydroxy-3-naphthanilide) (**35**). These arylamides are prepared by condensation of an arylamine with the important 2-hydroxy-3-naphthoic acid in the presence of phosphorus trichloride. 2-Hydroxy-3-naphthoic acid itself results from heating sodium 2-naphtholate in an atmosphere of carbon dioxide under pressure. When applied from alkaline solution these coupling components are loosely anchored to the fibre by virtue of their substantive character and are retained there until brought into contact with

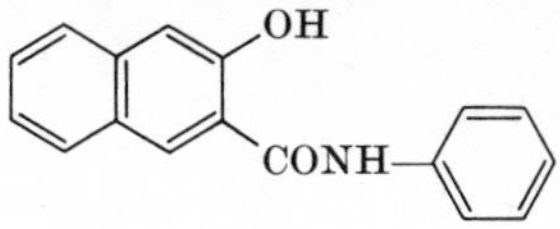

(35)

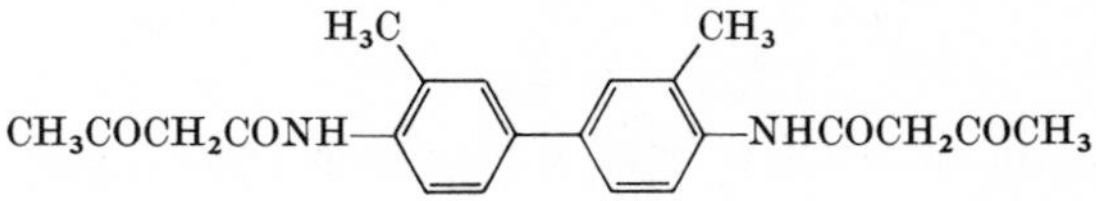

(36)

a solution of a diazonium salt prepared from an amine devoid of water-solubilizing groups. This results in the formation of an insoluble azo-compound on the fibre. Subsequently the particles are caused to aggregate within the fibre pores by boiling in hot soap solution. Naphtol AS and its analogues give chiefly orange and red shades with the chloro-, nitro-, methyl- and methoxy-substituted anilines commonly employed as diazo-components and blues with dianisidine, 4-aminodiphenylamine and 4-(benzoyl-amino)-2,5-dimethoxyaniline. Yellows were achieved with the introduction of Naphtol AT (**36**); browns result from heterocyclic analogues of Naphtol AS such as the arylamides of 2-hydroxy-carbazole-3-carboxylic acid. Diazonium compounds stabilized and marketed as their naphthalene-1,5-disulphonate and tetrachloro-zincate salts have freed the dyer from the necessity of diazotizing the arylamine component. Azoic dyes are superior to direct dyes in fastness but their application is more complex from the dyer's point of view. A far better solution to the problems encountered in the dyeing of cotton has been found in the reactive dyes, but since these are dominated by the chemistry of the reactive group rather than the properties of the chromophore, discussion of that type is deferred to a separate Section.

Simple Anthraquinone Dyes

As the name implies, these dyes are derivatives of anthraquinone (**37**). In this molecule the carbonyl groups of the central quinonoid unit function as electron-withdrawing substituents on the flanking benzenoid rings. When electron-releasing substituents are introduced into these rings, polar structures such as (**38**) are possible and the molecule shows the phenomenon of colour. Thus

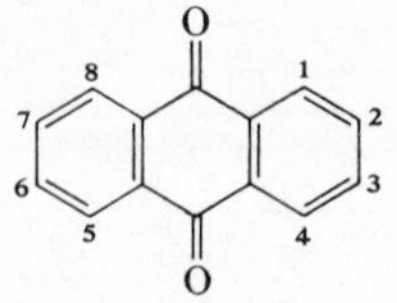

(37)

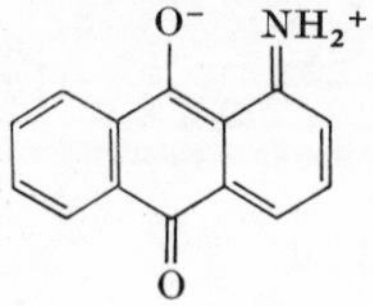

(38)

anthraquinone itself is only a very pale yellow, the 1-amino-compound is orange, the 1,5-diamino-derivative red and 1,4,5,8-tetra-aminoanthraquinone is blue.

Synthesis of these dyes usually starts from anthraquinone, which is itself obtained by the oxidation of anthracene or by Friedel–Crafts acylation of benzene with phthalic anhydride followed by cyclization of the resulting o-benzoylbenzoic acid. Nuclear substitution in anthraquinone is not easy because of the electron-withdrawing character of the carbonyl groups. Halogenation is so difficult as to be impracticable, and nitration under the forcing conditions required gives mixtures of isomers. Sulphonation is by far the most important reaction: it can be directed to produce either the 1-sulphonic acid, in the presence of mercury, or the 2-sulphonic acid in the absence of this catalyst. Further reaction introduces a second sulphonic acid into the other benzene ring, yielding isomeric mixtures. In the presence of mercury the second group is located at position 5 or 8, or in the absence of mercury at position 6 or 7.

In these compounds the sulphonic acid groups are labile; their displacement by heating with lime, to yield hydroxyanthra-

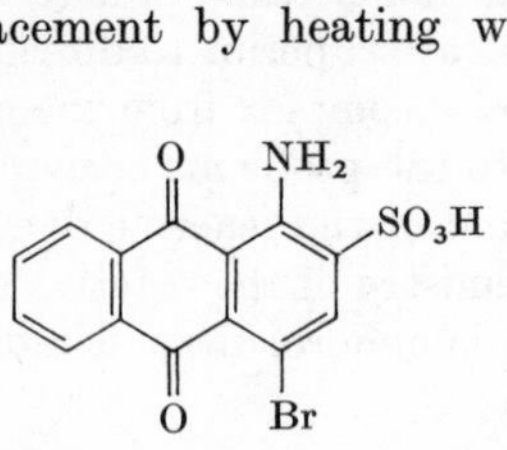

(39)

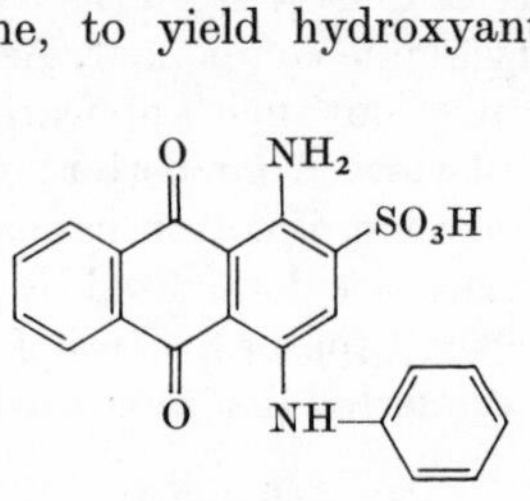

(40) Blue

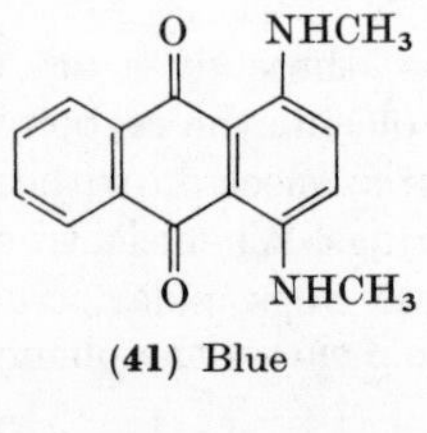

(41) Blue

quinones, or with ammonia under pressure, to yield amino-anthraquinones, are widely exploited in the preparation of dyestuffs intermediates. Fusion with alkali-metal hydroxides can result in further nuclear hydroxylation; this is utilized in the preparation of 1,2-dihydroxyanthraquinone (alizarin) by fusion of anthraquinone-2-sulphonic acid with potassium hydroxide. The introduction of an electron-releasing substituent makes further substitution of the anthraquinone nucleus easier. The order of stages in the synthesis of the key intermediate bromo-amino-acid **(39)** illustrate this. Anthraquinone is converted via the 1-sulphonic acid into 1-aminoanthraquinone which is then 2-sulphonated and finally 4-brominated. The ready copper-catalyzed condensation of this intermediate with amines is the main route to sulphonated anthraquinone dyes, for example the bright blue compound **(40)**.

Variations on the phthalic anhydride–benzene synthesis of anthraquinone are occasionally useful. An important intermediate prepared in this way is 1,4-dihydroxyanthraquinone (chrysazin) which results when the benzene is replaced by p-chlorophenol; hydrolysis of the chlorine atom occurs at the cyclization stage. Chrysazin is largely used as a source of simple 1,4-dialkylamino-anthraquinones, e.g. **(41)**, which result from its reaction with alkylamines.

The initial aim of anthraquinone research, synthesis of the naturally occurring alizarin, has now vastly widened to cover the exploitation of these dyes on all fibres. Whilst the full shade range can be reached by using these dyes it is in the blue-to-green portion that they have found their major use, since in this area they give a combination of brightness and light fastness unsurpassed by any other class of dye. Since the mechanisms for dye retention discussed under azo-dyes are independent of the chromophore they can be equally applied to the anthraquinone series. Thus simple water-insoluble compounds such as **(41)** are used as disperse dyes for synthetic fibres. Sulphonated water-soluble compounds, typically **(40)**, are used as acid dyes on wool, and cationic dyes for polyacrylonitrile arise when a quaternary ammonium grouping is located in a pendant side chain. Only on cotton have the simple non-reactive anthraquinone dyes found no use, owing to the completely non-substantive nature of the anthraquinone nucleus. Here, however, they have been employed in azoanthraquinone dyes such as **(42)** wherein a substantive yellow azo-dye is combined with a blue anthraquinone unit to yield a green direct dye as shown on p. 334.

(42) Green

Vat Dyes

Despite the failure of simple sulphonated anthraquinone dyes on cotton, more complex structures based on anthraquinone are extremely important in the dyeing of this fibre. These are the vat dyes. The salient feature of anthraquinone chemistry upon which these dyes depend is the ease with which this molecule is reduced (vatted) by sodium dithionite in dilute sodium hydroxide to give a solution of the disodium salt of the dihydroxy-compound (leuco-form). This is then readily re-oxidized to the insoluble quinone by air. In the case of anthraquinone itself the leuco-form has little substantivity and is of no use in the dyeing of cotton. More complex structures, however, possess marked substantivity and can be satisfactorily adsorbed on to cotton from their alkaline solution. Exposure to air regenerates the insoluble quinone on the fibre; the dyeing process is finally completed by treatment of the dyed fibre with boiling soap solution to aggregate the particles of dyestuff. The need for the dyer to carry out the reduction to the leuco-form has been circumvented by the marketing of the stable sulphuric acid esters of the reduced vat dyes. These water-soluble compounds are produced by reducing the vat dye in pyridine containing sulphur trioxide with a metal; after adsorption on the fibre they are readily re-oxidized to the insoluble quinone under acid conditions.

The simple useful vat dyes are exemplified by (43) (p. 336) where aminoanthraquinone units are joined via a dicarboxylic acid chloride into a substantive entity. Similar are the anthrimides, typically (44), formed by condensation of a chloroanthraquinone with a 1-aminoanthraquinone. Dyes of the latter type, besides being useful vat dyes in their own right, are important as inter-mediates in the preparation of the carbazole vat dyes, e.g. (45). Formation of the carbazole ring in these compounds is brought about by fusion of the anthrimide with aluminium chloride and sodium chloride; (benzoylamino)anthrimides cyclize more easily, treatment with concentrated sulphuric acid often sufficing. Far more complex dyes occur in this group. Thus the tetra-anthrimide obtained when 1-aminoanthraquinone is condensed with 1,4,5,8-tetrachloroanthraquinone cyclizes to the important Indånthrene Khaki GG. In dyes of this complexity there is often doubt whether formation of all the carbazole rings (four in the dye cited) is completed in the cyclization stage.

In a similar group, the anthraquinone azines, the anthraquinone units are joined together through a 1,4-diazine ring. The parent of these is the important indanthrone (46), obtained by fusion of

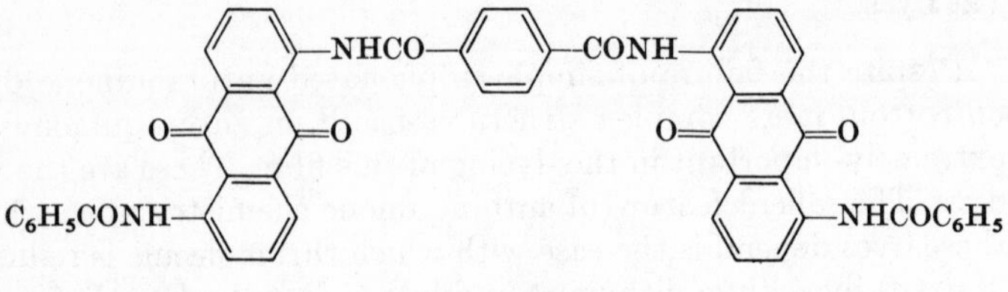

(43) Yellow

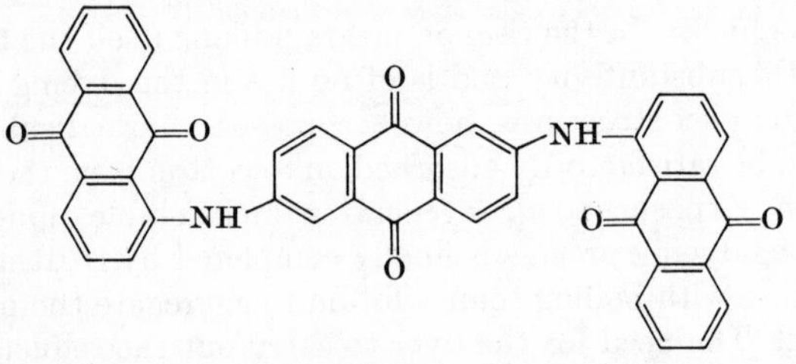

(44) Orange

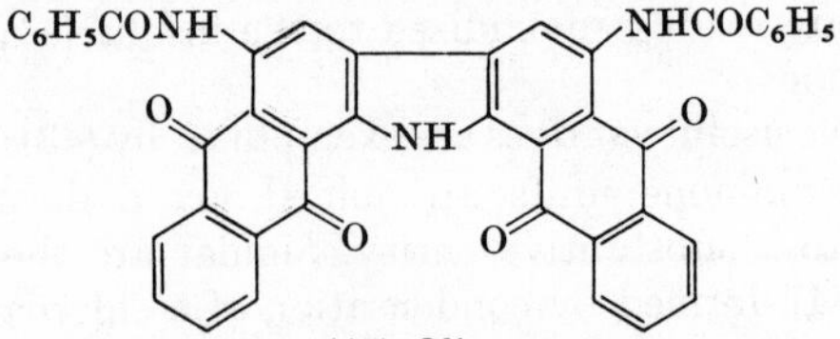

(45) Olive

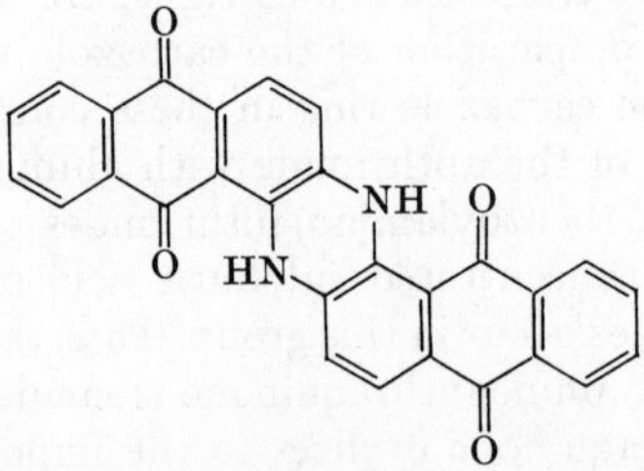

(46) Blue

2-aminoanthraquinone with potassium hydroxide. Numerous other heterocyclic ring systems including acridone and thioxanthone rings have been utilized in joining together anthraquinone residues but these are of lesser importance.

In the foregoing vat dyes, the individual anthraquinone units can be clearly distinguished. There are, in addition to these, further important groups of vat dyes containing anthrone (**47**, R = H) units. These units are condensed into a polycyclic aromatic structure wherein pyrene and perylene skeletons can be observed. Typical are the dibenzopyrenequinones (= dibenzochrysenes) (**48**) and the anthanthrones (**49**). The former are prepared by cyclization of 3-benzoylbenzanthrone (**51**, R = COC$_6$H$_5$) or 1,5-dibenzoyl-

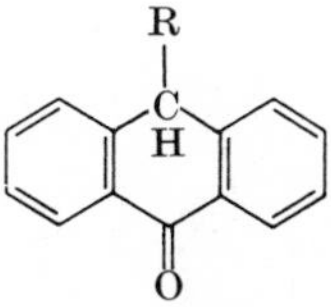

(**47**) 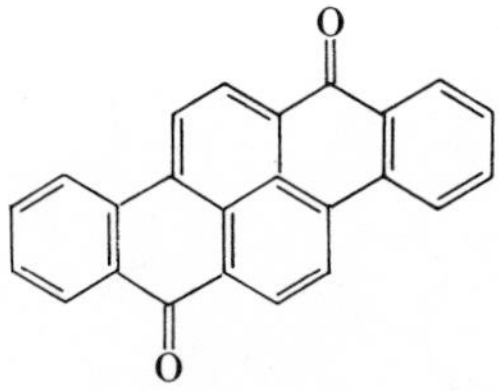   (**48**) Yellow 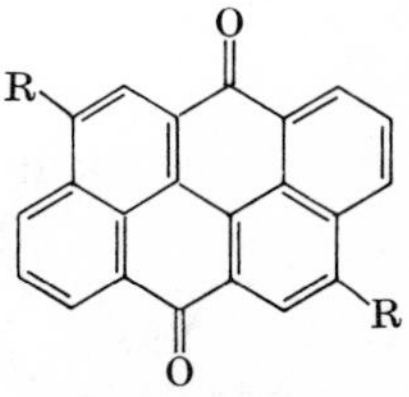   (**49**) Orange

naphthalene with aluminium chloride, both intermediates resulting from Friedel–Crafts benzoylation of the relevant hydrocarbon. Anthanthrone arises by cyclization of 1,1'-binaphthalene-8,8'-dicarboxylic acid in sulphuric acid; in itself it is of no practical value but its substitution products, e.g. (**49**, R = Br) are useful dyes. Anthrone units can also be distinguished in the pyranthrones

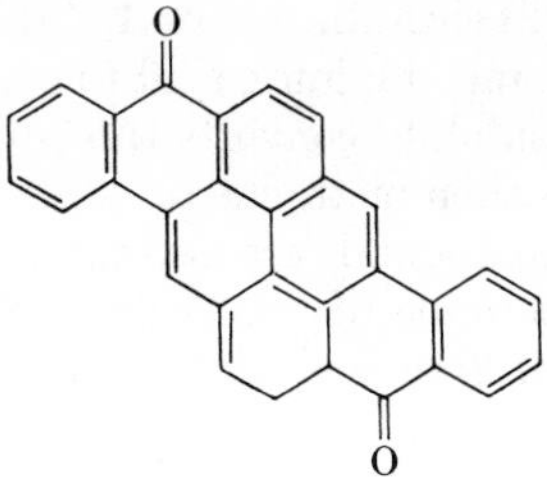

(**50**) Orange 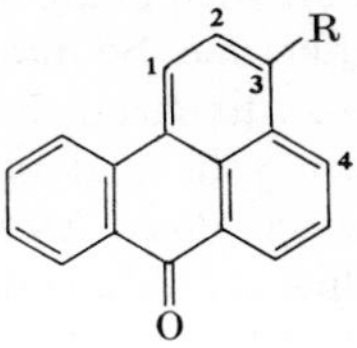   (**51**)

(**50**) which result when 2,2'-dimethyl-1,1'-bianthraquinone, prepared by the action of copper on 1-chloro-2-methylanthraquinone, is fused with potassium hydroxide.

A key intermediate in the preparation of important vat dyes containing a perylene skeleton is benzanthrone (**51**, R = H). This compound is readily prepared from anthraquinone by treatment in sulphuric acid with iron powder and glycerol. The initial reactions are reduction of the anthraquinone to anthrone and formation of acrolein from the glycerol. These materials then react either by condensation to give a hydroxy-ketone [**47**, R = CH(OH)CH=CH$_2$] which dehydrates and cyclizes to benzanthrone, or by α,β-addition of anthrone to acrolein to give (**47**, R = CH$_2$CH$_2$CHO) which cyclizes to yield, initially, a dihydrobenzanthrone. Alkali fusion of benzanthrone gives dibenzanthrone (**52**, R = H), the 4,4'-dibenzanthronyl being an intermediate product.

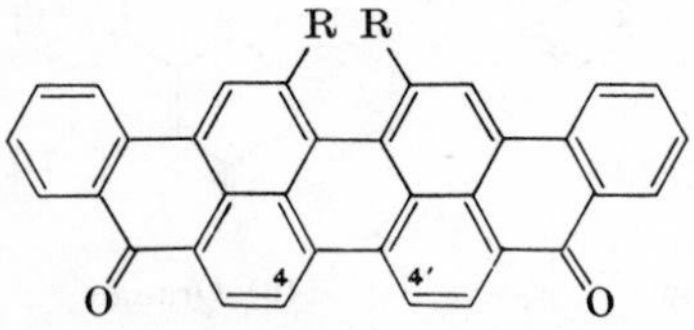

(52)	(53) Yellow

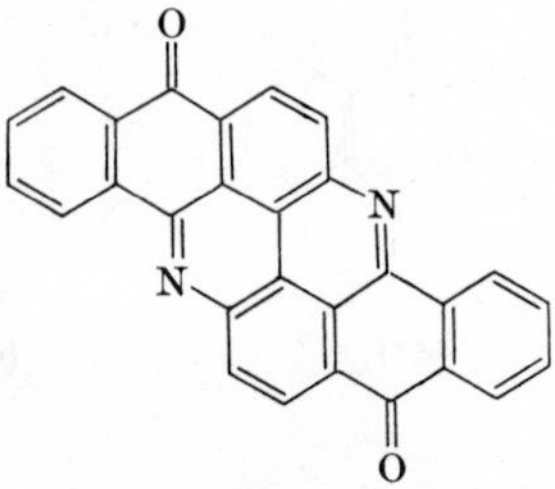

'Dibenzanthrone' (**52**, R = H) itself is an important blue vat dye and also the basis of numerous substituted derivatives which are of considerable value. Outstanding amongst these is the 16,17-dimethoxy-compound, Caledon Jade Green (**52**, R = OCH$_3$). This is prepared by oxidation of dibenzanthrone or 4,4'-dibenzanthronyl with manganese dioxide and sulphuric acid to give the 16,17-diketone. Sodium hydrogen sulphite converts this into the dihydroxydibenzanthrone, which is then methylated.

Heterocyclic analogues of these compounds are also extensively used as vat dyes. An important example of this is flavanthrone (**53**) which is obtained by fusion of 2-aminoanthraquinone with potassium hydroxide at high temperature.

Vat dyes cover the whole spectrum and generally give superb fastness properties. A defect with many of the yellow dyes is photo-tendering, a phenomenon whereby exposure of the dyed fabric to sunlight causes deterioration of the cotton fibre. Because of the multi-stage syntheses involved in their preparation they are costly and consequently their main use is in high-quality cotton textiles (e.g. curtain materials and furnishing fabrics) where outstanding light fastness is a necessity and the cost can be tolerated.

Indigoid and Thioindigoid Dyes

With the elucidation of the structure of the pre-eminent natural dye, indigo, its synthetic production became an automatic target for the early dyestuffs industry. Such was the importance attached to this venture that the German firm B.A.S.F. alone spent in excess of £1m. towards the end of the last century on achieving an economic synthesis. When this target was finally met the synthetic material rapidly ousted the natural product.

Two processes were developed by B.A.S.F. for the manufacture of indigo. In the original process, phenylglycine was cyclized by fusion with potassium hydroxide, yielding indoxyl, and this was then oxidized to indigo in alkaline solution. The second process substituted phenylglycine-2-carboxylic acid at the fusion stage, giving an increase in yield. Later work showed that fusion with sodamide gave an equally good yield from the cheaper phenyl-glycine; this became and has remained the major manufacturing route.

Indigo owes its deep blue colour to the contribution which polar forms such as (54) make to the molecular structure. It is applied by a vatting technique but differs from the anthraquinone vat dyes in that its almost colourless leuco-form is reasonably stable. This material is marketed as 'indigo white', thus relieving the dyer of the tedium of the reduction stage; a treatment with nitrous acid is frequently employed to re-oxidize this to the insoluble form. As with other dyes that are retained on the fibre by virtue of their insolubility, aggregation of the dye molecules is an important factor.

(54)

The blue colour of indigo is little affected by substitution and only a small number of halogenated derivatives have been of any technical significance, these also being blue in shade. However, the thio analogue (55), prepared from 2-carboxyphenylthio-glycollic acid by fusion with potassium hydroxide followed by oxidation is red in shade, being applied to cotton in the same way as indigo. Simple substituted derivatives of thioindigo are orange to red, but replacement of one or both of the benzene nuclei by naphthalene or anthracene allows browns, greens and greys to be achieved. In addition to thioindigo, the mixed structure (56) is also useful; further variations which have been exploited are the brominated isomer (57) and the thioindoxyl–acenaphthene-quinone condensate (58).

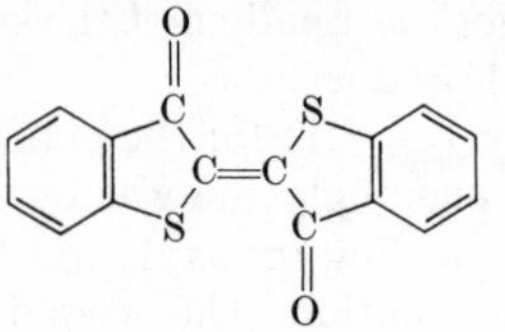

(55) Red

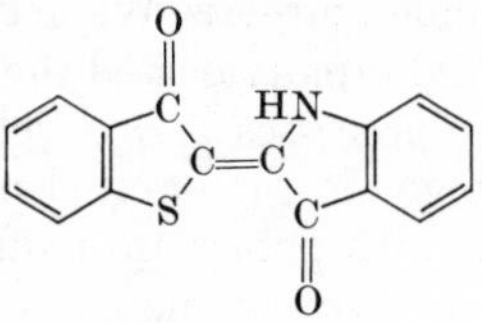

(56) Violet

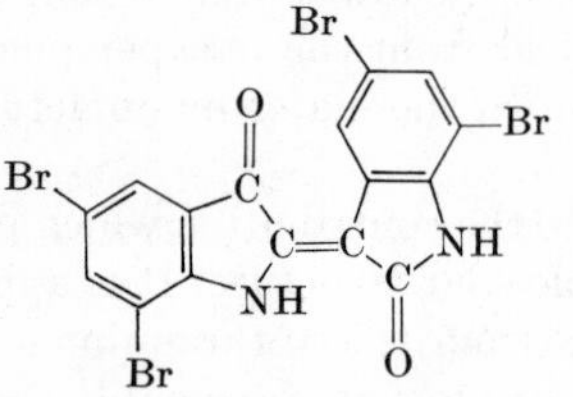

(57) Heliotrope

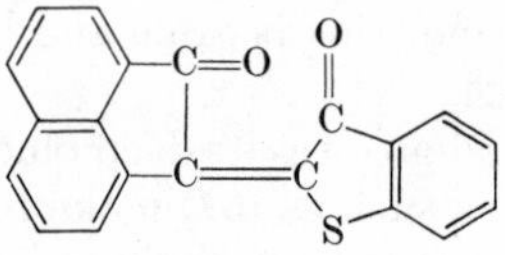

(58) Scarlet

Triarylmethanes

The triarylmethanes are amongst the oldest of synthetic dyes (Part 2, pp. 249–253). A typical structure (59) shows only one of the possible forms that can be written, the positive charge being delocalized over the two or three dialkylamino-groups located *para* to the central carbon atom. This extensive charge delocaliza-tion gives rise to brilliant violet, blue and green shades.

The main representative of the diamino-series is Malachite Green (59, R = H). This is prepared by condensation of benz-aldehyde with dimethylaniline in dilute hydrochloric acid, followed by oxidation of the resulting (60) with lead dioxide. Triaminotriarylmethanes, typically Crystal Violet [59, R = N(CH$_3$)$_2$], are prepared by phosgenation of dimethylaniline to give 4,4'-bis(dimethylamino)benzophenone, condensation of which with a third molecule of dimethylaniline yields the dyestuff. Use of N-phenyl-1-naphthylamine in the last condensation yields the important Victoria Blue. Related hydroxylated triarylmethane dyes such as Chrome Violet (61) are prepared by condensation of

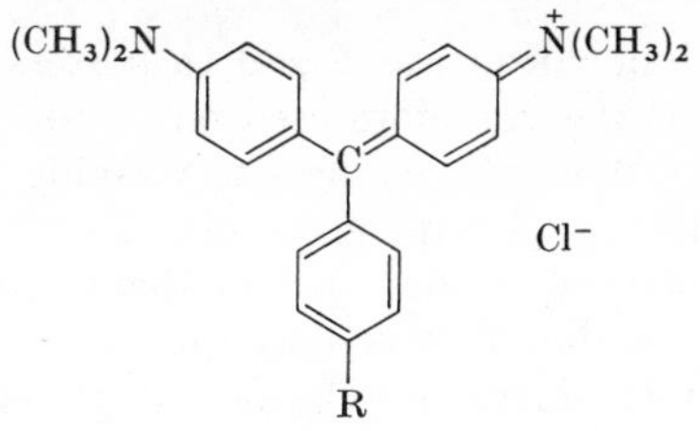

(59)

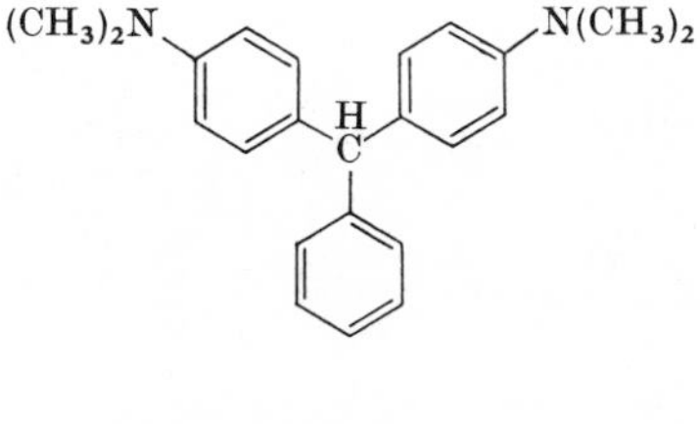

(60)

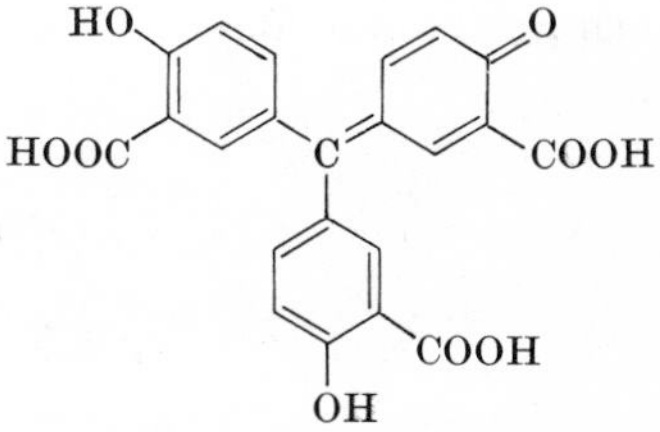

(61)

suitable intermediates in sulphuric acid in the presence of sodium nitrite as an oxidizing agent. Chrome Violet results from salicylic acid and formaldehyde.

The basic dyes of this class find some use on polyacrylonitrile, and their polysulphonated derivatives are applicable as acid dyes for wool. Their textile usage is, however, restricted because of poor light fastness and their main outlet as dyestuffs is in paper colouration where brilliance rather than permanence is required.

Phthalocyanine Dyestuffs

The macrocyclic pigment copper phthalocyanine (**77**; see p. 349) possesses a unique turquoise shade and superb light fastness. These attributes have made the preparation of dyestuffs based on this system obviously desirable, in particular for cotton dyes, where the bright shade would be very advantageous. The starting material for such preparations is often copper phthalocyanine itself since its durability enables vigorous reactions to be carried out on it without decomposition. Usually four substituents are introduced, one in each benzene ring. Moderation of the conditions enables mixtures containing less than four substituents per molecule to be prepared. Initially the mixture of sulphonic acids produced by sulphonation was found to possess some degree of substantivity and was therefore used on cotton as a direct dye. As with all direct dyes, the fastness to washing was limited and this approach did not solve the problem.

Much better fastness results by capitalizing the insolubility of copper phthalocyanine. This is accomplished in the dye Alcian Blue 8 GN by tetrachloromethylating the pigment and causing the chloromethyl groups to react with thiourea to give a tetra-isothiouronium salt. When printed on to the fibre and decomposed by heat, this produces an insoluble phthalocyanine pigment ingrained in the fibre.

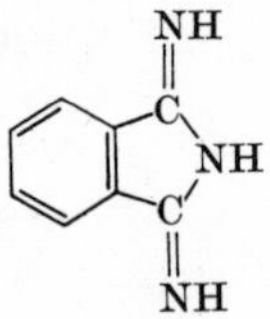

(62)

A different approach is utilized in the Phthalogens. These utilize di-iminoisoindoline (62) which is obtained as the sparingly soluble nitrate when phthalic anhydride is heated with urea, ammonium nitrate and ammonium molybdate. The free base arising after treatment with sodium hydroxide is mixed with a copper salt. When the mixture is printed on the cloth and heated, copper phthalocyanine is formed *in situ*.

Miscellaneous

In addition to the main chromophoric classes outlined above, numerous others exist but warrant only scant attention. The sulphur dyes, cheap products for unsophisticated markets, are complex compounds of ill-defined composition prepared by heating various organic materials with sulphur or a sulphur compound. In a typical case 2,4-dinitrophenol heated with sodium polysulphide solution yields Sulphur Black T. The dyes cover the whole shade range, are applied by a vatting technique and are retained on cotton by their insolubility. Aniline Black, a complex structure containing quinone imine units, produced on the fibre by oxidation of aniline, is still used in the dyeing of cotton.

Quinoline Yellow, obtained by sulphonation of the condensation product (63) of 2-methylquinoline with phthalic anhydride, is widely used in paper colouration. A few greenish-yellow cyanine dyes, e.g. (64) prepared by condensation of the appropriate aldehyde with ethyl cyanoacetate, are of value as disperse dyes for synthetic fibres and, after quaternization, as basic dyes for polyacrylonitrile. Individual xanthene dyes such as Rhodamine B (65) are used as fluorescent colours for paper; they are prepared by condensation of *m*-dialkylaminophenols with phthalic anhydride.

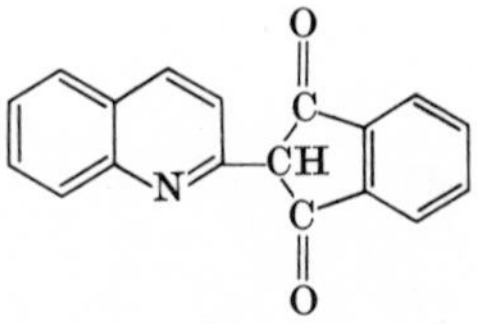

(63) 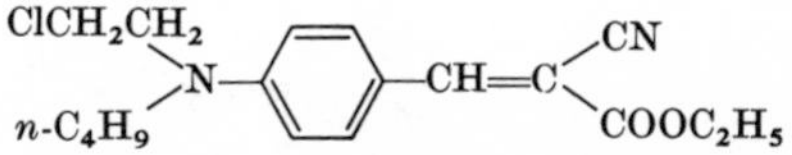   (64)

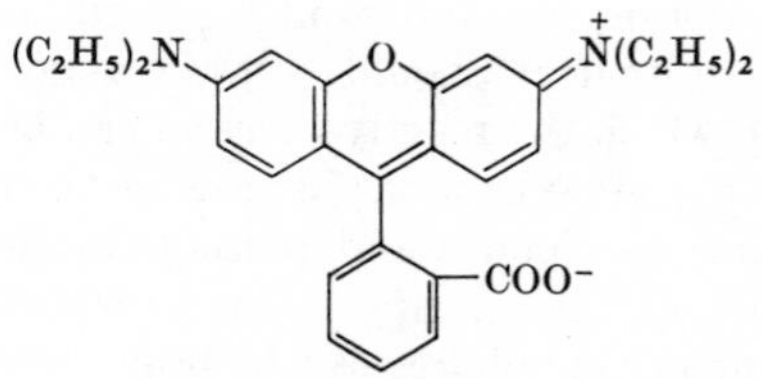

(65)

Reactive Dyes

The foregoing discussion of synthetic dyestuff classes has been organized according to the chromophore contained in the dye. Throughout, the particular problems posed in the dyeing of cotton have been tackled by utilizing features of the chromophore in one of two ways. Hydrogen-bonding has been exploited to hold soluble dyes or aggregates of soluble dye molecules on to the fibre as in the case of the direct dyes. Alternatively, formation of an insoluble dye on the fibre and aggregation of the particles has been relied on in the case of the azoic, vat, indigoid, thioindigoid, phthalocyanines and sulphur colours. The first method gives ease of application but limited fastness; the second can give much better fastness at the expense of a more difficult application process.

The major advance in post-war dyestuffs chemistry has been the introduction in 1956 of reactive dyes for cellulosic fibres. These dyes enabled excellent fastness to washing to be achieved from a water-soluble dyestuff by a simple application method. They function by the presence on the chromophore of a group that reacts with the fibre to form a covalent bond; this linkage retains the dye on the fibre during subsequent washing treatments. A

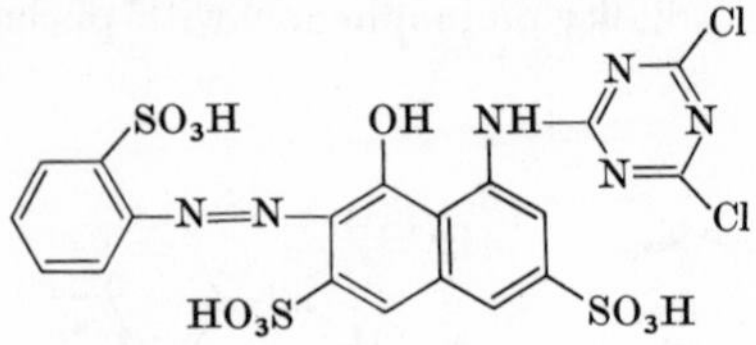

(66)

typical member of the first range of these dyes, the Procion M range, is **(66)** prepared by reaction of an aminoazo-compound with cyanuric chloride. No constraint is placed upon the chromophore other than that it should contain a group, usually an amino-group, to which the reactive system can be attached. Any class of dye can therefore be used; in practice the most important are the azos, anthraquinones and phthalocyanines. All carry a number of sulphonic acid groups to confer water-solubility and, since very simple chromophores can be used, bright clear shades are readily achieved.

The reactivity of these dichloro-*s*-triazinyl dyes is characteristic of nitrogen-containing six-membered heterocyclic compounds (cf. Part 3, Chapter 2). The reactivity of 2- or 4-halogenopyridines has been emphasized (Part 3, pp. 65–66), and the similar reactivity of pyrimidine, pyrazine and pyridazine has been described (Part 3, pp. 77–78). We should expect similar, if not greater, reactivity

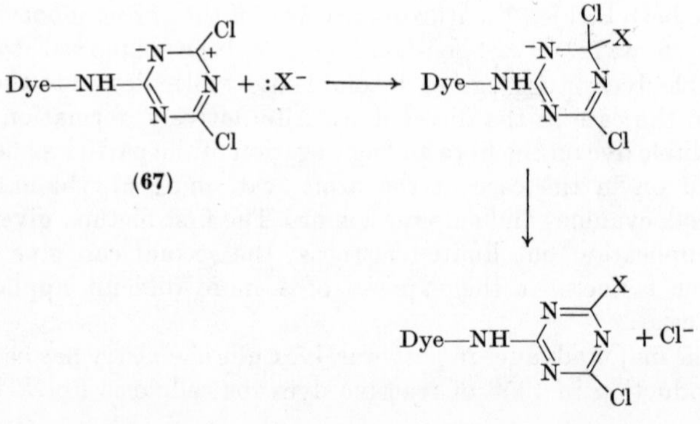

(67)

from a triazine (Part 3, p. 80). Nucleophilic attack by an anion X^- on a dye–cyanuric chloride compound (**67**) is shown. In this equation : X^- represents a nucleophile such as a hydroxyl ion or a cellulose O^- ion. It is important that the electronic influences make the reactive system susceptible to nucleophilic attack and not to attack by un-ionized species. Thus at neutral pH values the dye is quite stable. It can be dissolved in water, the cloth entered and salt added to the dye bath to exhaust the dye on to the fibre without any chemical reaction taking place. When these physical processes have taken place, sodium carbonate is added to the dye bath to raise the pH to about 10.5, causing nucleophilic attack to ensue at about 20°C. There are two possible modes of reaction. The attacking nucleophile can be either cellulose O^-, resulting in fixation to the fibre, or it can be a hydroxyl ion, resulting in hydrolysis and loss of the dye to the dyeing process. The rate constants for these two processes are roughly equal. However, the fact that the dye has been adsorbed by the fibre and is therefore concentrated in the fibre phase together with the roughly 25:1 preponderance of cellulose O^- ions over hydroxyl ions within the fibre causes fixation to predominate over hydrolysis. That portion of the dye which is hydrolyzed is washed out of the cloth at the end of the dyeing process.

Although this type of dye has two labile chlorine atoms it does not have a second chance of fixation if the first chlorine hydrolyses. The chloro-hydroxy-*s*-triazinyl entity produced ionizes under the alkaline conditions; the resulting O^- substituent feeds electrons back into the hetero-ring, stops the build-up of positive charge on the ring-carbon atoms and renders the remaining chlorine atom non-labile. This is an extreme example of how the second substituent can affect the reactivity of the chlorine atom in a mono-chloro-*s*-triazine. In less extreme cases, substituents that attract electrons preserve the reactivity and are themselves labile; 2,4-dinitrothiophenoxy, 2,4-disulphophenoxy and sulphonic acid groups are of this type; all are less accessible than the chloro-substituted triazines which remain the commercial choice.

Conversely, electron-releasing substituents such as amino, substituted amino and methoxyl diminish the reactivity of the remaining chlorine atom. Monochloro-*s*-triazines having these groups as second substituent require a temperature of above 60°C and often more alkaline conditions to undergo nucleophilic attack. Dyes of this type comprise the Procion H and Cibacron ranges which are applied by hot batchwise dyeing. These dyes can also be advantageously applied by printing methods. In these, the dye is

mixed with alkali, water and thickening agents to form a print
paste which is quite stable at normal temperatures. This paste is
printed on the cloth, and the impregnated cloth is heated to about
100°C, whereupon reaction with the fibre, together with some
hydrolysis, occurs. Unfixed dye is again washed out of the cloth
at the end of the process.

In both dyeing and printing applications it is, of course,
desirable that hydrolysis be kept to a minimum and as great a
proportion of the dye as possible fixed. In recent years the Procion
Supra (printing) and Procion HE (dyeing) ranges, both consisting
of dyes containing two reactive systems, have been marketed
with this specific aim in mind.

The numerous reactive systems available as alternatives to the
mono- and di-chloro-*s*-triazines fall into two classes. One is the
halogeno-heterocyclic type which includes, as well as the chloro-*s*-
triazines, the trichloropyrimidinylamino (Reactones and Drimar-
enes; **68**), the dichloropyridazone (Primazine P; **69**) and the
dichloroquinoxaline-6-carbonylamino (Levafix E; **70**) dyes. The

| (68) | (69) | (70) |

majority of these do not give the high level of reactivity shown by
the dichloro-*s*-triazines but are more comparable with the mono-
chloro-*s*-triazines. This is a consequence of their having only two
heterocyclic nitrogen atoms against the three ideally placed
nitrogen atoms in the *s*-triazine ring. In certain cases the level of
reactivity can be raised by a suitably placed electron-attracting
group, as in the 5-cyanodichloropyrimidinylamino dyes. Better
leaving groups than chlorine have also been exploited; fluorine
and methylsulphonyl groups are examples. A particularly effective
way of boosting the reactivity is by employing a quaternary
ammonium group as the leaving substituent; such compounds
readily result from reaction of monochloro-*s*-triazines with specific
tertiary amines, amongst which pyridine, trimethylamine and
1,4-diazobicyclo[2.2.2]octane are noteworthy. The second general
class of reactive dyes, typified by the Remazol dyes (**71**), contains a
2-sulphatoethyl sulphone group. In the presence of alkali this
splits out sulphuric acid, giving a vinyl sulphone in which the

electron movements (**72**) create a deficiency of electrons on the terminal carbon atom. Nucleophilic attack at this point followed by proton addition fixes the dye to the cellulose (**73**). Several other

Dye—$SO_2CH_2CH_2OSO_3H$

(**71**)

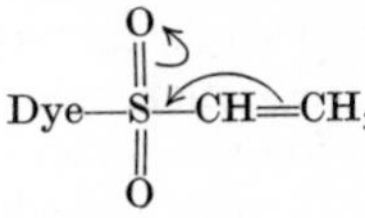

Dye—$\overset{\overset{O}{\|}}{\underset{\underset{O}{\|}}{S}}$—CH=CH$_2$

(**72**)

Dye—$SO_2CH_2CH_2O$—Cellulose

(**73**)

Dye—$NHCOCH_2CH_2OSO_3H$

(**74**)

systems, e.g. the 3-sulphatopropionylamino group in the Primazin dyes (**74**) function in a similar manner. The 2-sulphatoethyl sulphone system requires to be built into an intermediate rather than simply being attached to the chromophore in a final acylation stage. The steps entailed are conversion of, e.g., 3-nitrobenzene-sulphonic acid via the chloride into the sulphinic acid, reaction of this with 2-chloroethanol, reduction to the amine and sulphona-tion. This is circumvented in recent Remazol dyes which contain a 2-sulphatoethylsulphonylamino-group (as in **75**), introduced into an amino-containing dyestuff by reaction with 'carbyl sulphate'

Dye—$NHSO_2CH_2CH_2OSO_3H$

(**75**)

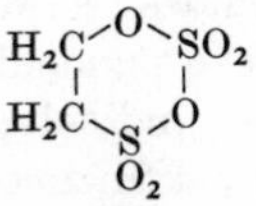

$\begin{array}{c} H_2C \\ | \\ H_2C \end{array} \overset{O}{\underset{S}{\diagdown}} \overset{SO_2}{\underset{O_2}{\diagup}} O$

(**76**)

(**76**), which is itself obtained by the action of sulphur trioxide on ethylene.

Whilst reactive dyes have made their main impact on the dyeing of cotton, the principle of retaining a dye on the fibre by formation of a covalent bond can be extended to other fibres. Wool and nylon, both of which contain amino-groups, are obvious candidates and ranges of reactive dyes for these fibres have been marketed. These are the Procinyl range of disperse dyes for nylon and the Procilan range of 1:2 metal-complex dyes for wool. Both contain reactive systems tailored to the particular needs of these fibres. Thus in the Procilan range the reactive unit is an acrylamido-group; this reacts sufficiently slowly to allow the dye to penetrate the fibre fully before becoming firmly attached.

PIGMENTS

The second major use of coloured organic compounds is as pigments, wherein they offer greater tinctorial power and softer texture than inorganic materials such as lead chromate. In pigmentation the coloured compound is bodily incorporated into the paint or plastic medium and becomes trapped therein when the medium hardens. Consequently no mechanism for ensuring uptake and retention is necessary. In common with dyestuffs, fastness to light is important but in the case of a pigment this requirement is more stringent, very high fastness to the effects of prolonged outdoor exposure to weather being a necessity. The remaining fastness requirements differ from those of dyestuffs. Fastness to heat is required so that the pigment may withstand high-temperature moulding in a plastic medium or stove-enamelling in a paint. Insolubility in organic solvents is needed to prevent migration into, for example, an oversprayed coat of white paint.

The readily accessible and cheap azo-compounds figure prominently as pigments, particularly in the yellow-to-red portion of the shade range. Typical diazo-components are 2,5-dichloro-aniline and 4-methyl-2-nitroaniline. Acetoacetarylamides, 1-aryl-3-methylpyrazol-5-ones, β-naphthol and 2-hydroxy-3-naphthoic arylamides serve as coupling components. Whilst the best monoazo-pigments have satisfactory light fastness, their resistance to heat and solvents is often deficient. Increasing the molecular size by utilizing a diamine, such as 3,3'-dichlorobenzidine, to prepare a bisazo-pigment is a prime means of improving these properties. This quest for greater insolubility forces the use of progressively more insoluble components, making a final aqueous diazotization and coupling increasingly difficult. To circumvent this, techniques involving a final insolubilizing stage carried out after the azo-linkage has been formed have been exploited. Thus, in the Chromophthal pigments an azo-carboxylic acid, e.g. 2,5-dichloroaniline→2-hydroxy-3-naphthoic acid, is converted into the acid chloride and this is condensed with a diamine such as benzidine in a high-boiling solvent, to yield a very insoluble bisamide.

In addition to these wholly organic azo-compounds, the insoluble calcium, barium and manganese salts of simple sulphonated azo-dyes find extensive use as cheap pigments. For less exacting applications, salts of basic triarylmethane dyes with phosphotungstomolybdic acid are also employed; these show much better fastness to light than the parent basic dyes.

Apart from these simple structures, the insoluble and durable copper phthalocyanine (**77**) displays ideal properties as a pigment.

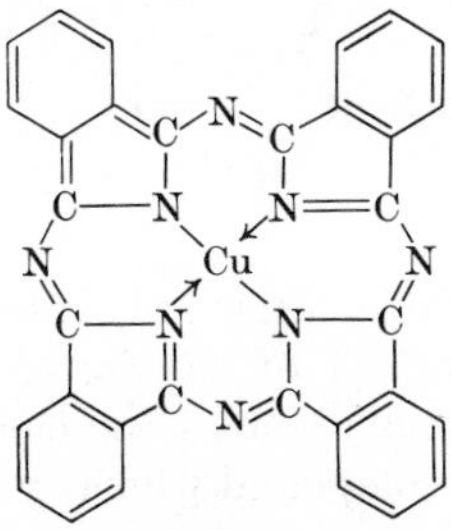

(**77**)

It is readily prepared by heating phthalic anhydride, urea, cuprous chloride and a catalyst in a high-boiling solvent and gives an attractive turquoise shade of outstanding fastness to heat, light and solvents. The shade variations that can be achieved by using this eminently suitable system are limited. A polychloro-compound, obtained by chlorinating copper phthalocyanine until 12–14 chlorine atoms per molecule have been introduced is outstanding as a green pigment; more yellow shades of green result when part of the chlorine is replaced by bromine.

The success of copper phthalocyanine and its halogenated derivatives as pigments has led to a search for other coloured heterocyclic compounds that could be exploited as pigments. These are especially desirable in the red part of the shade range since here the widely applicable azo-compounds often show their lowest light-fastness levels. Three major classes have emerged. The di(arylimides) (**78**) of perylenetetracarboxylic acid give a variety of shades in the red region, depending upon the exact nature of the aryl residue. These compounds can also be used as vat dyes on cotton.

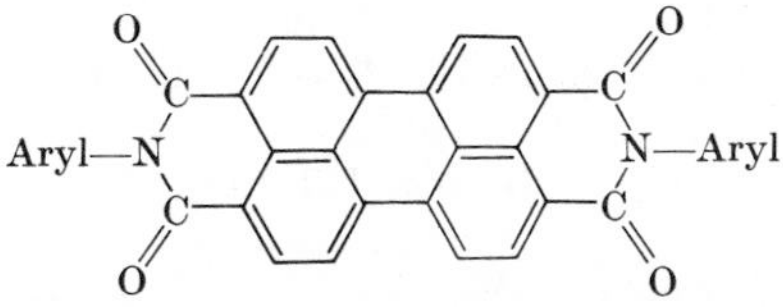

(**78**)

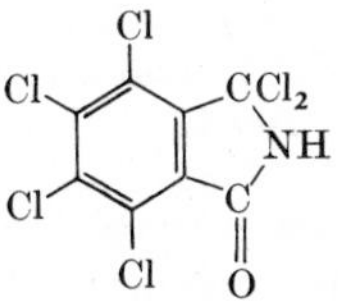

(**79**)

Orange, red and brown pigments also result when the phthalimide derivative (**79**) is condensed with a diamine to yield a tetrachloroisoindolinone (**80**). Thirdly, the polymorphic modifica-

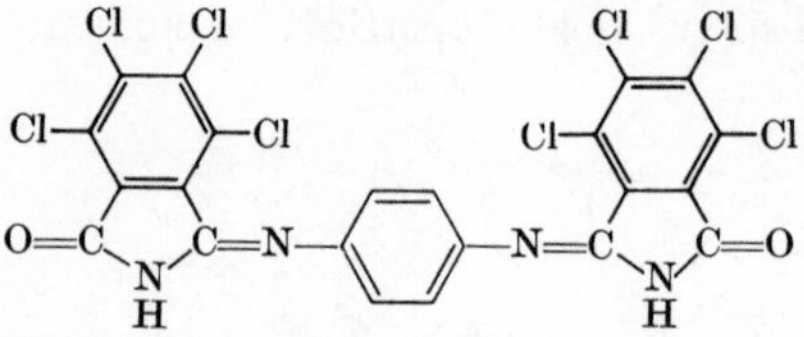

(**80**) Red 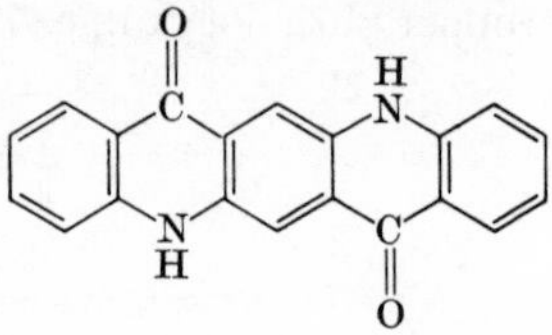   (**81**)

tions of lin-quinacridone (**81**) are valuable as red-to-violet pigments. The preparation of the latter compound from 2,5-dibromoterephthalic acid involves condensation with aniline to yield the 2,5-dianilino-compound which cyclizes to (**81**).

Nomenclature

Dyestuffs and pigments are marketed under brand names, usually registered trade marks of the manufacturing firm, which help indicate the particular application for which the product is suitable. Such brand names used in this Chapter are Alcian, Caledon, Procilan, Procion and Procinyl (ICI); Cibacron and Chromophthal (Ciba); Levafix and Phthalogen (Bayer); Primazin (B.A.S.F.); Reactone (Geigy); Drimarene (Sandoz); and Remazol (Hoechst). In this Chapter the names of chemicals are sometimes those current in the dyestuff industry rather than the modern systematic names, but in such cases the structures are evident from the formulae.

Bibliography

E. N. Abrahart, *Dyes and their Intermediates*, Pergamon Press Ltd., London, 1968.

H. E. Fierz-David and L. Blangey, *Fundamental Processes of Dye Chemistry*, Interscience, New York, 1949.

H. A. Lubs, *The Chemistry of Synthetic Dyes and Pigments*, Reinhold Publishing Corp., New York, 1955.

K. H. Saunders, *The Aromatic Diazo Compounds and their Technical Applications*, 2nd edn., Edward Arnold, London, 1949.

K. Venkataraman, *The Chemistry of Synthetic Dyes*, Vols. 1–4, Academic Press Inc., New York, 1952, 1970, 1971.

T. Vickerstaff, *The Physical Chemistry of Dyeing*, ICI Ltd., Oliver and Boyd, London, 1954.

H. Zollinger, *Azo and Diazo Chemistry, Aliphatic and Aromatic Compounds*, Interscience, New York, 1961.

CHAPTER 11

Pharmaceuticals

(N. F. ELMORE, G. R. BROWN, and A. T. GREER, Imperial Chemical Industries Limited, Pharmaceuticals Division, Alderley Park)

Introduction

Medicinal plants have been employed by physicians for centuries but it was not until the nineteenth century that alkaloids such as morphine were isolated and used as single substances. With the development of organic chemistry, synthetic compounds with useful properties (e.g. ether, 1846; aspirin, 1899; the barbiturates and arsenicals) became available. In 1932 the demonstration of the effectiveness of 'Prontosil Rubrum' (7), a red dyestuff (2',4'-diaminoazobenzene-4-sulphonamide) which was the first successful treatment of systemic Gram-positive bacterial infections in man and animals, provided a turning point which induced many chemical manufacturers to seek other compounds with such beneficial properties. The synthesis of organic medicinal products was greatly stimulated by the necessity during World War II for antimalarial compounds to replace quinine after the loss of routes to Indonesia. The development of penicillin at that time was fortuitous and less urgent. After the war the chemical industry underwent a tremendous expansion and successful drug research developed rapidly, first with steroids and antibiotics and subsequently with drugs for the treatment of the nervous and cardio-vascular systems. All of the main branches of organic chemistry are now involved in the search for new drugs and great ingenuity is currently employed in the synthesis of complex natural products, total synthesis of steroids, peptide hormones and prostaglandins.

The pharmaceutical industry is advancing on several major fronts, and has other broader interests such as vaccines, sterile drips, dressings, plaster casts, etc., which are outside the scope of this Chapter. Proprietary, or 'over-the-counter', medicines are sold through retail chemists and supermarkets without prescriptions by doctors. These are usually well-tried remedies for minor complaints such as colds, coughs, indigestion and headaches. This type of product, with a few exceptions such as antiseptics and antitussives, does not require novel organic constituents and sales are achieved by brand-name marketing skills (i.e. advertising) rather than by demonstration of the effectiveness of the remedy. The major preoccupation of the pharmaceutical industry is the discovery of novel medicinal products for the treatment of specific diseases and these drugs are usually referred to as 'prescription' or 'ethical' pharmaceuticals. Finally, the discovery of new products that promote the development and well-being of livestock is of great importance and in many ways parallels the work in human medicine since the problems are similar.

The discovery of ethical drugs is concentrated into two main areas, those that are active against infectious and non-infectious diseases. The medicinal chemist approaches these problems rationally by designing compounds based on his appreciation of related biochemistry or by preparing compounds that have a structural analogy to known active types. The compounds are tested by biochemists or biologists who will use, where appropriate, pathogens growing *in vitro* or *in vivo*, or models of diseased conditions in animals. When an 'active' compound is discovered, further evaluation in more refined systems will be carried out by the biologist whilst the chemist prepares related compounds termed 'analogues' to find the best of the series. Once one or two compounds have been selected, extended toxicology will begin. Regular dosing lasts for several months before the animals are examined in minute detail by toxicologists for undesirable abnormalities. In the event that no toxic effect is discovered, a limited clinical examination in humans (if appropriate) will commence once approval has been obtained by a Government body who will carefully sift through the toxicity and any other relevant data which have accumulated.

Several years will elapse between the discovery of useful biological activity in a compound and its sale as a drug. The utility of propanolol (see p. 406) was established late in 1962; the compound was launched in this country in August 1965 and its sale in the United States for use in cardiac arrhythmias was permitted in December 1967. By recent standards this is fast, since most registration authorities now require long-term animal toxicity data in order to get a better assessment of the benefit-to-risk ratio before administration to humans. Fison's anti-asthmatic drug sodium cromoglycate was discovered in 1965 and it was not approved for sale in the United States until February 1973. Recently the Federal Drug Administration has threatened seven-year efficacy trials with oral contraceptives before they can be marketed. Long trials will, of course, consume a significant proportion of the patent life of a compound.

Fison's 'Sodium cromoglycate' (Intal)

Structure–Activity Relationships

Since the beginning of medicinal chemistry scientists have endeavoured to find relationships between chemical structure and biological activity. Such an understanding would be beneficial in that it would lead to a rational design of new drugs having improved specificity, potency, stability and duration of action. Regrettably, most structure–activity relationships are recognized at a late stage and are only seen in retrospect.

A number of approaches are available for the organization of synthetic-chemistry programmes intended to lead to new drugs. Modification of known drugs or natural products has been fruitful (e.g. antibiotics and steroids) and has led to useful antibacterial and anti-inflammatory agents and oral contraceptives. Alternatively, one may take a small part of a known drug structure and build this unit into other molecules, then examine for a useful effect: compounds with the partial structure of cocaine retain useful anaesthetic activity. Elaboration of the structure of a known drug that has a potentially useful side effect can result in the potency of the latter being increased to a useful therapeutic level whilst the drug's original effect is suppressed. This is exemplified by the diuretic (substances which increase the rate of urine formation) sulphonamides which were developed as a result of the observation of this effect in patients who were being treated with the antibacterial agent sulphanilamide. Many examples exist of a chemical having been designed to interfere with a biological process: the drug amprolium prevents coccidiosis in chickens, turkeys and cattle by interfering with the metabolism of vitamin B_1 by parasitic coccidia and is toxic to the latter (and thus behaves as an 'antimetabolite'). In a less specific way, molecules have been designed that exert a specific chemical effect when inhibiting a biological process: alkylating agents need not be chemically related in order to inhibit the development of cancer, and molecules have been designed to complex (or chelate) with metals to exert a biological effect; ethacrynic acid, a diuretic drug, was designed during the preparation of compounds that would react with thiol groups on proteins. Biological activity can sometimes be discovered by the use of a hypothesis of a structure–activity relationship. Thus, a particular spatial separation of groups can be ascertained by making measurements on molecular models and this has been particularly useful in the preparation of molecules related to the analgesic alkaloid morphine. Attempts to describe receptor sites for particular types of biological activity have not been very successful for generating new active types. There has been some success with a completely different approach, the pursuit of novel chemistry which makes available for testing molecules with a shape and polarity never previously seen: an example here is the antianxiety drugs of the benzodiazepine type.

At this point it is pertinent to consider what can happen to a compound in an animal test system or in man. First, the compound may be absorbed and it will then have to pass through various membranes and systems to reach the site where it will act, probably inside a cell. During this process it can be lost by binding to protein or eliminated rapidly from the body. Alternatively, the compound may be transformed into another molecule (a metabolic process) which exerts the biological effect. Duration of action of the drug is further affected by its ability to withstand conversion into an inactive species (detoxification) or it may advantageously become dissolved in body fat, which can act as a reservoir. Clearly the polarity, shape and size of the molecule are important. Polar

effects are exemplified by the anticonvulsant primidone (**1**) which does not have the sedative activity of the related phenobarbitone

(2) and yet they differ only by one carbonyl group. Androsterone (3) is a potent androgen whilst the epimer (4), whose shape (but not size) is significantly different, is inactive. The introduction of

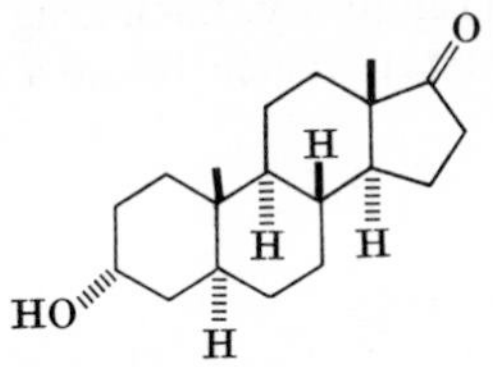

(3) 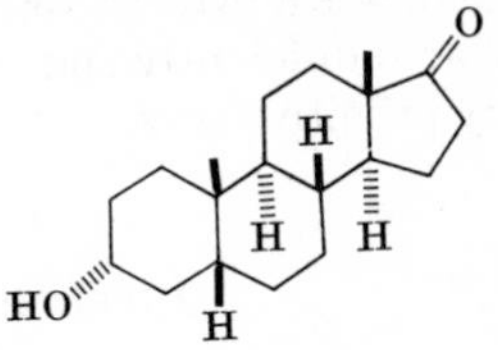   (4)

chemical groups may also modify physical properties and hence the level of activity. Thus, the extension or introduction of an alkyl chain increases lipid(fat)-solubility, and branching of the chain makes oxidative metabolism difficult; cycloalkyl groups make an alkyl chain more compact and facilitate van der Waals bonding. Halogen atoms also increase lipid-solubility and assist in the blockage of sites to metabolism (and hence inactivation) by hydroxylation. Thus, the oestrogenic hormone chlorotrianisine (5) has a much longer duration of action than diethylstilboestrol (6).

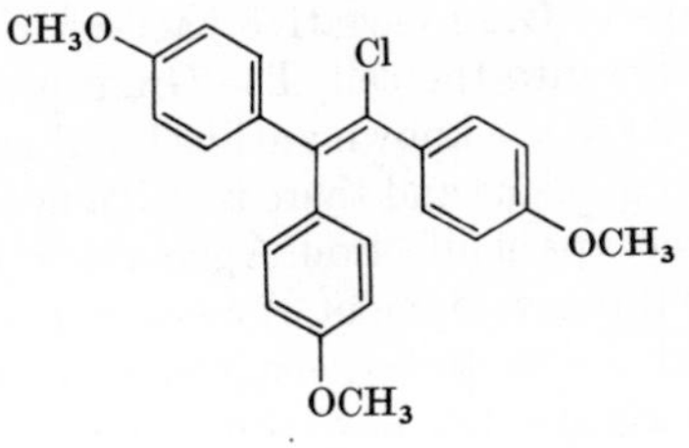

(5) 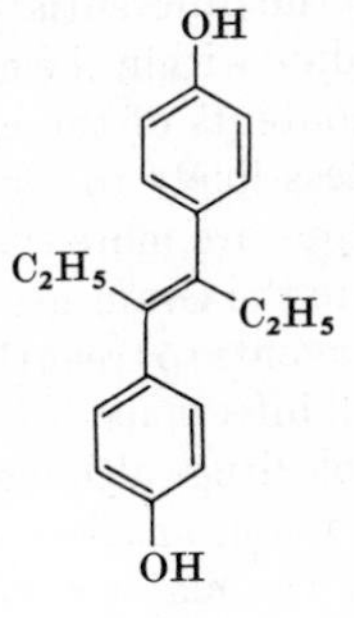   (6)

If a carboxyl group is replaced by sulphonic acid, water-solubility is enhanced. Carboxyl groups may be decarboxylated in biological systems and this can often be delayed by esterification; the polar nature of hydroxyl groups often represents a disadvantage and can be overcome by acylation or alkylation. The basicity of nitrogen atoms may be modified by varying the groups to which they are attached. Replacement of a benzene ring by a furan or thiophen ring is often a successful structural modification although furans often lack stability. When benzene is replaced by basic heterocycles, however, activity is frequently lost owing to the change in polarity of the molecule.

Although these generalizations about structure and activity are often valid, it is unwise to apply them rigorously. Each drug needs to be treated differently and our understanding of this area is still at a very primitive stage.

INFECTIOUS DISEASES

Antibacterial Drugs

The treatment of bacterial disease with chemical agents is a major achievement of the pharmaceutical industry. A range of synthetic drugs and antibiotics (products of microbial origin) is available which can control practically all types of bacterial infection. Bacteria have been arbitrarily divided into two classes by Gram: those bacteria that are stained blue by Gram's reagent (Crystal Violet and iodine) are termed Gram-positive and those that are not so stained are termed Gram-negative. There are several hypotheses to explain this behaviour; the most reasonable is that the differentiation is due to trapping of the Crystal Violet and iodine within the cells in the 'positive' bacteria whilst the high lipid contents of the cell walls of Gram-negative bacteria render them less likely to seal the dye into the cell. The Gram-positive organisms are more susceptible to the known antibacterial agents than are the Gram-negative organisms and there is still a need for better agents to treat the latter type of infection. A general problem with all infectious diseases is the development of resistance to the available drugs although this has sometimes been overcome by the use of a combination of two agents that have different modes of action. The major antibacterial agents in use will be described in the following Sections.

Synthetic Antibacterial Agents

Despite the discovery of a number of antiseptic compounds before 1930, none was found suitable for internal use until it was shown in 1932 that a red dyestuff 'Prontosil Rubrum' (7) controlled streptococcal and staphylococcal infections. Subsequently it was soon shown that (7) breaks down in the tissues to sulphanilamide (8) which is the active agent, and in a very short time every medically orientated institution in the world was said to be working on sulphonamides. Many drugs were developed as a result, and several of them are still in use despite the advent of antibiotics and the problems introduced by resistance. These compounds are used to treat respiratory and urinary-tract

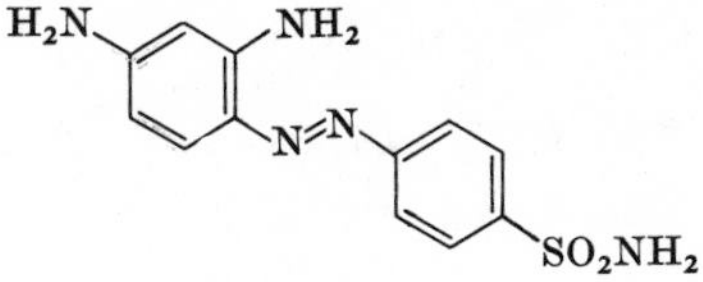

(7)

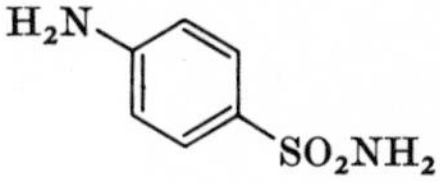

(8)

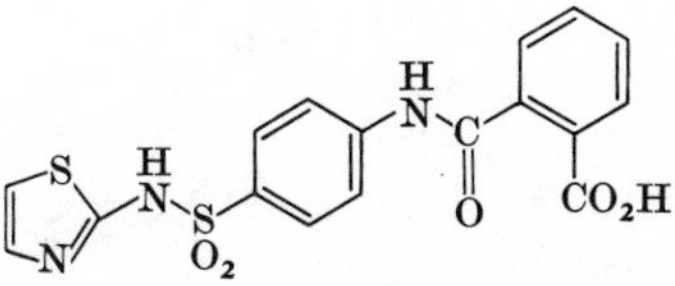

(9)

infections in particular, and a poorly absorbed derivative phthalyl-sulphathiazole (9) is used for enteric infections. The sulphonamides are used in combination with some other antibacterial agents which will be described below; examples of the type of synthetic methods employed in this series are shown in Schemes 1 and 2 for sulphadiazine (10) and sulphadimidine (11) (cf. Part 3, p. 120).

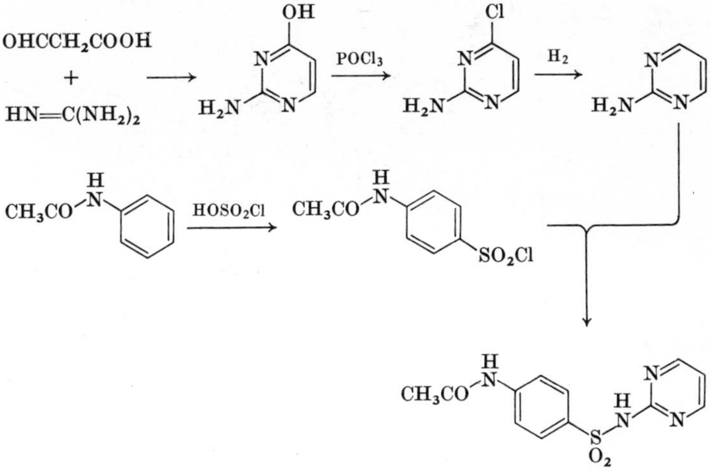

(10)

Scheme 1.

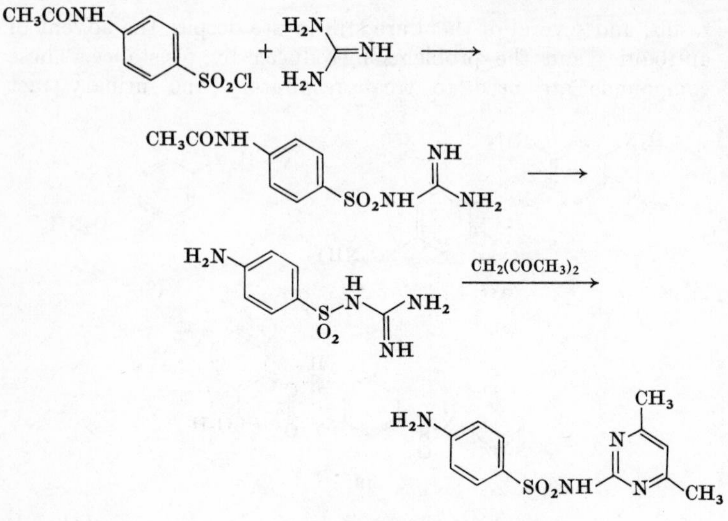

SCHEME 2.   (11)

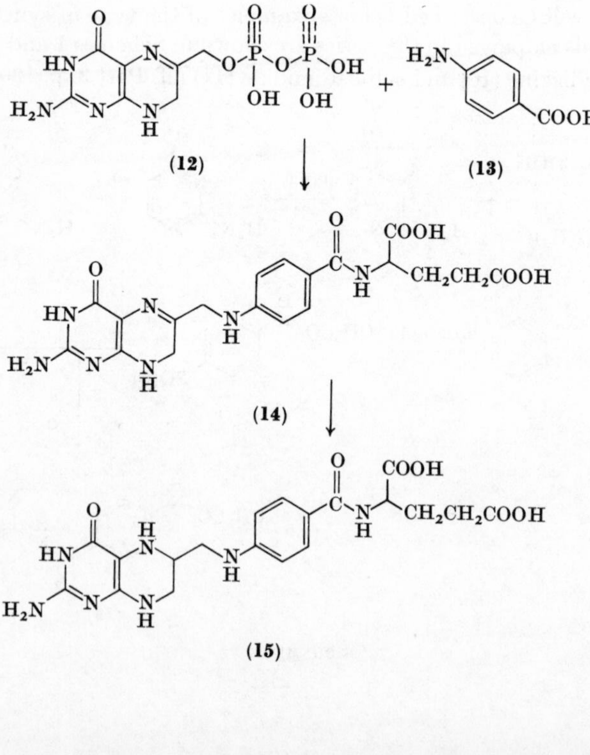

The sulphonamides are unique in that their mode of action is known at the enzyme level. They interfere with the conversion of the pyrophosphate ester (12) into dihydrofolic acid (14) by *p*-aminobenzoic acid (13) by forming spurious dihydrofolic acids whose significance in antibacterial action is not understood. Another potent antibacterial agent trimethoprim (16) inhibits the conversion of dihydrofolic acid (14) into the tetrahydro-derivative (15) and is synthesized by the conventional route shown.

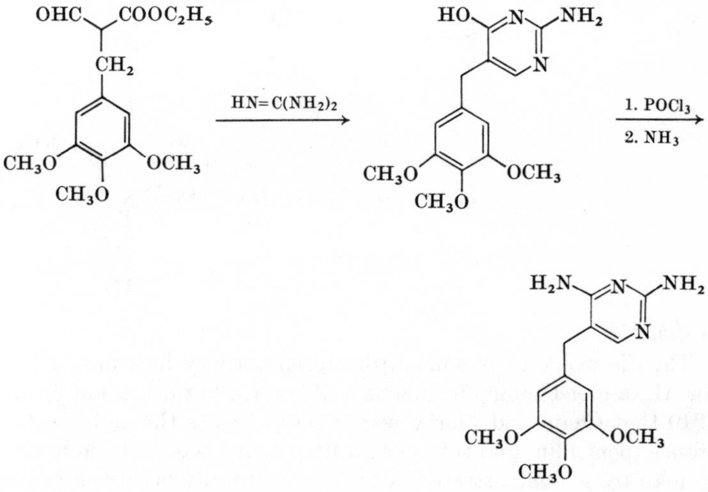

(16)

Two other antibacterial agents with good Gram-negative activity, in the treatment of the urinary tract, are nalidixic acid (17) (p. 360) and nitrofurantoin (18). Nalidixic acid is synthesized by condensation of 6-methyl-2-pyridinamine with diethyl ethoxy-methylenemalonate followed by ring closure, alkylation and

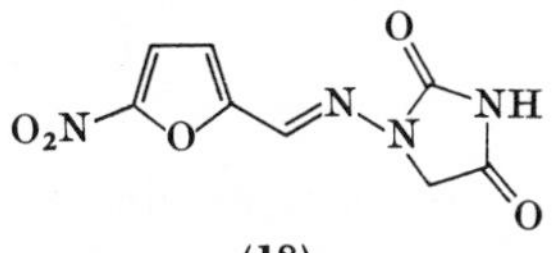

(18)

hydrolysis. Nitrofurantoin is prepared by nitration of furfur-aldehyde (which requires protection as the acetate), followed by condensation with *N*-aminoimidazolidine-2,4-dione.

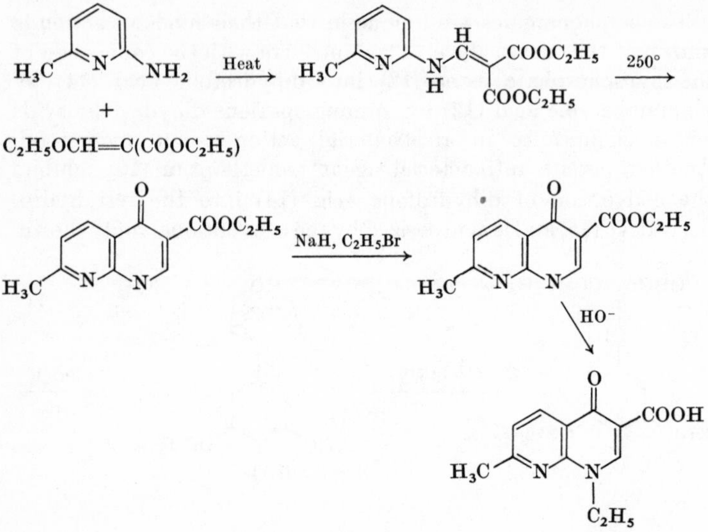
H3C
N
NH2
+
C2H5OCH=C(COOC2H5)
Heat
H
N
C
COOC2H5
COOC2H5
250°
O
COOC2H5
NaH, C2H5Br
HO−
COOH
C2H5

(17)

Antibiotics

The discovery of potent antibacterial activity in a mould by Sir Alexander Fleming in 1929 is well known, but it was not until 1940 that Chain and Florey were able to isolate the active substance (penicillin) and thus permit its potent bactericidal activity in man to be demonstrated. The seven naturally occurring penicillins are all amides derived from a common amine (cf. Part 4, p. 315). Benzylpenicillin (**19**) is a typical example and still retains a valuable position in therapy because of its cheapness and low toxicity.

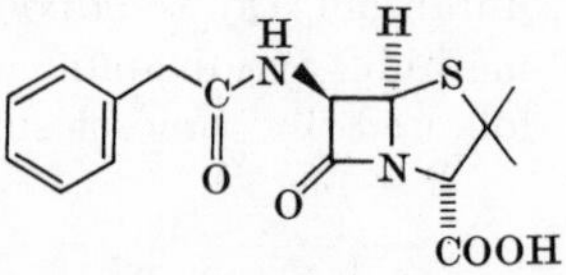
H
H
N
C
S
O
N
O
COOH

(19)

Although penicillins are manufactured by fermentation processes, a great deal of work has been carried out, notably by Henery-Logan and Sheehan, to synthesize penicillins by routes that could yield new compounds with even greater utility. There is much current interest in molecular modification of the penicillin structure. A major break-through in the penicillin saga was the discovery by the Beecham group that fermentations carried out in the absence of the side-chain (acyl) precursor afforded 6-amino-penicillanic acid (**20**). Over 2000 acyl derivatives of this acid have been made and one of the most useful of these is ampicillin (**22**)

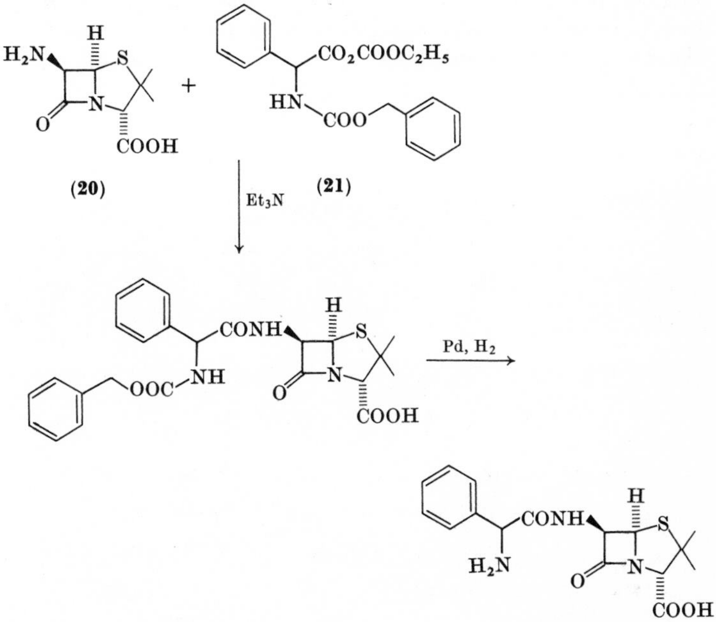

whose advantages are its broad-spectrum activity, its administration by the oral route, and its (relative) acid-resistance. It is *not*, however, resistant to the enzyme penicillinase and cannot be used with resistant staphylococcal infections; but some other semi-synthetic penicillins such as methicillin (**23**) do not suffer this disadvantage.

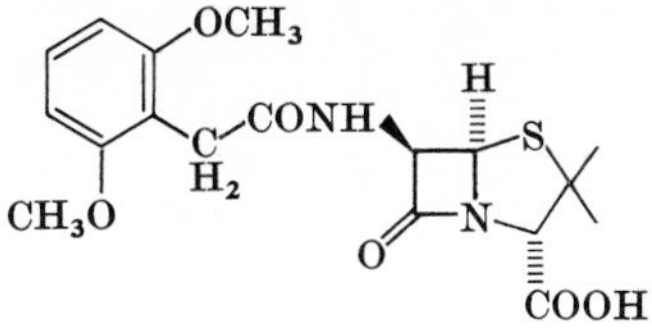

Another series of antibiotics chemically related to the penicillins are the cephalosporins. Cephalosporin C (**24**) is a fungal metabolite; it can be degraded to 7-aminocephalosporanic acid (**25**) which can then be acylated to give a range of products similar to those obtained from 6-aminopenicillanic acid (**20**); cephalothin (**26**) and cephaloridine (**27**) are prepared from (**25**) and are particularly useful for patients with respiratory or urinary-tract infections who are sensitive to penicillins.

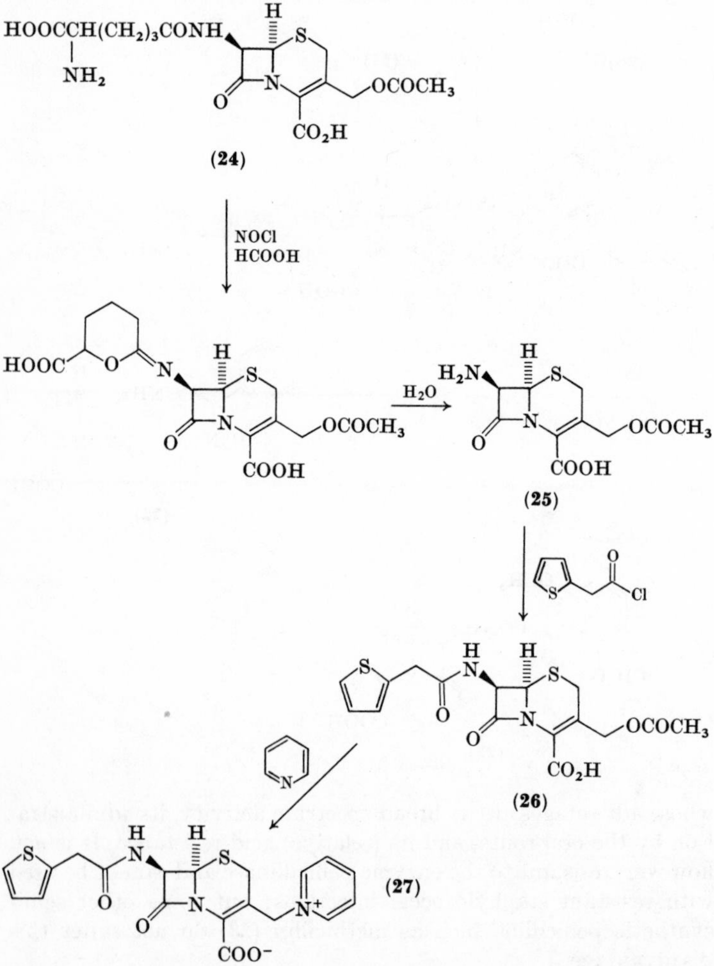

Chloramphenicol (**28**), an antibiotic which was initially widely used, is now restricted for use against typhoid fever because of serious side effects.

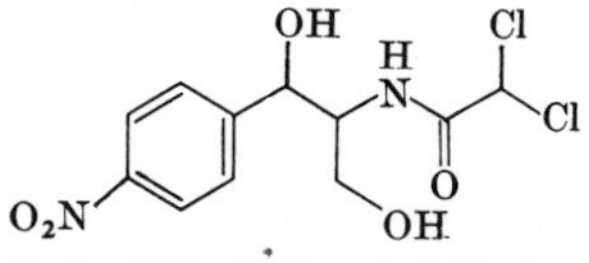

(28)

The tetracycline group of antibiotics (Part 4, pp. 198–207) have the broadest spectrum of antibacterial activity yet discovered and they are consequently widely used, particularly against pneumonia, Gram-negative infections and rickettsial diseases. They are fungal metabolites and some products are derived from molecular modification of the natural products. Prominent amongst compounds on the market are these shown in Table 11.1.

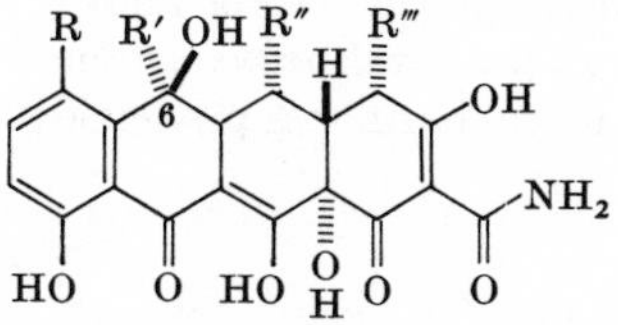

(29)

Table 11.1. Some members of the tetracycline group

	Substituents in (**29**)			
	R	R′	R″	R‴
Tetracycline	H	CH_3	H	$N(CH_3)_2$
Chlorotetracycline	Cl	CH_3	H	$N(CH_3)_2$
Oxytetracycline	H	CH_3	OH	$N(CH_3)_2$
Desmethylchlortetracycline	Cl	H	H	$N(CH_3)_2$
Methacycline	H	$=CH_2$	H	$N(CH_3)_2$
		at position 6		
Minocycline	H	CH_3	OH	$NHCH_3$

The term 'macrolide' is used to describe a large class of large-ring lactone antibiotics obtained from *Streptomyces* species. The best-known member of this group is erythromycin (**30**) (Part 4, p. 209) which has an antibacterial spectrum similar to that of penicillin and is useful in penicillin-sensitive patients and has been particularly effective against diphtheria.

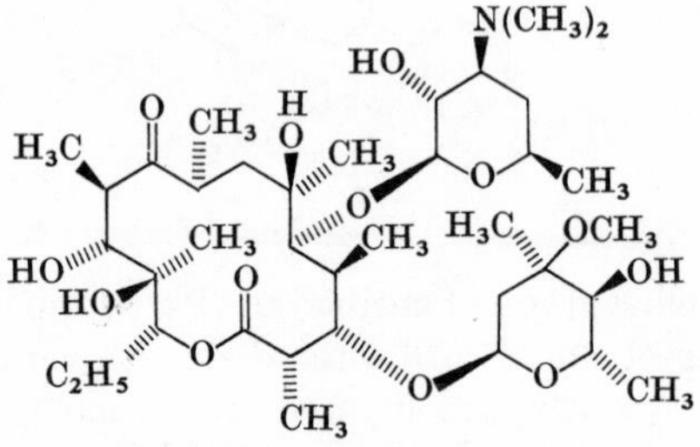

(**30**)

Lincomycin (**31**) is not a macrolide but has a similar spectrum of biological activity. Fusidic acid (**32**) is claimed to be outstanding for treatment of staphylococcal infections and although it is a steroid it differs from the better-known hormonal steroids in having a *trans-syn-trans-anti-trans* arrangement of the rings and in consequence has no hormonal properties.

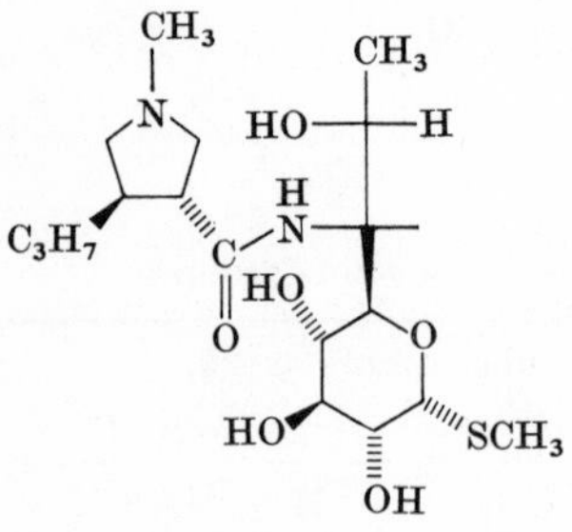

(**31**) 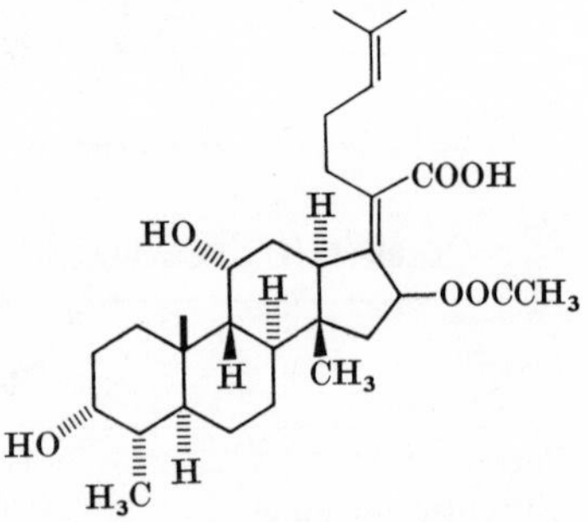   (**32**)

Antimycobacterial Agents

Mycobacteria are responsible for tuberculosis and leprosy; isoniazid (**33**) has pride of place among the antituberculosis drugs since it combines high potency with low toxicity and cheapness and can be administered orally. Its synthesis from γ-picoline is straightforward.

Two second-line drugs which are related to isoniazid in structure but not cross-resistant (i.e. it will attack infections resistant to isoniazid), are ethionamide (**34**) and its homologue prothionamide (**35**).

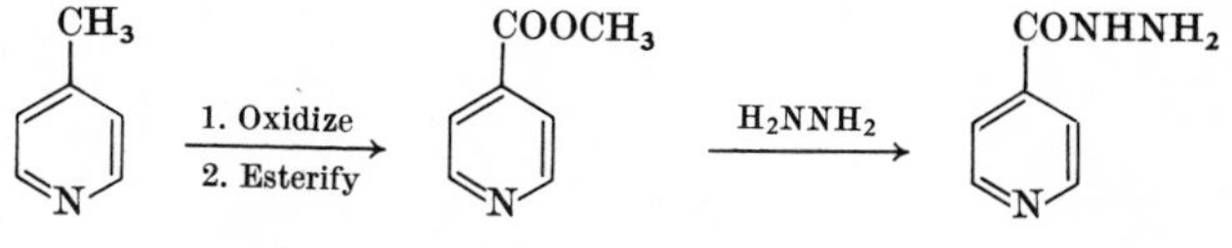

(**33**)

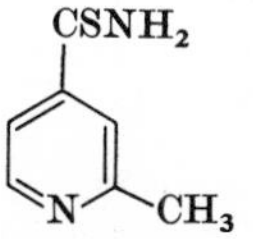

(**34**) 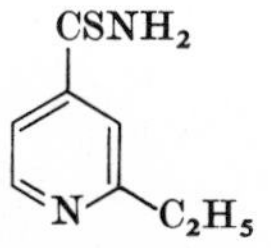   (**35**) 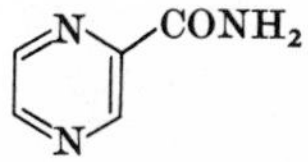   (**36**)

Pyrazinamide (**36**) is also closely related to isoniazid; it is prepared from quinoxaline. There is a rapid development of drug resistance in tuberculosis and, to overcome this, 'combination therapy' (a mixture of drugs) is often employed. The antibiotic streptomycin (**37**) (Part 4, p. 78) was the first successful drug to be used against tuberculosis and is often used in combination with *p*-aminosalicylic acid (**38**). Thioacetazone (**39**) owes its place in therapy to the fact that it is relatively cheap and does not need to be given in such enormous doses; it is synthesized from toluene.

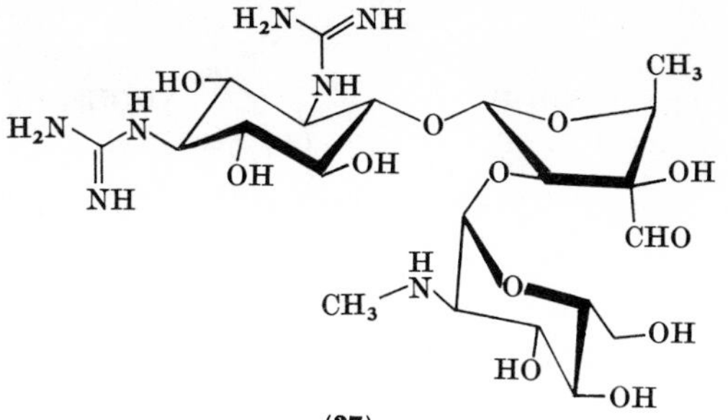

(**37**)

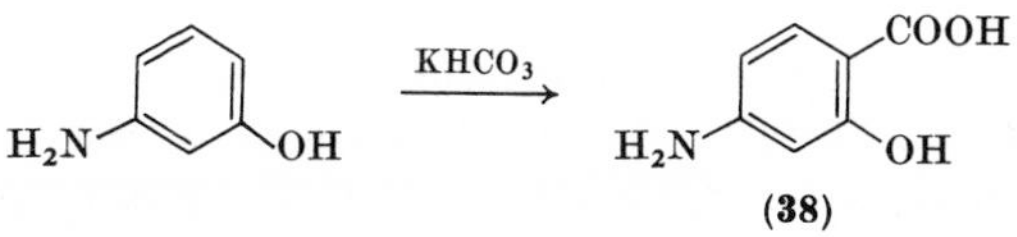

(**38**)

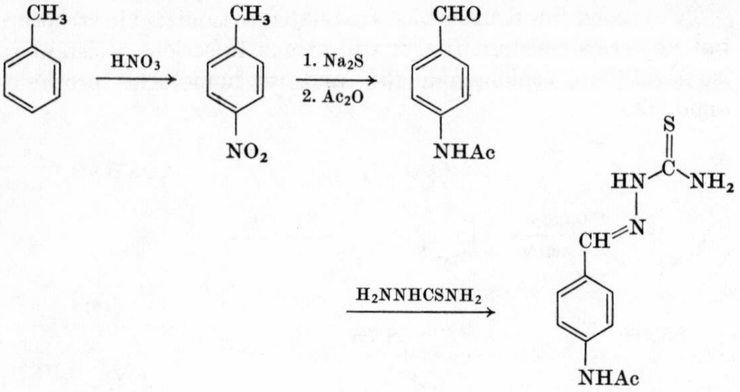

(39)

Cycloserine (**40**) is an antibiotic that has been studied extensively against mycobacteria; on its own it is inadequate but in combinations it was effective in treating tuberculosis in humans. Its synthesis from serine by the Merck group is shown.

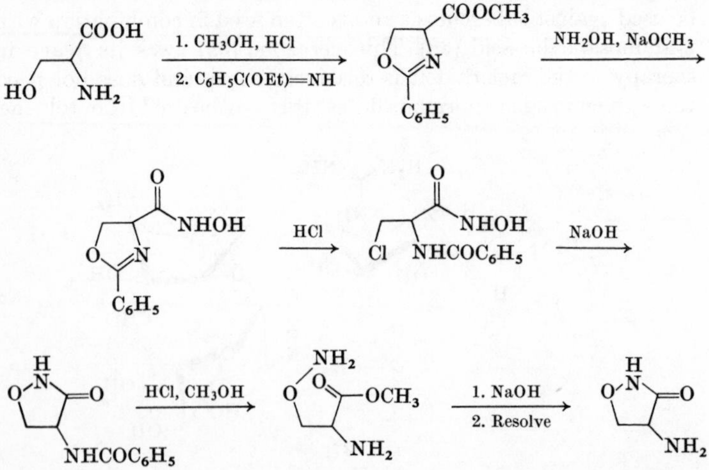

(40)

When a random selection of compounds was screened at Lederle Laboratories against tuberculosis in mice, activity comparable with that of streptomycin was found in the diamine (**41**). A large number of analogues were prepared and the pattern of activity implicated metal chelation. This could be direct, e.g. by interfering with a metal-containing enzyme, or indirect by facilitating drug transport. When the activities of the amines were correlated with

$(CH_3)_2CHNH(CH_2)_2NHCH(CH_3)_2$

(41)

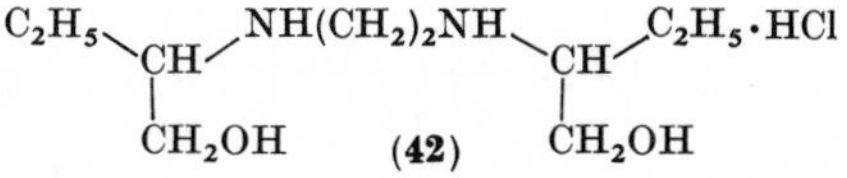

(42)

their abilities to chelate there was an indication that more 'open' chelates containing terminal hydroxyl groups would be of interest and these were made; this reasoning led to ethambutol (42). The stereospecificity of its antimycobacterial activity is striking since the *dextro*-isomer is twelve times more active than the *meso*-isomer and 200–500 times more active than the *laevo*-isomer.

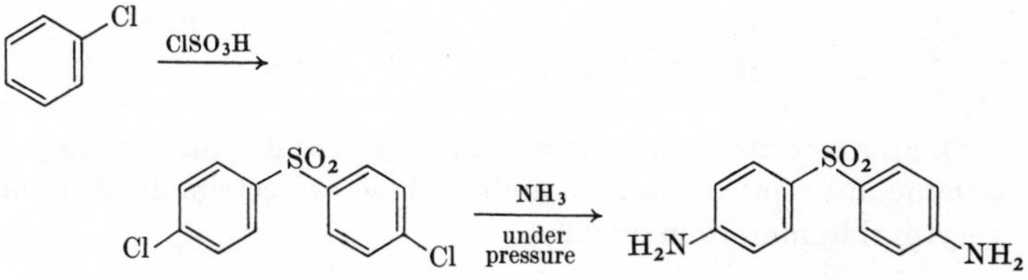

+ other isomers

(43)

Leprosy is also caused by mycobacterial infection, and the drugs used for its treatment are also active against, but not clinically suitable for, tuberculosis. Dapsone (43) is synthesized from chlorobenzene and chlorosulphonic acid.

Antiseptics

The words antispetic, disinfectant and germicide all signify an agent that kills microbes on contact. Drugs in this category are applied locally and must have properties in addition to anti-bacterial action, such as freedom from tissue damage, dyeing properties, etc. Many simple compounds are germicidal, e.g. ethanol, formaldehyde and iodine. Products are marketed for specific users such as hospitals, farmers, housewives, or veterinary surgeons, and the extent of their success is often related to formulation, packaging and the size of the advertising budget rather than their bactericidal merit.

Lister in 1867 first used phenol in antiseptic surgery and phenols are still in antiseptic use today. Resorcinol is used in shampoos and a chloroxylenol (**44**) and hexachlorophane (**46**) [prepared from (**45**)] are general antispetics. Thymol (**47**) a natural product, is used in throat lozenges and gargles, as are trihalogenophenols.

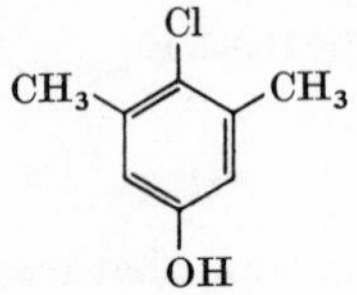

(44)

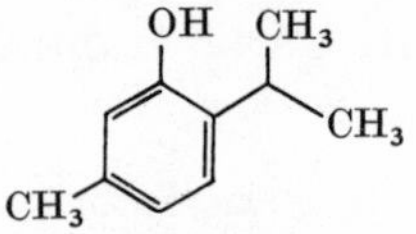

(47)

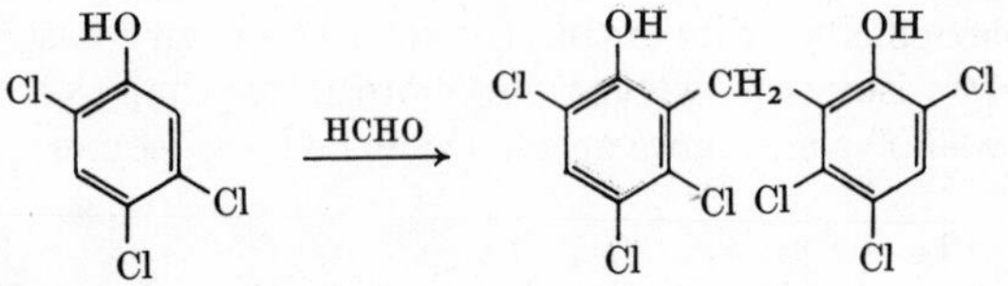

(45)

(46)

Quaternary ammonium compounds represent a further class of organic antiseptic. Notable among these are cetrimide (**48**) and cetylpyridinium bromide (**49**).

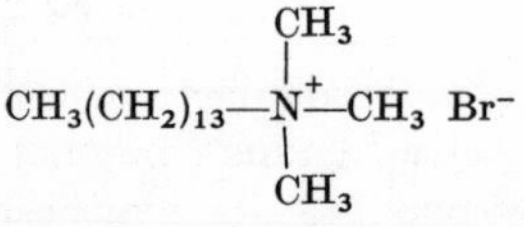

(48)

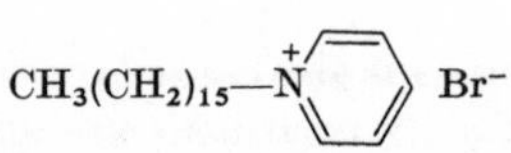

(49)

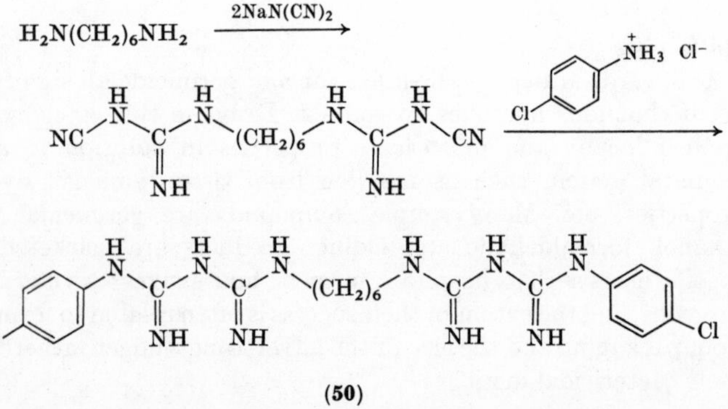

(50)

Chlorhexidine **(50)** is also a widely used antiseptic; it is synthesized from hexane-1,6-diamine dihydrochloride and sodium dicyanimide.

Antiviral Agents

Our knowledge of viruses has increased rapidly during the past twenty years but the introduction of antiviral agents has made little progress. *N*-Methylisatin 3-thiosemicarbazone **(51)** is useful for the prevention of smallpox and has encouraging activity against vaccinia virus. Two close analogues **(52)** and **(53)** have no

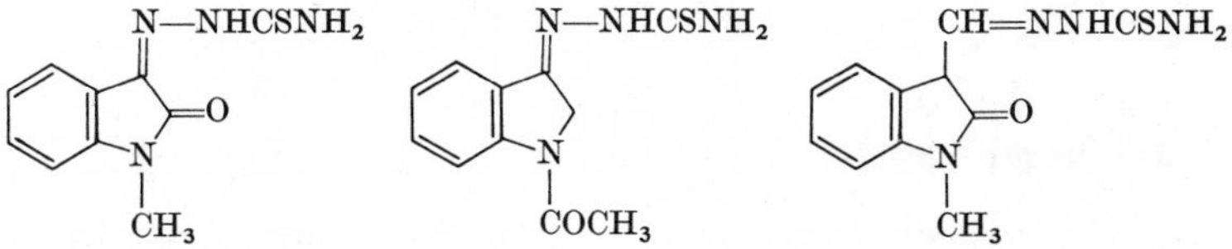

(51)   **(52)**   **(53)**

antiviral activity. Adamantanamine **(54)** prevents some strains of influenza from developing and is active against the rubella (measles) virus; this compound was the first drug discovered by du Pont and its development cost at least 20×10^6. Many other alicyclic amines have been prepared and tested; the less-hindered amines represented by **(55)** are claimed to have activity against influenza virus in mice. 2′-Deoxy-5-iodouridine **(56)** has been used

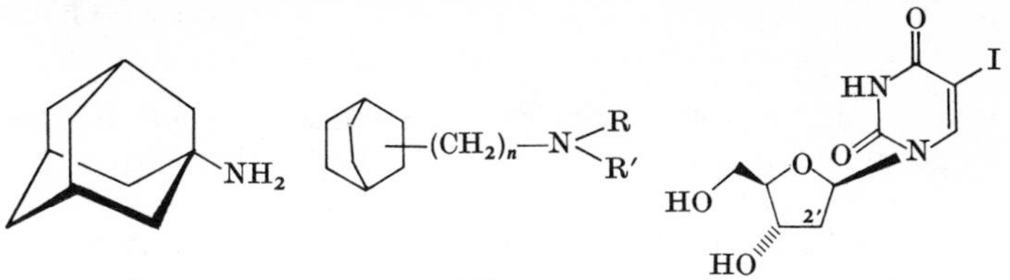

(54)   **(55)**   **(56)**

in the treatment of herpes virus. As in the treatment of cancer, many analogues of the natural pyrimidine and purine nucleosides have been prepared as potential antimetabolites, i.e. compounds that proceed down the same biochemical pathway as the natural compound but are unable to perform the function of the latter, in this case the synthesis of DNA and RNA polymerase. Compounds with both adamantane and nucleoside components have been prepared. The fluorouridine (**57**) has activity against vaccinia virus *in vitro* but the thymidine derivative (**58**) is inactive.

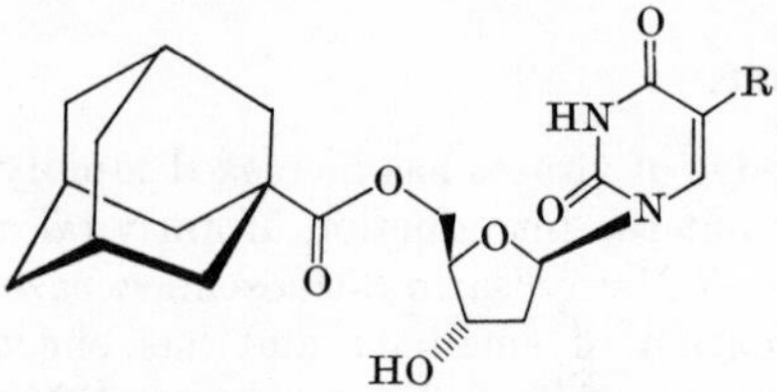

(**57**) R = F.   (**58**) R = CH$_3$.

Antifungal Agents

Fungi include yeasts and moulds; the number of species is enormous and whilst some are beneficial others are responsible for parasitic diseases. Many phenols are antifungal and their activity

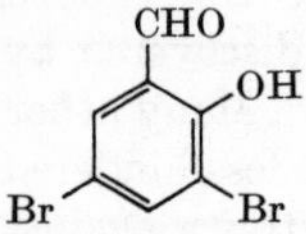

(**59**) 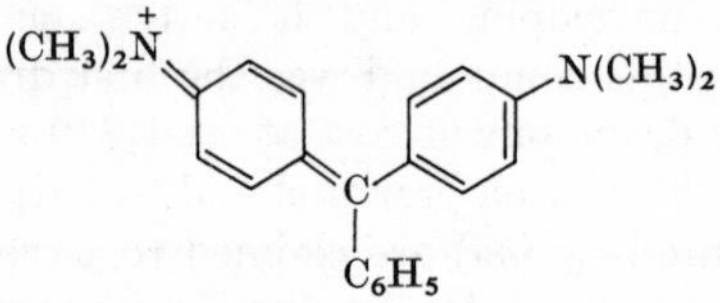   (**60**) 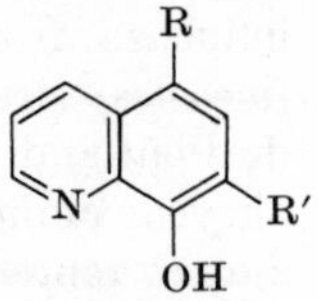   (**61**) R = R' = H
                                                              (**62**) R = Cl, R' = I

is frequently improved by halogen or alkyl substituents, e.g. (**59**). Triphenylmethane dyes (Part 2, pp. 249–253; this Part, p. 340) such as Malachite Green (**60**), Gentian Violet and Basic Fuchsin have been applied topically to certain infections but the disadvantages of these compounds is obvious. 8-Hydroxyquinoline (8-quinolinol) (**61**) is the active ingredient in several preparations whose effect is bacteriostatic and fungicidal; its activity is attributed to its ability to chelate metals since none of the isomeric hydroxyquinolines is active. Halogenation as in (**62**) gives improved activity. Some heavy-metal ions (Ag$^+$, Hg^{2+}, Ca^{2+} and Zn^{2+}) have long been used against agricultural fungi, and a combination of heavy-metal ions with an organic component as in zinc caprylate (octanoate) (63), zinc or copper undec-10-enoate (64) and sodium ethyl mercurithiosalicylate (65) produces selective

$$[CH_3(CH_2)_6COO]_2Zn \qquad [CH_2{=}CH(CH_2)_8COO]_2Zn$$

(63)   (64)

COONa
SHgC$_2$H$_5$

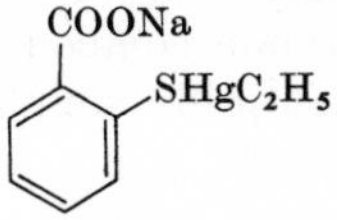

(65)

toxicity and permits their use against some animal and human infections. Various dithiocarbamates such as sodium dimethyl-dithiocarbamate (66) have been used as agricultural fungicides for forty years (see Chapter 12) and the broad antimicrobial activity in this class of compound has led to tolnaftate (67) for use as a topical antifungal agent. Lauroyl peroxide (bisdodecanoyl peroxide) (68) finds a use against *Epidermophyton* fungal species which cause the so-called 'athlete's foot'.

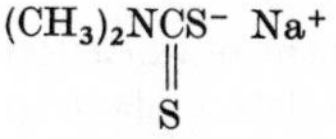

$$(CH_3)_2NCS^- \ Na^+$$
$$\overset{\|}{\underset{S}{}}$$

(66)

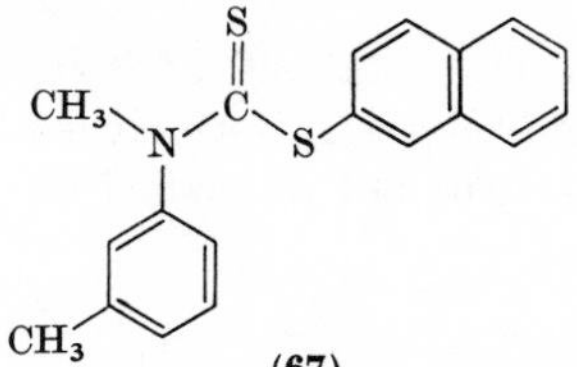

(67)

$$[CH_3(CH_2)_{10}COO]_2$$

(68)

Pride of place in antifungal chemotherapy belongs to the antifungal antibiotics. The polyene antibiotics contain a large lactone ring with several conjugated double bonds. These compounds are frequently antifungal and present a formidable challenge to the chemist engaged in organic synthesis. Pimaricin (69) is an example of this type.

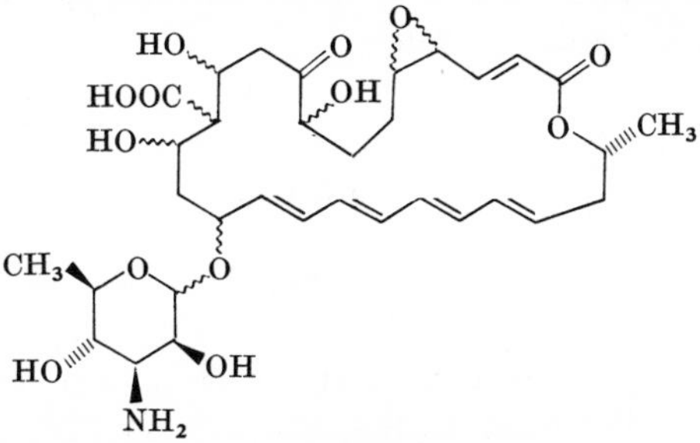

(69)

Griseofulvin (**70**) was discovered in 1939 and has been the subject of a vast amount of work (Part 4, p. 177). It is active orally (but not topically) against human fungal infections. The racemic compound has been synthesized stereospecifically in a beautifully simple synthesis by Stork and Tomasz (see Part 4, p. 179).

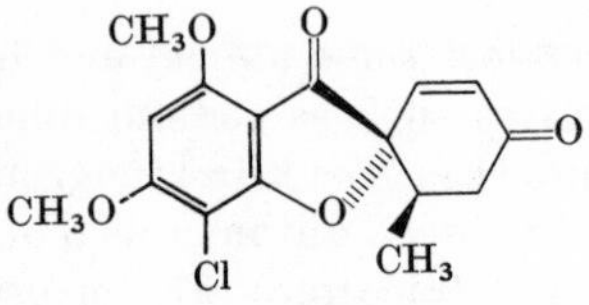

(**70**)

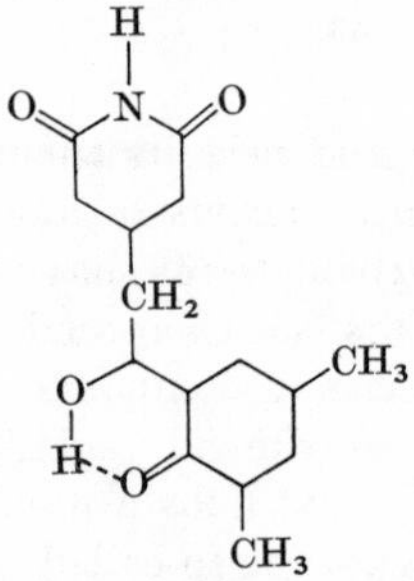

(**71**)

Cycloheximide (**71**) is a metabolite of *Streptomyces griseus* which inhibits fungi but not bacteria; it is widely used for the eradication of fungal plant pathogens but has failed to establish a place in human medicine.

Anticancer Drugs

Cancer is a group of diseases that can affect different organs and body systems. Uncontrolled, abnormal cell division takes place, normally at a rate greater than that of most normal body cells. Medicinal agents for the treatment of cancer began with the 'nitrogen mustards', folic acid antimetabolites ('antagonists') and steroid hormones in the 1940's.

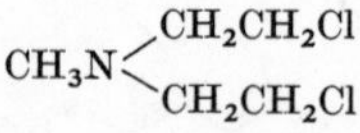

(**72**)

Of the drugs with no known specific action (and there are many!) the alkylating agents are a large and useful group. 'Nitrogen mustard' (**72**) is the first and best known, and several clinically useful drugs have emerged from the hundreds of compounds that have followed its discovery. There is much speculation on the mode of action of these compounds, whose indiscriminate chemical reactivity makes them difficult to study. A lot of evidence indicates that DNA is the cell component most sensitive to attack by these agents and that the 7-position of guanine is the primary site of action; alkylation leads to cleavage of the basic component as shown.

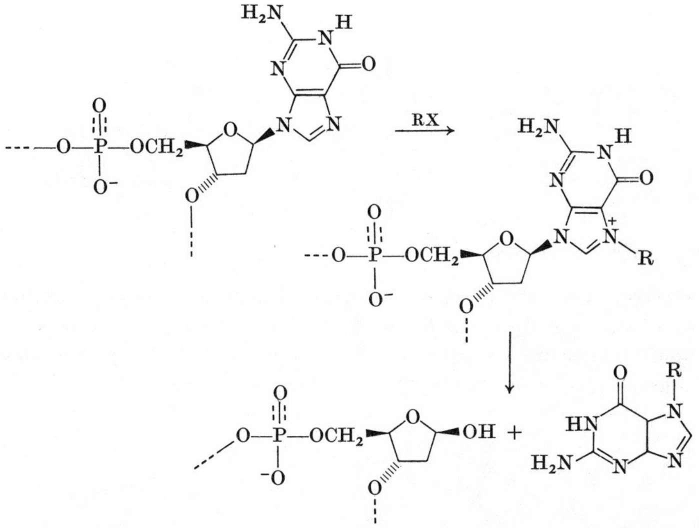

The most active agents are bifunctional and this is explained by their ability to cross-link strands of DNA. Aziridines are very effective alkylating agents and the *sym*-triazine (**73**) was the first

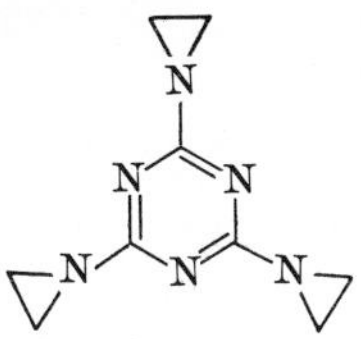

(**73**) 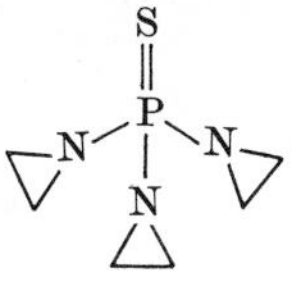   (**74**)

such compound found to be suitable for oral administration. Attachment of the aziridine component to organic molecules of all shapes and sizes has been tried and the phosphoramides represented by thiotepa (**74**) are particularly important.

Antimetabolites interfere with the formation or utilization of normal cellular metabolites; this interference can result from inhibition of an enzyme (or enzymes) or from incorporation as a fraudulent building block into proteins or nucleic acids. Pyrimidine and purine antimetabolites have been comprehensively studied. 5-Azauracil (**75**) and 5-azaorotic acid (**76**) are able to inhibit adenocarcinoma (malignant growth in skin cells) but not leukaemia. They are both inhibitors of the enzyme that converts orotic acid (**77**) into orotidylic acid (**78**). 8-Azaguanine (**79**) is a guanine

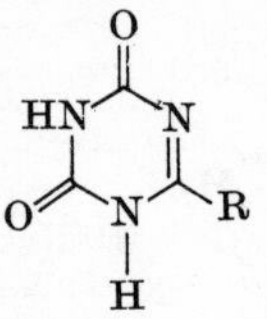

(**75**) R = H
(**76**) R = COOH

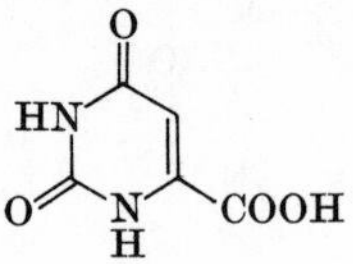

(**77**)

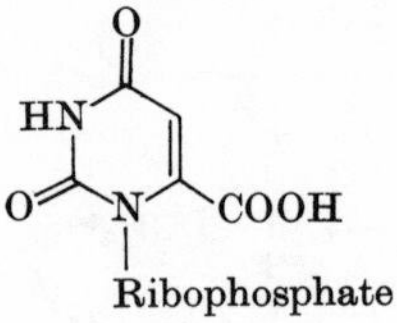

(**78**)

analogue wherein nitrogen has been introduced in place of carbon and was the first (1949) nucleic-acid antimetabolite to show definite antitumour activity. The purine is not the active principle since it has been shown that it needs to be converted into the

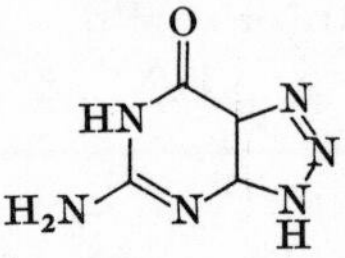

(**79**)

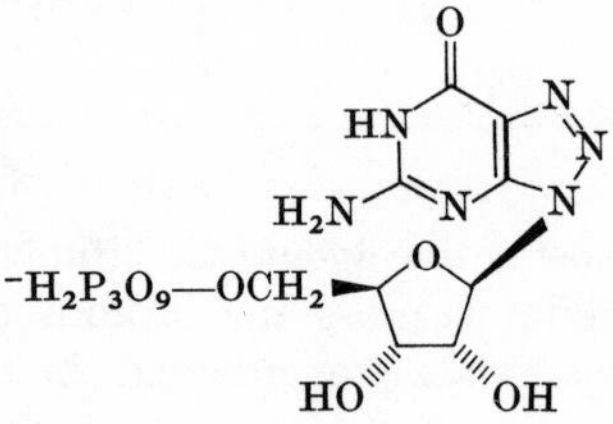

$-H_2P_3O_9$—OCH_2

(**80**)

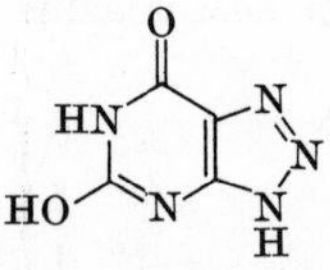

(**81**)

triphosphate (**80**) which is presumably incorporated into messenger RNA and inhibits cell growth. The enzyme guanase converts 8-azaguanine (**79**) into 8-azaxanthine which is devoid of antitumour activity; hence tumours that have low guanase content should be most responsive to (**81**), but this is not the case and one concludes that the mode of action is complex.

Amino-acid antimetabolites have been examined with disappointing results. Ethionine (**82**), the ethyl counterpart of methionine, inhibits certain tumours in rats but not in humans. The same is true of (**83**) (the selenium analogue of cysteine) and the tyrosine analogue betazine (**84**). Antagonists of the metabolites involved in the synthesis of nucleic acids are typified by azaserine

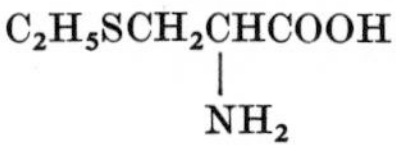

C2H5SCH2CHCOOH
NH2

(**82**)

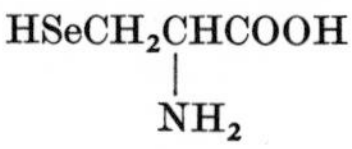

HSeCH2CHCOOH
NH2

(**83**)

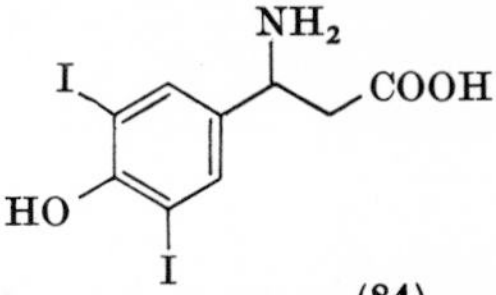

NH2
I
COOH
HO
I

(**84**)

(**85**) which was first isolated from a *Streptomyces* culture filtrate and later synthesized. Azaserine interferes with the processes in which glutamine (**86**) is involved.

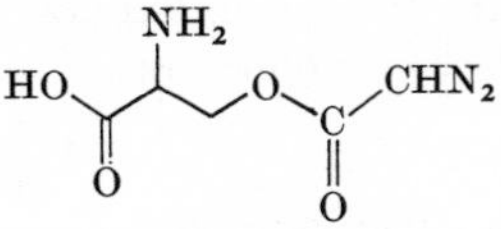

NH2
HO
O
C
CHN2
O
O

(**85**)

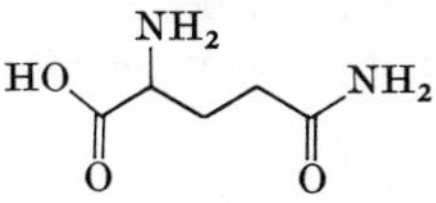

NH2
HO
NH2
O
O

(**86**)

Folic acid (Part 4, pp. 419, 436–437) antagonists are of significance since the derived co-enzymes are involved at three points in the nucleic acid biosynthetic pathway. Aminopterin (**87**) and its *N*-methyl analogue methotrexate (**88**) are two potent antifolic agents with a high degree of anticancer activity; their biological activity is due to their ability to prevent cells from reducing folic and dihydrofolic acid to tetrahydrofolic acid.

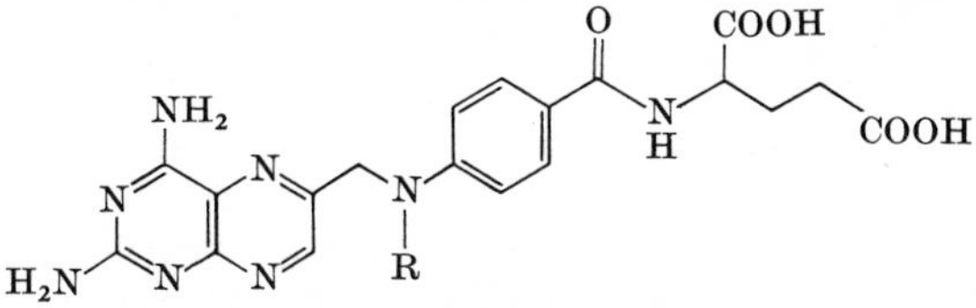

O
COOH
NH2
N
H
COOH
N
N
N
R
H2N
N
N

(**87**) R = H
(**88**) $R = CH_3$

Vitamin and co-enzyme antagonists have been studied and inhibition of tumours has been observed in laboratory animals but a clinically useful drug has yet to appear. The derivative (**89**) of riboflavine (Part 4, pp. 14 and 419) was effective in mice but did not appear as a typical riboflavin antagonist when studied in the usual test system *Lactobacillus casei*.

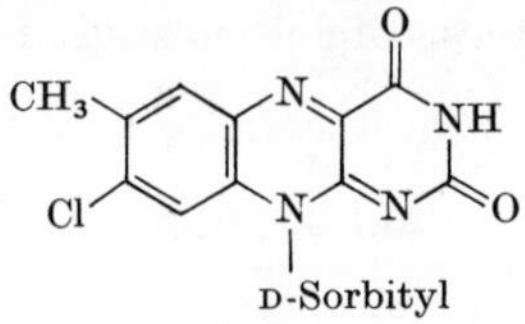

(**89**)

Antiparasitic Drugs

Parasitic diseases in humans are especially prevalent in tropical and sub-tropical countries and have accounted for the deaths of many millions of people. Parasitic diseases in animals are world-wide and are responsibile for severe economic loss.

Antimalarial Drugs

Malaria is a formidable global problem since some 400 million people live in areas where there is no eradication programme. The first reported remedy was obtained from the powdered roots of the Chinese plant *Dichroa febrifuga* where the active constituent is the alkaloid febrifugine (**90**). Cinchona bark yields the alkaloid quinine (**91**) (Part 4, pp. 376–379) whose fascinating history goes

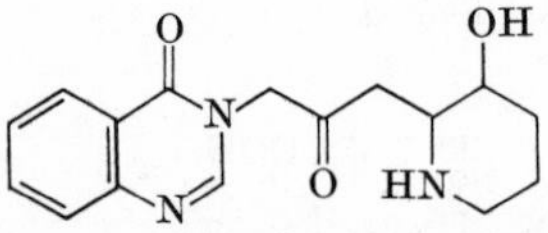

(**90**) 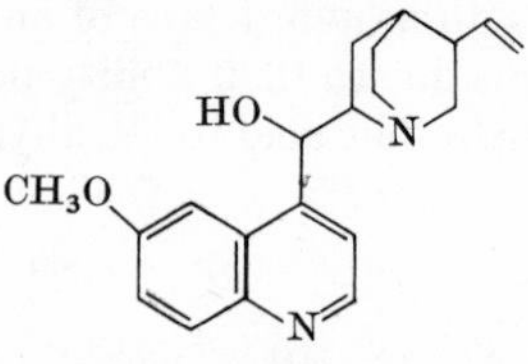   (**91**)

back to 1630 and whose structure was finally proved when it was synthesized by Woodward and Doering in 1944. When supplies of quinine from Java and Sumatra were cut off during World War II, German chemists started an intensive research programme, using the dye Methylene Blue (**92**), which had been shown to have some antimalarial activity, as their starting material. Structural modifications of this molecule indicated that a (dialkylamino-alkyl)amino side-chain was a necessary requirement for high

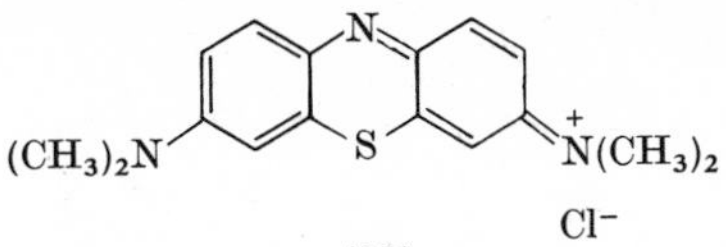

(92)

activity, and in 1932 these researches led to the first synthetic antimalarial drug pamaquine (**93**). Its toxicity and its ineffective-

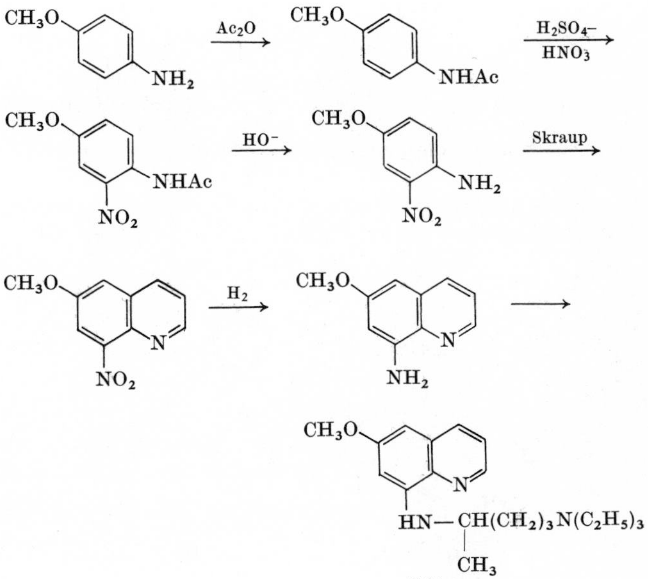

(93)

ness against acute attacks led to the replacement of pamaquine by mepacrine (**94**), which was widely used up to the outbreak of World War II.

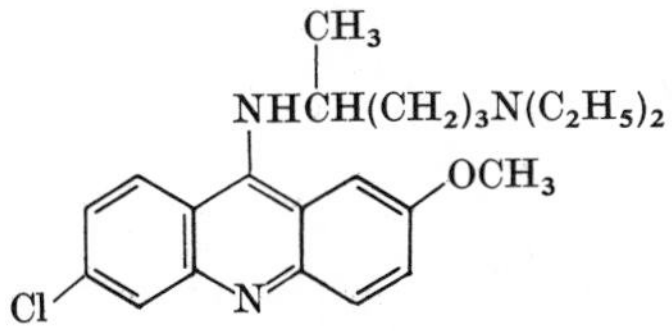

(94)

(95)

(96)

(97)

The potential of 4-aminoquinolines as antimalarials was not fully recognized until the early part of World War II. One of the most important antimalarial drugs chloroquine (**95**) comes from this group and can be synthesized from *m*-chloroaniline.

An interesting range of antimalarial drugs developed at ICI by Curd and Rose during World War II were the biguanides, represented by proguanil (**96**); the latter is metabolized to the dihydrodiazine (**97**), which is also antimalarial and whose close resemblance to the more accessible 4-amino-5-arylpyrimidines was recognized and led to pyrimethamine (**98**) as the most important of this group.

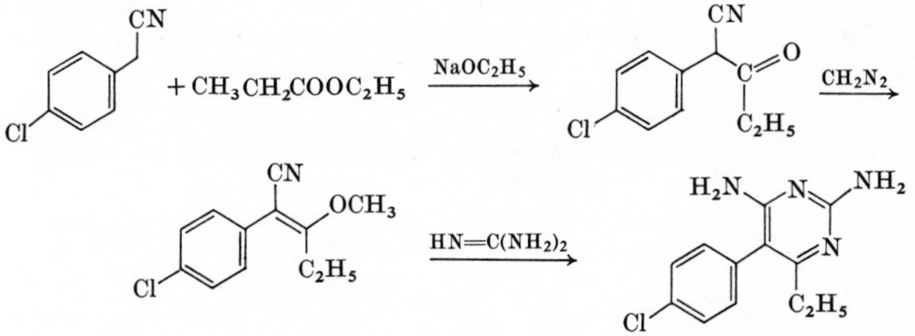

(98)

In recent years resistance to antimalarial drugs has become an increasingly serious problem which has been highlighted by the Vietnam war, and interest in quinine has been rekindled. In 1969 Roche announced that they had developed an industrial synthesis of it.

Antiamoebic Agents

Amoebic dysentery is considered by many people to be a tropical disease but, although more frequently seen in warm climates, it is by no means confined to those areas. The condition in man is caused by *Entamoeba histolytica* in the bowel and extraintestinal sites; the discovery that emetine (**99**) (Part 4, pp. 388–390) has

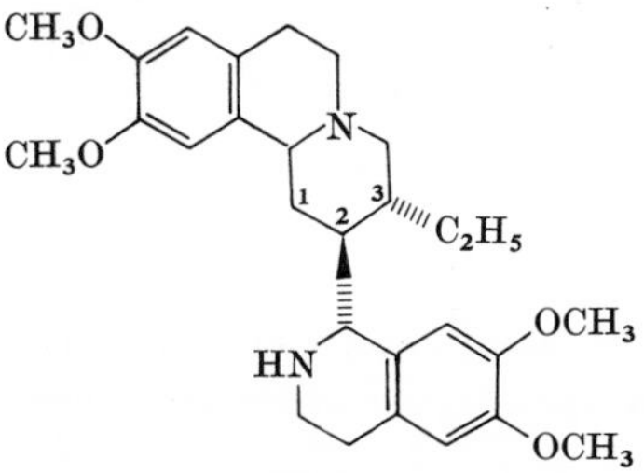

(99)

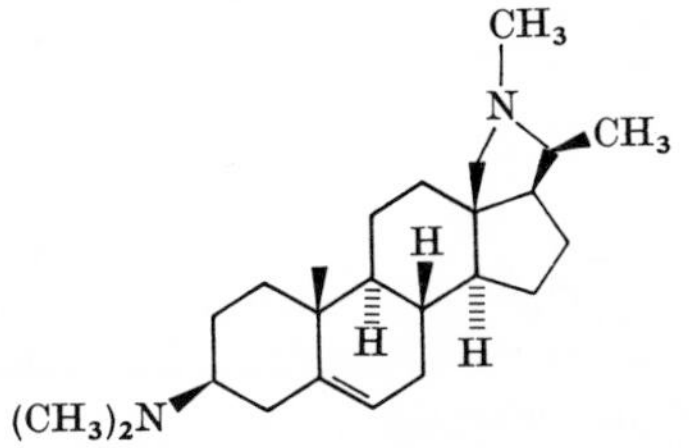

(100)

specific antiamoebic properties catalysed the search for agents which are more effective and less toxic. Only 2,3-didehydroemetine is sufficiently interesting to be considered in this context. There are several total syntheses of such compounds.

Conessine (**100**) from the bark of *Holarrhena antidysenterica* has also been used in the treatment of amoebic dysentery. There are two interesting total syntheses of this molecule by D. H. R. Barton and W. S. Johnson.

Many antibacterial antibiotics are effective against amoebiasis in animals and man. Chlortetracycline (cf. **29**), erythromycin (**30**) and oxytetracycline (cf. **29**) have all been shown to be effective and the most recent antibiotic to be introduced is paramomycin (**101**). There is a wide range of synthetic products for the treatment

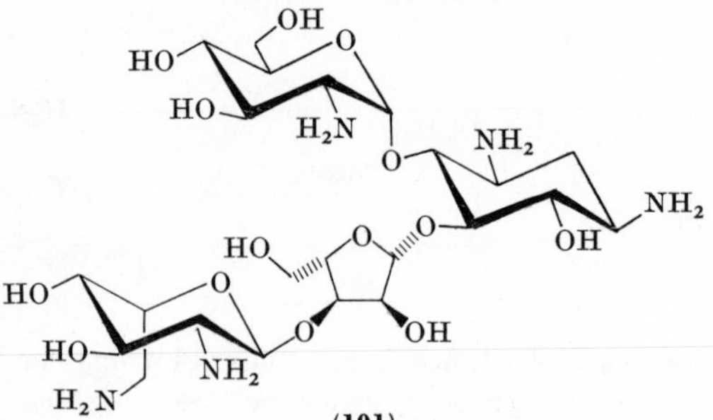

(**101**)

of amoebiasis. These compounds are not in themselves completely satisfactory, frequently because of their toxic effects on the host. Such agents have utility against resistant strains and are sometimes used in combination with emetine (**99**). Iodinated 8-hydroxy-quinolines have been used; chiniofon (**102**) was first synthesized in 1892 and was used as an antiseptic and 'universal cure-all' before it was discovered in 1921 to be effective in amoebiasis.

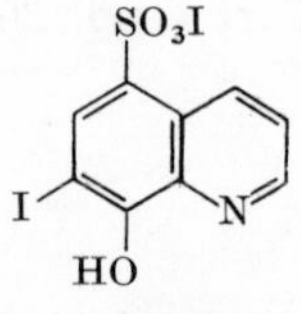

(**102**) 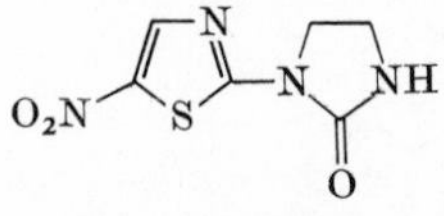   (**103**)

The use of nitro-heterocycles such as niridazole (**103**) against amoebiasis and schistosomiasis (cf. p. 384) was first described in 1964.

Chloroquine (**95**), whose effectiveness against malaria has been described above, has found extensive use also against amoebic infections.

Antitrypanosomiasis Agents

Trypanosomiasis (sleeping sickness) in Africa is caused by the bites of tsetse flies; in South America, Chaga's disease is transmitted by triatomid bugs. These diseases are dangerous and affect humans and animals.

The selective staining of parasites by different dyes prompted Ehrlich to examine hundreds of these compounds as antitrypanosomiasis agents and he was remarkably successful; Trypan Blue (**104**) and Afridol Violet (**105**) had good activity but

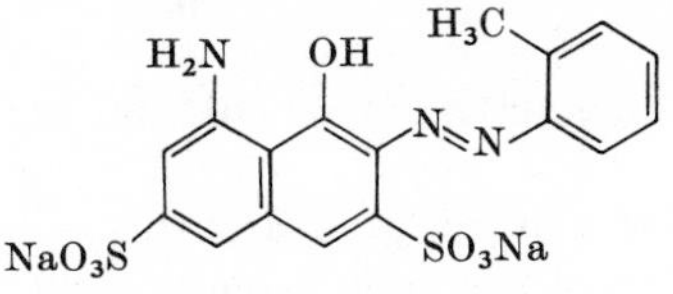

(**104**)

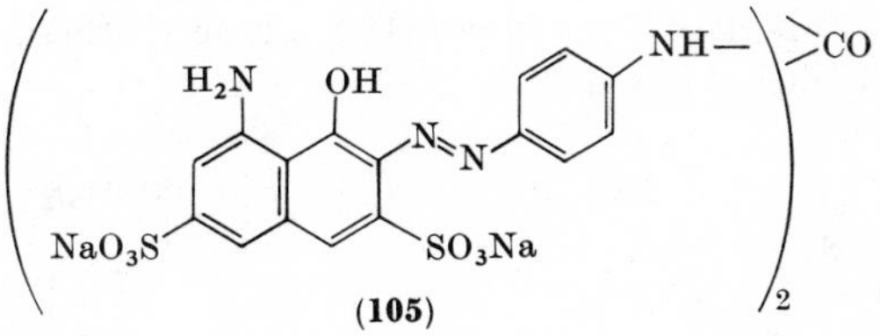

(**105**)

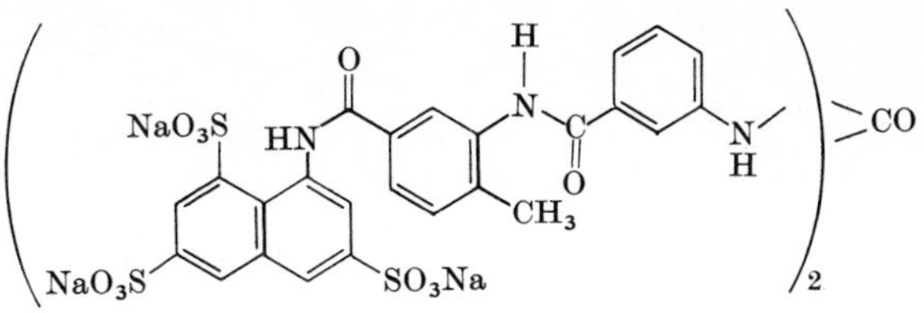

(**106**)

suffered from the obvious disadvantage of colouring tissues and body fluids. It was reasoned by chemists at Bayer that, if the action of the dye is due to fixation by the parasite, compounds with such an affinity but lacking the chromophoric azo-group could well have the same effect. Eight years of work were rewarded by the drug suramin (**106**) in 1920.

Trypanosomes in the animal host consume vast quantities of glucose, and marked hypoglycaemia is observed during the terminal stages of infection. In an attempt to 'starve' the parasite —a rational approach to chemotherapy—the hypoglycaemic agent synthalin A (**107**) was tried, and found to be effective. *In vitro* experiments, however, indicated that the drug has a direct effect on the trypanosome, independent of any effect it may exert through the host. Subsequently a group of aromatic diamidines

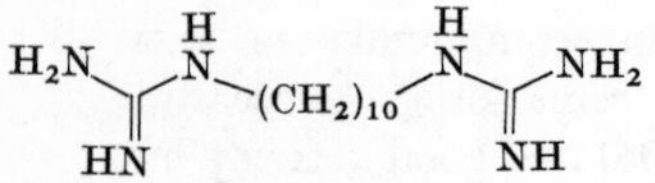

(**107**)

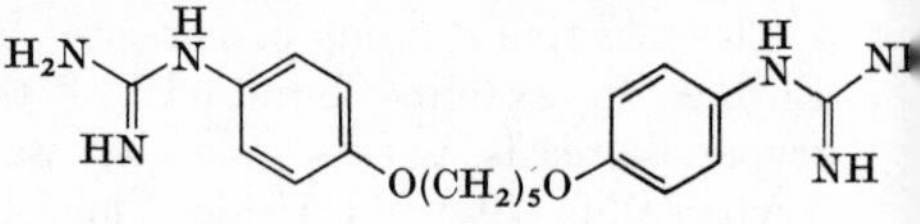

(**108**)

was prepared and, among those active, one, pentamidine (**108**), is still employed clinically.

Some acridines show activity in this area, and this is increased by quaternization. Phenanthridines are isomeric and possess quinoline and isoquinoline units which are associated with compounds with high antitrypanocidal activity; among active compounds in this class is dimidium (**109**) whose synthesis starts from 2-aminobiphenyl (see p. 383).

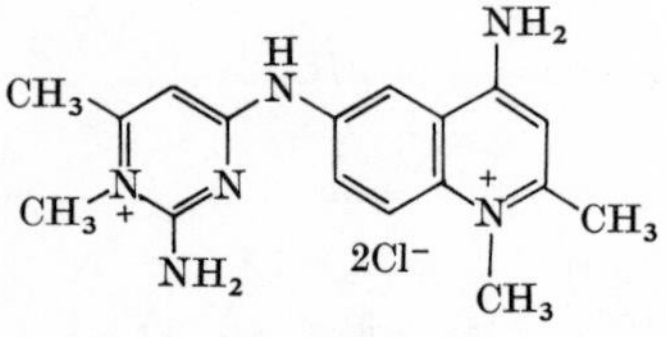

(**110**)

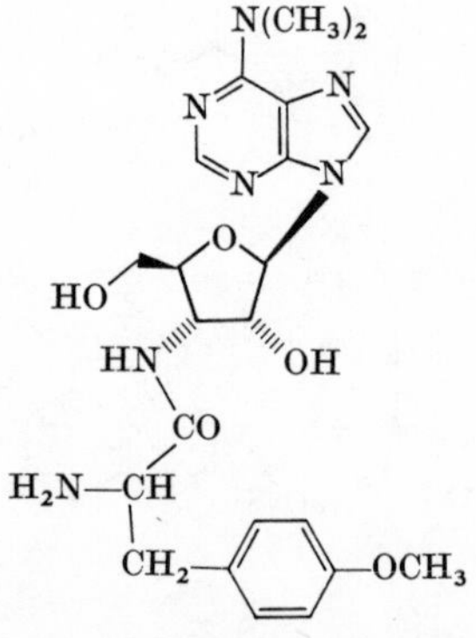

(**111**)

In an attempt to simplify the synthesis, some derivatives of 4-aminoquinoline were prepared. The most notable compound to arise from this work was quinapyramine (**110**) whose synthesis is similar to that of the antimalarial quinolines.

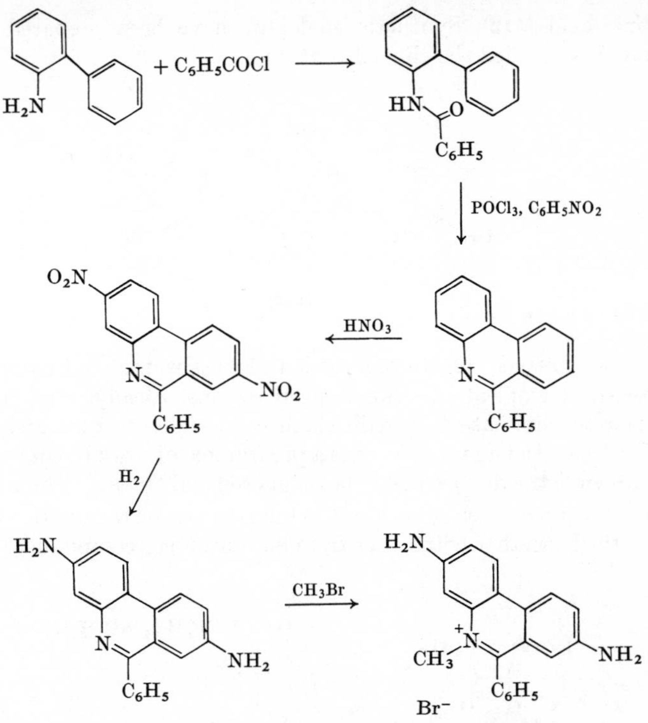

(109)

Understandably, many antibiotics have been screened against trypanosomes. One compound with a broad trypanocidal spectrum is puromycin (**111**) which suffers from the rapid development of resistance and was synthesized by the late B. R. Baker (Lederle) in 1954.

Anthelmintics

These are compounds that destroy parasitic worms (helminths) or remove them from the host. Of the three main classes of helminth, the tapeworms have received least attention because of their relative medical and economic insignificance, especially when compared with nematodes and trematodes.

Extract of male fern *Aspidum oleoresin* has been used for the removal of tapeworms and the active constituent is thought to be filixic acid (**112**). Synthetic analogues have been prepared but none has reached the clinical trial stage.

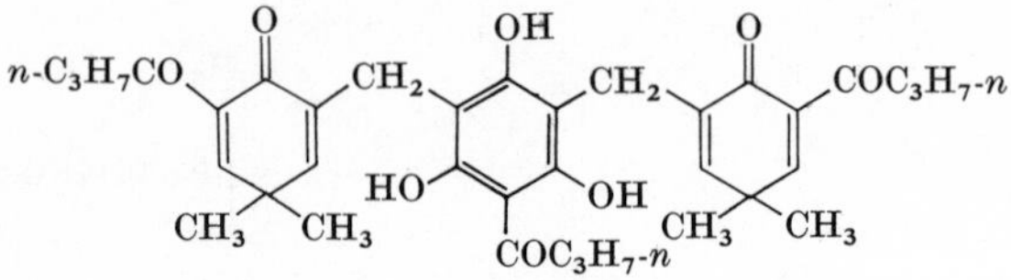

(**112**)

After malaria, schistosomiasis is the most widespread parasitic disease of tropical and sub-tropical regions. Ideally, one would prefer to eliminate the snail which carries the disease, and the World Health Organization has a programme of work to that end; meanwhile the disease has to be diagnosed and treated in humans. The first agent to show an effect in humans was tartar emetic (**113**) and this, together with other trivalent antimony compounds, has

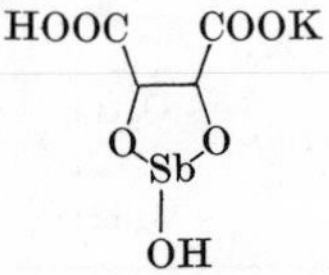

(**113**) 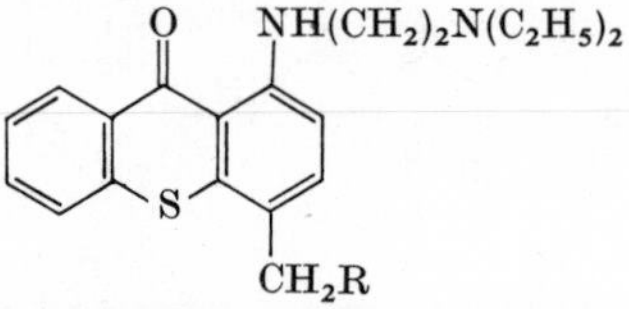   (**114**) R = H;   (**115**) R = OH

continued to be a most effective agent for treatment. Lucanthone (**114**) was the first metal-free compound to show clinical activity and until recently no related compound had improved activity over it, but metabolic studies have yielded hycanthone (**115**) and preliminary reports tell of high activity in human schistosomiasis.

Schistosomes are human blood flukes (trematodes), and liver-fluke infections of man and his animals are of great medical and economic significance. Here again, snails serve as hosts (vectors) and, fortunately, several chemotherapeutic agents are available. Carbon tetrachloride and hexachloroethane have been used in sheep and cattle and a new series of salicylanilides, e.g. oxyclozonide (**116**), has been marketed. Phenothiazine (**117**) has been in use for over 30 years and is manufactured cheaply from diphenylamine and sulphur with a catalytic amount of iodine.

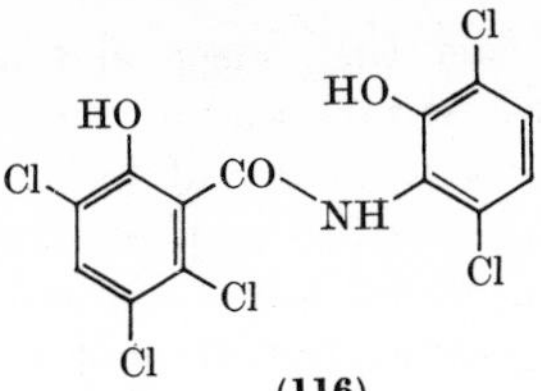

(116)

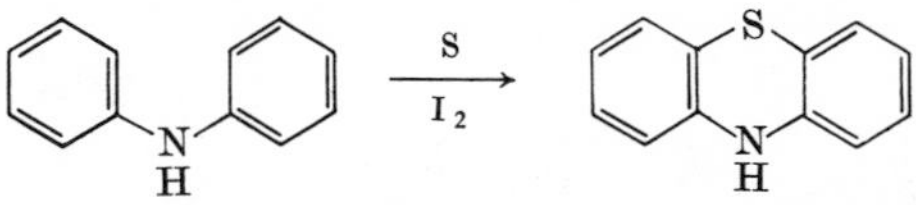

(117)

Some very potent compounds are now available for the treatment of parasitic worm (nematode) infections. The synthesis of thiabendazole **(118)** discovered by Merck in 1961 is shown.

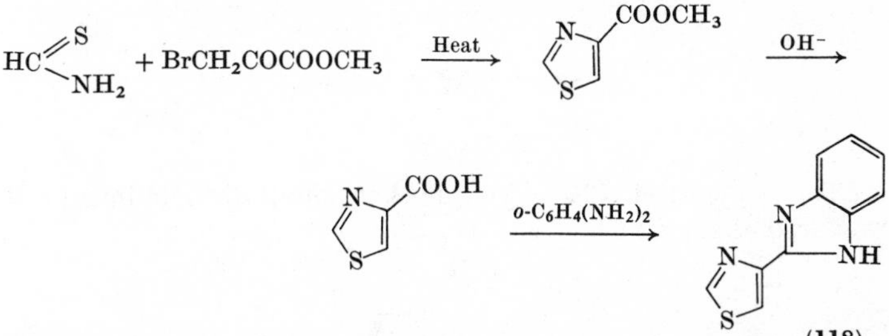

(118)

Tetramisole **(121)** (p. 386) is an outstanding anthelmintic discovered by Janssen in Belgium. The initial compound of interest was the thiophen derivative **(119)** which was active in sheep and poultry but not in the rats and mice that are the normal laboratory test animals. The inference drawn from this observation was that compound **(119)** was not active but that it was metabolized to an active compound, and the animal specificity described above is

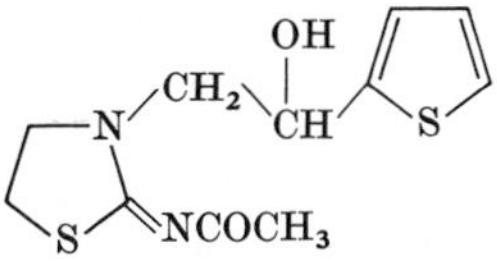

(119) 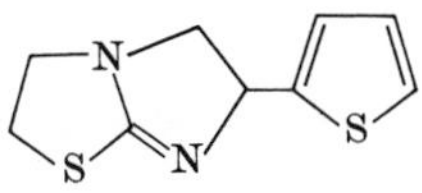   **(120)**

explained by the fact that sheep and poultry can profitably metabolize (**119**) whilst rats and mice cannot. In the event, the metabolite (**120**) was isolated and has high activity which is improved further by substituting phenyl for the thienyl group, as in (**121**). Tetramisole is optically active and the biological activity resides in the *laevo*-isomer; resolution on a manufacturing scale is now carried out to remove and recycle the inactive *dextro*-isomer.

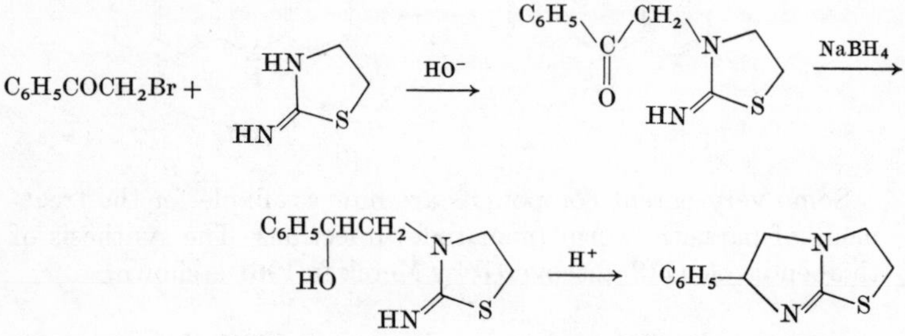

$C_6H_5COCH_2Br +$

Synthesis of tetramisole.

Pfizer's pyrantel (**122**) is yet another potent anthelmintic; it is prepared as shown.

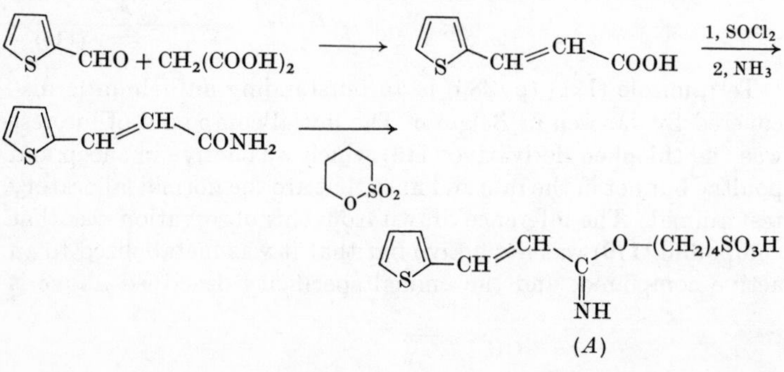

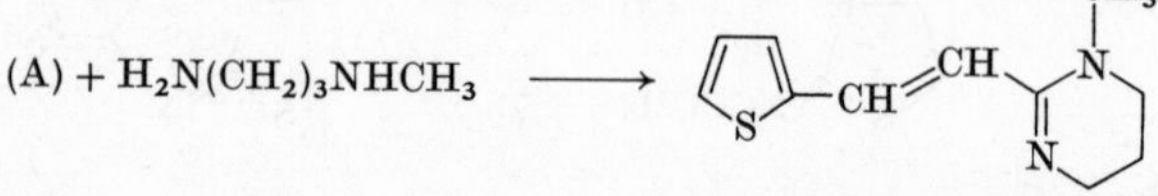

$(A) + H_2N(CH_2)_3NHCH_3 \longrightarrow$

Coccidiosis

This parasitic disease affects most animals and birds and its prevention is of greatest importance in intensive breeding units in the poultry industry. Early treatment of coccidiosis was with sulphur or borax, but the sulphonamides were the first drugs to be used successfully against the disease, and they are still important at the present time. The most commonly used are sulphaquinoxaline (**123**) and sulphadimidine (**11**).

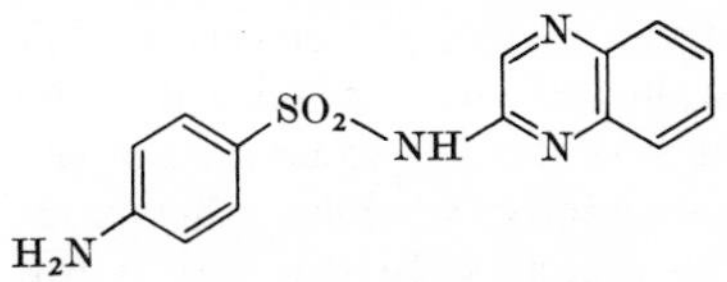

(**123**)

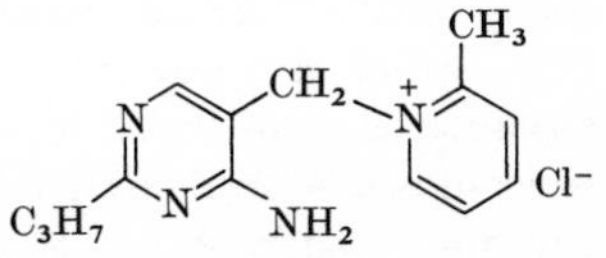

(**124**)

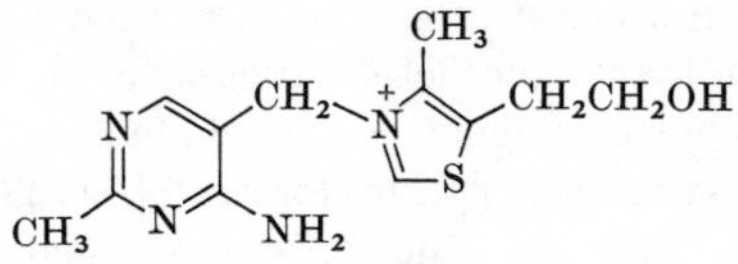

(**125**)

Since *p*-aminobenzoic acid is antagonized by the action of sulphonamides as both antibacterial and anticoccidial agent, one could expect other *p*-aminobenzoic acid antagonists to have anticoccidial properties. This is in fact the case and both pyrimethamine (**98**) and the dihydrodiazine (**97**) are active. Other vitamin antagonists have been successfully employed as preventative (but not curative) agents. Hence, amprolium (**124**) is a useful drug owing to its action as a thiamine (**125**) antagonist.

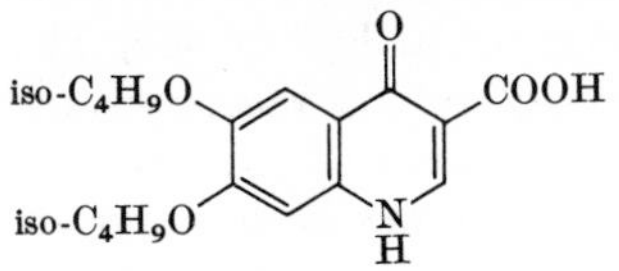

(**126**)

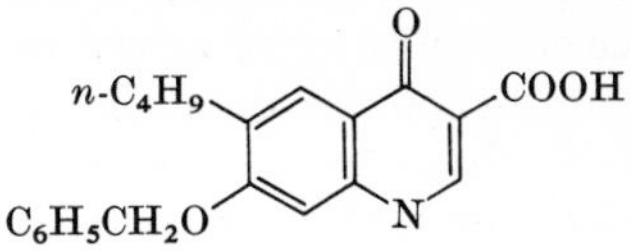

(**127**)

Some quinolin-4(1*H*)-one-3-carboxylic acids, of which buquinolate (**126**) and methyl benzoquate (**127**) are examples, are very potent coccidiostats. These compounds are not antibacterial and show a strong similarity to the antibacterial agent nalidixic acid (**17**), which is not anticoccidial.

NON-INFECTIOUS DISEASES

Antifertility Agents

A comprehensive study of the inhibition of ovulation by steroids which suppress the oestrus cycle during pregnancy (progestins, progestational agents) by Rock, Pincus and Garcia in 1957 revealed that ovulation could be abolished at will for as long as desired; this work led to the development and sale of a series of orally active steroidal contraceptives, known collectively as 'The Pill', which are now in world-wide use and represent a major advance in medicine. Some of the first steroids to be used in this work possessed inherent oestrogenic activity whilst others were contaminated with oestrogenic by-products. In this way it was found that the presence of an oestrogen was essential for successful contraception. A number of progestin–oestrogen combinations are now available that are completely successful in inhibiting conception, so that choice of agent can be made solely on the degree to which side effects occur. Ethynyloestradiol (**128**) or its methyl ether mestranol (**129**) are the only oestrogens employed in currently available preparations and are synthesized from oestrone.

Two of the most important contraceptive agents, norethisterone (**131**) and norethynodrel (**130**), can be synthesized from oestrone by routes that illustrate the value of protecting groups and selective reactions in steroid chemistry. Norethisterone acetate (**132**) is also in common use, as are lynoestrenol (**133**) and ethynodiol diacetate (**134**). Contraceptive activity is not confined to steroids lacking a methyl group at C-19 ('nor-steroids'), and compounds that more closely resemble progesterone (Part 4, pp. 261 and 262), such as methoxyprogesterone acetate (**135**) and megestrol acetate (**136**), are in use but chlormadinone (**137**) has been withdrawn because it produced tumours in dogs (p. 390). The current concern about the relation between oral contraceptives and thromboses in young women has further stimulated the search for non-steroidal contraceptives but as yet none has been marketed. Methods for introducing the methyl group or the chlorine atom at position 6 are discussed in Part 4, pp. 266–267.

Steroidal Anti-inflammatory Drugs

In a search for new hormones, Kendall's group at the Mayo Clinic in Minnesota isolated five compounds from adrenal cortical extracts, and one of these, cortisone (**139**) showed hormone-like

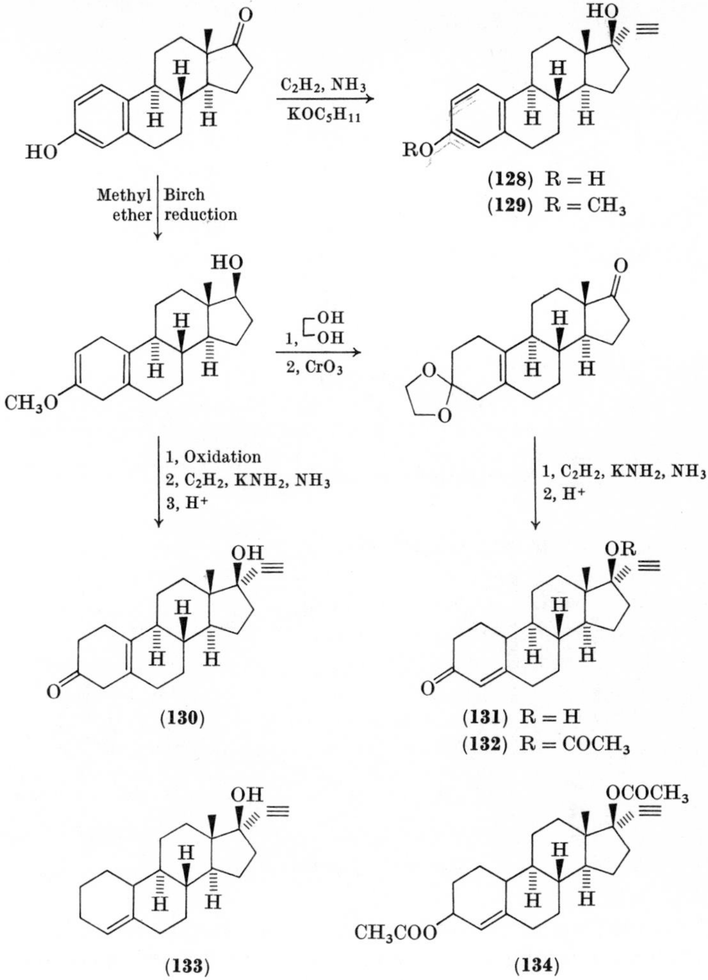

Some antifertility agents.

activity in rats. Exploitation of this biological activity was dependent upon compound availability, so in 1942 seven groups of American chemists and others in Switzerland began work on syntheses of the compound to produce material for evaluation. Most syntheses commenced with the bile acid deoxycholic acid (138) which involved, amongst other changes, translocation of oxygen from C-12 to C-11 (see p. 390).

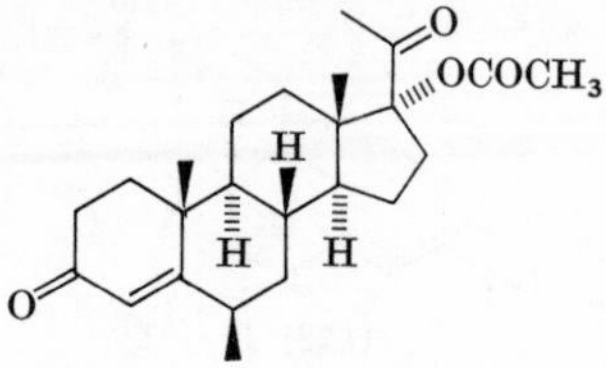

(135)

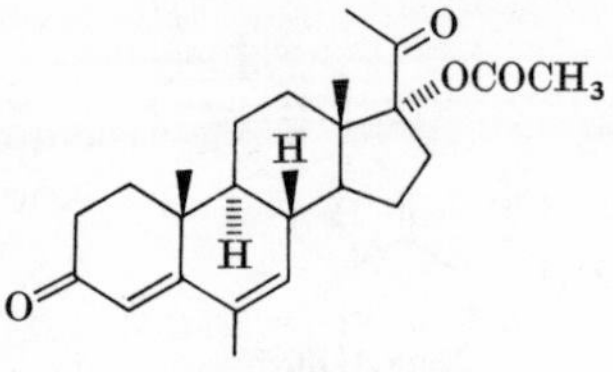

(136)

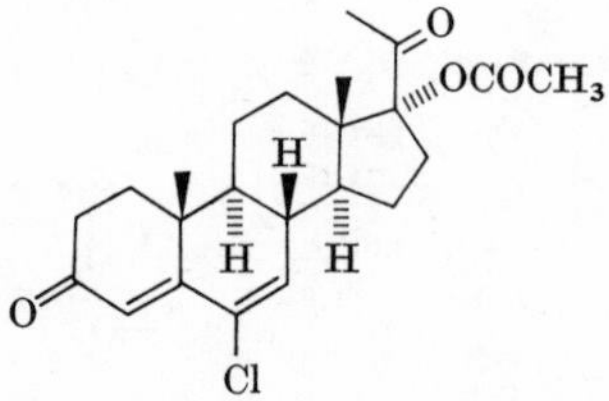

(137)

Some further antifertility agents.

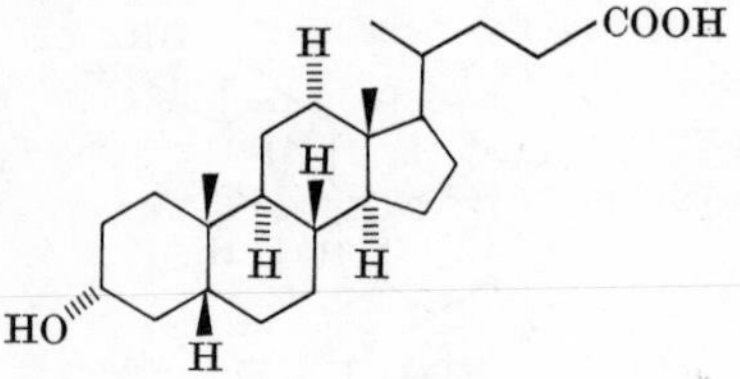

(138)

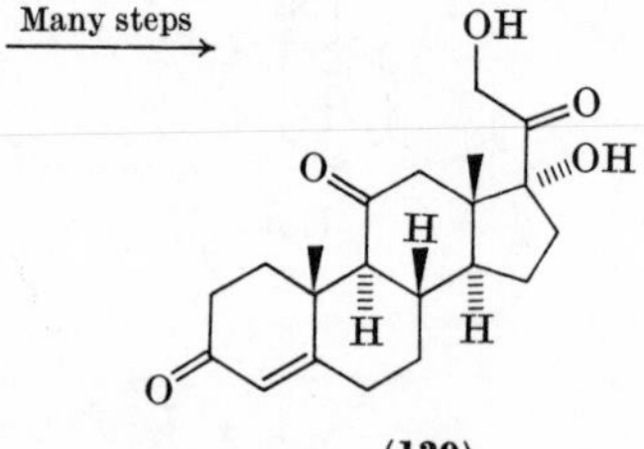

Many steps

(139)

Synthesis of cortisone.

In 1949 the use of cortisone for the treatment of rheumatoid arthritis, a painful and crippling disease involving inflammation of the joints, was described; understandably this set off an intensive search for other steroids with anti-inflammatory activity. There was great concern about the inadequacy of the bile-acid route for meeting the requirements, as the best process required thirty stages with an overall yield below 1%. The prescribed daily dose for arthritic patients was 100 mg and, since cortisone led to alleviation of the symptoms and not a cure, protracted use was inevitable. There was clearly a great need for alternatives to routes that did not require bile acid intermediates. The development of a method for the introduction of an oxygen function at C-11 in a relatively abundant steroid having no substituent in ring C (e.g. progesterone) was desirable. Other approaches considered were total synthesis (unlikely to be feasible on a large scale) and utilization of hecogenin (**140**), then a rare compound but soon found in abundance. All approaches were eventually successful and the totally unexpected microbiological oxidation of steroids at C-11 led to the route from progesterone as the only synthetic one

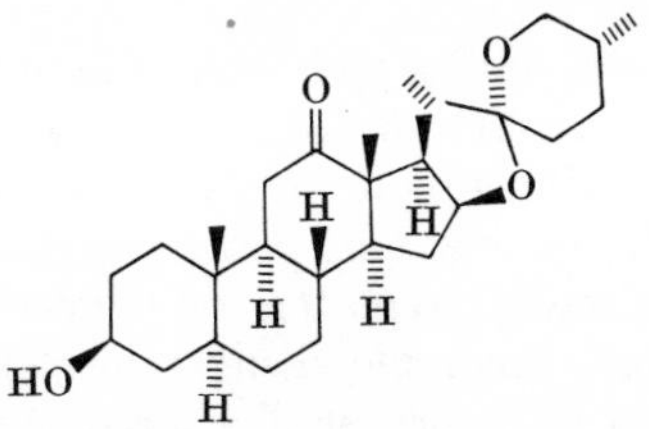

(**140**)

developed for practical production. Oxidation at C-11 and reduction of steroids by yeast extracts were discovered in 1937; the first introduction of oxygen at C-11 by a biochemical process was in 1949 and this involved the perfusion of cortexone (**141**) in serum through an isolated adrenal gland—hardly the makings of a plant process! The Upjohn group achieved a major breakthrough when they set out with the deliberate intention of finding a micro-

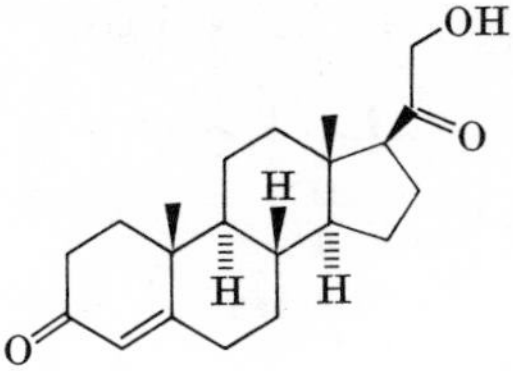

(**141**) 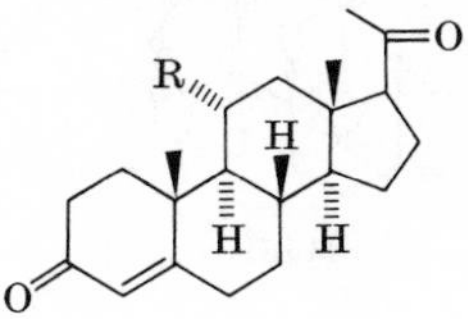   (**142**) R = H;   (**143**) R = OH

organism in soil that would perform the oxidation and were successful. A culture of *Rhizopus arrhizus*, isolated as a result of exposing an agar plate on one of their laboratory window-ledges in Kalamazoo, Michigan, converted progesterone (**142**) into 11α-hydroxyprogesterone (**143**) in yields which were eventually raised to 50%. Better yields were obtained with *Rhizopus nigricans* and, since progesterone is easily obtained from the naturally occurring and abundant diosgenin (**144**) (Part 4, pp. 259 and 263), this new oxidation formed the basis of an attractive four-step route to cortisone.

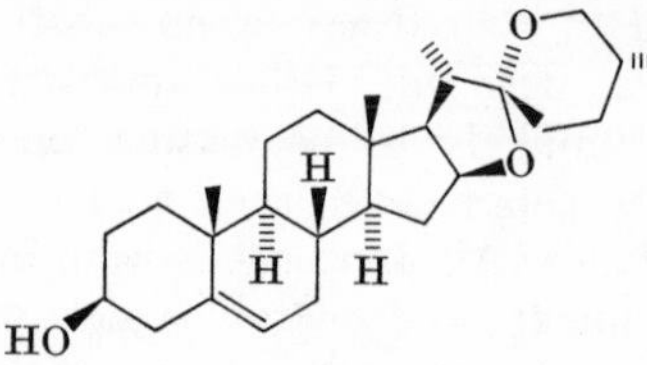

(144)

The corticosteroids have a wide spectrum of biological activity and the small concentrations maintained by the adrenal cortex regulate a number of metabolic processes. The administration of high levels of cortisone over long periods often causes serious side effects, e.g. skin atrophy, hypertension (elevated blood pressure) and hyperglycaemia (high blood-sugar levels). For this reason medicinal chemists have prepared a large number of cortisone analogues with the intention also of increasing potency. 9α-Fluoro-cortisone (145) has about ten times the activity of cortisone but has such high salt-retaining properties that it is now used thera-peutically for that purpose (mineralocorticoids). Subsequent developments have brought forth prednisone (146) and predniso-lone (147). These seemingly minor structural changes have

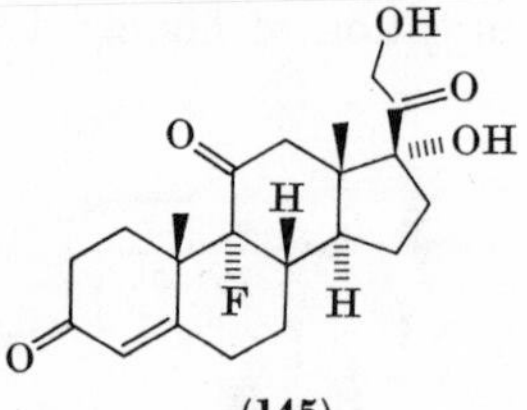

(145)

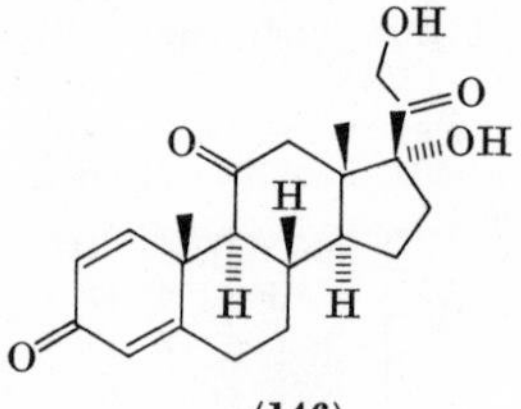

(146)

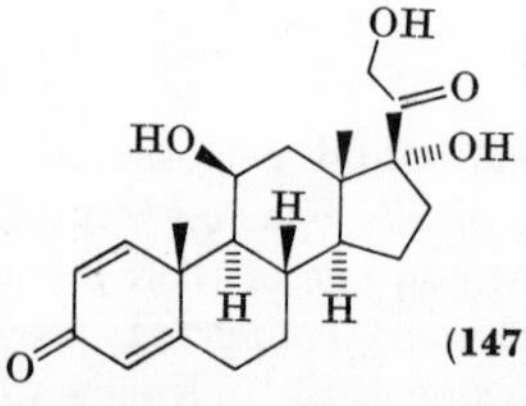

(147)

brought several-fold increases in anti-inflammatory potency without increasing the salt-retaining effects. The potential hazards of cortisone therapy, coupled with the fact that termination of the treatment brings its own difficulties, have divided the opinions of physicians on the long-term use of these compounds.

Non-steroidal Anti-inflammatory Drugs

Interest in these compounds has increased recently and medicinal chemists have attempted to find compounds whose activity is comparable with the cortisone group of compounds but lack their side effects. Early compounds that were synthesized and effective were called analgesic–antipyretic (i.e. pain-relieving–fever-reducing) before tests were developed that could distinguish the anti-inflammatory components from other effects; diseases in this group include rheumatoid arthritis, rheumatic fever and osteoarthritis and they are second only to diseases of the heart as causes of limited physical activity. A compound more effective than aspirin against rheumatoid arthritis has yet to be discovered. Most of the drugs to be described are antipyretic and analgesic; the latter is particularly useful in the treatment of arthritis.

Salicin was introduced for the treatment of acute rheumatism (rheumatic fever) by Maclagen in England in 1874. Now, one hundred years later, we still do not understand the mode of action of this or any other drug with the inflammation site. Salicylates are preferred for the alleviation of rheumatoid arthritis and aspirin is frequently used in combination with an antibiotic for the treatment of rheumatic fever. Aspirin (acetylsalicylic acid) is synthesized from phenol by the Kolbe–Schmitt process (Part 2, p. 218); despite its effectiveness and prolonged use, side effects such as gastrointestinal bleeding and hypersensitivity occur sometimes and cannot be ignored; the use of 'soluble', 'buffered' or 'enterically coated' aspirin has not altogether overcome these side effects. Flufenisal (**148**) is a salicylate described recently by Merck and is said to have four times the anti-inflammatory potency of aspirin with less gastrointestinal irritation; its synthesis from an accessible biphenyl is shown on p. 394 and includes an interesting extrusion of sulphur dioxide.

(**149**) (**150**)

The nitrogen-containing analogue (**149**) has comparable activity whilst (**150**) is essentially inactive.

(148)

It is advantageous to divide the anti-inflammatory pyrazolones into two groups, the pyrazol-5-ones typified by antipyrine (151) and amidopyrine (152) and the more important pyrazolidine-3,5-diones whose principal member is phenylbutazone (153). The clinical use of (152) has declined in recent years owing to an increasing awareness of its side effects and now is frequently used as a reference standard for comparison with new candidates.

The interesting biological activity of the acidic phenylbutazone (153) was discovered accidentally. It was found to be analgesic and antipyretic and was combined with the basic amidopyrine (152) in the hope that together they would have an increased solubility and greater potency. In 1952 there was a report that the antirheumatic activity of the combination was greater than that due to the amidopyrine content and that phenylbutazone was on its own a striking reliever of arthritic conditions. Its synthesis starts, as shown, from hydrazobenzene.

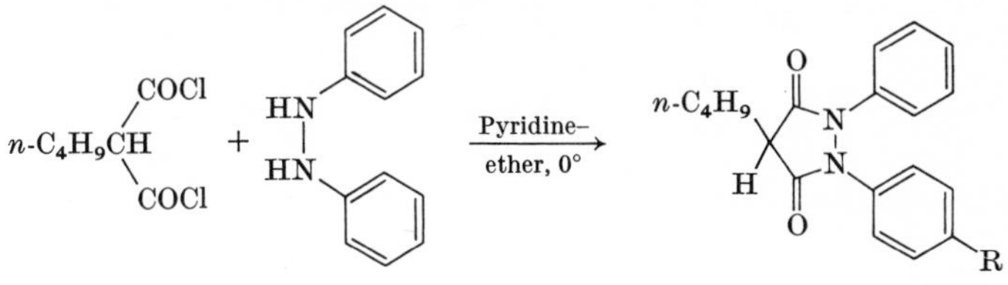

(**153**) R = H
(**154**) R = OH

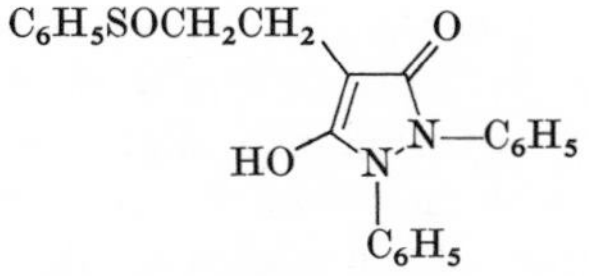

(**155**)

A metabolite of (**153**) is oxyphenylbutazone (**154**) which has equal anti-inflammatory activity and lower toxicity. Another analogue, sulphinpyrazone (**155**), is a potent drug against gout; there is a close relationship between the pK_a of these compounds and their biological activity.

Compounds derived from anthranilic acid represent a different class of anti-inflammatory drug; these compounds could be looked upon as analogues of the salicylates and three interesting compounds, mefenamic acid (**156**), flufenamic acid (**157**) and 'CI-583' (**158**) have been described by the Parke-Davis group. The most active of the series is (**158**) which is said to have fifteen times the potency of phenylbutazone and 150 times that of aspirin.

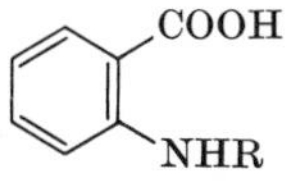

R = 2′,3′-$(CH_3)_2C_6H_3$ (**156**)
R = 3′-$CF_3C_6H_4$ (**157**)
R = 2′,3′-Cl_2,3-$CH_3C_6H_2$ (**158**)

The heterocyclic alkanoic acids are a group of antiarthritic compounds where there is currently the most effort and interest. The potent drug indomethacin (**159**) was first described by Merck chemists in 1963. The synthesis of indoles as anti-inflammatories was prompted by the finding that rheumatic patients excrete tryptophan (**160**) metabolites and there has, in addition, been much speculation on the role of serotonin (**161**) in the inflammatory process.

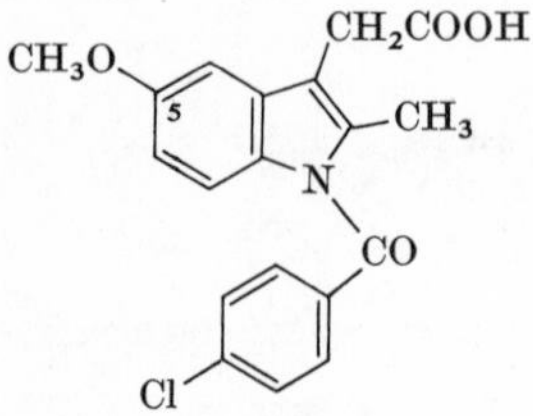

(**159**) 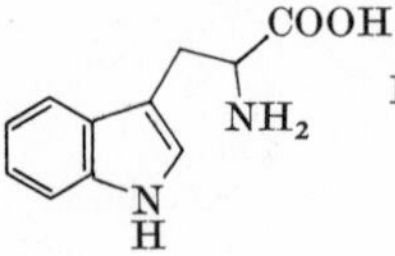   (**160**) 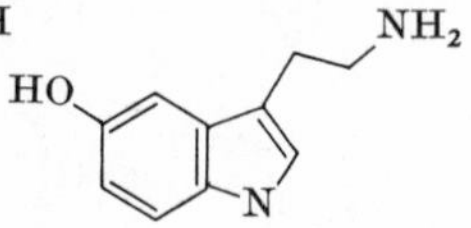   (**161**)

In the indomethacin series there is a close relationship between pK_a and biological activity. Deacylation produces essentially inactive compounds; substitution of *p*-chlorobenzyl for *p*-chlorobenzoyl reduces activity slightly whilst 5-demethylation reduces it more markedly. A study of the related α-substituted propionic acids (e.g. **162**) has shown that after resolution into its two optical isomers the anti-inflammatory activity resides only in the (+)-enantiomer. The preferred conformation of the *p*-chlorophenyl group is *cis* to the methoxyphenyl portion of, yet not coplanar with, the entire indole nucleus. A closely related indene (**163**) is half as active as indomethacin whilst the *trans*-isomer (**164**) has

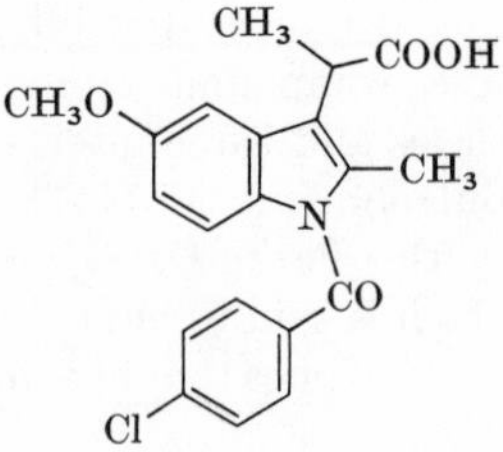

(**162**) 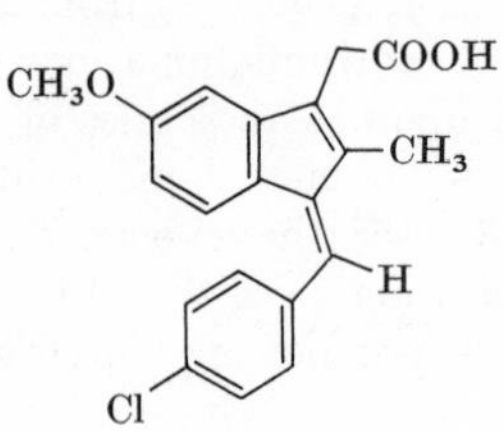   (**163**)

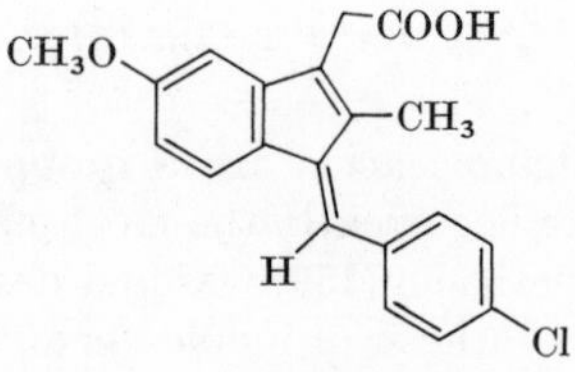

(**164**)

only 10% of the activity of (162). Indomethacin is effective against rheumatoid arthritis and is of value in gout and osteoarthritis; its side effects include headaches, dizziness, dyspepsia and gastro-intestinal upset but the incidence has been reduced by more conservative dosage.

Interest in arylacetic acids was stimulated by the discovery by Boots that ibufenac (165) had 2–4 times the potency of aspirin. The compound was withdrawn after signs of liver damage (hepato-toxicity) in man and was replaced by the corresponding α-substi-tuted propionic acid ibuprofen (166); unlike other α-substituted

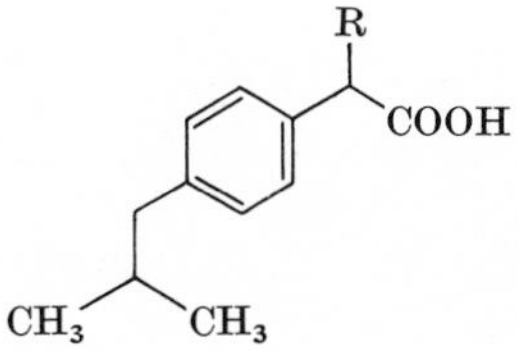

(165) R = H
(166) R = CH₃

propionic acids with anti-inflammatory activity, both enantio-mers of (166) are active. Substituted 1- and 2-naphthylacetic acids, e.g. (167), have been developed by Syntex over a number of years and this compound was launched in the U.K. in 1973. A typical synthesis is shown.

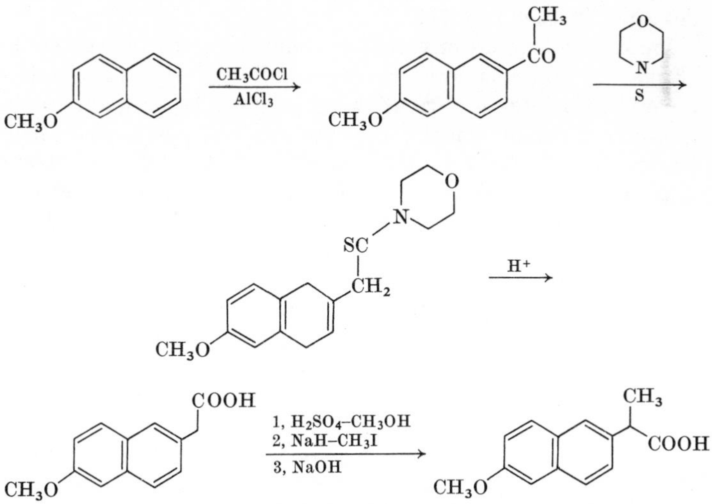

(167)

Heterocyclic acetic acids distinctly different from indomethacin abound in the patent literature. Fenclozic acid (168) was withdrawn from clinical trial because its side effects were similar to those of ibufenac. This compound is of particular interest since it is metabolized in the rat to (169) wherein the halogen had migrated to an adjacent position. This hydroxylation of *para*-substituted benzene derivatives with halogen migration is known as the NIH shift since it was discovered in 1966 by Daly and his co-workers at the National Institutes of Health at Bethesda, Maryland.

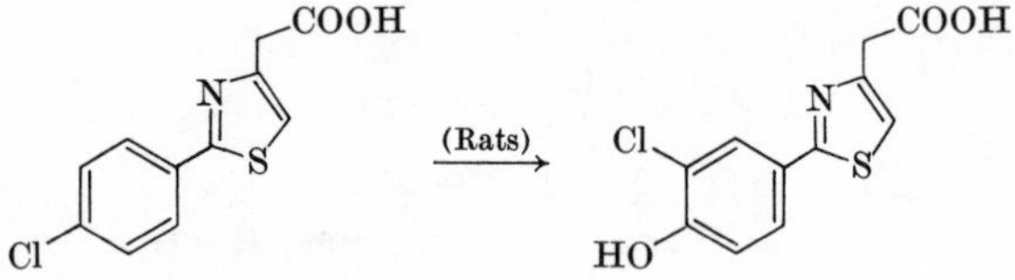

Coagulants and Anticoagulants

Disruption of blood circulation may result in either thrombosis or haemorrhage; in the latter there is an abnormal tendency to bleed; several antihaemorrhagic drugs, both naturally occurring and synthetic, have been introduced. The most important natural product in this respect is vitamin K whose chemistry has been described in Part 4 (p. 271).

Snake venoms are mixtures of compounds that have specific effects on some factors involved in blood clotting. Modern separative techniques are allowing progress to be made in the isolation of venom constituents: proteins and polypeptides have been identified. The coagulant properties of the venoms of several different snakes have been used in therapy and it is evident that these fascinating agents have untapped potential.

An interesting coagulant is carbazochrome salicylate (170) which is a complex of adrenochrome monosemicarbazone and sodium salicylate. Adrenochrome is an oxidation product of the important amine adrenaline. The drug (170) is non-toxic and can be taken orally or by intramuscular injection for post-operative haemorrhage in, e.g. tonsillectomy.

Protamine sulphate is a protein of low molecular weight found in the sperm of some fish and has weak anticoagulant properties. It is well tolerated (i.e. has no adverse side effects) and has been in use for over 20 years.

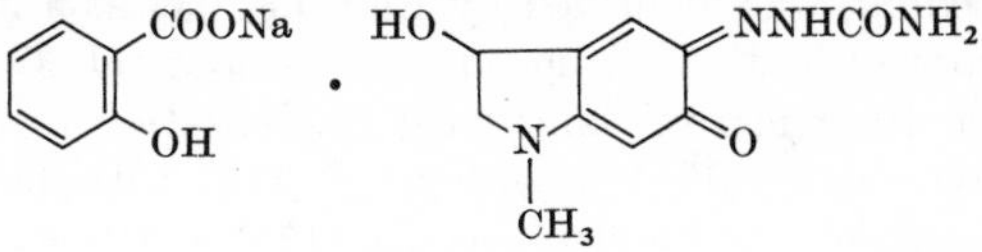

(170)

ε-Aminocaproic acid (6-aminohexanoic acid) is readily obtained from its lactam (cf. Chapter 7), an important nylon intermediate which is in turn obtained by Beckmann rearrangement of cyclohexanone oxime. This compound is useful in controlling serious haemorrhage. Two related compounds (171) and (172) discovered respectively by Kabi (Sweden) and Merck have recently been described; (171) is ten times as potent as, and (172) sixty times as active as ε-aminocaproic acid. A number of compounds are

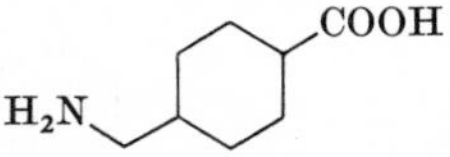

(171) 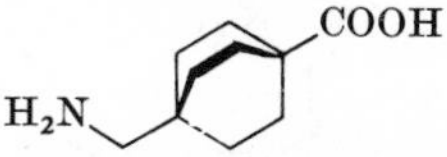   (172)

capable of accelerating coagulation when applied to bleeding surfaces; such topical agents do not control severe bleeding; they include gelatin foam, oxidized cellulose, thrombin and many preparations from animal sources.

Anticoagulants prolong the time taken for coagulation and are classified according to their mode of action and administration. There are two large groups, the direct-acting heparin and the indirect (oral) agents.

Heparin has been known since 1916 when it was discovered by an undergraduate at Johns Hopkins Medical School, Baltimore, Maryland. It is a sulphuric acid ester of a polysaccharide that consists of alternating D-glucuronic acid and 2-amino-2-deoxy-D-glucose units. The smallest unit possessing anticoagulant activity has a molecular weight of about 8000. Protamine sulphate is a potent heparin antagonist, i.e. it reverses the effect of heparin. For a long time heparin was considered to be the ideal anticoagulant but its main disadvantage was that it had to be taken parenterally (by injection) which made it particularly inconvenient for long-term therapy. Two new groups of compounds were subsequently discovered that were active *via* the oral route. In the early 1920's there was a serious outbreak of sweet clover disease among cattle in North Dakota and Canada. The toxic substance, which caused severe bleeding, was identified by Link in 1939 as a dicoumarol (**174**) whose 4-hydroxycoumarin component (**173**) is easily synthesized from methyl salicylate; condensation of (**173**) with formaldehyde gave (**174**). Reaction of (**173**) with benzylidene-acetone gave the well-known warfarin (**175**) which is effective as a rat poison since the animals bleed to death.

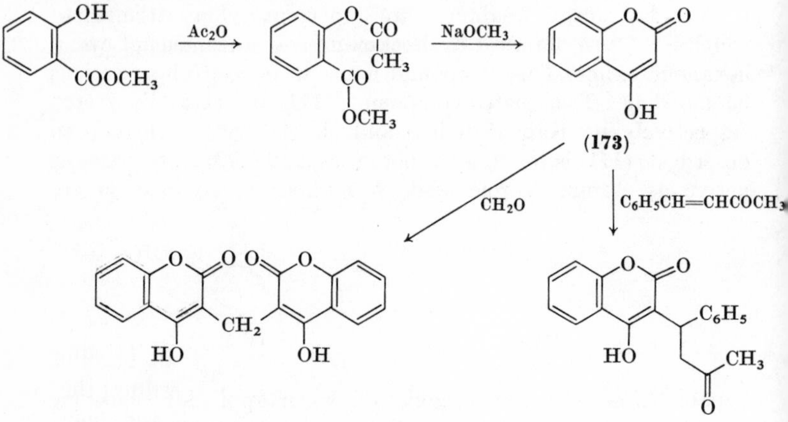

(**173**)

(**174**)   (**175**)

Anticoagulants of the other group are the structurally similar indane-1,3-diones. Two compounds, phenindione (**176**) and diphenadione (**177**), are established. They are synthesized by classical routes.

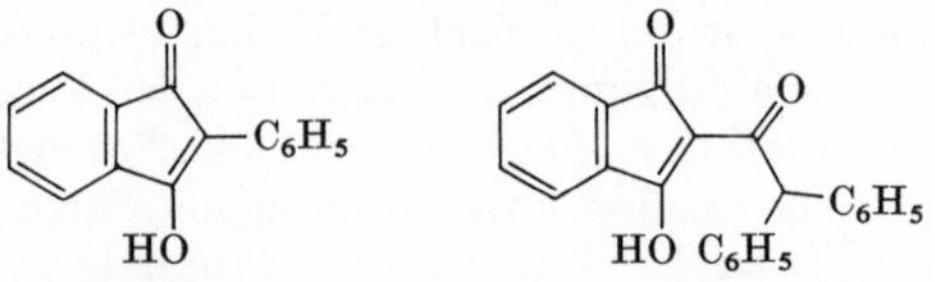

(**176**)   (**177**)

Hypocholesteraemic Drugs

These drugs are concerned with the lowering of cholesterol levels in blood vessels. The non-steroidal agents are frequently simple molecules. When the biosynthesis of cholesterol was under investigation (Part 4, pp. 217–226, 247–251) farnesyl pyrophosphate (178) was oxidized to the acid (179). Such a compound, which is not on the pathway between mevalonic acid (Part 4, pp. 219–220) and squalene is able to inhibit that transformation and therefore has potential as an agent to prevent cholesterol formation. So far there appears to have been little research in

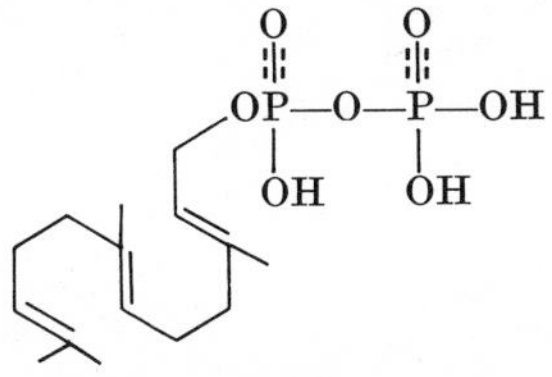

(178) 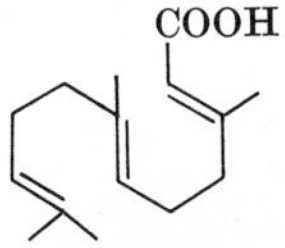　　**(179)**

this area orientated towards the discovery of sesquiterpenoid hypocholesteraemic agents. However, several analogues of mevalonic acid have been prepared with a view to their use as inhibitors of the conversion of acetate into cholesterol. The biochemical inhibiting (antimevalonic) activity of (180) is significant, as is that of the interesting fluorinated analogues (181) and (182), and developments in this area are eagerly awaited.

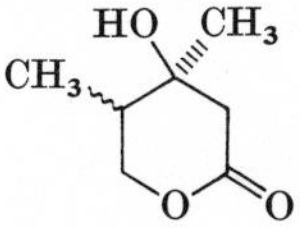

(180) 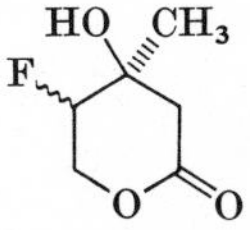　　**(181)** 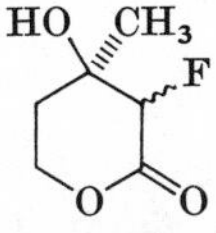　　**(182)**

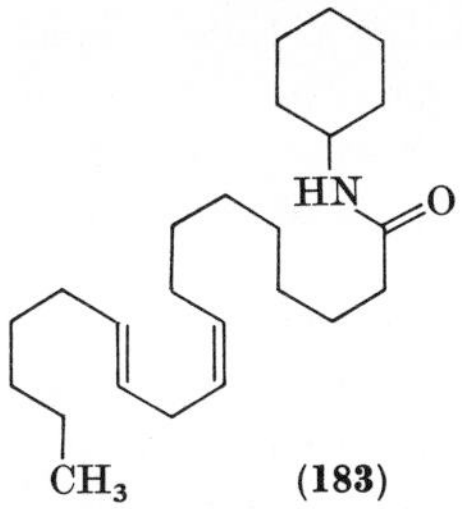

(183)

Several linoleic (Part 4, pp. 150 *et seq.*) derivatives that are believed to inhibit cholesterol absorption have been prepared. The *N*-cyclohexylamide (183) is particularly effective. Studies with aryl- and aryloxy-alkanoic acids have also led to clinically useful compounds. Some 20 years ago evidence was published of the hypocholesteraemic activity in a series of alkanoic acids and esters of which (**184**) was initially most interesting; but subsequently among the many analogues prepared there was greater activity in (**185**) and (**186**). Later, two ICI workers, Waring and Thorp, prepared and examined a selection of phenoxy-analogues leading to clofibrate (**187**), which is readily prepared from ethanol, acetone, chloroform and *p*-chlorophenol. Clofibrate has become an extremely successful agent in this area.

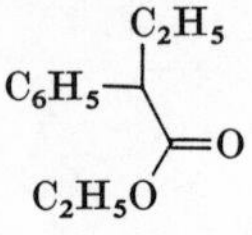

(**184**) 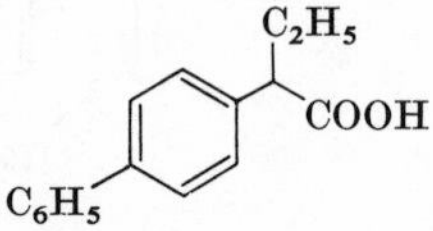   (**185**) 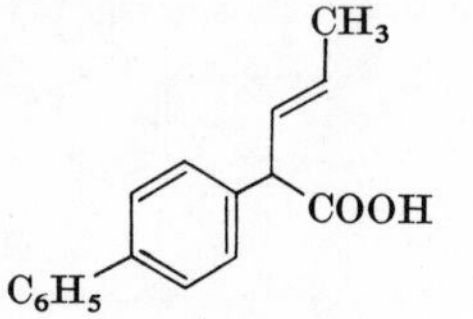   (**186**)

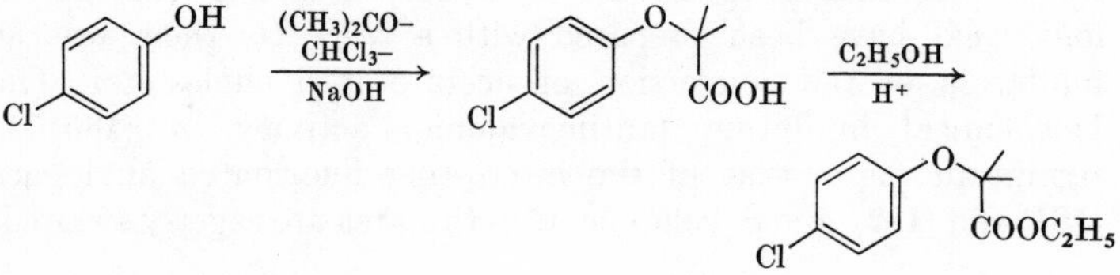

(**187**)

Hypoglycaemic Agents

Diabetic patients maintain abnormally high levels of glucose in their blood. None of the antidiabetic agents currently in use was available 20 years ago. At that time therapy was dietary or injection of insulin, a polypeptide hormone which performs several functions of which causing increased glucose utilization is of relevance here.

When a potential antityphoid compound (**188**) was reported in 1942 to produce hypoglycaemia in humans, the prospects for a new oral agent were improved, but this discovery was apparently obscured by World War II since the clinical effectiveness of the related carbutamide (**189**) was not reported until 1955. The latter is a potent but somewhat toxic compound and, of the hundreds of follow-up compounds, tolbutamide (**190**), acetohexamide (**191**), chlorpropamide (**192**) and tolazamide (**193**) are clinically in use in the U.S.A.

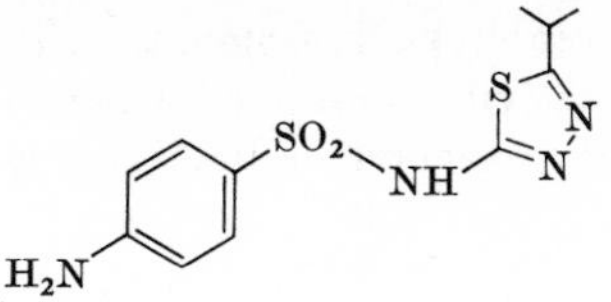

(188)

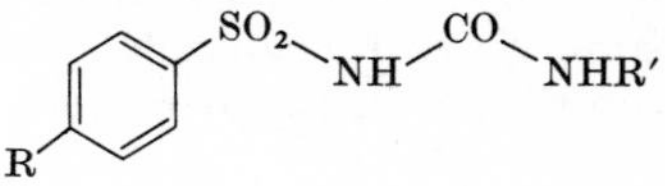

(189) R = NH₂, R′ = C₄H₉
(190) R = CH₃, R′ = C₄H₉

(191) R = CH₃CO, R′ = ⬡

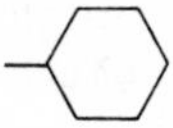

(192) R = Cl, R′ = C₃H₇

(193) R = CH₃, R′ = —N⬡

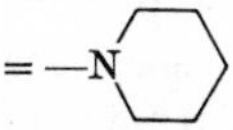

One of the earliest reports of oral hypoglycaemic activity was in guanidine (194); its high toxicity led to the development of the more acceptable synthalins A and B (195) whose effectiveness was limited by liver and kidney toxicity. Phenformin (196) was a major advance, but its cyclic analogue (197) was only weakly active. A series of quaternary pyridinium salts (198) have been

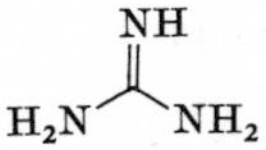

(194)

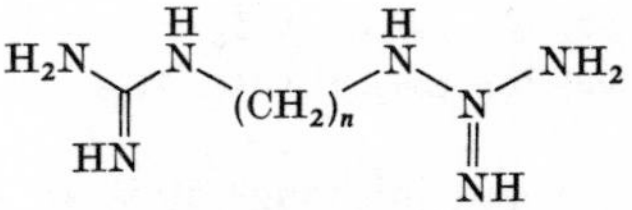

(195A) $n = 10$
(195B) $n = 12$

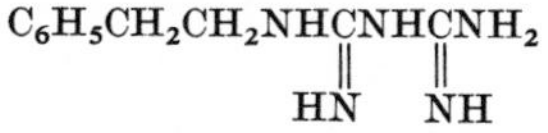

$C_6H_5CH_2CH_2NHCNHCNH_2$
  $\|$ $\|$
  HN   NH

(196)

$C_6H_5CH_2CH_2NH$ ⬡ NH_2

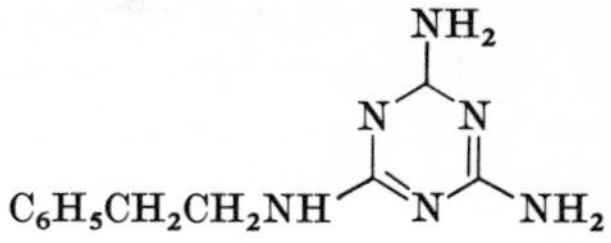

(197)

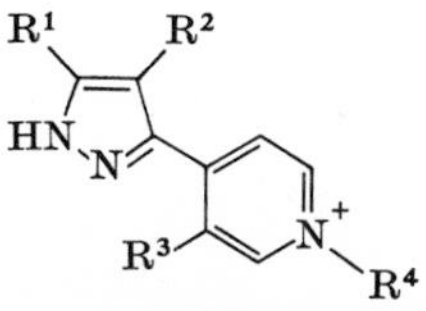

(198)

reported by the Lederle group to have oral hypoglycaemic activity and further studies have shown that activity is retained when an isoxazolyl, 1,2,4-oxadiazolyl, thiazolyl, oxazolyl or indolyl ring is substituted for the pyrazolyl unit.

Antihypertensive Drugs

High blood pressure (hypertension) is a sign of a fundamental disturbance and is in itself not a specific disease. When the disorder is due to cardiovascular, renal, neurogenic or endocrine causes, therapy is directed at the primary defect. However, in the majority of cases, the primary cause cannot be defined and it is for the treatment of these people that antihypertensive drugs are used.

Such sedatives and tranquillizers as phenobarbitone, phenothiazines and meprobamate (199) have been studied in hypertension but mental depression and other undesirable effects severely restrict their usefulness. Mebutamate (200) is a close relative of

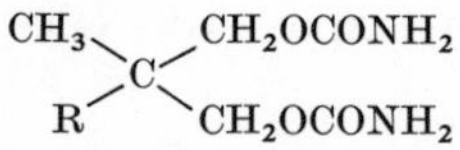

(199) R = $CH_3CH_2CH_2$—
(200) R = $CH_3CH_2CH(CH_3)$—

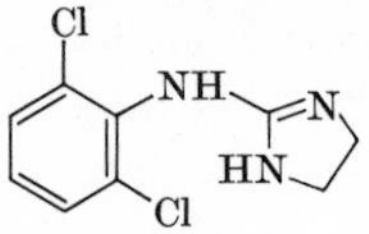

(201)

meprobamate and, although it is an effective antihypertensive agent in the dog, its usefulness in man is open to controversy. On the other hand, there is little doubt at present that clonidine (201) is a very potent compound which produces a sustained decrease in blood pressure and heart rate lasting for several days. Clonidine acts on the central nervous system and it is of interest that its effects were first noticed when it was used as an additive in shaving lotions; it has a distinct advantage in that it can be applied orally or intravenously. The effects of hydrazinophthalazines on the cardiovascular system have been known since the early 1950's; the most extensively studied compounds in this series are hydralazine (202) and dihydralazine (203), whose synthesis from

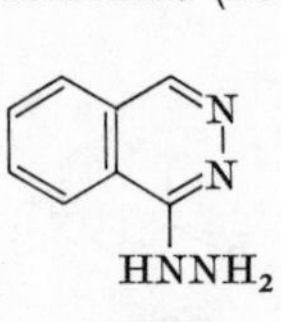

HNNH$_2$

(202)

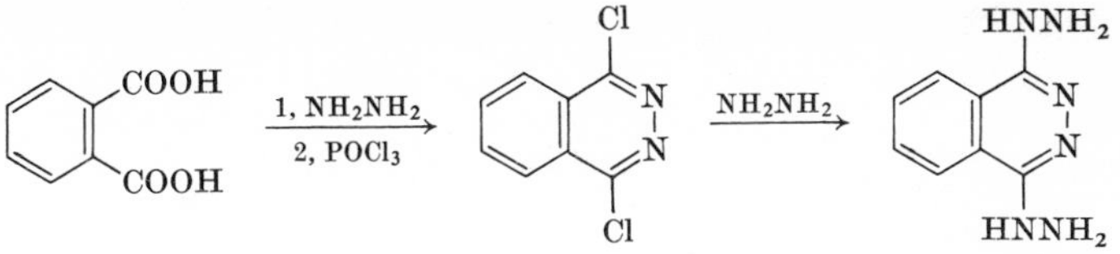

(203)

phthalic acid is straightforward as shown. The use of these compounds in man is restricted by their toxicity, erratic behaviour and some unpleasant side effects.

Decarboxylation of dihydroxyphenylalanine (dopa) by dopa-decarboxylase is inhibited by 'α-methyl dopa' (204); the latter is

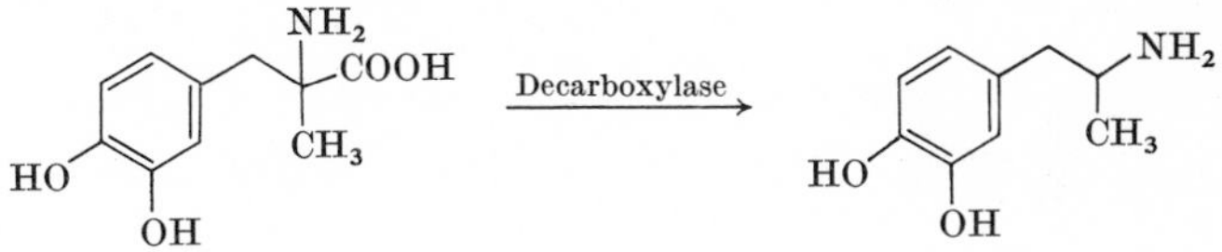

(204)

a potent antihypertensive agent in man and its effect is thought to be due to its ability to act as a substrate for the enzyme rather than an enzyme inhibitor.

A recent highly significant advance in antihypertension is the development of compounds that are able to suppress responses to certain nervous stimuli and are termed 'adrenergic blocking agents'. These compounds compete with catecholamines at their receptor sites and protect the heart against excessive stimulation. Some of these compounds are also effective against angina (wherein physical or emotional stress drives the heart excessively, so that it beats faster and more forcefully, more oxygen is required, the demand exceeds the supply and pain results) and also against heart arrhythmias (irregular heart beat) by allowing the heart to beat more slowly, less forcibly and with greater efficiency. The first of these compounds to be studied was xylocholine (205) (p. 406) which was originally synthesized as a cholinergic agent, that would show its effect by substituting for acetylcholine in transmission at nerve junctions. Xylocholine is, in addition, a potent local anaesthetic but had unacceptable side effects and pointed the way to the preparation of bretylium (206), a much improved drug whose own shortcomings were poor oral absorption and a rapid development of tolerance. Guanethidine (207) is the forerunner of a large group of compounds that show beneficial antihypertensive effects.

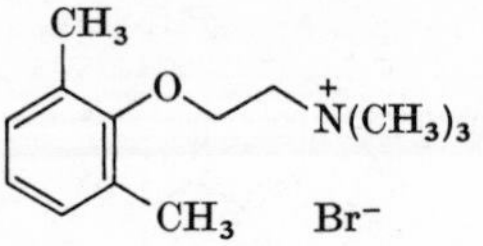

(205)

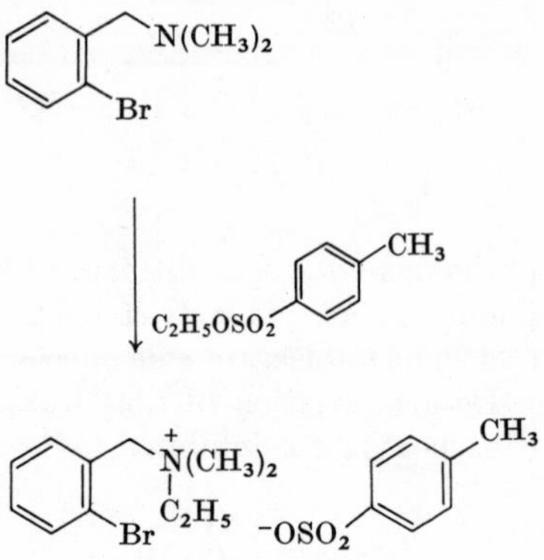

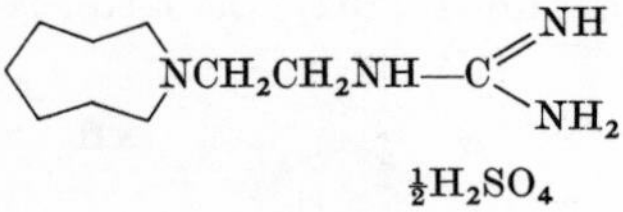

(207)   (206)

A most significant development in this area is represented by the phenethanolamines as typified by pronethalol (**208**) and the phenoxypropanolamines, e.g. propranolol (**209**).

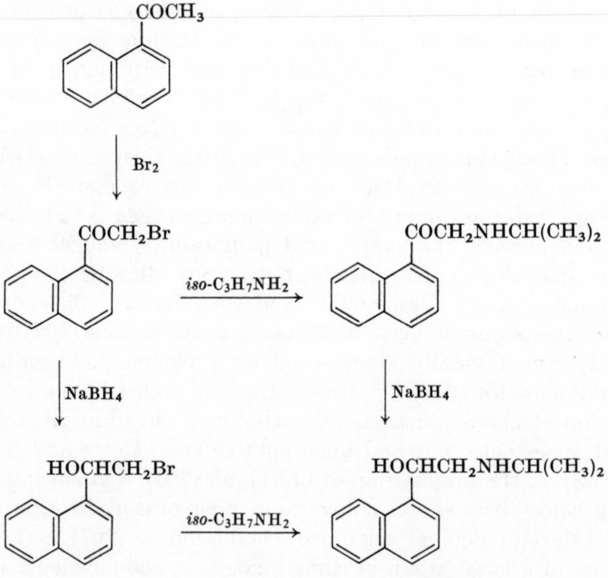

(208)

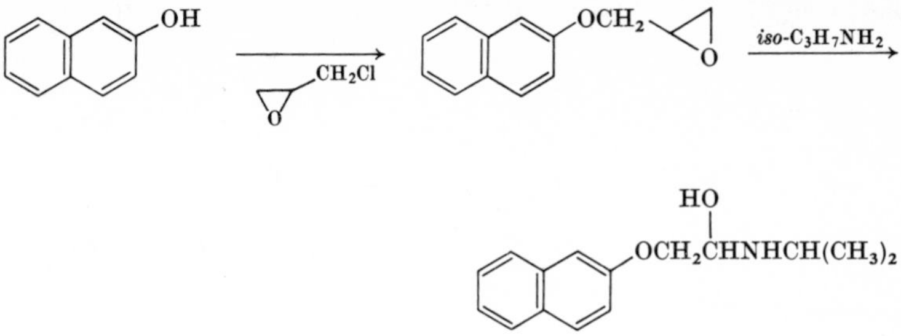

(209)

Central Nervous System Drugs

During the past 20 years the use of sleeping pills, sedatives and tranquillizers has become commonplace in the developed countries, total sales being second only to those of antibiotics. Besides these better-known medicinal agents, appetite suppressants are widely used and physicians are able to prescribe selective agents for the control of the less common but unpleasant diseases such as epilepsy and Parkinson's disease. The safe anaesthetics now available have extended the scope of surgery and increased the success rate of many operations. An important target for the pharmaceutical industry has been the development of a non-addictive analgesic to replace morphine, and to some extent this has been achieved. Despite the wide range of drugs in use there is constant search for new and improved agents, particularly for the treatment of mental illness, and the new areas being explored include memory loss. The main drugs in use in these areas are described in the following Sections.

Tranquillizers

Up to 1953 the only effective treatments available for psychotic disorders were psychotherapy (which is of limited value in excited patients), electroconvulsive therapy and brain surgery. The discovery of chlorpromazine (**211**) and related phenothiazine drugs has revolutionized the treatment of mental disease with its ability to calm excited patients and its significant effect on schizophrenia. These compounds were discovered by structural modification of the diphenylmethane antihistamines (**210**) which were known to have sedative side effects; over forty phenothiazines are currently

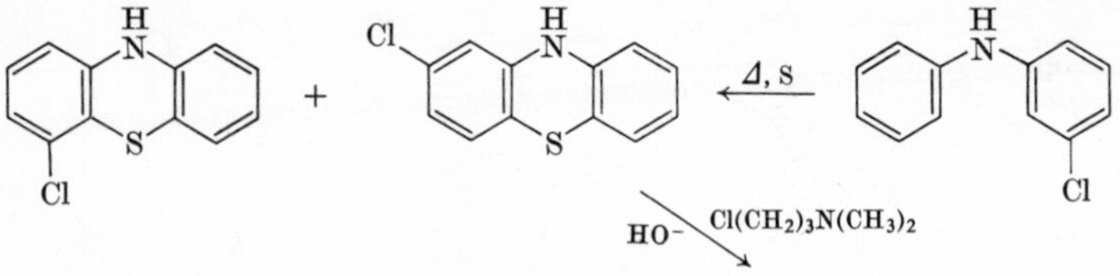

$(C_6H_5)_2CHOCH_2CH_2N(CH_3)_2$

(210)

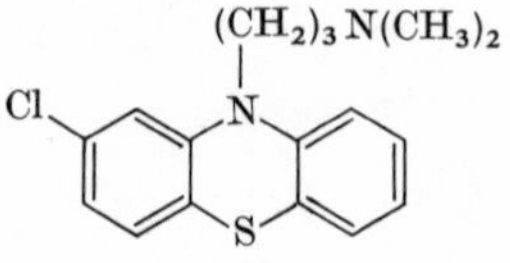

(211)

marketed. Chloropromazine (211) can be synthesized as shown, starting with an Ullmann reaction between *o*-chlorobenzoic acid and *m*-chloroaniline; the insertion of sulphur takes place mostly in the required sense. For substituted phenothiazines that are sensitive to basic conditions, the final alkylation can be achieved in other ways, e.g. by elimination of carbon dioxide from a carbamate as illustrated.

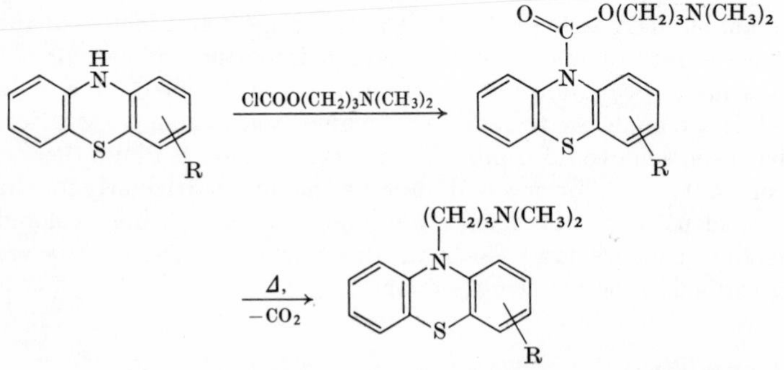

An alternative series of tranquillizers that are useful for patients who cannot tolerate the phenothiazines are the butyrophenones, notably haloperidol (212). These compounds were discovered during a research programme intended to improve the analgesic (pain-relieving) potency of a series of 4-phenylpiperidines related to the drug pethidine (213). The synthesis of haloperidol gives scope for variation of the substitution pattern in both rings and over 5000 compounds have been tested.

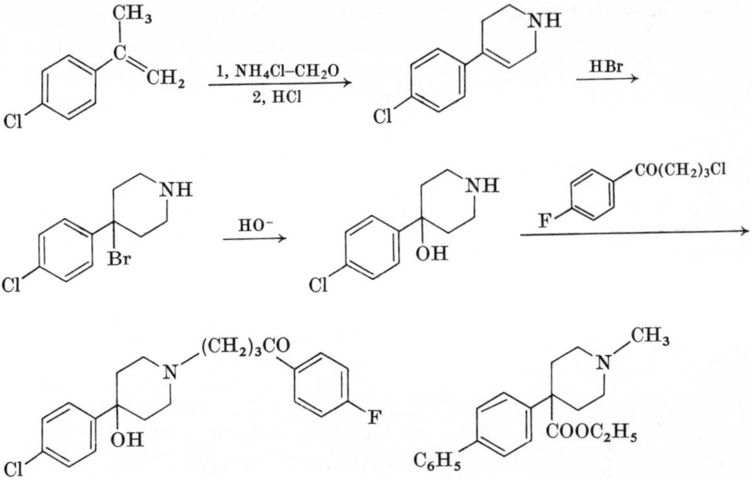

(212)   **(213)**

A further tranquillizer of considerable importance is reserpine **(214)** (Part 4, pp. 395–404) which is extracted from the Indian plant *Rauwolfia serpentina*, extracts of which had been known to have tranquillizing properties and had been used in folk medicine for over three centuries. Reserpine was initially expected to be of great importance in psychotic therapy and the chloro-analogue

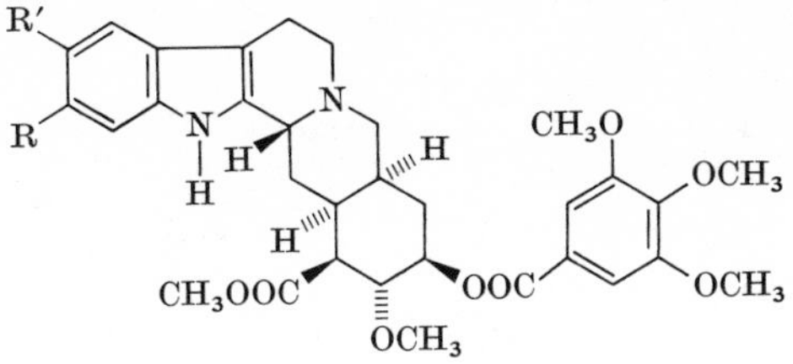

(214) R = CH$_3$O, R′ = H.   **(215)** R = H, R′ = Cl

(215) was made by total synthesis by Roussel in France. These compounds do not appear to have fulfilled their initial promise.

Antidepressants

The discovery of the phenothiazines encouraged biologists to test other compounds prepared as potential antihistamines and in this way a group of aminoalkyliminobibenzyls were found which, although having weak tranquillizing effects, were useful in the treatment of depression. The compound selected for clinical trial was imipramine (**216**). A closely related agent, amitriptyline (**218**), is also available for the treatment of depression and is interesting

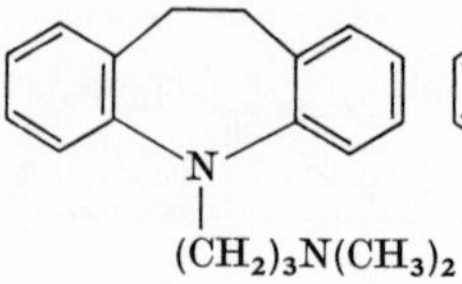

(**216**) 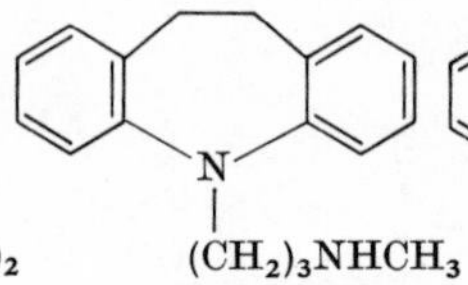   (**217**)

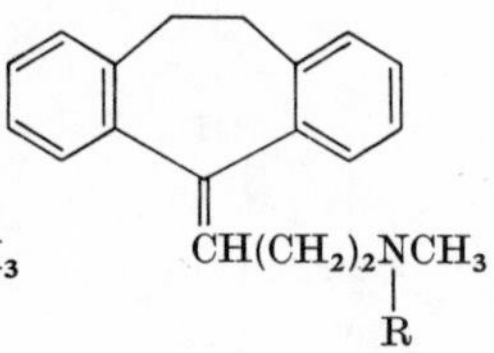

(**218**) R = CH$_3$; (**219**) R = H

since it does not carry a ring nitrogen atom. In an attempt to improve the slow onset of action of these compounds the two demethyl compounds (**217**) and (**219**) were prepared and they, too, have a place in antidepressant therapy.

Before the discovery of these tricyclic antidepressants it had been discovered that the mood of tubercular patients was improved after treatment with iproniazid (**220**). It was soon shown that this class of compound inhibited the enzyme monoamine oxidase, the system responsible for the control of amines in the brain, and a range of this type of compound was soon marketed,

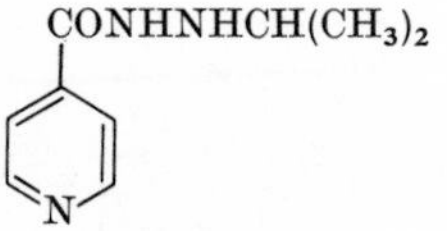

(**220**)   (**221**)

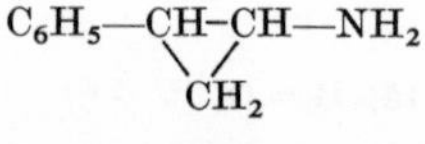

(**222**) 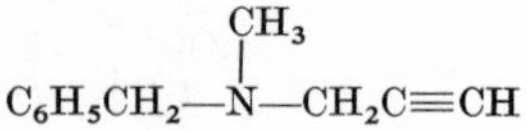   (**223**)

but they showed toxic effects. The tricyclic compounds were preferred with the result that the pyridine derivatives were withdrawn. Three compounds currently available are phenelzine (**221**), tranylcypromine (**222**) and pargyline (**223**) and they illustrate the fact that useful biological activity can be found in quite simple structures.

Analgesics

Drugs that relieve pain can be divided into classes on the basis of their biological effect: the mild analgesics, e.g. aspirin and *p*-acetamidophenol, and the potent analgesics such as morphine (**224**) (Part 4, pp. 343, 362–366, 369–372). The intensive search to obtain a drug without morphine's side effects (addiction, nausea, respiratory depression) can only be briefly discussed but the most useful compounds are described. The first of these, pethidine (**213**), has been mentioned already. It was initially claimed to be superior

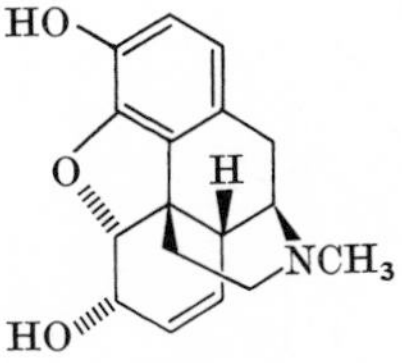

(**224**)

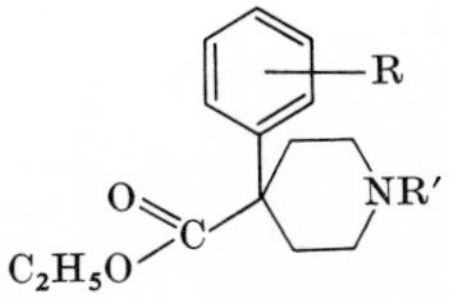

(**213**)

R = H, R′ = CH₃

to morphine but experience has shown that it has all of the disadvantages of morphine therapy. Two compounds that have a more striking resemblance to morphine have had some commercial success; phenazocine (**225**) is more potent than morphine but shares its side effects, whereas pentazocine (**226**) has the important advantage of being non-addictive whilst being less potent than morphine. The benzomorphan ring system in these compounds can be prepared from alkylpyridines (see p. 412); the cyclization gives a racemic mixture of two isomers and the analgesic activity of pentazocine rests mainly in the laevorotatory one.

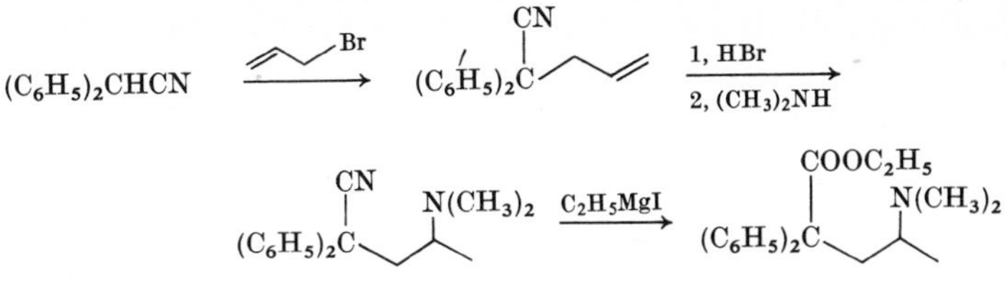

(**227**)

A further interesting analgesic, methadone (**227**), was synthesized by German chemists and came into use at the end of

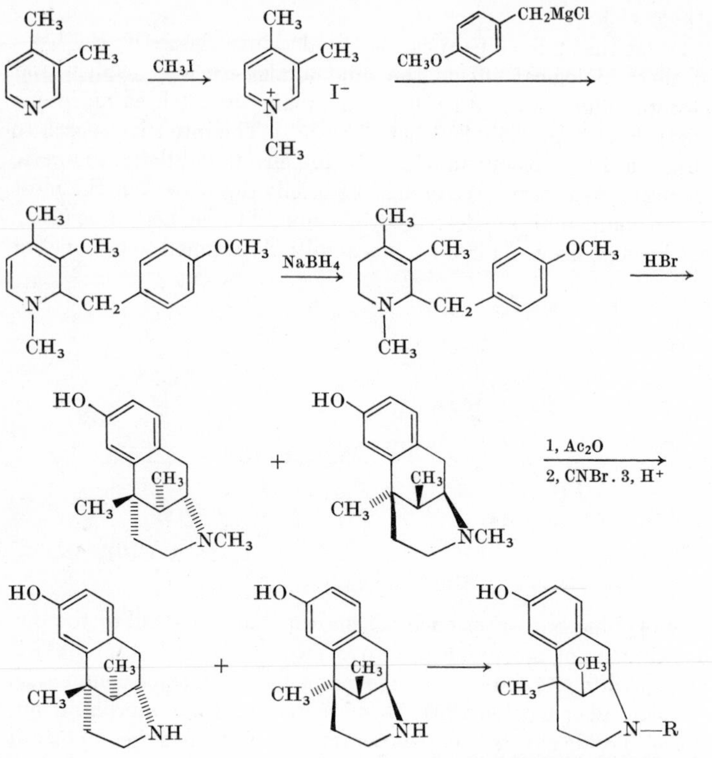

(225) R = CH₂CH₂C₆H₅
(226) R = CH₂CH=C(CH₃)₂

World War II. Despite its apparent lack of structural similarity to morphine it was derived from pethidine and has the identical biological activity spectrum of morphine. It is used as an antidote to the withdrawal symptoms exhibited by heroin addicts. One of the many syntheses of methadone is shown on the preceding page.

Antianxiety Agents

Anxiety is an emotional state which is difficult to define but is characterized by worry and tension. It can last for a long time and patients are often best treated with a mild sedative. The first of these was mephenesin (228), introduced initially as a muscle relaxant and synthesized from glycidol and o-cresol. Attempts to

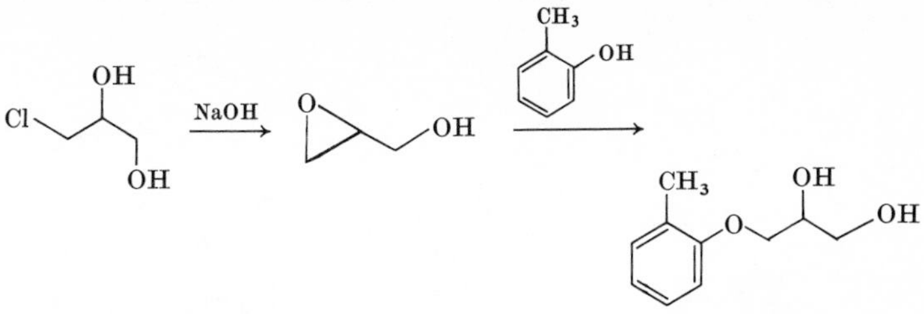

(228)

increase its duration of action led to the discovery that carbamate ester groups were a useful addition to the structure and this produced meprobamate (**199**) as a useful daytime sedative.

The benzodiazepine antianxiety agents discovered by Roche have largely replaced meprobamate. They were discovered by routine screening procedures after their preparation by an unexpected ring-enlargement (p. 414): chlorodiazepoxide (**229**) was obtained rather than the expected methylaminomethyl compound (**230**). Another useful compound of this type is diazepam (**232**), which, like (**229**), is made from the versatile 2-amino-5-chlorobenzophenone (**231**) as shown. Oxazepam (**234**) was made by Wyeth and can be prepared by Polonowski rearrangement of the oxide (**233**) or by the alternative route from (**235**) which is also obtained from (**231**). [For (**231**) see p. 414.]

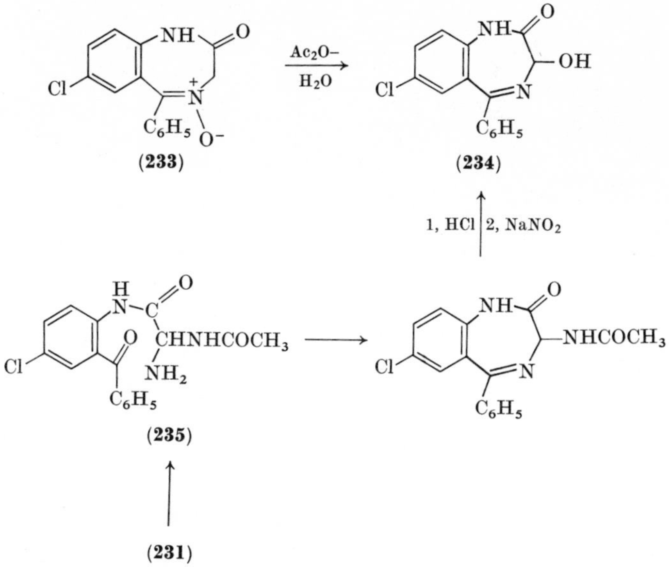

(233) **(234)**

1, HCl | 2, NaNO₂

(235)

(231)

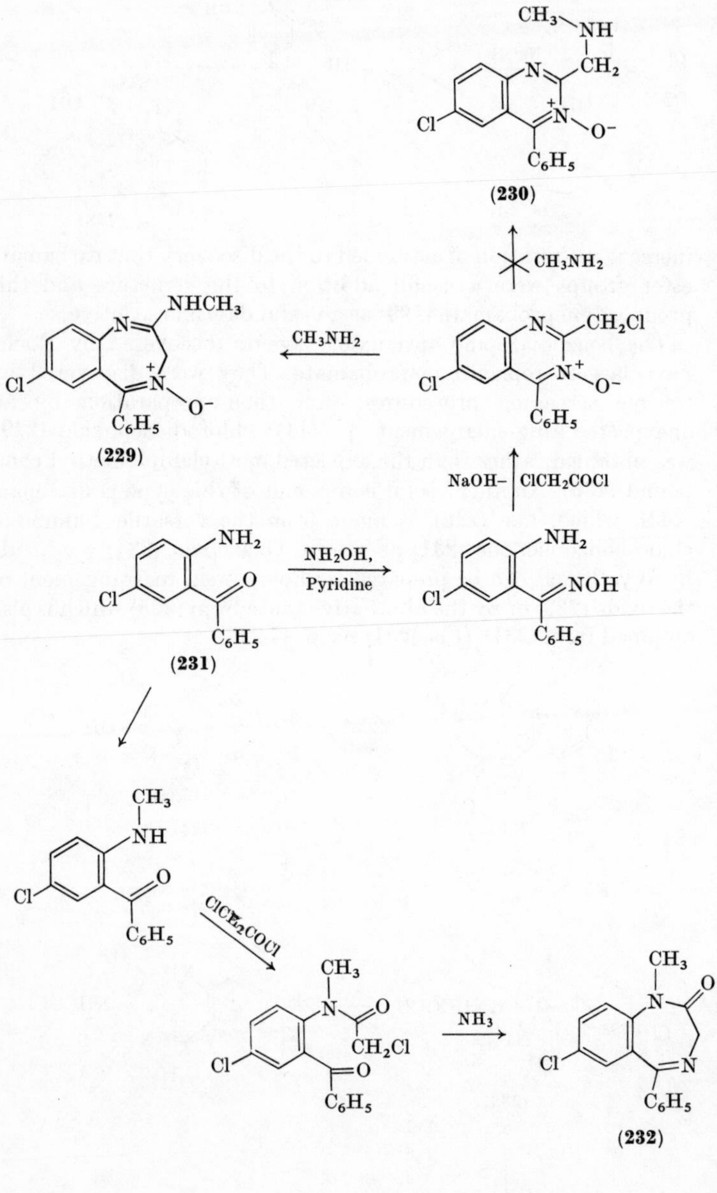

CH3
NH
N
CH2
N
O−
Cl
C6H5
(230)
CH3NH2
NHCH3
N
CH3NH2
N
CH2Cl
Cl
N
O−
Cl
N
O−
C6H5
C6H5
(229)
NaOH– ClCH2COCl
NH2
NH2
NH2OH,
Pyridine
O
Cl
Cl
NOH
C6H5
C6H5
(231)
CH3
NH
O
Cl
C6H5
ClCH2COCl
CH3
CH3
N
O
N
O
NH3
CH2Cl
Cl
O
Cl
N
C6H5
C6H5
(232)

Hypnotics and Sedatives

The barbiturates are amongst the oldest drugs in use today although their use as sedatives has been superseded by the benzodiazepines. Emil Fischer discovered the hypnotic potency of 5,5-diethylbarbituric acid (240) in 1903, having noted that the hypnotic compounds (236) and (237) both contained a tetrasubstituted carbon atom. Fischer's original synthesis enables a range of barbiturates such as phenobarbitone (241) to be prepared (cf. Part 3, p. 120).

$(CH_3)_2C(SO_2C_2H_5)_2$

(236)

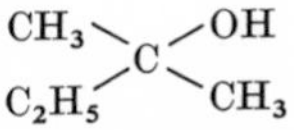

$CH_3 \diagdown \atop C_2H_5 \diagup C \diagup OH \atop \diagdown CH_3$

(237)

$Cl_3CCH(OH)_2$

(238)

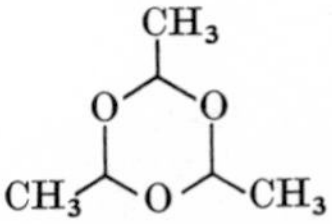

(239)

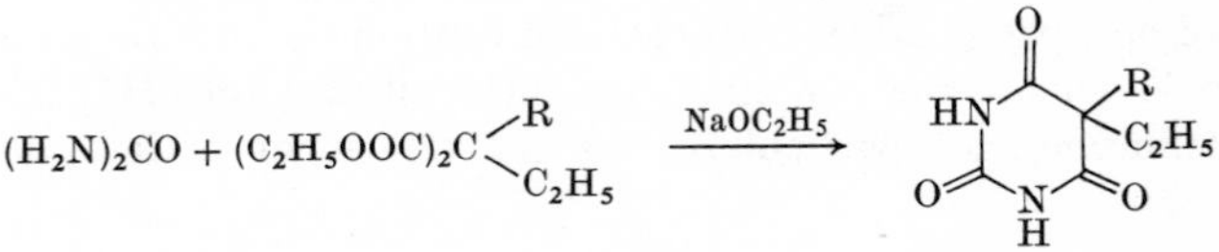

$(H_2N)_2CO + (C_2H_5OOC)_2C \diagup^R_{\diagdown C_2H_5}$ $\xrightarrow{NaOC_2H_5}$

(240) R = C_2H_5
(241) R = C_6H_5

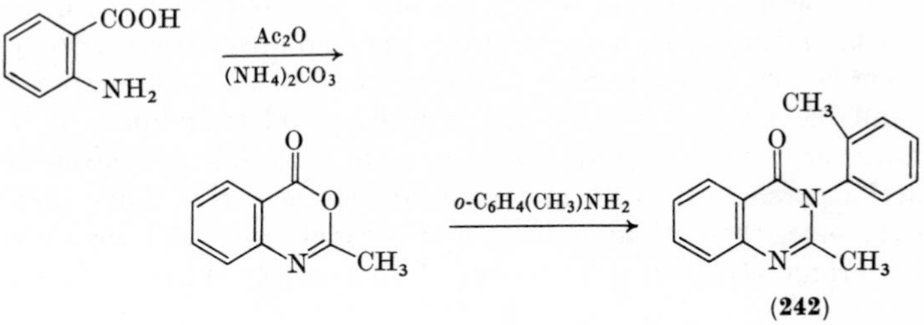

A number of other hypnotics of varying structure are known: chloral hydrate (238) and paraldehyde (239) have been in use for some time, but methaqualone (242) is a more recent discovery and is prepared from anthranilic acid and *o*-toluidine.

Anticonvulsants

Epilepsy is a collective term for a group of convulsive disorders, the most common being petit mal epilepsy which is observed as brief attacks of unconsciousness, sometimes with jerking of the body, and grand mal epilepsy which embraces major convulsions.

Phenobarbitone (**241**) is still used for the latter but a related compound, primidone (**1**), is less soporific and is superior in epilepsy. Diphenylhydantoin (**244**) is the other standard drug used for grand mal.

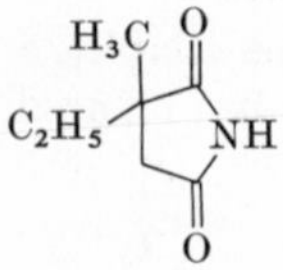

(**243**) 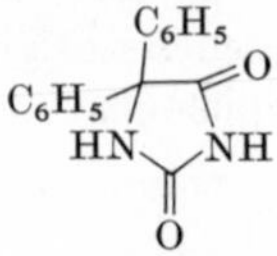   (**244**) 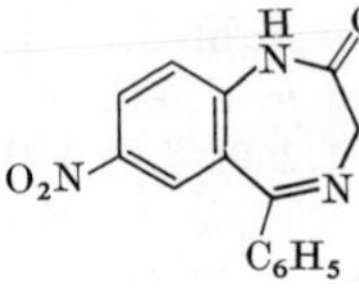   (**245**)

The most satisfactory drug for the treatment of petit mal is ethosuximide (**243**); the introduction of new agents for the treatment of epilepsy is taken with extreme caution because they are used over long periods of time. Recently nitrazepam (**245**) has become prominent, particularly for the treatment of infantile seizures.

Anaesthetics

Gaseous anaesthetics such as nitrous oxide and volatile organic compounds are used in present-day surgical anaesthesia. Diethyl ether is now used only where modern anaesthetic facilities are not available because it does not require special techniques; it is, however, flammable and there is a risk of explosion; moreover, its action is slow in onset and in disappearance. Halothane (**246**) overcomes these disadvantages and is widely used; it is prepared by bromination or chlorination of the corresponding trifluorohaloethane.

$$ClCH_2CF_3 \xrightarrow{Br_2} BrClCHCF_3 \xleftarrow{Cl_2} BrCH_2CF_3$$

(**246**)

For convenient induction of anaesthesia intravenous anaesthetics are now used and these are usually short-acting barbiturates such as thiopental (sodium) (**247**). Local anaesthetics such as cocaine (**248**) (the tropane alkaloid; Part 4, p. 349) has given way to a range of synthetic benzoic esters such as lidocaine (**249**) and procaine (**250**).

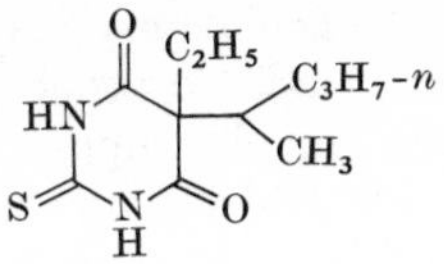

(247)

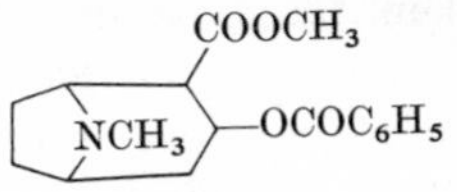

(248)

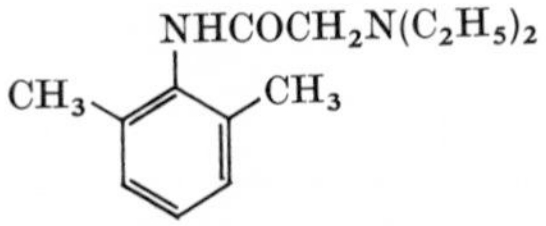

(249)

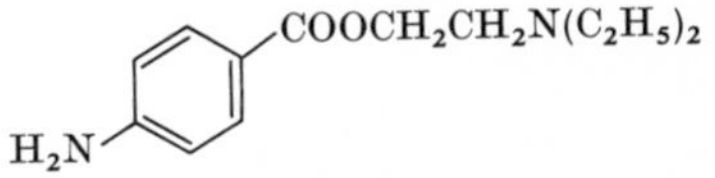

(250)

Central Nervous System Stimulants

The xanthine alkaloids found in coffee and tea, for example caffeine (251), bear a certain structural similarity to the barbiturates but exert the opposite biological effect. Caffeine (Part 3, p. 114; Part 4, p. 418), a weak stimulant, is used in proprietary headache preparations mixed with aspirin or *p*-acetamidophenol.

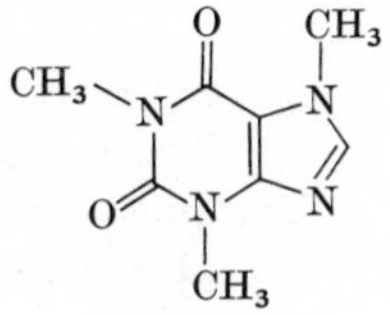

(251)

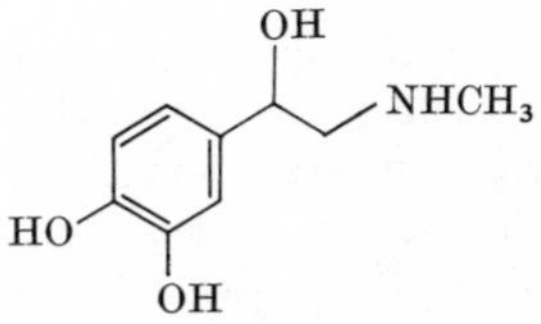

(252)

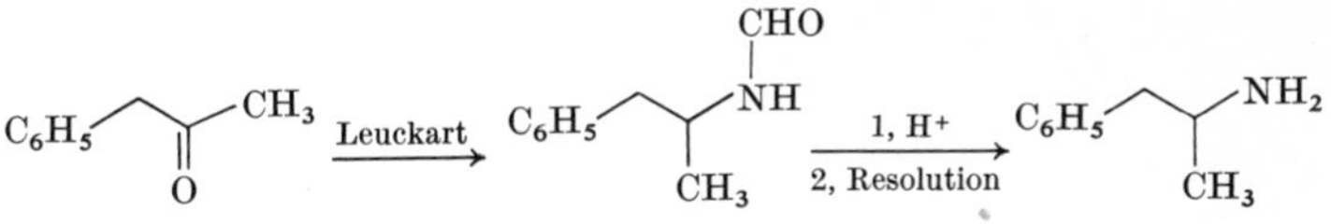

(253)

Structural modification of the adrenal hormone adrenaline (252) (Part 4, p. 127 *et seq.*) led to the discovery of amphetamine (253), a more potent stimulant. The undesirable properties of the amphetamines have, however, restricted their use to appetite suppression, and their abuse represents a serious sociological problem. Obese patients find that these compounds make a diet more tolerable and attempts have been made to reduce the stimulant properties whilst preserving the appetite suppression; examples derived from this approach are phentermine (254), diethylpropion (255) and phenmetrazine (256). A very widely used compound fenfluramine (257) is claimed not to be a stimulant and to increase the uptake of sugar by muscles.

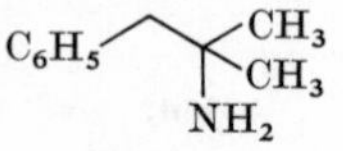

(254)

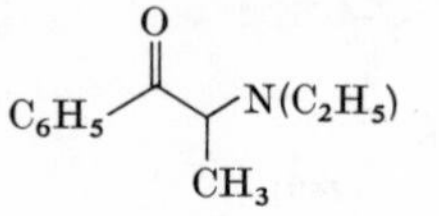

(255)

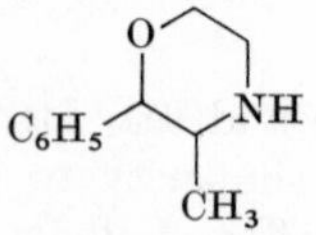

(256)

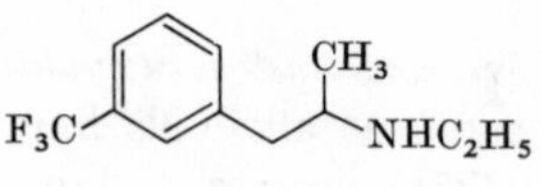

(257)

Drugs for Use against Parkinson's Disease

Some success has been achieved in the treatment of the trembling and rigidity symptoms of Parkinson's disease. A variety of agents is in use and prominent among these are trihexyphenidyl (258), which can be synthesized in one step by a Grignard reaction, and orphenadrine (259) (see facing page).

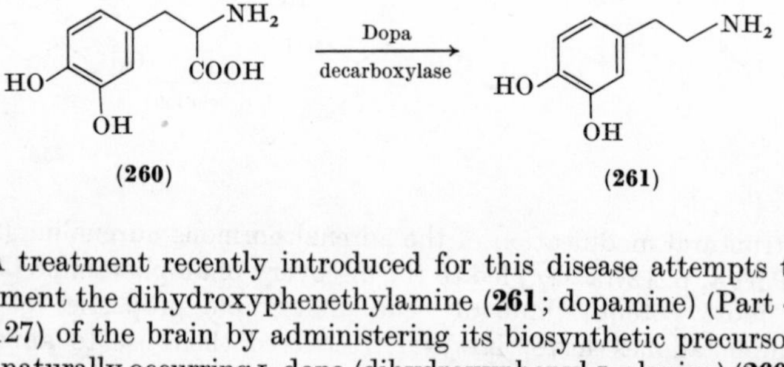

(260)

(261)

A treatment recently introduced for this disease attempts to augment the dihydroxyphenethylamine (261; dopamine) (Part 4, p. 127) of the brain by administering its biosynthetic precursor, the naturally occurring L-dopa (dihydroxyphenyl-L-alanine) (260).

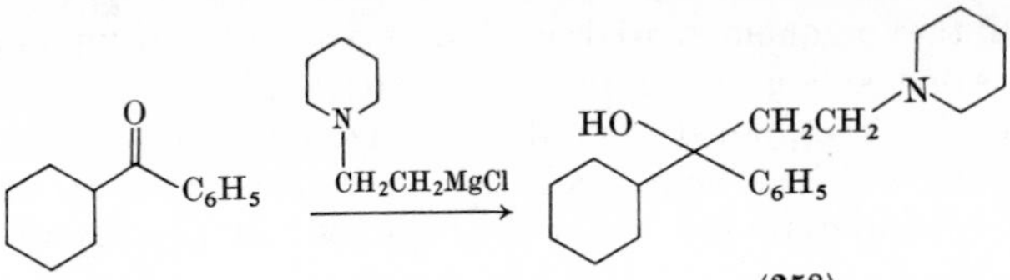

(258)

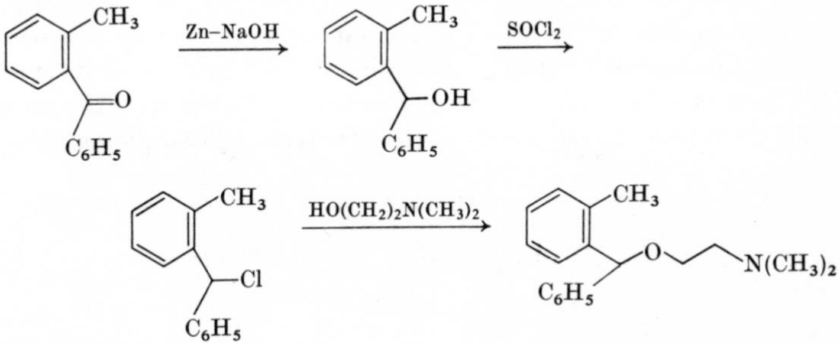

(259)

Recent work, aimed towards enhancement of activity with diminution of side effects, involves the administration of dopa with a compound that inhibits its decarboxylation in cerebral tissue but not in the central nervous system. Two such compounds (262) and (263) are under study. There is also much current interest in the use of 'α-methyldopa' (204) in Parkinson's disease.

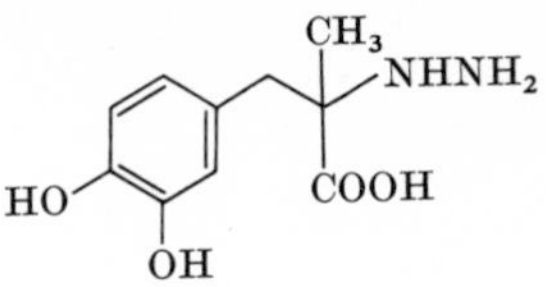

(262)

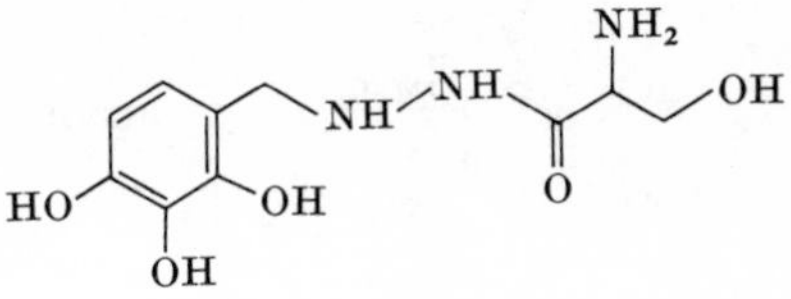

(263)

Diuretic Agents

Diuretic agents are used to increase the flow of urine; the resulting decrease in body fluid is useful in the treatment of a number of unrelated diseases, including high blood pressure and kidney malfunction.

Mercurous chloride has been known since the sixteenth century to exert a diuretic effect but it was not until 1920 that it was discovered that organomercurials used in the treatment of syphilis had the same action. Intensive investigation of this type of compound, however, failed to produce a safe, orally active drug. Similarly the xanthine bases, caffeine and theophylline, have a long history as diuretic agents but in this series high potency was not encountered.

It was the study of the side effects of the antibacterial agent sulphanilamide (**8**) which showed that diuretic activity was conferred by the sulphonamide grouping and led to the discovery of the potent orally active drugs. The first of these, acetazolamide (**264**), was prepared as one of a series of heterocyclic sulphonamides,

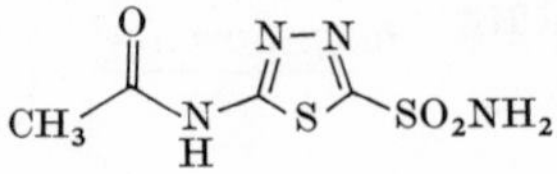

(**264**)

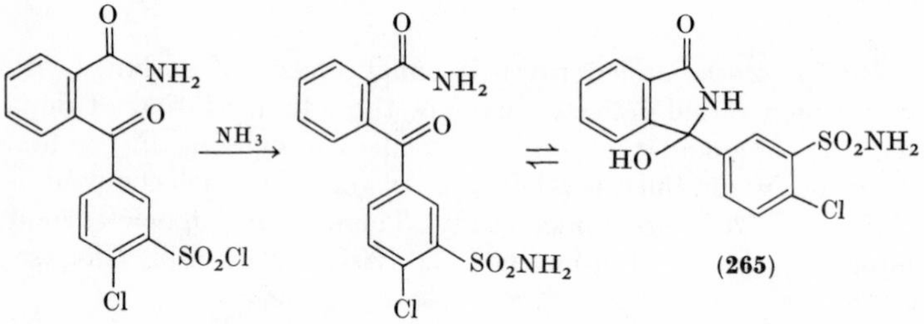

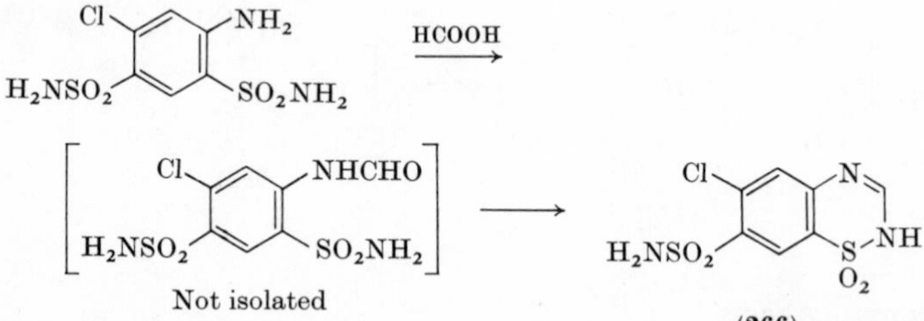

but utility in long-term treatment, such as for congestive heart failure, was limited by side effects and only moderate potency. Further investigation of the sulphonamide 'lead' brought forth chlorthalidone (**265**), a compound currently in clinical use. The observation that the presence of acylamino-groups gave diuretics with higher activity than the parent substance led Sprague to the preparation of chlorothiazide (**266**).

From a continuation of this line of investigation it was found that the hydrothiazines were another potent series of compounds, and cyclopenthiazide (**267**) is used in therapy. The most prescribed compound in the sulphonamide series is frusemide (**268**), a mono-sulphonamide.

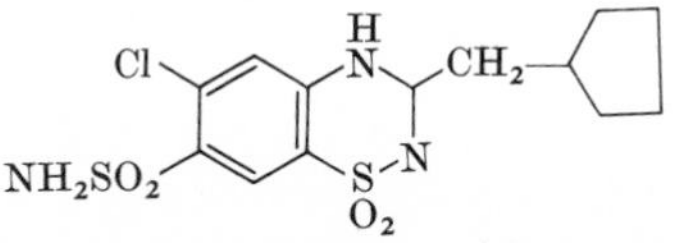

(**267**) 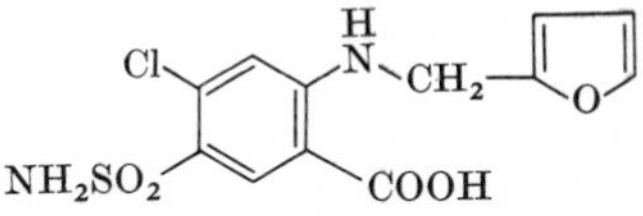   (**268**)

A completely new series of potent compounds was found as part of a search for compounds that would block thiol groups as the organomercurials do. Ethacrynic acid (**269**) rivals frusemide in potency; its synthesis demonstrates the use of the Mannich reaction to prepare an $\alpha\beta$-unsaturated ketone.

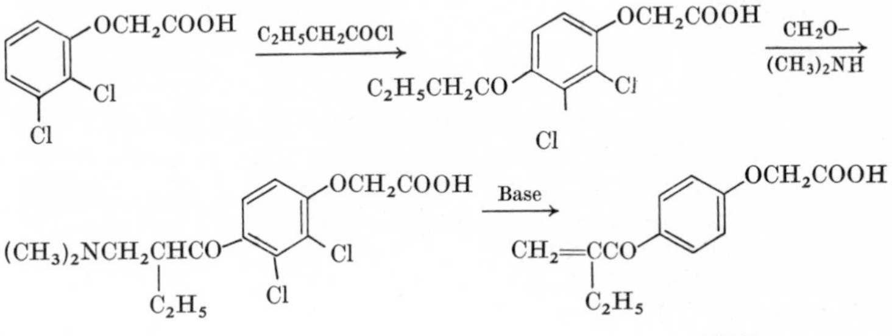

(**269**)

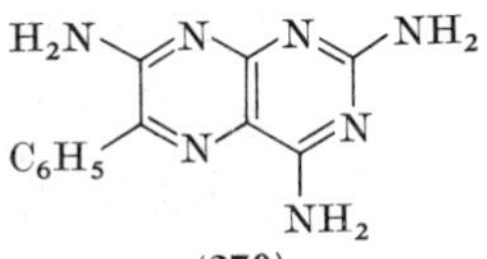

(**270**)

Routine screening of compounds by Merck, Sharpe and Dohme revealed that triamterene (**270**) was of value in combination with other diuretic agents in that it reduced loss of potassium, which

follows undesirably from administration of the sulphonamide and
thiazide diuretics.

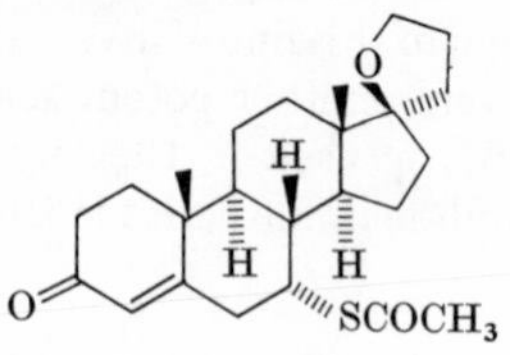

(271)

The discovery that aldosterone (Part 4, p. 262) was the hormone
controlling potassium secretion and sodium reabsorption led to
attempts to prepare compounds that would block its synthesis in
the body and possibly its action. It is believed that both these
objectives have been attained with the expensive diuretic
spironolactone (**271**).

Bibliography

A. Burger, *Medicinal Chemistry*, 3rd edn., Wiley–Interscience, New
York, 1970.

J. P. Remington, *Remington's Pharmaceutical Sciences*, 14th edn.,
Mack Publishing Co., Easton, Penn., 1970.

L. S. Goodman and A. Gilman, *The Pharmacological Basis of Thera-
peutics*, 4th edn., Macmillan, New York, 1970.

R. Slack and A. W. Nineham, *Medical and Veterinary Chemicals*,
Pergamon Press, Oxford, 1968.

S. Martindale, *The Extra Pharmacopoeia*, 26th edn., The Pharmaceutical Press, London, 1972.

The Merck Index, 8th edn., Merck and Co., Rahway, New Jersey, 1968.

A. Albert, *Selective Toxicity and Related Topics*, 4th edn., Methuen,
London, 1968.

W. A. Sexton, *Chemical Constitution and Biological Activity*, 3rd edn.,
Spon, London, 1963.

R. E. Kirk and D. F. Othmer, *Encyclopaedia of Chemical Technology*,
2nd edn., Wiley, New York, 1963–1972.

V. Drill, *Drill's Pharmacology in Medicine*, 4th edn., McGraw-Hill,
New York, 1971.

CHAPTER 12

Chemicals for Agriculture

(I. T. KAY, B. K. SNELL, and C. D. S. TOMLIN, Plant Protection
Limited, Jealott's Hill Research Station)

It is estimated that at the present time some 15,000 people die
every day from starvation and yet the World population is
increasing and is expected to double in the next 40 years. In this
situation loss of 10% of field crops and stored foodstuffs by insect
attack and poor crop yields as the result of plant disease can only
be regarded as catastrophic. Disease borne by insects (flies,
mosquitoes, etc.), often attacking those already suffering from
malnutrition, takes a further toll of mankind. The chemical
industry, besides providing the inorganic nitrogen and phosphorus
necessary to keep the land fertile, makes a big contribution to
agriculture through plant protection and pest control. The present
Chapter deals first with the latter, i.e. with insecticides. The second
Section discusses fungicides and other plant-protecting chemicals.
The third Section deals with herbicides, and the fourth deals with
plant growth regulators. Although hand and mechanical weeding
has always been an integral part of farming practice, current
economics and the growth in crop acreage have resulted in a situation where traditional methods are no longer practical. Apart from
the obvious disadvantage of crop adulteration, the growth of
weeds in cultivated crops is undesirable for a number of other
reasons. Weeds, if allowed to flourish, will compete with crops for
nutrients and for water; often they are faster growing and successfully crowd out the crop plant which is deprived of sunlight. The
feasibility of a weed-free field has made farmers more weed-
conscious and today a standard of crop purity is aimed for that
would not have been possible some decades ago. Whilst these
considerations are not so compelling to the domestic gardener, he
has not been reluctant to use the spin-off of agrochemical research

to relieve him of many of the traditional chores in the garden. Moreover, the use of selective herbicides on lawns has made possible a higher standard of lawn that was practicable before.

Insecticides

The search for synthetic insecticides begun in the 1930's requires the setting up of biological screens to evaluate insecticidal activity for many thousands of test chemicals. This in turn involves rearing insects of many species together, and their host plants, in the laboratory. The insects are then sprayed with the test chemicals at measured dose rates, generally of the order of a few hundred parts per million of active ingredient in the formulation. For a contact poison the formulation has to ensure that the test insect is thoroughly wetted. For a residual (stomach) poison the formulation must give a uniform covering of the waxy leaf surface. Even when the poison is applied directly to the insect, many factors can affect the performance: these include its rate of penetration to the site in the insect where the biochemical lesion will occur, and its rate of detoxification by enzymes during its transportation to that site. Efficiency in blocking some vital process at its site of action is essential for good activity.

The use in the field of a chemical having these favourable factors is further complicated by, for example, its stability to rain and sunlight (persistence), and its toxicity to other forms of life, including man (selectivity). Finally, the cost of the chemical is important since it must not, of course, exceed the value of the crop saved by controlling the insect pests.

As a result of these stringent requirements only one chemical in ten-thousand tested reaches a late stage in development as a marketable insecticide, and the cost of developing an insecticide today is of the order of one million pounds (sterling).

Chlorinated Hydrocarbons

Early research, involving the random screening of many thousands of synthetic compounds, established that many organic compounds containing chlorine in the molecule showed weak insecticidal activity. This work culminated in the early 1940's in the discovery of DDT (*d*ichloro*d*iphenyl*t*richloroethane). This is a powerful insecticide having activity against a broad spectrum of insects and little apparent toxicity to mammals. DDT is capable of cheap production by acid-catalyzed condensation of chlorobenzene and chloral (Part 2, p. 213):

DDT

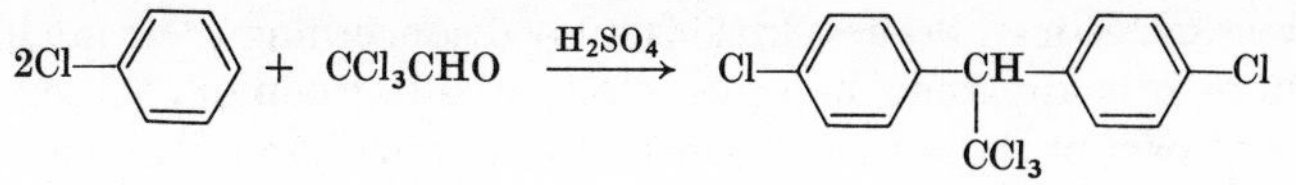

After World War II, and encouraged by the commercial success of DDT, a spate of industrial research led to the development of a whole series of chlorinated hydrocarbon insecticides including BHC (hexachlorocyclohexane) and the 'cyclodiene' group of

BHC

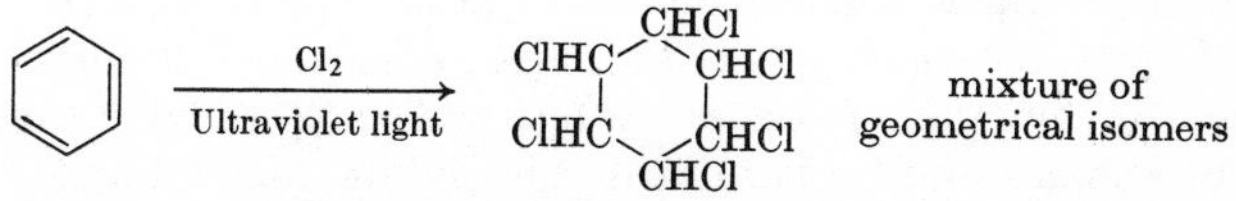

insecticides derived by Diels–Alder addition of various dieneophiles to hexachlorocyclopentadiene, including aldrin, dieldrin and endosulfan.

Aldrin Dieldrin

Endosulfan

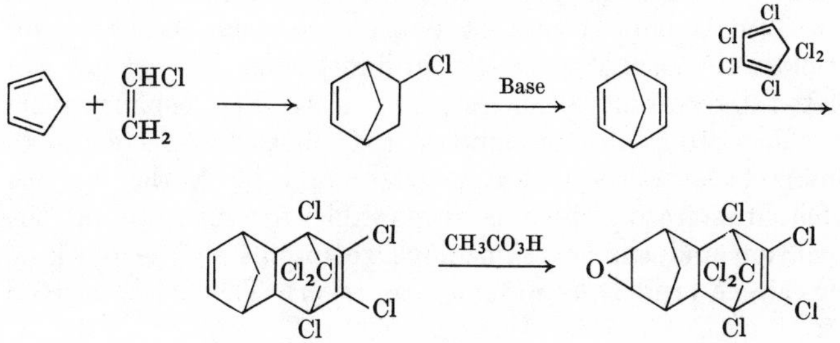

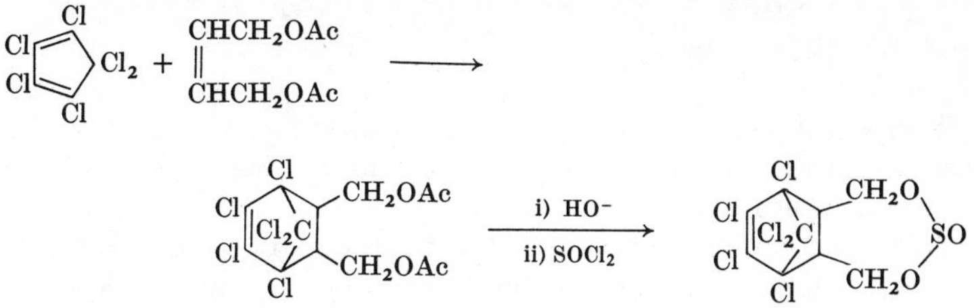

Except that these compounds owe their insecticidal activity to their ability to interfere with the transmission of nerve impulses in insects, possibly by upsetting sodium or potassium ion balance across nerve membranes, little is known concerning their mode of action. It is apparent, however, that for this group of compounds no one part of the molecule constitutes a lethal grouping of atoms or toxophore. Rather, the structure of the molecule as a whole, and its conformation, determines its toxicity—as if it were a key unlocking some Pandora's Box of biochemical disorders.

The hopes that these inexpensive chemicals would offer a panacea to the problems of insect control were dashed by the growing realization that the factors making them so desirable as insecticides made them undesirable on ecological grounds. Their great persistence, biological availability and fat-solubility cause them to accumulate in the ecosystem and to be stored in the body fat of animals and birds towards the end of the food chain. This property has culminated (1970) in the banning of the use of many of these insecticides, particularly for situations where alternative chemicals can be used.

Organophosphorus Esters

Almost simultaneously with the discovery of DDT, it was found that certain organic esters of phosphoric acid had powerful insecticidal activity. Within certain limitations this activity can be related to the ability of the esters to behave as phosphorylating agents. This group of compounds is toxic both to insects and to mammals because they phosphorylate and block the enzyme acetylcholinesterase, which is responsible for maintaining the organization and transmission of nerve impulses. The blocking action can be crudely represented as in (*a*). The anion derived

$$\text{enzyme—CH}_2\text{O}^- + \overset{\overset{\displaystyle O}{\|}}{\underset{\underset{\displaystyle OR}{|}}{P}}\diagdown_L^{OR} \longrightarrow \text{enzyme—CH}_2\text{O}\overset{\overset{\displaystyle O}{\|}}{\underset{\underset{\displaystyle OR}{|}}{P}}\diagdown OR + L^- \quad (a)$$

from a primary hydroxyl group attached to the enzyme behaves as a nucleophile towards the phosphorus ester, displacing a leaving group L.

One of the first compounds of this group to be used widely as an insecticide was parathion. Here the resonance-stabilized *p*-nitrophenoxide anion provides a suitable leaving group, thus conferring acylating properties upon the ester. Unlike the chlorinated hydrocarbons, therefore, these insecticides have a part of the molecule that is essential to their activity (the toxophore) with another part of the molecule L that has the correct

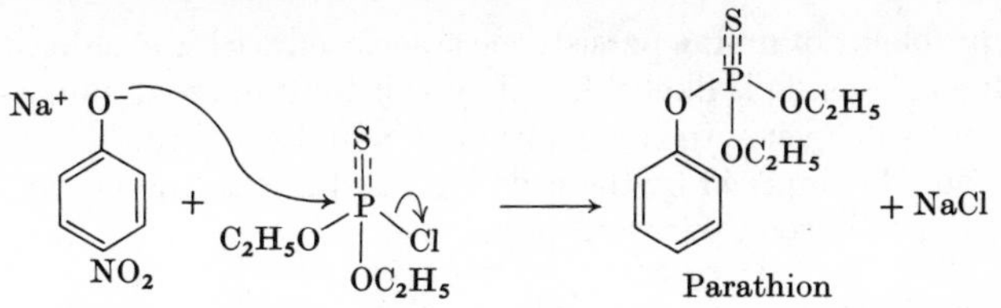

Parathion

electronic properties to act as a carrier and that is capable of wide variation. The scope of the variation is further illustrated by the insecticides dichlorvos, malathion and diazinon. Early examples

$$CCl_3CHO + (CH_3O)_3P \xrightarrow{\text{Heat}} CCl_2{=}CHOP(OCH_3)_2 + CH_3Cl$$

(cf. Part 3, p. 241)   Dichlorvos

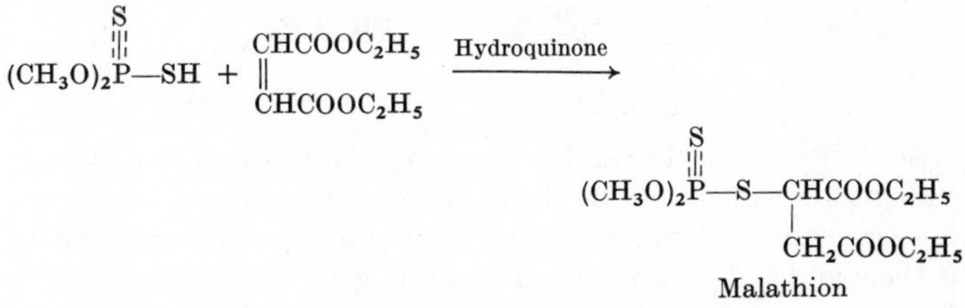

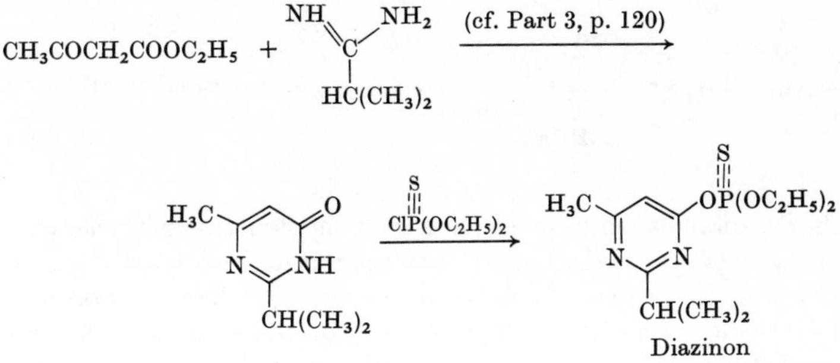

Malathion

Diazinon

of organophosphorus insecticides, e.g. parathion, were highly toxic to mammals as well as to insects. However, some of the more recently discovered members of this class, e.g. malathion, have little toxicity to mammals, although in many cases the reasons for this selectivity are not understood.

The organophosphorus insecticides, being generally biodegradable by hydrolysis of the phosphate ester toxophore, pose few of the problems of undue persistence associated with the chlorinated hydrocarbons. It is likely that they will continue to dominate the field for some years to come, although a limitation to their usefulness may be imposed by the growing resistance of insects to them that has followed their widespread use.

Carbamates

Insecticides of this group were developed from the 1950's onwards, the lead for work in this area being generated from consideration of the mode of action of the toxic alkaloid

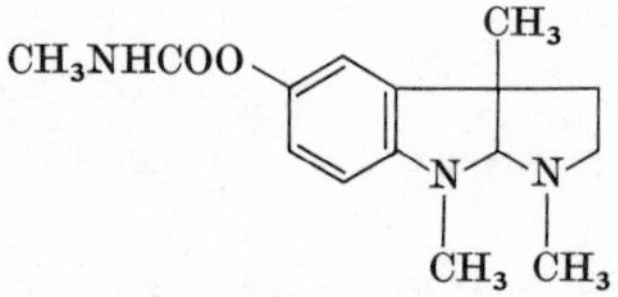

(1)

physostigmine (1). Research with physostigmine and its synthetic analogues established that, like the organophosphorus compounds, these carbamic esters are inhibitors of acetylcholinesterase owing to their ability to transfer a carbamoyl group (the toxophore) to the active site of the enzyme, illustrated in (b). Unlike the case of

the organophosphorus esters, however, the structure of the leaving group RO^- is critical in determining insecticidal activity. In practice, three main types of carbamic ester have emerged as insecticides of economic importance: phenol carbamates, heteroaromatic carbamates, and oxime carbamates. These are exemplified by carbaryl, pirimicarb and aldicarb.

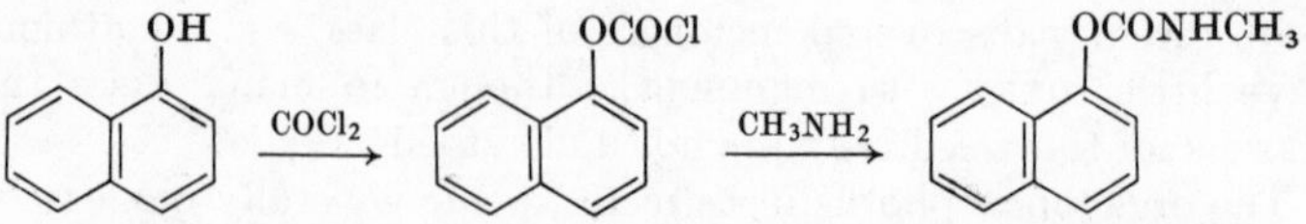

Carbaryl

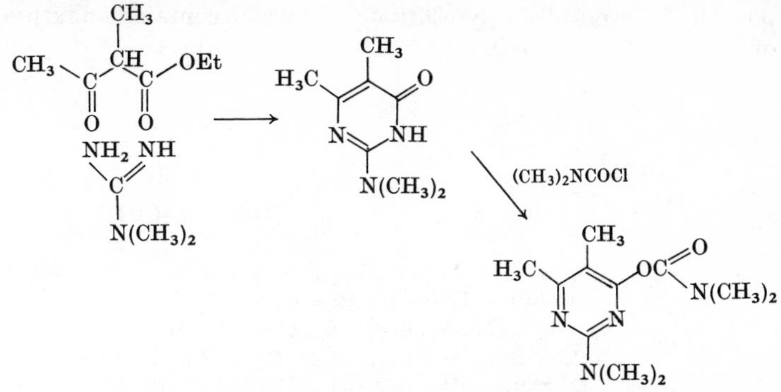

Pirimicarb

$(CH_3)_2C=CH_2 \xrightarrow{NOCl} ClC(CH_3)_2CH=NOH \xrightarrow{CH_3SNa}$

$CH_3SC(CH_3)_2CH=NOH \xrightarrow{COCl_2}$

$CH_3SC(CH_3)_2CH=NOCOCl \xrightarrow{CH_3NH_2} CH_3SC(CH_3)_2CH=NCONHCH_3$

Aldicarb

One of the problems associated with carbamates is that many insecticides of this group tend also to be highly toxic to mammals. Nevertheless, certain of these compounds have been developed commercially since they can offer advantages over other insecticides, as, for example, in the control of soil nematodes (microscopic eel-worms that attack the developing roots of plants) by aldicarb.

Natural Products as Insecticides

A number of natural products obtained as plant extracts have found some use as insecticides, and these include nicotine, derris (containing rotenone; cf. Part 3, p. 87), and the extracts of the

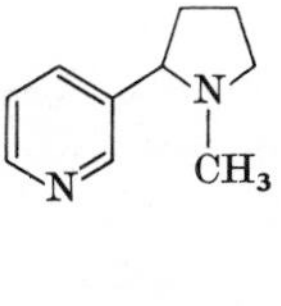

Nicotine 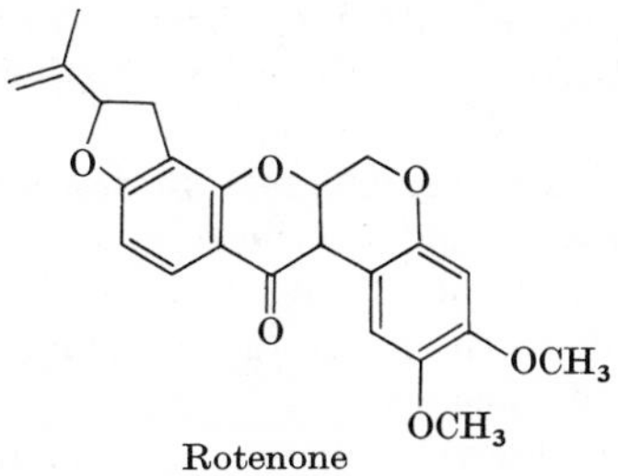   Rotenone

powdered flowers of the pyrethrum plant which contains a mixture of pyrethrins and cinerins.

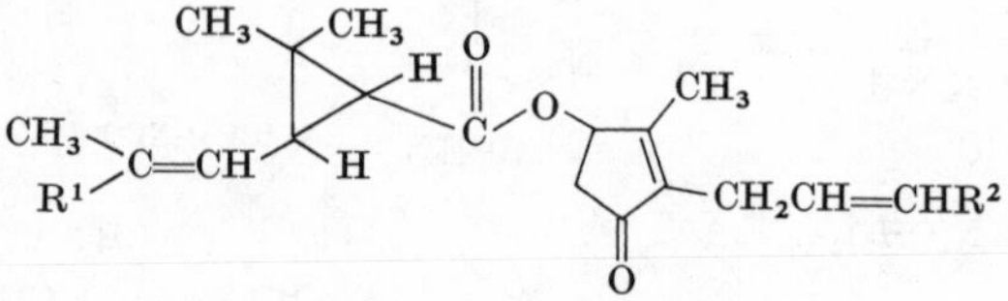

Pyrethrin I: $R^1 = CH_3$, $R^2 = CH{=}CH_2$.
Pyrethrin II: $R^1 = COOCH_3$, $R^2 = CH{=}CH_2$.
Cinerin I: $R^1 = CH_3$, $R^2 = CH_3$.
Cinerin II: $R^1 = COOCH_3$, $R^2 = CH_3$.

Of greatest commercial importance are the pyrethrum extracts which, because of their spectacular 'knock-down' action on flies and their low toxicity to mammals, have found their principal outlet as aerosols for domestic use. Their use against crop pests is restricted by their lack of persistence in the field and by their high cost relative to other insecticides.

Future Trends

Future trends in insect control will probably arise from a more profound interpretation of the fact that what the chemical industry sells to the farmer or grower is the *effect* of insect-free crops. Thus it may not be necessary to use chemicals that have a direct toxic action upon the pests. Chemicals that, for example, deter insects from feeding upon the crop (repellants or anti-feedants), attract the insect away from the crop (sex- or feeding-attractants), or prevent the insects from breeding large populations (chemicosterilants) may be equally effective. Indeed there is good reason to believe that the reduced disturbance to the ecology arising from more selective techniques would provide other, hidden, advantages: natural insect predators of the pest species could be permitted to survive, whereas the use of the present broad-spectrum insecticides with their 'over-kill' tends to eliminate these and other beneficial insects too.

In this connexion great interest has recently been generated by the isolation and identification of an insect hormone that regulates insect metamorphosis. Treating insects with an excess of the 'juvenile hormone' at an early stage in their development distorts their life history and they remain in a juvenile (larval) form instead of developing via pupation into adults. Despite a demanding isolation procedure (from the *Cecropia* moth) with its attendant bioassay, and the subsequent isolation of only a few milligrams of the juvenile hormone, its structure was established as methyl 10,11- epoxy- 7- ethyl - 3,11- dimethyltrideca - 2,6 - dienoate (**2**) (cf. Part 4, p. 243).

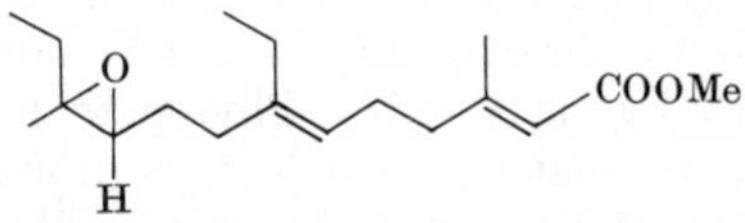

(**2**)

Such has been the interest generated by this hormone that in the four years elapsed since the elucidation of its structure no less than fifteen synthetic routes to it have been reported. Many of these are extremely elegant and employ reactions of high stereoselectivity and stereospecificity.

Whether or not compounds showing these types of biological activity can be used as practical means of insect control remains to be seen, but certainly at this time over twenty major chemical Companies are actively engaged in research on the biological testing and synthesis of analogues of the hormones.

Fungicides

The devastating effects of plant disease have been recorded since the early days of civilization. In the nineteenth century, two major calamities showed the need to control epidemics of plant disease. One, the Irish potato famine of 1845, described recently by Cecil Woodham-Smith in *The Great Hunger*, was caused by the potato blight fungus, *Phtophthora infestans*. The other, the total destruction of the coffee crop in Ceylon, was caused by the coffee leaf rust fungus, *Hemileia vastatrix*.

The scientific study of plant disease began in the last century with the pioneer work of Prevost, De Bary, Pasteur and others. The role played by fungi, bacteria and viruses in causing plant disease has become recognized over the past 150 years.

Many plant diseases can be held in check by cultural practices. These methods include crop rotation, selection of suitable sites, destruction of weeds which harbour parasites, and removal of diseased plant parts. The production of disease-resistant varieties by the plant breeder is an ideal control method in theory, but not many varieties remain disease-resistant because of evolution or selection of new virulent types of the pathogen. If these methods fail to control disease, crop protection chemicals are used.

Surface Fungicides

A definition of a fungicide as 'A chemical or physical agent that kills, or inhibits, the development of fungus spores or mycelium' was given in 1943 by the American Phytopathological Society.

The majority of fungicides that have been used commercially are *surface fungicides*. By correct placement of the chemical on the seed or foliage, a protective cover is formed that prevents or retards the growth of the invading fungus.

Inorganic fungicides will be described briefly first because of their historical importance. Copper-containing fungicides include Bordeaux mixture, a complex of copper sulphate ($CuSO_4$) and slaked lime [$Ca(OH)_2$], which was used by Millardet in 1885 to control vine downy mildew. It has been used widely ever since, though less corrosive substitutes such as copper oxychloride [$3Cu(OH)_2 \cdot CuCl_2$] have been found. Elemental sulphur has been used for over 100 years to control powdery mildew fungi. Fungicidal polysulphide derivatives include 'lime sulphur', prepared by boiling a suspension of lime with flowers of sulphur. Lime sulphur, still commonly used, gives reasonably effective disease control. Sulphur products can, however, cause plant damage, some apple and pear varieties being particularly susceptible.

Mercury compounds include mercurous chloride (calomel, Hg_2Cl_2), used as a root dip to control clubroot of brassicas (*Plasmodiophora brassicae*); and a wide variety of organomercurial compounds, R—Hg—X, where R is alkyl or aryl, and X is an organic or inorganic acid group (e.g. $C_6H_5HgOCOCH_3$), is widely used as cereal seed dressing. These are cheap and give extremely good control of 'bunt' and 'covered smuts' of cereals, diseases which once were common in the United Kingdom. Very low rates of application, about 2 g of mercury per acre, are used. Most mercurial preparations are toxic to mammals and birds. It is almost certain, therefore, that their use will cease provided that effective, less toxic, substitutes can be found.

Derivatives of dithiocarbamic acid (NH_2CSSH) were discovered just before World War II to be important organic fungicides (they were originally prepared as 'accelerators' for the vulcanization of rubber). An exhaustive study of their potential as agricultural fungicides has been made, but mention will be made here of only two of the important groups. A typical example, thiram (tetramethylthiuram disulphide), is prepared from simple precursors as shown in (c). The fungicide nabam is prepared from

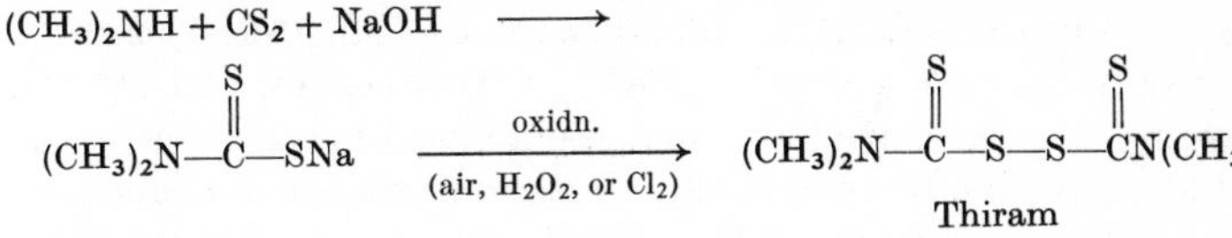

$$(CH_3)_2NH + CS_2 + NaOH \longrightarrow \tag{c}$$

1,2-diaminoethane; it is oxidized in dilute solution by exposure to air, and is therefore converted into either the more stable zinc (zineb) or manganese (maneb) derivative.

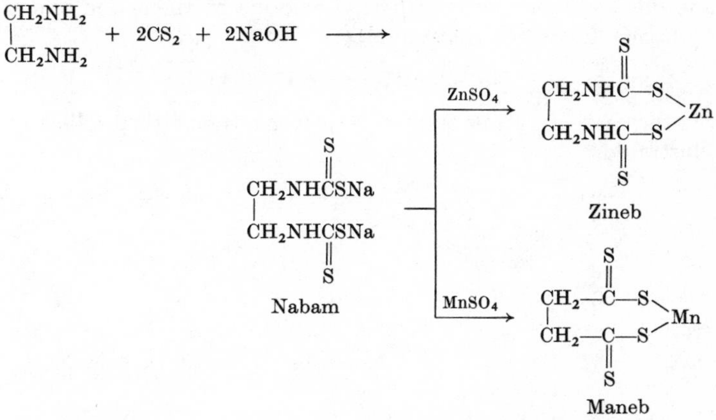

The dithiocarbamates are cheap and effective fungicides and are still popular. Thiram is used, for example, to control grey mould (*Botrytis cinerea*) on strawberries and lettuces; maneb and zineb to control potato blight.

Many compounds containing the trichloromethylsulphenamide ($>$N—S—CCl$_3$) group are potent fungicides. It has been suggested that their fungitoxicity depends on direct reaction with thiol groups present in fungi:

$$>NSCCl_3 + 2RSH \longrightarrow \ >NH + RSSR + CSCl_2 + HCl$$

or that thiophosgene liberated in this reaction is itself fungitoxic. This is an oversimplified picture, of course, and precise details of the mode of action are difficult to obtain. However, the recognition that the NSCCl$_3$ group was responsible for the fungicidal activity of captan (it was suggested at one time that the imide was the toxic portion) led to the development of a number of related compounds, such as folicid, captan and dichlofluanid. These compounds are effective, broad-spectrum fungicides used to control apple scab and grey mould of strawberries, and as seed dressings for the control of many seed and soil-borne diseases.

Sulphenyl halides used in the synthesis of these compounds are obtained simply from carbon disulphide, e.g. trichloromethyl-sulphenyl chloride by use of chlorine:

$$CS_2 + 3Cl_2 \xrightarrow{\;I_2\;} Cl_3CSCl + SCl_2$$

and dichlorofluoromethylsulphenyl chloride by the action of dry hydrogen fluoride on that product:

$$Cl_3CSCl + HF \longrightarrow FCl_2CSCl$$

Syntheses of the three fungicides just mentioned then follow as illustrated.

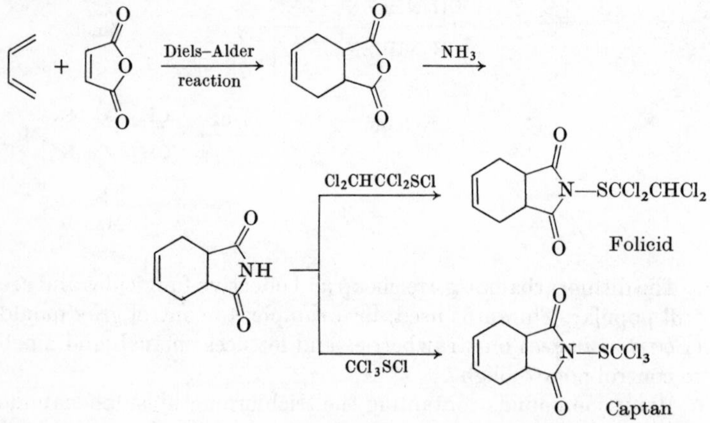

$$(CH_3)_2NH + SO_2Cl_2 \longrightarrow (CH_3)_2NSO_2 \cdot Cl \xrightarrow[\text{NaOH}]{C_6H_5NH_2}$$

$$(CH_3)_2NSO_2NHC_6H_5 \xrightarrow{ClSCFCl_2} (CH_3)_2NSO_2N(C_6H_5)SCFCl_2$$

$$\text{Dichlofluanid}$$

Guanidine derivatives are also important and *n*-dodecyl-guanidine acetate (dodine) is a cationic wetting agent used to control apple scab. It is prepared by a conventional guanidine synthesis from dodecylamine and cyanamide. n-Dodecylguanidine is also used as its alkyl sulphate salts, which are prepared by reaction of n-dodecylamine with an S-alkylisothiouronium alkyl sulphate.

$$n\text{-}C_{12}H_{25}NH_2 + NH_2CN + CH_3COOH \longrightarrow$$
$$[n\text{-}C_{12}H_{25}NHC(NH_2){=}NH_2]^+ \; CH_3COO^-$$
$$\text{Dodine}$$

$$NH_2CSNH_2 + SO_2(OR)_2 \longrightarrow$$
$$[RSC(NH_2){=}NH_2]^+ \; ROSO_2O^- \xrightarrow[-RSH]{n\text{-}C_{12}H_{25}NH_2}$$
$$[C_{12}H_{25}NHCH(NH_2){=}NH_2]^+ \; ROSO_2O^-$$
$$\text{Dodine alkyl sulphate}$$

Systemic Fungicides

Surface fungicides do not eradicate established infections, nor do they enter the plant. They do not, therefore, protect new growth from fungal attack, and as surface deposits they are subject to weathering. Repeated applications of fungicide are necessary to maintain disease control throughout the growing season, and this is uneconomic in labour, machinery, and chemical. Furthermore, pathogens such as *Verticillium* wilts, which invade the vascular materials within the plant, cannot be reached or controlled by surface fungicides. These limitations of surface fungicides showed the need for systemic fungicides, namely compounds that enter and move freely within the plant to eradicate established infections and protect the plant, including new growth, from invading pathogens. Although the search for systemic fungicides began in the 1940's, only within the last few years have a number of such compounds been developed by chemical Companies in the United Kingdom, the United States, Germany and Japan.

In 1966 it was found that the 1,4-oxathiin derivative carboxin gave excellent control of loose smut of barley (*Ustilago nuda*). This fungus, being within the seed, cannot be controlled satisfactorily by the surface-applied organomercurial fungicides. The sulphone analogue oxycarboxin is effective against rust on cereals and ornamentals. The systemic properties of these compounds have been demonstrated, for instance, by control of rust disease on leaves of bean plants following soil treatment. One synthesis of carboxin (**3**) is formulated below and peracid is used to convert it into oxycarboxin (**4**) (oxidation of sulphide to sulphone).

$$CH_3COCH_2CONHC_6H_5 + SO_2Cl_2 \longrightarrow CH_3COCHClCONHC_6H_5$$

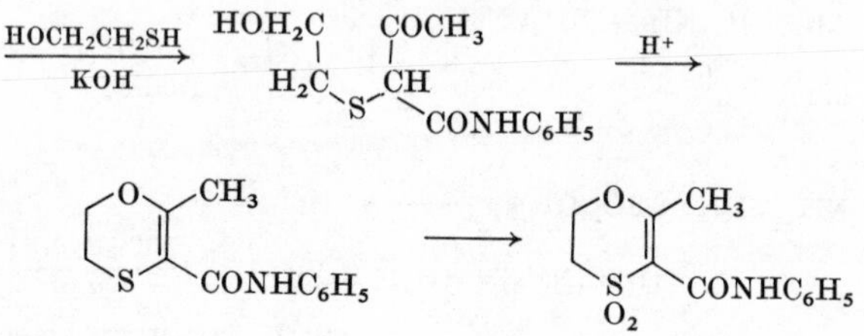

(**3**)   (**4**)

Another aliphatic heterocyclic ring system occurring in fungicides is the morpholine ring of tridemorph (**5**), a systemic compound under development for the control of powdery mildew of cereals (*Erysiphe graminis*), being prepared by a two-stage process from tridecylamine.

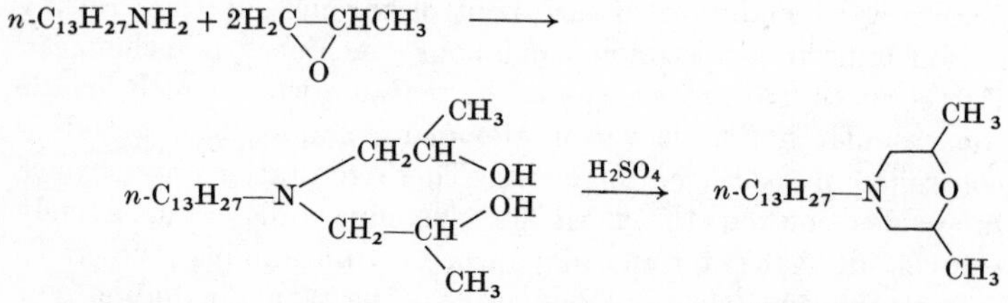

(**5**) Tridemorph

The first aromatic heterocyclic compounds we shall consider are the pyrimidines. Dimethirimol, discovered in 1965, is used to control powdery mildew (*Sphaerotheca fuliginea*) of the leaves and stem of cucurbits (i.e. cucumbers, marrows, etc.). The compound is effective as a spray, but also a single application of the compound to the soil gives disease control on cucumbers for up to eight weeks; traditional surface fungicides must be sprayed at intervals of about ten days and still give unsatisfactory disease control. Dimethirimol is prepared by condensation of *N,N*-dimethylguanidine (obtained from dimethylamine and cyanamide) with ethyl α-*n*-butylacetoacetate (cf. Part 3, p. 120).

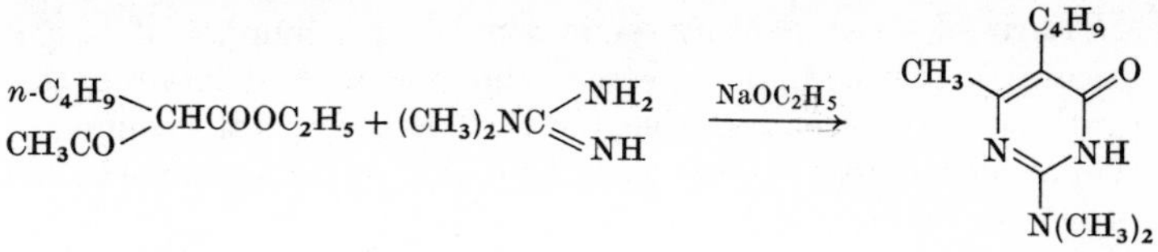

Dimethirimol

The related pyrimidine ethirimol, when applied to barley as a seed dressing, gives control of powdery mildew throughout the growing season. Its synthesis is similar to that of dimethirimol, but in this case the desired product must be separated from the isomeric *N*-ethyl-oxopyrimidines which are not fungicidal.

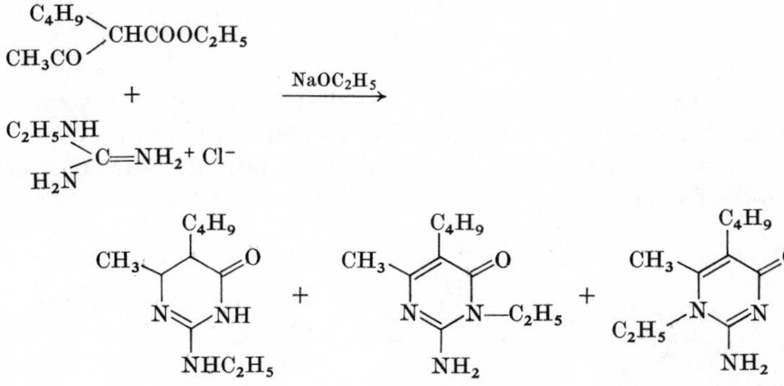

Ethirimol (Minor product) (Trace)

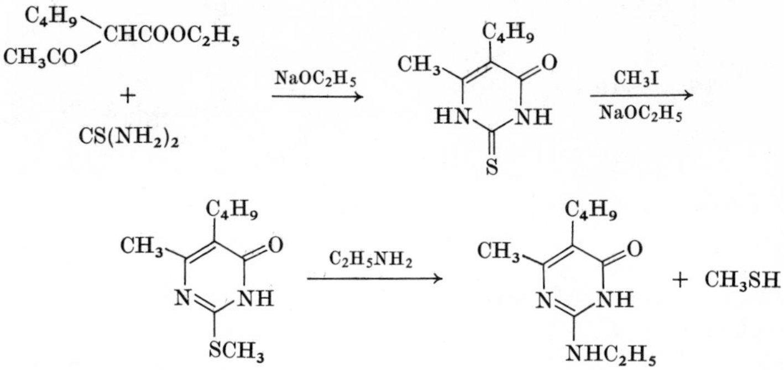

Ethirimol

The unambiguous synthesis of ethirimol shown above is less attractive commercially as the toxic methanethiol liberated in the final stage must be trapped effectively.

Triarimol gives good systemic control of a number of fungal diseases, including both apple scab and mildew. Full details of this synthesis are not yet available, but the final stage is the condensation of 5-bromopyrimidine with 2,4-dichlorobenzophenone in the presence of *n*-butyl-lithium:

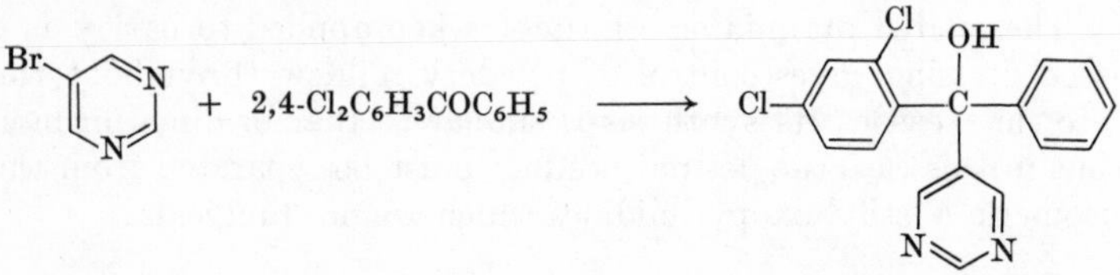

Triarimol

The benzimidazole ring occurs in benomyl which is highly active against a wide range of foliar diseases and soil-borne pathogens. One efficient synthesis is illustrated in the formulae shown in (*d*) on the facing page.

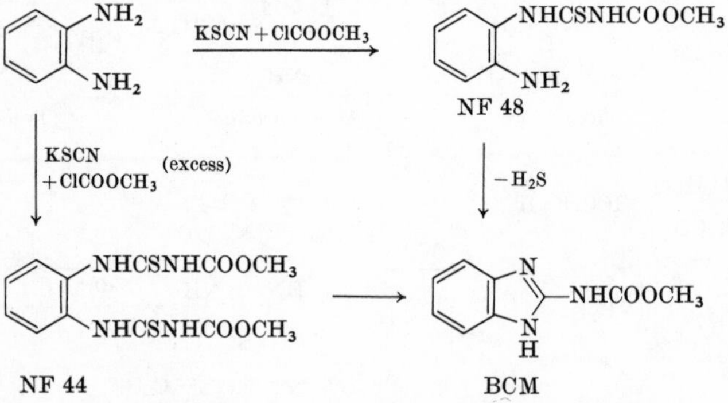

Two thioureidoformates (NF44 and NF48) prepared from *o*-phenylenediamine and methyl isothiocyanatoformate (S=C=NCOOCH$_3$; generated *in situ* from potassium thiocyanate and methyl chloroformate) have similar fungicidal activity to that of benomyl. It is known that these compounds are readily converted into methyl benzimidazol-2-ylcarbamate (BCM), one of the degradation products derived from benomyl; these three fungicides could, therefore, have the same mode of action.

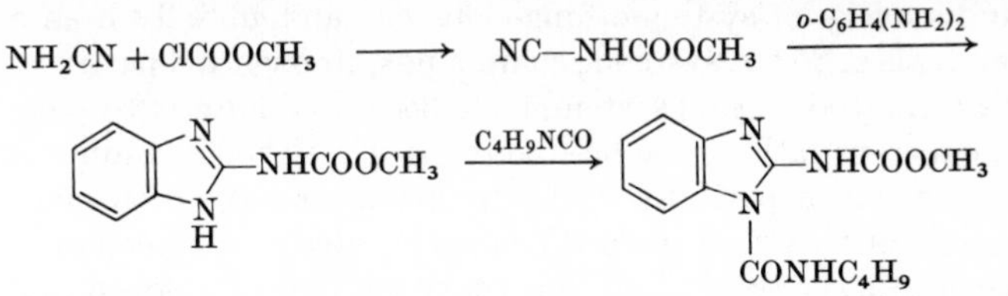

$$NH_2CN + ClCOOCH_3 \longrightarrow NC{-}NHCOOCH_3 \xrightarrow{o\text{-}C_6H_4(NH_2)_2}$$

(d)

BCM   Benomyl

Antibiotics

Since the discovery of penicillin attempts have been made to control plant diseases by the use of antibiotics, but very few have been developed for practical use. Streptomycin, one of the first antibiotics introduced into agriculture, was used in the United States to control fireblight of pears (*Erwinia amylovora*). This antibiotic is recommended in the United Kingdom for the control of downy mildew of hops; to prevent contamination of the harvested hops by such a potent antibiotic, final application should be made at least eight weeks before harvest. In Japan, more extensive use has been made of antibiotics in agriculture, particularly for the control of blast (*Piricularia oryzae*) and bacterial leaf blight (*Xanthomonas oryzae*) on rice; blasticidin S, a cytosine derivative isolated from *Streptomyces griseochromogenes*, gives excellent control of rice blast; it occasionally causes plant damage and has been replaced to some extent by newer antibiotics, including kasugamycin, which is an extremely effective fungicide safe to plants and mammals.

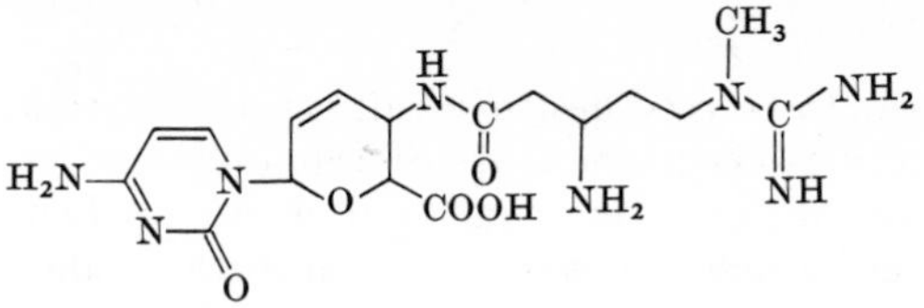

Blasticidin S

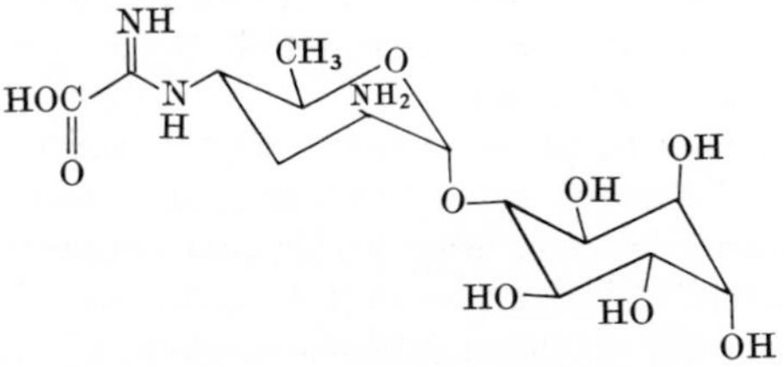

Kasugamycin

The discovery of systemic fungicides has undoubtedly been a major advance. So far, systemic compounds, transported passively in the xylem (root→leaf) system, have been used most effectively on annual crops rather than on woody plants. Although a number of unsuccessful or partially successful attempts have been made to control dutch elm disease (*Ceratosystis ulmi*) by injection of chemicals into the trunk, fungicides have yet to be developed that are transported satisfactorily within shrubs and trees. The discovery is also awaited of fungicides that will enter sprayed leaves and move throughout the plant in the phloem system.

A few cases are known of the development of fungicide-resistant strains of plant-pathogenic fungi; and with current development of systemic fungicides which, in some cases, may have a very specific mode of action rather than being general enzyme-inhibitors, acquired resistance to fungicides may become a more serious problem. We have only very limited knowledge of the factors that influence movement of chemicals within plants, and of the mechanism of selective fungicidal activity. Much remains to be learned of the physiology and biochemistry of plant pathogenic fungi and of the specific host–parasite relation involved. The majority of fungicides used today have been discovered fortuitously or as the result of large-scale screening programmes, but the day may well come when it will be possible to design molecules on a more rational basis.

Herbicides

The first weedkillers to be used destroyed all vegetation (total herbicides). Some, such as arsenic derivatives were quite dangerous; others, such as creosote were industrial waste products applied in what were massive doses compared with modern herbicides. The earliest herbicide still in general use is sodium chlorate ($NaClO_3$) which is employed as a total herbicide for paths. It is, however, easily leached out of soil and is not effective for so long as more recent 'residual' herbicides.

The first organic herbicides to be discovered were the dinitroalkylphenols, probably the most important one today being dinoseb. They act on plants by interfering with oxidative phosphorylation and the synthesis of ATP (see Part 4, p. 20). As this biochemical process is similar in plants and mammals it is not surprising that this class of herbicide is rather toxic, and nowadays their use is declining.

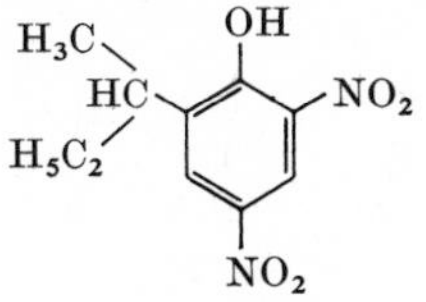

Dinoseb

The first selective herbicide to be used was sulphuric acid, which despite its obvious unpleasant properties was widely used before World War II in cereal crops. The narrow, waxy cereal leaves allow the spray to run off and are unharmed, whilst broad-leaved weeds (dicotyledons) not only receive more spray but also lose it less readily by run-off. This can be regarded as physiological selectivity.

Biochemical selectivity was first found in the so-called hormone weedkillers. In 1928 the first plant hormone, indolylacetic acid (IAA) was discovered. This has a very powerful action in stimulating the growth of plants. It might be thought that applying large

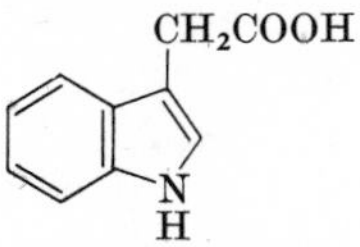

IAA 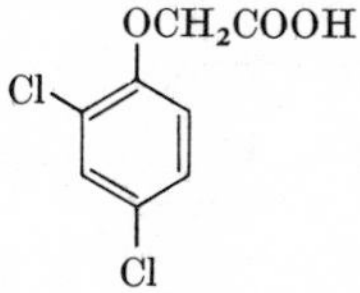   2,4-D 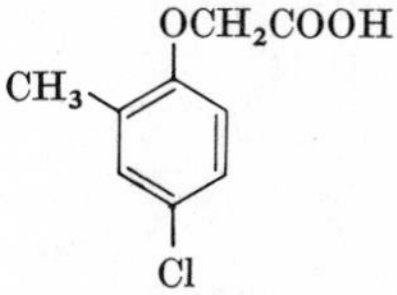   MCPA

amounts of IAA to a plant would cause it to grow excessively and die as a result; however, plants are able to regulate metabolically the amount of IAA they contain and as a result the chemical is ineffective as a weedkiller. The hormone weedkillers, discovered in 1940, are synthetic analogues of IAA that the plant is unable to metabolize rapidly; plants treated with these compounds do die as a result of uncontrolled and grossly distorted growth. The most important herbicides of this type are 2,4-dichlorophenoxyacetic acid (2,4-D) and 2-methyl-4-chlorophenoxyacetic acid (MCPA). For reasons that are still not fully understood these chemicals are much more effective in killing dicotyledons (broad-leaved plants) than monocotyledons (e.g. grasses) and they are therefore widely used for the control of broad-leafed weeds in cereal crops. The homologous 2-(aryloxy)propionic acids show similar herbicidal activity; mecoprop is used for killing chickweed (*Stellaria media*) and cleavers (*Galium aparine*) which are resistant to the phenoxyacetic acids.

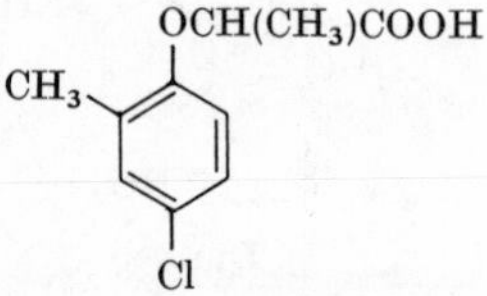

Mecoprop

The isomeric straight-chain isomers of mecoprop are, however, inactive. It has been found that compounds of the general structure $ArO(CH_2)_nCOOH$ are herbicides if n is odd but are inactive if n is even. This is explained by a biochemical process known as β-oxidation in which methylene groups β to the carboxyl group are oxidized to a carbonyl group, the resulting compound then readily cleaving to the corresponding acid with two fewer carbons (Part 4, p. 155). For example, MCPB is active because in the plant

MCPB

MCPA

it breaks down to MCPA. Some plants are unable to cause β-oxidation and are therefore resistant to the phenoxybutyric acids even though they are killed by the corresponding phenoxyacetic acids. The phenoxybutyric acids are, therefore, used to exploit these differences and find application in legumes such as peas.

Recent evidence suggests that the full explanation of selectivity is more complicated than this. Some plants, although able to carry out β-oxidation, are capable of even faster aliphatic homologation, probably by addition of malonate; consequently, phenoxybutyric herbicides are effectively removed as long-chain phenoxyaliphatic acids.

The phenoxyacetic acids are manufactured by reaction of the sodium salt of the appropriate phenol with chloroacetic acid. 2,4-Dichlorophenoxybutyric acid is made by reaction of the phenol with butyrolactone.

Other hormone-type weedkillers in use include a series of benzoic acids of which 2,3,6-trichlorobenzoic acid (TBA) and 3,6-dichloro-2-methoxybenzoic acid (dicamba) are representative; also the heterocyclic analogue picloram. They are manufactured by the routes indicated. All these organic acids kill plants by distorting their growth although the detailed symptoms vary

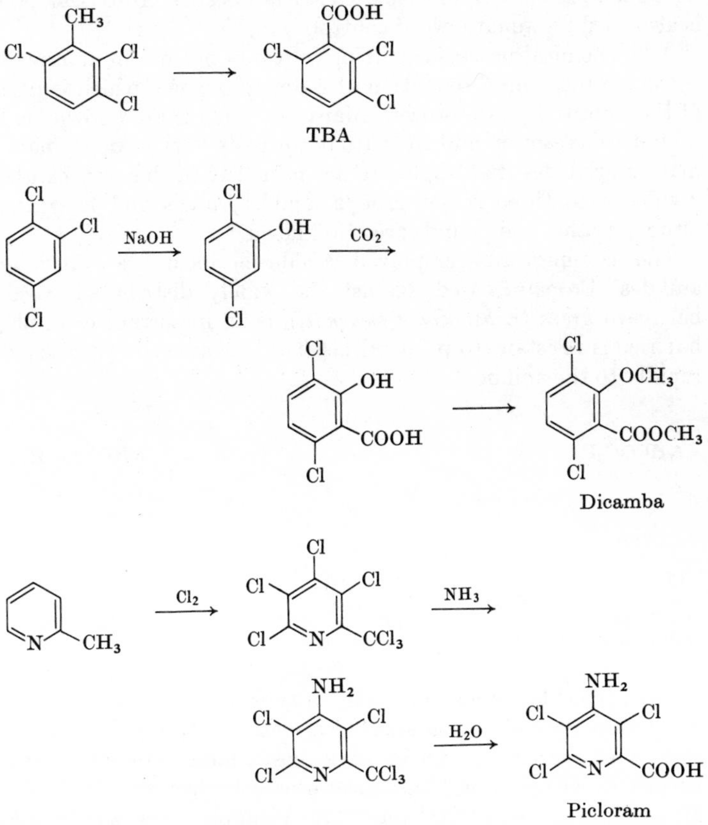

TBA

Dicamba

Picloram

from one compound to another. It has been suggested that a molecule with this type of herbicidal effect must meet three structural criteria: a negatively charged atom; an atom with a partial or full positive charge situated 0.55 nm from the negative site; and a lipophilic group coplanar with the positive centre. The anions of the compounds just discussed fit these requirements, the chlorine atoms producing a partial positive charge inductively at a site on the benzene ring, and the lipophilic group being the ring itself. The elaboration of theories such as these should allow the more rational design of future herbicides.

The first compounds to show selective herbicidal activity against grasses were the simple halogenated acids trichloroacetic acid (TCA) and 2,2-dichloropropionic acid (dalapon). They are used for controlling perennial grasses such as couch grass, and against wild oats, but they cannot be used safely in cereal crops. Dalapon is also used for aquatic weed control.

A biochemical process occurring in plants but not in mammals is the photochemical oxidation of water to oxygen, which is part of the photosynthetic process. Many compounds are known that inhibit this reaction and they are not usually very toxic to mammals. Herbicides that appear to act primarily in this way can be divided into three major groups, amides, ureas and triazines, although other compounds are also known.

The commercially employed amide herbicides are usually anilides. Propanil, used against the widely distributed weed, barnyard grass (*Echinochloa crusgalli*), is an important example, but rice is resistant to propanil because it is able to hydrolyze it rapidly to the aniline.

$NHCOC_2H_5$

Cl

Cl

Propanil

$R'NHCONR_2$

(6)

$NHCON(CH_3)_2$

Cl

Monuron

The herbicidal ureas are usually of type (6) where R is an alkyl group and R' is either an alkyl or a methoxy group. By varying these and the aryl group in analogues a large number of urea herbicides with a range of physical properties has been produced. Monuron is one of several important examples. The amides and the ureas are produced by conventional chemical methods, e.g. propanil from 3,4-dichloroaniline and propionyl chloride, and monuron from p-chlorophenyl isocyanate and dimethylamine.

Most of the herbicidal triazines have the structure (7), where X and Y are either alkylamino- or dialkylamino-substituents and Z is usually Cl, OCH_3 or SCH_3, although recently triazines bearing other groups have been introduced. Early and still very important triazines include simazine and atrazine which are used particularly for weed control in maize. The high resistance of maize to atrazine and simazine has been found to be due to the presence in maize of a compound that rapidly removes the chlorine atom hydrolytically, resulting in the detoxification of the herbicide.

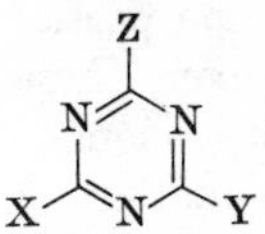

(7) 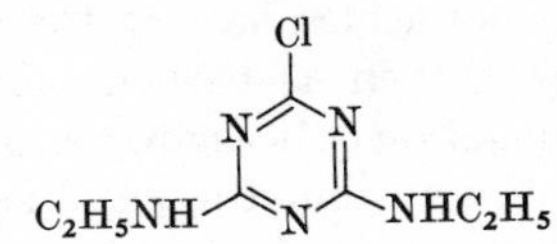   Simazine 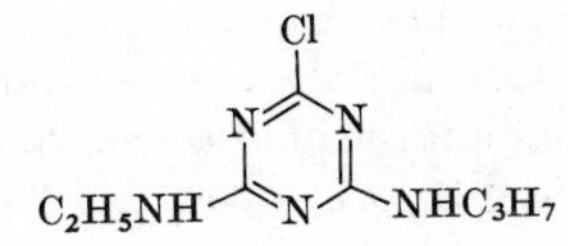   Atrazine

The triazines are made by reaction of cyanuric chloride with the appropriate nucleophile. Since each substituent introduced is electron-releasing, the remaining chlorine atoms become deactivated towards nucleophilic attack, and stepwise substitution is easy, as illustrated here for the synthesis of desmetryne.

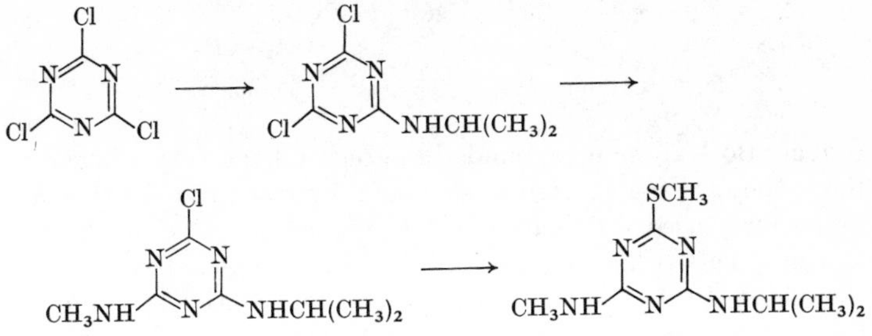

Desmetryne

It will be seen that all of the herbicides inhibiting the photochemical oxidation of water described have a common structural feature, namely a group (A) where X is an atom with a lone pair of

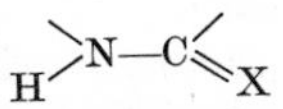

electrons (either N or O). It is thought that this part of the molecule is the toxophore responsible for binding to an enzyme involved in the photochemical oxidation, thereby inhibiting it, but the details of the process are not understood. In recent years, a number of more complex structures bearing this toxophore have appeared, notably a group of uracils of which bromacil is an important

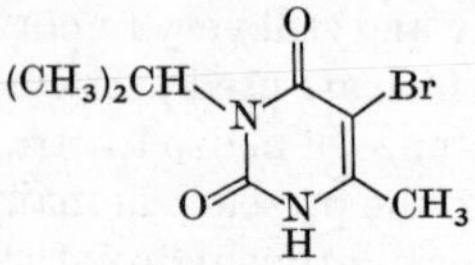

Bromacil

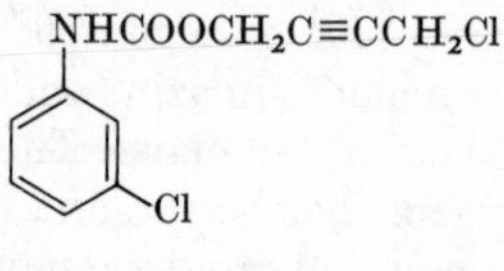

Barban

example. Another group of herbicides bearing the same partial structure are the carbamates, such as barban; however, these inhibit the photochemical oxidation only slightly and it is thought that these compounds kill plants by inhibiting cell division. Barban is used for wild-oat control in cereals. It is prepared by reaction of the corresponding alcohol with *m*-chlorophenyl isocyanate.

Two important herbicides with a novel mode of action and unique properties were discovered by Imperial Chemical Industries Limited in 1956; these were the bipyridylium salts paraquat and

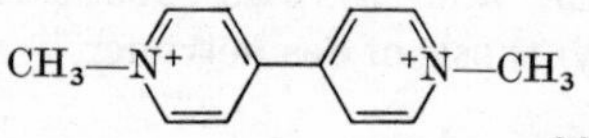

Paraquat

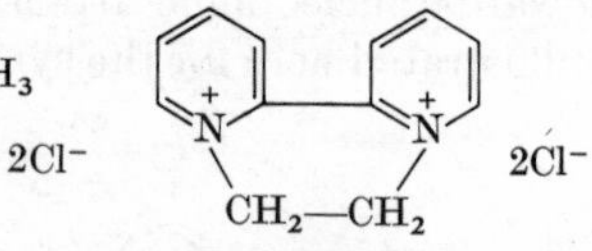

Diquat

diquat. Both these compounds kill green foliage very efficiently. Since they are inactivated as soon as they reach the soil they are extensively used for clearing a field immediately before or after sowing. This practice can, in certain circumstances, enable ploughing to be replaced by simpler cultivation techniques, thereby maintaining the soil structure and minimizing erosion. The bipyridylium herbicides are also used for winter stubble clearing, for weed control in plantations and orchards and for many other agricultural purposes. Diquat is also used for crop desiccation (removing unwanted leaves to facilitate harvesting) and for aquatic weed control.

The bipyridylium salts act by becoming involved in the primary chain of oxidation–reduction reactions which occur in photosynthesis. Paraquat is reduced to the resonance-stabilized radical and this is then re-oxidized by atmospheric oxygen, forming a peroxide species that is highly toxic to plant cells.

Paraquat is manufactured by reduction of pyridine to its radical ion by means of sodium in liquid ammonia; this species dimerizes exclusively in the 4-position to give a dihydrobipyridyl, which is

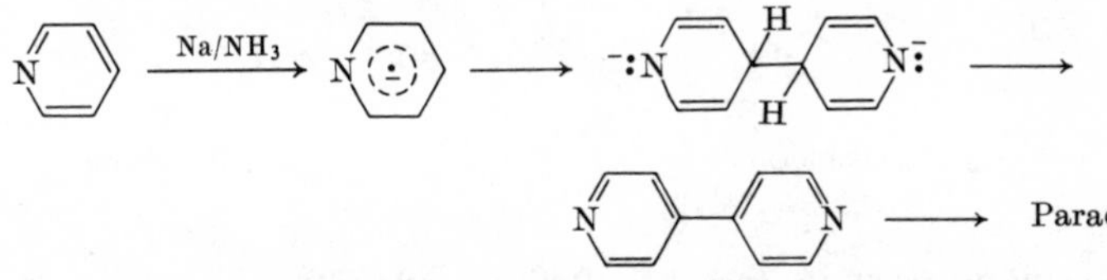

then oxidized and quaternized. The process is relatively complex and reflects the growing sophistication possible in herbicide synthesis.

There are a number of important herbicides not falling into any of the groups discussed above. Among these, amitrole should be mentioned; it is a total herbicide and is believed to act by preventing protein synthesis. Dichlobenil is a pre-emergence herbicide,

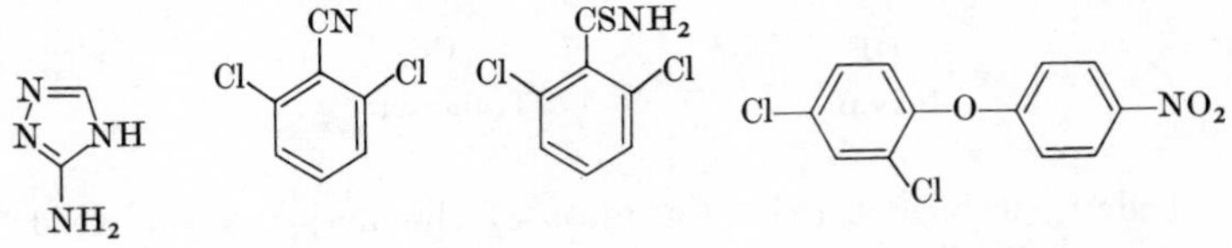

Amitrole   Dichlobenil   Chlorthiamid   Nitrofen

that is to say, the chemical is applied before the weed seeds have germinated; it is also used as an aquatic herbicide; it is rather volatile, and a large number of less volatile derivatives have been tried as substitutes in order to retain the chemical in the soil for longer periods. Chlorthiamid, the H_2S adduct of dichlobenil, is used commercially; it slowly liberates dichlobenil in the soil. Nitrofen is also used for pre-emergent weed control in brassica (rape, kale, cabbage, etc.) and other crops. A number of chemically related compounds have been announced recently. Pyrazone is one of a range of N-phenylpyridazinones; it is used particularly in sugar-beet, a plant which is able to detoxify the chemical. Pyrazone acts by inhibiting photosynthesis and may be related to the other groups of herbicide that inhibit the photochemical oxidation of water discussed above. It is prepared by reaction of phenyl-hydrazine with mucochloric acid, followed by ammonolysis of the product.

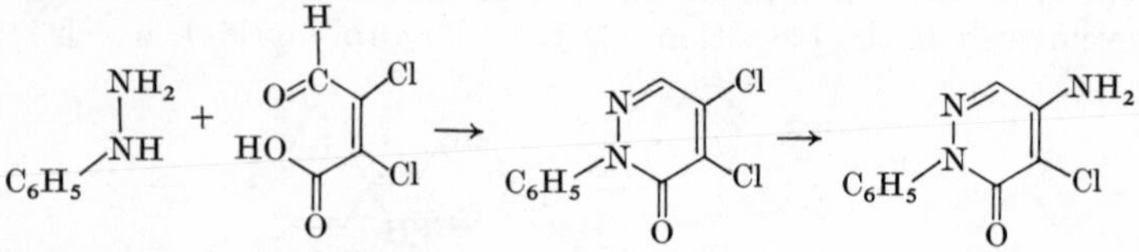

Mucochloric acid

Pyrazone

Ioxynil is used in cereals, often mixed with phenoxy acid herbicides. Its biochemical mode of action has been much discussed; it has been found to uncouple oxidative phosphorylation and may therefore be related to that of the nitrophenols. Trifluralin is used for pre-emergence grass weed control in cotton.

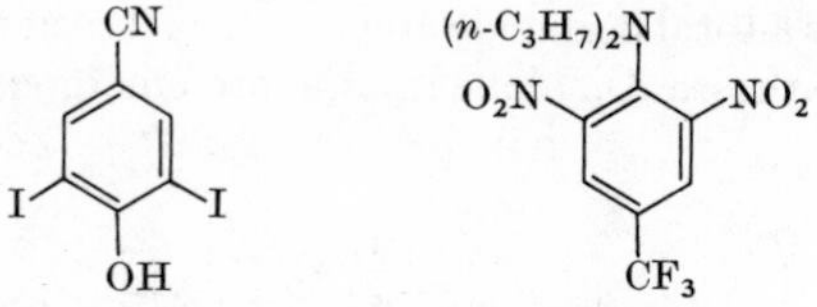

Ioxynil

Trifluralin

Today, herbicides exist for most of the major world weed problems. To discover new compounds that are herbicides is not sufficient; they need to be better than an existing compound already in use—better by being cheaper, safer on the crop or safer to handle, or by having just the right degree of persistence for the job in hand. The weed flora of cultivated land in the U.K. has changed dramatically in the last 25 years. Largely as a result of the use of hormone weedkillers, the broad-leaved weeds, once common in cereals, are now rare; but in their place a new range of weeds, more like the crop, is developing; the biggest weed problem in cereals today is wild oats (*Avena fatua*). To solve these problems a considerable research-based industry has arisen to discover and evaluate new herbicides. It is less than 40 years ago that the first organic herbicides were introduced. These and later compounds were discovered largely as a result of empirical testing with little scientific rationale. Although we may still expect discoveries to arise occasionally 'by chance', it is becoming necessary today to design new herbicides by considering the fundamental biochemistry of the plant system and predicting the physicochemical properties that a compound requires in order to enter a plant through roots and leaves and to be translocated through the plant.

Growth regulators

All the remarkable physiological changes occurring during the life of a plant are mediated by chemicals. One such compound, indolyacetic acid, IAA, has been mentioned above (p. 441). The possibility of using synthetic chemicals to induce desirable physiological changes in plants has been considered for a long time and is currently receiving increasing attention. Among the effects considered desirable, particularly in an age of increasing mechanization of harvesting, are uniformity of fruit size and maturity, loosening of fruit at harvest time and alteration of plant shape to facilitate growth and/or harvesting. It would also be useful if chemicals could be found that, in conjunction with fertilizers, increased crop yield. Other useful properties are described below.

1-Naphthylacetic acid and 3-indolylbutyric acid (IBA), both analogues of the natural hormone IAA, are used to promote root growth in plant cuttings.

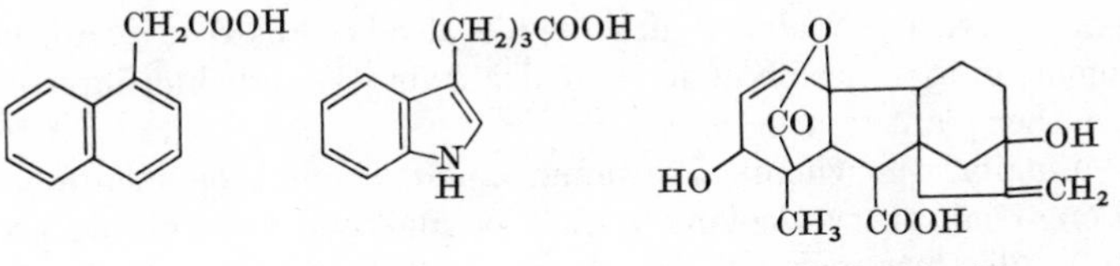

1-Naphthylacetic acid   IBA   Gibberellic acid

The second group of natural plant hormones to be discovered were the gibberellins. One of these, gibberellic acid, is used for a variety of purposes, the most important of which is inducing germination in barley during brewing.

The simplest plant hormone is ethylene; this has the various effects of inhibiting growth, accelerating abscission (loss of leaves, flowers and fruit), ripening and stimulating flowering according to the growth stage of the plant. In orchards a 'chemical formulation' of ethylene in the form of 'Ethrel' is used to promote ripening and loosening of fruit. A new application of increasing importance is the stimulating of latex flow in rubber trees. The compound fragments as indicated, liberating ethylene within the plant.

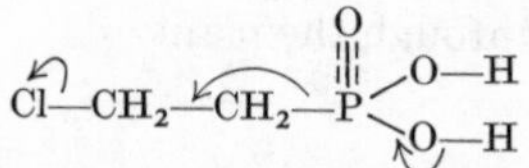

Ethrel

$ClCH_2CH_2\overset{+}{N}(CH_3)_3\ Cl^-$

Chloromequat chloride

There are various other types of plant hormone, analogues of which may well lead to commercial application in the future.

One of the most important growth-regulators in terms of quantities used is chloromequat. It has the effect of shortening plant stems and is used on cereals to prevent the crop bending under its own weight (lodging). Larger amounts of fertilizer can then be applied, resulting in increased yields.

In horticultural practice, a succinamic acid (B9) is used to stunt plants such as chrysanthemums, making them more suitable for

$CH_2CONHN(CH_3)_2$
CH_2CO_2H

B9

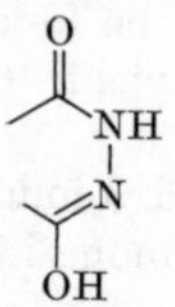

Maleic hydrazide

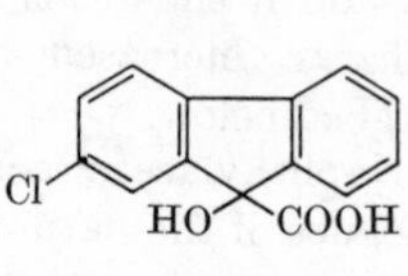

Chlorflurecol

pot cultivation. Maleic hydrazide has also been used as a stunting agent, in this case on grass in roadside verges; and chlorflurecol is another plant retardant.

Finally, the chemical pruning agents should be mentioned. Long-chain fatty alcohols (mainly octanol and decanol) are used to inhibit terminal bud growth and hence stimulate branching in many plant species. They are also used on tobacco to stop sucker growth and have been applied to chrysanthemums.

The use of growth regulators in agriculture is at present in its infancy: it is not difficult to visualize the enormous importance that such compounds could assume in the future.

Bibliography

Insecticides

R. C. Reay, *Insects and Insecticides*, Oliver and Boyd, Edinburgh, 1969 (paperback).

R. D. O'Brien, *Insecticides—Action and Metabolism*, Academic Press, London, 1967.

Bibliography (contd.)

C. E. Berkoff, 'The Chemistry and Biochemistry of Insect Hormones,' *Quart. Rev.*, 1969, **23**, 372.

M. Jacobson and P. G. Crosby, *Naturally Occurring Insecticides*, Marcel Dekker, New York, 1971.

Fungicides

E. C. Large, *The Advance of the Fungi*, Jonathan Cape, London, 1940.

C. Woodham-Smith, *The Great Hunger*, Hamish Hamilton, London, 1963.

B. E. J. Wheeler, *An Introduction to Plant Diseases*, Wiley, London, 1969.

R. Wegler (Ed.), *Chemie der Pflanzenschutz- und Schadlings-bekämpfungsmittel*, Vol. 2, Springer-Verlag, Berlin, 1970.

Herbicides

L. J. Audus (Ed.), *The Physiology and Biochemistry of Herbicides*, Academic Press, London, 1964.

J. D. Fryer and R. J. Makepeace (Eds.), *Weed Control Handbook*, 2 vols., Blackwell, Oxford, 1972.

D. D. Davies, J. Giovanelli and T. Ap Rees, *Plant Biochemistry*, Blackwell, Oxford, 1964.

G. E. Fogg, *The Growth of Plants*, Penguin, Harmondsworth, 1966.

CHAPTER 13

Detergents

(K. JONES, Unilever Research, Port Sunlight Laboratory)

The Detergency Process

It is a relatively simple matter to define detergency. It could, for example, be defined as 'the act of cleaning the surface of a soiled material in a liquid bath in which a solute (or solutes) which aids this cleaning—detergents—are dissolved'. Unfortunately, the simplicity ends there. Many widely different processes contribute to the cleaning, and their relative importance depends very greatly on the nature of the substrate, the nature of the dirt to be removed, and the cleaning conditions (detergent concentration, temperature, degree of agitation). One has only to consider the everyday detergency operations carried out by a housewife— washing clothes, washing dishes, cleaning a kitchen floor and, of course, washing herself—to realize how varied is the range of substrates which are cleaned and how diverse is the nature of the dirt on them.

Most important detergency systems use water as solvent and, for simplicity, the discussion in this Chapter is restricted to detergents designed for use in water.

Any successful detergent system must do two things: it must detach the dirt from the surface to be cleaned; and it must disperse or dissolve the dirt in the wash liquor in such a way that the cleaned substrate can be separated from the wash liquor without the dirt being redeposited on it. The key to both these requirements is the nature of the interfaces between the substrate, the dirt, and the wash liquor. A fully formulated detergent system functions by modifying the properties of these interfaces, thus changing the energy of the interactions between the dirt and the substrate. To do this, the detergent formulation must contain materials that are adsorbed at these interfaces, i.e. it must contain surface-active molecules. These are termed surfactants and can be represented by a general formula RX where R is a hydrophobic tail group, in almost all cases a hydrocarbon residue, and X is a hydrophilic head group such as —COO⁻. Depending on the nature of R and X, such molecules may be either water-soluble or oil-soluble and the balance between the hydrophilic and the hydrophobic groups largely determines the detergent properties.

The hydrocarbon chain of a water-soluble surfactant is not entirely comfortable in bulk aqueous solution. This is not because of repulsion between the hydrocarbon chain and water. In fact, the van der Waals attraction of water for the hydrocarbon chain is slightly greater than the attraction between paraffin chains but both these attractions are much lower than the water–water attraction. Liquid water consists of a three-dimensional hydrogen-bonded network of water molecules, the hydrogen bonds being continually broken and formed. A paraffin chain modifies the interactions of the water molecules near it so as to give a different kind of structure. Under most circumstances there is a tendency to restore the bulk water structure, and the surfactant molecule always tends to take up a position in which its chain is at least partially removed from the bulk solution. The most obvious way for it to effect this is for surfactant molecules to congregate at the interface between its aqueous solution and air, and also at interfaces with oil and with certain solids if these are present in the system. This adsorption at the interfaces has a quite drastic effect on the interfacial properties, particularly interfacial tensions, which are considerably lowered. For example, the surface tension of a 10^{-3}-molar surfactant solution at 20°C is about 0.3–0.4×10^{-3} N cm⁻¹, which compares with 0.73×10^{-3} N cm⁻¹ for pure water. As the total surfactant concentration is increased, a concentration is eventually reached where the surfactant molecules have to find another way of removing their chains from bulk solution and they begin to form more or less spherical aggregates with their hydrophobic tails on the inside and their polar head-groups outside. Such aggregates are termed micelles and the concentration at which they begin to be formed is called the critical micelle concentration (CMC). Below the CMC the thermodynamic properties of surfactant solutions are approximately those expected of dilute solutions of unassociated molecules. Above the CMC colligative properties such as osmotic pressure change with concentration of surfactant much more slowly than would be predicted, because the surfactant molecules in excess of the critical micelle concentration exist in the associated form as micelles. As each micelle consists of many molecules (usually 20–200) the transition from relatively ideal to very non-ideal behaviour is sharp. Other, non-thermodynamic, properties—conductivity, diffusion, etc.—change in an analogous manner.

Effect of Surfactants on Dirt Removal

In order to explain how surfactants assist dirt removal, it is necessary to consider the factors that hold the dirt on the substrate and the processes that occur when the dirt is removed. Dirt can contain liquid oily materials which can flow and hence change their shape during the removal process (e.g. human sebum, fatty food residues on dishes) and solid particulate materials which have to be removed as they are, without deformation (e.g. carbon, metal oxides). Whilst many factors are common to the removal of both types of soil, there are also important factors that are not common to them. First, consider a dry fabric soiled with particulate dirt: the dirt is held on the surface in this case by van der Waals attractive forces. When the fabric is immersed in an aqueous solution and agitated, the dirt particles will tend to be shaken away from the fabric. The surface of the fabric previously in contact with the dirt, and the parts of the dirt particles previously in contact with the fabric, can then be wetted. Once in contact with the aqueous medium, these surfaces and, of course, the rest of the surface of the dirt and the fabric will acquire an electric charge as a result of preferential adsorption of ions from solution, or ionization of groups present in the surfaces. Anions are more readily adsorbed than cations and hence most materials become negatively charged. These surface charges are partially balanced by a diffuse area of charge of opposite sign caused by concentration of the counter-ions in the aqueous phase in the immediate vicinity of the surface. Hence, each surface builds up an electrical double layer and, as the charges on the surface of the fabric and the dirt particles are normally both negative, these double layers usually repel one another, aiding removal of the dirt.

If an anionic surfactant is present in the solution, as is the case in most practical washing systems, it will, for reasons given above, be adsorbed on the surface of the dirt particles and the fabric. This anionic adsorption occurs even though the surfaces are already negatively charged; it increases the charge on the surface of both dirt and fabric which, in turn, considerably increases the double-layer repulsion and thus clearly enhances dirt removal.

Separation of the dirt particle and the fabric generates two new solid–liquid interfaces from the previous particle–fabric interface. Adsorption of the detergent at these new interfaces considerably

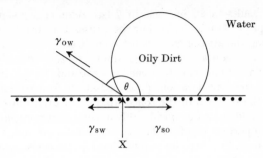

Figure 13.1. Wetting angle on a soiled surface. γ_{sw}, γ_{so} and γ_{ow} are, respectively, substrate–water, substrate–oil and oil–water interfacial tensions.

reduces their surface energy and this decreases the work required for their generation, i.e. decreases the amount of work required to remove the dirt.

In the case of liquid dirt, additional dirt removal processes—rolling up, solubilization and formation of mixed phases—can occur.

The rolling-up process can occur when a substrate surface is initially wetted by oily dirt. On immersion in a detergent solution, preferential wetting of the substrate by the aqueous phase forces the oil to roll up. This is illustrated in Figure 13.1 which represents a drop of oily dirt on a substrate immersed in water. The forces acting at a point (X) where the oil-drop surface meets the substrate are the substrate–water, substrate–oil and oil–water interfacial tensions (respectively γ_{sw}, γ_{so} and γ_{ow}). The oil–substrate contact angle is θ. These are related as shown in eqn. (1).

$$\gamma_{sw} = \gamma_{so} + \gamma_{ow} \cos \theta \qquad (1)$$

Rearranging this gives:

$$\cos \theta = \frac{\gamma_{sw} - \gamma_{so}}{\gamma_{ow}} \qquad (2)$$

If an oil-insoluble surfactant is dissolved in the aqueous phase it will be adsorbed at the oil–water and substrate–water interfaces and hence decreases both γ_{sw} and γ_{ow}. If $\cos \theta$ is negative, as in the case chosen in Figure 13.1, both these changes will decrease $\cos \theta$, i.e. increase θ. This is a grossly over-simplified treatment of the phenomenon, but a full analysis confirms that addition of sur-

factants results in an increase in θ (see references given in the
bibliography). Thus, the surfactant causes the dirt to roll up,
diminishing the area of contact between dirt and substrate. Even
if the area of contact is reduced almost to zero ($\theta \approx 180°$), the
van der Waals attraction between the drop and the substrate still
exists and the considerations described above for particulate dirt
apply to the removal of the rolled-up dirt. Clearly, however, the
rolling-up phenomenon considerably aids the dirt removal.

Many important dirts are fairly polar semisolids and, for these
to be removed by the rolling-up mechanism, their viscosity has to
be reduced so that they can flow sufficiently rapidly for their shape
to change during the washing period. High-temperature washing
can clearly achieve this. Certain surfactants also penetrate such
dirt and form complexes or cause phase changes that radically
alter the dirt's physical properties. These changes in physical
properties can also contribute significantly to the ease of dirt
removal.

In the dirt-removal processes considered so far, the surfactants
function by virtue of their adsorption at interfaces. In contrast,
solubilization of oily dirt depends on the ability of detergents to
form micelles, for these provide a non-polar environment in the
aqueous medium in which oily materials can dissolve. In most
detergent applications, solubilization is relatively unimportant.
The surfactant is used at low concentrations (e.g. 0.01–0.3% in
domestic fabric washing) and the ratio of surfactant to oily dirt
is so low that only a fraction of the oil can be solubilized. One
notable exception to this is personal washing with toilet soaps.
Concentrations of up to 10% can occur during hand-washing for
example, and solubilization can contribute appreciably to detergency under these conditions.

Dirt Redeposition

Once dirt has been detached from a substrate it has to be
prevented from redepositing if the detergency process is to be
successful. From the previous Section, it is clear that in most
detergency operations the majority of the dirt is not dissolved
but is dispersed in the wash liquor; so, unless some barrier to
redeposition exists, collisions between substrate and dirt will
inevitably cause partial redeposition.

Surfactants adsorbed on the surface of the dirt and the cleaned
substrate help to oppose redeposition in a number of ways. As
explained above, the electrical-double-layer repulsions between
the dirt particles or droplets and the cleaned substrate are con-

siderably increased by the adsorption of charged surfactants, and this repulsion, as well as aiding dirt removal, presents a barrier to redeposition. The dirt and the substrate usually adsorb the surfactant molecules with their tails inward and their heads outward. The polar head groups are strongly hydrated and the resultant rigid hydration shells function as a further barrier to redeposition. The adsorbed surfactants also lower the surface energy of the dirt and substrate, which decreases the driving force for recombination.

All these effects improve the situation but redeposition is still an important practical problem, particularly in the rinse stage of the fabric-washing operation when the surfactant adsorbed on the dirt and the substrate is partially desorbed into the rinsing water. Antiredeposition agents are added to fabric-washing products to suppress redeposition. By far the most important of these is sodium carboxymethylcellulose (cellulose with an average of ca. 0.5 of the —OH groups in each glucose unit substituted by —OCH_2COONa) which suppresses redeposition of dirt on to cotton. There is still some controversy as to how this material works but, as it is an anionic polyelectrolyte, it seems most likely to function by adsorption on to the dirt and/or the cotton, hence intensifying the double-layer repulsions.

Builders

The discussion so far has been limited to the effect of surfactants in de-ionized water. Surfactant solutions containing other electrolytes, particularly salts with multivalent cations, behave rather differently and this has to be taken into account when considering practical washing systems.

Domestic fabric washing, probably the most important detergency process, is a case in point. Many homes are supplied with hard water containing Ca^{2+} and Mg^{2+} ions, and a successful fabric-washing product must function efficiently in such hard water. The hardness ions influence washing efficiency adversely in a number of ways. First, the calcium salts of many surfactants are only sparingly soluble in water, and the surfactant would be at least partially precipitated and hence ineffective in hard water. Secondly, clothes are often soiled with sebum containing fatty acids; in hard water these are converted into insoluble calcium salts which are not washed off the clothes. Thirdly, and possibly most important, increasing the electrolyte concentration in the wash liquor decreases the double-layer repulsions. Multivalent electrolytes are particularly potent in this respect and their presence at concentrations found in hard water has a significant deleterious effect on redeposition and dirt removal.

A successful fabric-washing formulation must therefore effectively remove free Ca^{2+} and Mg^{2+} from solution. Compounds termed builders are added to the formulations to achieve this removal. These builders either sequester or precipitate the hardness ions. Typical effective materials are ethylenediaminetetra-acetic acid, sodium polyacrylate and, by far the most important builder used commercially to date, sodium triphosphate ($Na_5P_3O_{10}$).

As stated at the beginning of the Chapter, many processes contribute to detergency, and numerous diverse substrates and dirts have to be dealt with. This complexity makes a 'General Theory of Detergency' impossible to formulate. The above description is intended only to show some of the more important ways in which surfactants aid the detergency process. More rigorous and detailed accounts of detergency are to be found in the references given in the bibliography.

Soaps and Soap Products

Soaps derived from natural oils and fats are the classical surfactants. Soap has been used at least since biblical times and, although its use has declined since World War II, it is still used to make the majority of the world's toilet tablets and, in the underdeveloped countries, hard soap bars are still the most important products used for fabric washing. Also a substantial proportion of the fabric-washing powder used in the United Kingdom is soap-based.

Natural oils and fats are mixtures of triglycerides, i.e. glycerol esterified by fatty acids (carboxylic acids with long hydrocarbon chains, Part 1, p. 154; Part 4, p. 150). Soaps are the salts of these fatty acids and are produced by saponification of these glycerides with alkalis, usually sodium hydroxide.

$$
\begin{array}{lll}
CH_2OOCR^1 & R^1COONa & CH_2OH \\
| & + & | \\
CHOOCR^2 + 3NaOH \longrightarrow & R^2COONa & + CHOH \\
| & + & | \\
CH_2COOR^3 & R^3COONa & CH_2OH \\
\end{array}
$$

The most important raw materials for soapmaking are tallow (produced by rendering the body fat from cattle and sheep) and coconut oil. Table 13.1 gives the approximate composition of the fatty acids combined with glycerol in these mixed glycerides.

Table 13.1. Fatty-acid distribution in tallow and coconut oil

Acid	Structure	Proportion (%) contained in	
		coconut oil	beef tallow
Caprylic	$CH_3(CH_2)_6COOH$	8–9	
Capric	$CH_3(CH_2)_8COOH$	7–10	
Lauric	$CH_3(CH_2)_{10}COOH$	45–51	
Myristic	$CH_3(CH_2)_{12}COOH$	16–18	2–4
Palmitic	$CH_3(CH_2)_{14}COOH$	7–10	29–32
Stearic	$CH_3(CH_2)_{16}COOH$	1–3	19–26
Oleic	cis-$CH_3(CH_2)_7CH$= $CH(CH_2)_7COOH$	5–8	40–43
Linoleic	cis,cis-$CH_3(CH_2)_4CH$= CH—CH_2—CH= $CH(CH_2)_7COOH$	1–3	2

Other materials are also used (depending on availability and price), e.g. palm oil (predominantly C_{16} and C_{18}), palm kernel oil (composition similar to coconut oil). The oils may, if required, be treated before saponification, e.g. bleached with acid fuller's earth to remove coloured impurities or, in the case of oils with a high content of unsaturated glycerides, partially hydrogenated (Part 1, p. 156) to improve colour and stability.

Saponification

A satisfactory soapmaking process must not only saponify the glycerides, it must also produce the soap with a low water content, remove coloured impurities and separate the by-product glycerol in such a way that it can be conveniently recovered.

The classical method, evolved over many years and still in use to produce a significant amount of the world's soap, involves treating the fat with sodium hydroxide in open pans equipped with steam-heating and facilities for phase separation. In one variant of this procedure, the fat charge is saponified under carefully controlled conditions of temperature, water content and alkali content to give a product containing only a very small excess of alkali. Dry salt is then added and the mixture separates into two phases, a 'soap curd' phase and a 'lye', the latter being a thin aqueous salt solution containing about half the glycerol. The phases are separated and water and salt are added to the soap phase so as to separate more of the glycerol in a second lye. Finally, a critical amount of water is added to the soap curd (after separation of the second lye) and a 'nigre', a thin phase containing coloured impurities and some soap, settles out. This is separated off to leave a 'neat soap' layer, containing about 30% of water, which is suitable for processing into finished products.

There are numerous variants of this type of process, all of which are time-consuming because of the large number of sequential operations and the long settling times required. A number of more rapid continuous processes have been developed and these have, in many cases, superseded the pan method. Some are similar in principle to the pan approach, involving alkaline saponification and phase separations with salt. Others use an entirely different approach, namely high-temperature (ca. 250°C)/high-pressure $(4-5 \times 10^6 \text{ Nm}^{-2})$ hydrolysis of fats with water in counter-current tower reactors, separation of the resultant crude fatty acids from the aqueous glycerol, purification of the fatty acids by distillation, and neutralization of the purified fatty acid with sodium hydroxide or other alkali.

Preparation of Finished Soap Products

Toilet Bars. The neat soap from the saponification contains about 30% of water. The water content is first reduced to ca. 12%. Next, minor additives (e.g. perfumes, preservatives such as ethylenediaminetetra-acetic acid, whiteners such as TiO_2, or colourants and, in some cases, germicides) are added and mixed in thoroughly by working the soap usually on multistage roll-mills which give homogeneous soap chips. These chips are processed in a 'plodder' (a screw extruder like a large mincing machine) which compresses the chips and extrudes them through a die to give a continuous bar of soap from which the finished soap tablets are stamped. Toilet bars usually contain 20–50% of coconut-oil soap and 50–80% of tallow soap and may also contain up to 10% of un-neutralized fatty acid. The more expensive coconut-oil soap is necessary to give a product with good lathering and satisfactory dissolution properties. Such soaps are very satisfactory when used in soft water, but in hard water they give undesirable 'scum' (insoluble calcium soap). Toilet bars based on synthetic surfactants with soluble calcium and magnesium salts have been recently developed and are used to a significant extent in some areas of the world, but they have not so far displaced soap-based toilet bars from their dominant position.

Fabric-washing Powders. Typical formulations consist of

sodium soap (50–60%, mostly tallow-based), sodium perborate (8–20%), sodium silicate (3–6%) and small amounts of lather-boosters, sodium carboxymethylcellulose, fluorescent whitening agents (see p. 483), perfumes and water. Sodium carbonate and/or sodium phosphates may also be present. The formulations are sold in the form of spray-cooled powders prepared by mixing the formulation ingredients (excluding any heat-sensitive materials such as sodium perborate), superheating the resultant slurry under pressure, and spraying the superheated slurry through nozzles into a tower in which the water flashes off from the resultant droplets as they fall under gravity This gives the familiar powder granules into which are dry-mixed the heat-sensitive ingredients to give the finished powder.

Soap powders do not need added builders as the Ca^{2+} in hard water is precipitated as calcium soap. This means that sufficient product must be used both to precipitate the Ca^{2+} and to provide a sufficiently high concentration of additional surfactant for satisfactory detergency. Thus in hard water soap powders must be used at fairly high concentrations.

Such products once dominated the fabric-washing market but they have now been almost completely displaced in most world markets by products based on alkylbenzenesulphonates (see p. 462). One notable exception is the United Kingdom, where soap powders still have an important share of the fabric-washing market.

Other Uses for Soaps. Soaps have been displaced by synthetic surfactants in many applications, but they are still used in, for example, textile finishing, emulsion polymerization, cosmetics, polishes and emulsion paints.

Some Problems Associated with Soap Products

Although soap-based products are highly satisfactory in many respects they have a number of drawbacks, the most familiar being the undesirable scum which is produced when they are used in hard water.

The availability and, in consequence, the price of natural oils and fats varies unpredictably, and in many cases supplies cannot be increased easily. For example, the supply of coconut oil depends on climatic conditions and is very variable. Also, its production can only be increased by laying down new coconut-palm plantations, which is obviously a much slower process than erecting a new chemical plant. Tallow is a by-product of beef production and hence the amount available for soapmaking is governed directly by the amount of beef produced. Good-quality tallow is also edible and, whilst soapmaking uses poorer grades, increase in the world population with a consequent increase in demand for food and detergent products would intensify the competition for high-grade tallows unless alternative raw materials were used in detergents.

In addition, the distribution of chain lengths of natural fatty acids available to the formulator of detergent products is quite limited. Most oils and fats have chain-length distributions peaking at either C_{12} or $C_{16/18}$ and if a particular detergency operation required, say, a C_{14-16} distribution, such a soap simply could not be produced economically from natural raw materials.

All these reasons, and others, have contributed to the decline of soap products. The use of synthetic surfactants that are based on cheap and readily available petrochemical raw materials and which, in many applications, are more effective than soaps has grown at a considerable rate since World War II.

Synthetic Surfactants

Surfactants are usually classified, according to the charge carried by their head group, into anionic surfactants, e.g. $CH_3(CH_2)_{11}OSO_3^-Na^+$, non-ionic surfactants, e.g. $CH_3(CH_2)_{11}$-$(OCH_2CH_2)_8OH$, and cationic surfactants, e.g. $[CH_3(CH_2)_{17}]_2$-$N^+(CH_3)_2Cl^-$. There are very many examples within each class, and only materials of actual or potential industrial importance are considered in this Chapter. More comprehensive accounts are to be found in the references given in the bibliography.

Anionic Surfactants

The anionics are by far the most important surfactants and 1969 world production was close to a million tons. The materials of greatest industrial importance contain C_{10-15} saturated hydrocarbon chains attached directly or indirectly to sulphonate or sulphate groups, and the major uses are in domestic fabric-washing and dishwashing products.

Biologically Hard Branched-chain Alkylbenzenesulphonates. The first synthetic anionic surfactants to be commercially exploited on a large scale were based on alkylbenzenes produced by tetramerizing propylene and adding the resultant mixture of highly branched dodecenes to benzene. Both the propylene

$$4CH_3CH=CH_2 \xrightarrow[\text{catalyst}]{\substack{\text{Phosphoric} \\ \text{acid-based}}}$$

[structures of propylene tetramer and dodecylbenzene with Friedel-Crafts alkylation]

$$\xrightarrow[\text{HF or AlCl}_3]{C_6H_6}$$

tetramerization and the Friedel–Crafts alkylation involve intermediates with carbonium ion character (Part 2, p. 111) in which carbon-skeletal rearrangements and hydride shifts can occur. The above formulae simply gives structures representative of the numerous highly branched propylene tetramers and dodecylbenzenes formed.

These dodecylbenzenes can be converted into sodium dodecylbenzenesulphonate in very high yield. This is a good surfactant

$$C_{12}H_{25}C_6H_5 \xrightarrow[\text{(2) NaOH}]{\text{(1) SO}_3 \text{ or oleum}} p\text{-}C_{12}H_{25}\text{—}C_6H_4\text{—}SO_3Na$$

with excellent foaming properties and, since it is derived from cheap and readily available raw materials, apparently fulfilled all the technical and economic requirements for a synthetic surfactant. As its calcium salt is much more soluble than calcium soap, it requires a builder to be effective when used for fabric washing in hard water. Sodium triphosphate ($Na_5P_3O_{10}$) is an efficient cheap builder, and fabric-washing powders based on sodium dodecylbenzenesulphonate and sodium triphosphate, widely introduced after World War II, had almost completely replaced soap-based fabric-washing powders in many important world markets by the end of the 1950's. At that time, however, it became apparent that propylene-tetramer-based materials have one very important defect. They are biologically 'hard', i.e. they degrade biologically rather slowly. This meant that a significant proportion of the sodium alkylbenzenesulphonate entering domestic drains survived sewage treatment and was discharged into rivers and lakes. This resulted in unpleasant foams during sewage treatment and on rivers. These foams were not only aesthetically undesirable; they interfered severely with the sewage treatment and also inhibited the normal uptake of oxygen by natural waters. Consequently these biologically hard surfactants had to be withdrawn in densely populated areas.

Biologically Soft Straight-chain Alkylbenzenesulphonates. The slow rate of biodegradation of propylene-tetramer-based alkylbenzenesulphonates is caused by the highly branched alkyl group which they contain. Sodium alkylbenzenesulphonates with straight-chain alkyl groups are much more rapidly biodegraded, i.e. they are biologically 'soft'. A number of routes to straight-chain alkylbenzenes were developed and commercialized in the early 1960's and the hard propylene-tetramer-based materials have now been replaced by soft alkylbenzenesulphonates in most of the major world markets. These straight-chain sodium alkylbenzenesulphonates are cheap and efficient and are by far the most successful synthetic surfactants to date. Although they are now being challenged by a number of newly developed materials (see p. 467) it seems likely that they will continue to be important for quite a number of years.

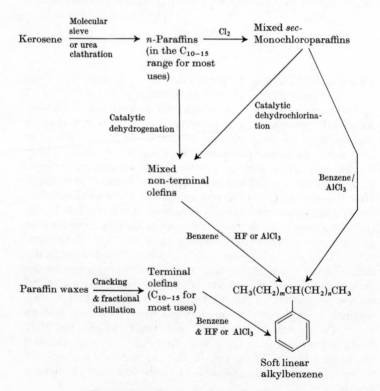

Figure 13.2.   Routes to linear alkylbenzenes.

The long straight-chain alkyl group in these soft alkylbenzene-sulphonates is derived either from *n*-paraffins in the C_{10-15} range or from paraffin waxes (C_{15-35}). The *n*-paraffins are isolated from kerosene petroleum fractions by molecular-sieve techniques or urea-clathration methods; both methods give straight-chain paraffins of greater than 90% purity. The paraffins are converted either into monochloroparaffins or into non-terminal olefins which are then used to alkylate benzene under Friedel–Crafts conditions. Paraffin waxes are cracked to give a range of terminal olefins of widely different chain length; this mixture is fractionated to give the C_{10-15} fraction which is used to alkylate benzene. A number of industrial approaches are summarized in Figure 13.2 and further details of the steps involved are given below.

The paraffin chlorination is a radical reaction which gives random substitution (Part 1, p. 17; Part 2, p. 132). Chlorination of dodecane, for example, gives almost equal amounts of 2-, 3-, 4-, 5- and 6-monochlorododecane together with smaller amounts of the 1-isomer. Only 20–30% of the paraffin is chlorinated, as higher degrees of chlorination give a significant amount of poly-substitution; this could not be tolerated as it would give an inferior final product containing appreciable amounts of disodium diphenylalkanedisulphonate and other monosulphonates with indan (**1**) and tetralin (**2**) structures. The chloroparaffin–paraffin

(**1**)   (**2**)

mixture from the chlorination is fed to the alkylation step where it is treated with a large excess of benzene in the presence of $AlCl_3$-based catalysts. The excess of benzene and paraffin are distilled off and recycled to the appropriate stages of the process, and the alkylbenzene is recovered by distillation. This recycle procedure is simpler than separating the paraffin–chloroparaffin mixture before alkylation as the boiling points of an *n*-paraffin and its monochloro-derivatives are much closer than the boiling points of an *n*-paraffin and the phenylalkanes derived from it. Because of the low conversion in the paraffin chlorination, the chlorination–dehydrochlorination gives 20–30% of random non-terminal olefin in unchanged paraffin. The alternative paraffin dehydrogenation has similarly to be operated at low paraffin conversion to give a high yield of non-terminal mono-olefins substantially free from dienes. Thus, both routes involving non-terminal olefins generate an olefin–paraffin mixture, and recycle systems similar to that used for the chloroparaffin route are operated.

The properties of the straight-chain sodium alkylbenzene-sulphonates depend both on the chain length of the alkyl group and the position of the benzene ring on the alkyl chain. In particular, the properties of 2-phenylalkanesulphonates are significantly different from those of the isomers with the benzene ring near the centre of the chain, e.g. the latter are better foamers.

The isomer distribution in the alkylbenzene is determined partly by the isomer distribution in the olefin or chloroparaffin used as alkylating agent but is largely controlled by the alkylation conditions and catalyst. Olefins are isomerized by HF; and AlCl$_3$ can isomerize olefins, chloroparaffins and the product alkyl-benzenes. The extent of these isomerizations can be manipulated by varying the reaction conditions: for example, alkylation of benzene with dodec-1-ene can be made to yield products with 2-phenyldodecane contents ranging from 14% (HF catalysis, 55°C, hexane diluent) to 44% (AlCl$_3$ catalysis, 0–5°C). Alkyl-benzenes with low 2-phenyl isomer contents are generally considered best and a number of Companies have developed their individual approaches for controlling the isomerizations so as to minimize 2-phenyl content.

Sulphonation. Alkylbenzenes are sulphonated commercially either with oleum or with gaseous sulphur trioxide. Both reagents give very high yields of predominantly *para*-monosulphonate.

The oleum process, which can be carried out in simple equipment, was the first method used and is still employed to a considerable extent. The alkylbenzene is treated with an excess of oleum (usually ca. 20% of SO$_3$ in 100% H$_2$SO$_4$) to give a >95% yield of sulphonate. Sufficient water is then added to the reaction mixture to dilute the sulphuric acid present to about 71%. At this concentration, the aqueous sulphuric acid forms a second phase which can be separated to leave only ca. 10% of sulphuric acid in the alkylbenzenesulphonic acid phase. Neutralization with aqueous NaOH after phase separation gives a slurry containing 40–55% of sodium alkylbenzenesulphonate and 5–6% of sodium sulphate, this mixture being suitable for processing into the majority of finished detergent products.

Sulphur trioxide sulphonation avoids the excess of sulphuric acid and has now become the preferred method. The reaction of SO_3 with alkylbenzenes is extremely rapid and highly exothermic. An excess of SO_3 has to be avoided as this gives badly discoloured products. In modern sulphonation plants, the gaseous SO_3, diluted with air (ca. 3–10% of SO_3 by volume), is passed concurrently over a film of alkylbenzene flowing down a heat exchanger surface. This gives efficient absorption of SO_3 from the gas phase and enables the heat of reaction to be efficiently removed. Such a system enables the SO_3 : alkylbenzene ratio and temperature to be controlled accurately and gives a high-quality sulphonic acid product virtually free from sulphuric acid. The sulphonic acid is usually neutralized with aqueous sodium hydroxide to give a 40–60% slurry suitable for processing into finished detergent products (see p. 483).

Other Synthetic Anionic Surfactants. From the point of view of surfactancy, the benzene ring in alkylbenzenesulphonates is not strictly necessary. It is really present in the molecule to provide a convenient means of attaching the hydrophilic sulphonate group to a hydrophobic alkyl chain. A number of other methods for attaching long alkyl chains to anionic groups have been developed and are detailed below. Most of the anionic surfactants produced by these methods could give performance comparable to that of alkylbenzenesulphonates in fabric washing and/or other detergency applications, and each of them has its own technical and economic advantages and disadvantages. In the future, some of these alternative surfactants may well displace part of the alkylbenzenesulphonate now used, but it is doubtful whether any one of them will reach the dominant position that alkylbenzenesulphonates have occupied for the last 15 years.

(a) Olefin-sulphonates. In the early 1960's good-quality long-chain terminal olefins began to appear in commercial quantities. Two routes are used for their manufacture. The first, the cracking of paraffin waxes which was mentioned above in connexion with alkylbenzene production, gives cheap olefins of fairly good quality, which contain diene and naphthenic impurities. The second route, based on the telomerization of ethylene with triethylaluminium or similar catalysts, gives very good quality olefins which are rather more expensive than the cracked-wax materials. Both routes produce a wide range of olefins and the price of the C_{10-18} 'detergent-range' materials depends on the

value of the olefins of chain length less than C_{10} and greater than C_{18} which are inevitably also produced.

The telomerization route involves a two-stage reaction. In the first stage, ethylene reacts with triethylaluminium at moderate temperature and fairly high pressure (e.g. 100°C and 10×10^6 Nm^{-2}), a growth reaction (*a*) occurring. Once the desired chain-

$$(CH_3CH_2)_3Al + (p + q + r)H_2C{=}CH_2 \longrightarrow \begin{array}{l} CH_3CH_2(CH_2CH_2)_p \\ CH_3CH_2(CH_2CH_2)_q{-}Al \\ CH_3CH_2(CH_2CH_2)_r \end{array} \quad (a)$$

length distribution has been reached, the reaction conditions are changed, either by substantially increasing the temperature (e.g. to 200–300°C) or by introducing a displacement catalyst such as colloidal nickel, so as to give a displacement reaction (*b*). This

$$\begin{array}{l} CH_3CH_2(CH_2CH_2)_p \\ CH_3CH_2(CH_2CH_2)_q{-}Al + 3H_2C{=}CH_2 \longrightarrow \\ CH_3CH_2(CH_2CH_2)_r \end{array}$$

$$\begin{array}{c} CH_3CH_2(CH_2CH_2)_{p-1}CH{=}CH_2 \\ + \\ CH_3CH_2(CH_2CH_2)_{q-1}CH{=}CH_2 + (CH_3CH_2)_3Al \quad (b) \\ + \\ CH_3CH_2(CH_2CH_2)_{r-1}CH{=}CH_2 \end{array}$$

liberates long-chain olefins and regenerates the triethylaluminium which is recycled. The net reaction is thus the conversion of ethylene into a mixture of long-chain olefins. In a simple 'single-growth', 'single-displacement' system, this mixture contains the various olefins in proportions that fall on a Poisson distribution. Various multigrowth, multidisplacement systems have been evolved to narrow this distribution, but the operation of these to produce solely detergent-range olefins appears to be too costly and hence olefin manufacturers using the telomerization route still have to face the problem of disposing of olefins of a wide range of chain lengths.

Olefins are sulphonated with sulphur trioxide in film reactors (see p. 467). The reaction is very complex, giving alkenesulphonic acids and sultones as major products together with minor amounts of sultone-sulphonic acids, alkenedisulphonates and other products. Neutralization/hydrolysis of this sulphonation product with aqueous sodium hydroxide gives a mixture of sodium alkene-sulphonates (ca. 60%), sodium hydroxyalkanesulphonates (ca. 22%) and disulphonates (ca. 18%). The formation of the major products can be rationalized by the following Scheme.

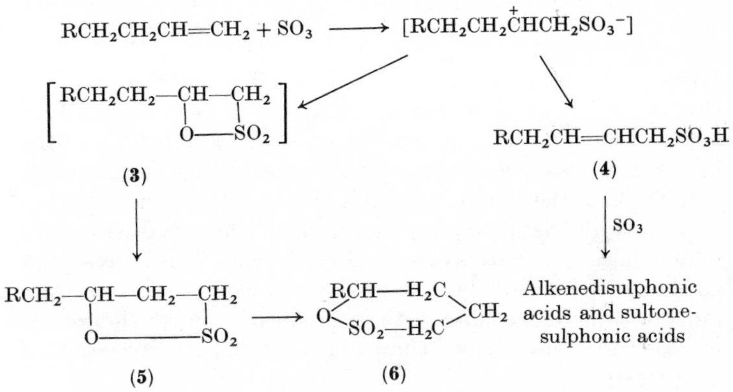

$$RCH_2CH_2CH{=}CH_2 + SO_3 \longrightarrow [RCH_2CH_2\overset{+}{C}HCH_2SO_3{}^-]$$

$$\left[\begin{array}{c} RCH_2CH_2{-}CH{-}CH_2 \\ \quad\quad\quad | \quad\quad | \\ \quad\quad\quad O{-\!-}SO_2 \end{array}\right]$$

(3)

$$RCH_2CH{=}CHCH_2SO_3H$$

(4)

$$\Big| SO_3$$

$$\begin{array}{c} RCH_2{-}CH{-}CH_2{-}CH_2 \\ \quad\quad | \quad\quad\quad\quad | \\ \quad\quad O{-\!-\!-\!-\!-}SO_2 \end{array}$$

(5)

$$\longrightarrow \quad O{\Big\langle}\begin{array}{c} RCH{-}H_2C \\ \\ SO_2{-}H_2C \end{array}{\Big\rangle}CH_2$$

(6)

Alkenedisulphonic acids and sultone-sulphonic acids

In the neutralization/hydrolysis step, the sulphonic acids are simply neutralized and the sultones are converted into mixtures of hydroxyalkane- and alkene-sulphonates, e.g. from (5) $RCH{=}CHCH_2CH_2SO_3Na$, $RCH_2CH{=}CHCH_2SO_3Na$ and $RCH_2CH(OH)CH_2CH_2SO_3Na$. Sodium 2-hydroxyalkane-1-sulphonate, presumed to be derived from the β-sultone (3), has been isolated from the neutralized hydrolysis products from sulphonations carried out under mild conditions; this is the only concrete evidence for the presence of this presumably unstable sultone, which has not been isolated from the sulphonation product. Under normal industrial sulphonation conditions in a film reactor, the 3- and 4-hydroxyalkanesulphonates, from the sultones (5) and (6), respectively, are the major hydroxy-mono-sulphonate products. Part of the unsaturated sulphonic acid (4) produced reacts with a further molecule of sulphur trioxide to give an analogous series of disulphonate products.

The complex mixture of sulphonates produced from these reactions, termed α-olefin sulphonate, is rapidly biodegradable and is an excellent surfactant.

(*b*) *Alkyl sulphates*. These were amongst the first synthetic surfactants to be used in detergent products and were originally prepared by sulphating long-chain alcohols produced by hydrogenation of esters of fatty acids, e.g.:

$$CH_3(CH_2)_{16}COOR \xrightarrow[\substack{Cu \\ chromite}]{H_2}$$

$$CH_3(CH_2)_{16}CH_2OH \xrightarrow[\text{2, NaOH}]{\text{1, SO}_3} CH_3(CH_2)_{16}CH_2OSO_3Na$$
$$+ ROH$$

Natural oils and fats can either be reduced directly to give fatty alcohols and glycerol or transesterified with a short-chain alcohol before reduction. This process is still operated but two other industrial routes to long-chain primary alcohols are now available.

In the first, the aluminium alkyls produced by telomerization of ethylene with triethylaluminium (see p. 467) are oxidized with air to alkoxides which are hydrolysed to alcohols. This process gives good-quality alcohols but, as in the long-chain olefin process, a range of different chain-length is produced. Unlike the α-olefin process, the aluminium compounds cannot be recycled (cf. Chapter 4).

The second route uses hydroformylation (Chapter 6) in which an olefin reacts with carbon monoxide and hydrogen in the presence of a cobalt-based catalyst. The primary products are aldehydes but the process conditions can be arranged to reduce these *in situ*

$$CH_3(CH_2)_9CH{=}CH_2 \xrightarrow[\text{Co}_2(CO)_8]{H_2{-}CO}$$

$$[CH_3(CH_2)_9CH(CH_3)CHO + CH_3(CH_2)_9CH_2CH_2CHO] \xrightarrow{H_2}$$
$$CH_3(CH_2)_9CH(CH_3)CH_2OH + CH_3(CH_2)_9CH_2CH_2CH_2OH$$

to alcohols. The above reaction scheme is simplified, as the hydroformylation catalysts also isomerize the olefins. Terminal olefins are hydroformylated faster than non-terminal olefins and hence, despite the isomerization, the alcohol produced consists predominantly of the straight-chain alcohol and the 2-methyl-alkan-1-ol shown in the Scheme. Smaller amounts of other branched alcohols (2-ethyl-, 2-propyl-, etc.) are formed from the non-terminal olefins produced by the isomerization. The product composition depends greatly on reaction conditions and can be varied within the range 45–90% of straight-chain alcohol and 10–55% of branched-chain alcohol.

Sulphur trioxide is the most convenient sulphating agent, and reaction of primary alcohols with this in, for example, film reactors (see p. 467) gives the sulphate in up to 95% yields. Sulphuric acid and chlorosulphonic acid can also be used for the sulphation but are less convenient in practice.

Primary alkyl sulphates, which are rather more expensive than alkylbenzenesulphonates, have only been used to a limited extent. The related alkyl ether sulphates, e.g. (7), prepared by condensing primary alcohols with ethylene oxide (see p. 474) and sulphonating the resultant ethoxylates, are widely used in dishwashing liquids.

$$CH_3(CH_2)_{11}(OCH_2CH_2)_3OSO_3Na$$
$$\text{(7)}$$

Secondary alkyl sulphates are prepared by treating long-chain olefins with 90–95% sulphuric acid, as illustrated in the following Scheme (this is simplified and does not show the isomeric *sec*-alkyl

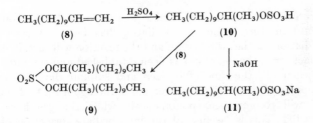

sulphates produced by rearrangement of the intermediate carbonium ions). The side reaction producing the dialkyl sulphate (9) presents a problem: it can be only partially suppressed by using an excess of sulphuric acid. Two approaches for dealing with this have been evolved. In one, a solvent is added and the reaction conditions are adjusted so that the half-ester (10) crystallizes and is not available for reaction with a second olefin molecule. In the second, an excess of sodium hydroxide is added to the sulphation product, and the alkaline solution is heated; this hydrolyzes the dialkyl sulphate to give the required monosulphonate (11) and secondary alcohol which is extracted with petroleum solvent and used in other petrochemical processes.

These surfactants are used industrially and in some domestic liquid detergent products.

Alkanesulphonates. Linear paraffins have been mentioned above as raw materials for the production of linear alkylbenzene-sulphonates. They can also be converted directly into surfactants of general formula $RR'CHSO_3Na$ by reaction with sulphur dioxide and oxygen. This reaction, termed sulphoxidation, is a

radical-chain process and is initiated either by ultraviolet or
γ-irradiation (Part 1, p. 179). The following sequence of reactions
explains the observed features of the sulphoxidation of *pure*
n-paraffins (RH represents the paraffin):

$$\begin{align}
\text{RH} &\xrightarrow{\text{Initiation}} \text{R}\cdot \\
\text{R}\cdot + \text{SO}_2 &\longrightarrow \text{RSO}_2\cdot \\
\text{RSO}_2\cdot + \text{O}_2 &\longrightarrow \text{RSO}_2\text{OO}\cdot \\
\text{RSO}_2\text{OO}\cdot + \text{RH} &\longrightarrow \text{RSO}_2\text{OOH} + \text{R}\cdot \\
\text{RSO}_2\text{OOH} &\longrightarrow \text{RSO}_2\text{O}\cdot + \cdot\text{OH} \\
\text{RSO}_2\text{O}\cdot + \text{RH} &\longrightarrow \text{RSO}_2\text{OH} + \text{R}\cdot \\
\cdot\text{OH} + \text{RH} &\longrightarrow \text{H}_2\text{O} + \text{R}\cdot \\
\text{RSO}_2\text{OOH} + \text{H}_2\text{O} + \text{SO}_2 &\longrightarrow \text{RSO}_2\text{OH} + \text{H}_2\text{SO}_4
\end{align}$$

The reaction has one remarkable feature, namely that once a
certain limiting amount of radiation has been absorbed, the
irradiation can be switched off and the reaction will sustain itself.
In fact, once the reaction has been established and the radiation
switched off, the flow of oxygen and sulphur dioxide can be
stopped and the reaction can remain dormant for up to 2–3 days
and will recommence spontaneously when the gas flows are
restarted. This is because of the intermediate formation of the
peroxysulphonic acids, which act as chemical initiators. A mixture
of secondary alkanesulphonates is produced and the production
of disulphonic acids, which are too polar to be good surfactants,
is suppressed by operating the reaction with a large excess of
paraffin, which is phase-separated from the sulphonic acid
product and recycled.

Olefinic or branched-chain paraffin impurities in the paraffin
inhibit the reaction. This is usually attributed to their scavenging
the reactive radicals to give allylic or tertiary radicals which are
too stable to participate efficiently in the chain process. One way
to overcome this is to start the reaction with specially pure
paraffin and, once it is well established, to change to the paraffin
of lower industrial purity which is normally given by, e.g., urea
clathration.

These secondary alkanesulphonates are readily biodegradable
and perform well in a number of detergency applications.

Addition of sodium hydrogen sulphite to terminal olefins gives
sodium primary alkanesulphonates:

$$\text{RCH}=\text{CH}_2 + \text{NaHSO}_3 \longrightarrow \text{RCH}_2\text{CH}_2\text{SO}_3\text{Na}$$

This is also a radical-chain process, usually initiated by peroxy-compounds (e.g. *tert*-$C_4H_9OOCOC_6H_5$), and the main reactions involved in the addition are:

$$HSO_3^- + X\cdot \longrightarrow XH + SO_3^- \cdot$$
$$SO_3^- \cdot + RCH{=}CH_2 \longrightarrow R\dot{C}HCH_2SO_3^-$$
$$R\dot{C}HCH_2SO_3^- + HSO_3^- \longrightarrow RCH_2CH_2SO_3^- + SO_3^- \cdot$$

The major problem associated with this process is the provision of a reaction medium in which both the non-polar olefin and the sodium hydrogen sulphite are soluble: aqueous lower alcohols (e.g. propan-2-ol) are the most suitable; they do not give single-phase systems at the start of the reaction but both the reactants have sufficient solubility in the aqueous phase for efficient reaction.

The sodium salts of these terminal sulphonates are only sparingly soluble in water and this limits their use, but they can be used in conjunction with other surfactants.

Non-ionic Surfactants

Non-ionic surfactants are composed of a long hydrophobic alkyl group connected to a highly polar neutral group. The polar group must be sufficiently hydrophilic to take the hydrophobic group into aqueous solution. Sugar units such as glucose can do this and, for example, *n*-dodecyl β-glucoside (**12**) functions as a non-ionic surfactant.

$$HO\text{---}\!\!\!\overset{\textstyle OH}{\diagdown}\!\!\!\overset{O}{\diagup} \qquad HO\text{---}\!\!\!\overset{}{\diagup}\!\!\!\overset{}{\underset{OH}{\diagdown}}\!\!\!\overset{O}{\diagdown}(CH_2)_{11}CH_3$$

(12)

Some industrially produced non-ionic surfactants are in fact based on sugars and related polyols but by far the most important class are those represented by the general structure (**13**) in which the hydrophilic group is a polyoxyethylene chain, R contains a long hydrocarbon chain and X is a linking group such as —O— or —COO—. These are produced by treating a substrate RXH with

$$RXH + n\text{-}H_2C\!\!\underset{O}{\overset{}{\diagdown\!\!\diagup}}\!\!CH_2 \longrightarrow RX[CH_2CH_2O]_nH$$
(13)

ethylene oxide. Such reactions are termed ethoxylations and the products are mixtures of materials with different numbers of oxyethylene (—OCH_2CH_2—) groups attached to the hydrophobic

R group; n in formula (**13**) represents the average number of oxyethylene groups attached to the hydrophobe. As described below, the distribution of oxyethylene chain lengths is determined both by the properties of RXH and the ethoxylation conditions. Non-ionic surfactants are liquids or waxy solids and their properties differ in some important respects from those of charged surfactants. They produce lower surface tensions at equivalent concentration and give CMC's lower than those of charged surfactants containing the same hydrophobic group; this is because the electrical repulsions occurring between the polar groups of charged surfactants oriented either at interfaces or in micelles do not arise with non-ionic surfactants which are therefore more readily adsorbed at interfaces and aggregate more easily in micelles.

These surfactants also show unusual solubility characteristics. If the temperature of an aqueous solution of a polyoxyethylene non-ionic surfactant is raised, a temperature termed the cloud point is reached at which the solution becomes turbid. Above the cloud point, a surfactant-rich second phase separates, leaving very little surfactant in the water; below the cloud point, most non-ionic surfactants are miscible in all proportions with water. The cloud point of a given surfactant depends on the structure of its hydrophobic group and the number of oxyethylene units it contains. Longer hydrocarbon chains obviously require more oxyethylene units to take them into solution, e.g. a n-decyl chain attached to three oxyethylene units is virtually water-insoluble whilst the same chain attached to four oxyethylene units is water-miscible at room temperature. For a n-hexadecyl chain the solubility transition occurs at 5–6 units. The explanation generally accepted is that the non-ionic surfactant is taken into solution below its cloud point by hydration of its polyoxyethylene chain and that, as the temperature is raised, a point (the cloud point) is reached where this rather loosely bound hydration sheath is sufficiently broken down to render the surfactant insoluble.

The Ethoxylation Reaction. Ethylene oxide does not react with protic substrates unless a basic or acidic catalyst is present or the substrate is itself basic (e.g. an amine).

(a) *Base-catalysed ethoxylation.* Because of its strained three-membered ring, ethylene oxide is susceptible to nucleophilic attack and the following reactions occur during condensation of a substrate RXH with ethylene oxide in the presence of a basic catalyst B^- as illustrated in reactions (3)–(9).

$$RXH + B^- \xrightleftharpoons{K_0} RX^- + BH \tag{3}$$

$$RX^- + H_2C\overset{\displaystyle\diagdown\diagup}{\underset{O}{}}CH_2 \xrightarrow{k_1} RXCH_2CH_2O^- \; (14) \tag{4}$$

$$RXCH_2CH_2O^- + RXH \xrightleftharpoons{K_1} RXCH_2CH_2OH + RX^- \tag{5}$$

$$RXCH_2CH_2O^- + H_2C\overset{\displaystyle\diagdown\diagup}{\underset{O}{}}CH_2 \xrightarrow{k_2} RXCH_2CH_2OCH_2CH_2O^- \; (15) \tag{6}$$

$$RXCH_2CH_2OCH_2CH_2O^- + RXH \xrightleftharpoons{K_2}$$
$$RXCH_2CH_2OCH_2CH_2OH + RX^- \tag{7}$$

$$RXCH_2CH_2OCH_2CH_2O^- + RXCH_2CH_2OH \xrightleftharpoons{K_3}$$
$$RXCH_2CH_2OCH_2CH_2OH + RXCH_2CH_2OH \tag{8}$$

$$RXCH_2CH_2OCH_2CH_2O^- + H_2C\overset{\displaystyle\diagdown\diagup}{\underset{O}{}}CH_2 \xrightarrow{k_3}$$
$$RXCH_2CH_2OCH_2CH_2OCH_2CH_2O^- \text{ etc.} \tag{9}$$

The substrate RXH first reacts with the catalyst to give the corresponding anion RX^- (eqn. 3). This then reacts with ethylene oxide to give the alkoxide (**14**) (eqn. 4) which reacts either with a further molecule of ethylene oxide to give the alkoxide (**15**) containing two polyoxyethylene groups (eqn. 6) or with the substrate to regenerate RX^- (eqn. 5). The alkoxide (**15**) undergoes analogous reactions, the first set of which are shown (eqns. 7–9), and the process continues, generating compounds with longer and longer polyoxyethylene chains until all the ethylene oxide is consumed.

The reversible reactions (eqns. 3, 5, 7, 8, etc.), which involve only proton-transfer, are fast. The S_N2 ring-opening reactions (eqns. 4, 6, 9, etc.) are much slower. The product distribution depends on the amount of catalyst used, the acidities of RXH and $RX[CH_2CH_2O]_nH$, which determine K_0, K_1, etc., and the nucleophilicities of the anions RX^- and $RX[CH_2CH_2O]_n^-$ which determine the rate constants k_1, k_2, etc.

The reaction of the catalyst itself with ethylene oxide, omitted from the above scheme, can sometimes be important. Most industrial ethoxylations use basic catalysts, usually sodium hydroxide, at ca. 0.5 mole-%. When alcohols are used as sub-strates, HO^- cannot function directly as a catalyst because it is insufficiently basic to remove a proton from the alcohol. In fact, it first reacts with ethylene oxide to produce the anion of ethylene glycol which is a much stronger base and can catalyse the alcohol ethoxylation:

$$HO^- + H_2C{-}CH_2 \longrightarrow HOCH_2CH_2O^-$$
$$\underset{O}{\bigvee}$$

$$HOCH_2CH_2O^- + ROH \rightleftharpoons HOCH_2CH_2OH + RO^-$$

The resultant ethylene glycol can undergo a series of reactions analogous to the above set to give polyethylene glycols as by-products:

$$HOCH_2CH_2OH + n\text{-}H_2C{-}CH_2 \longrightarrow HO[CH_2CH_2O]_{n-1}H$$
$$\underset{O}{\bigvee}$$

Water in the raw materials can clearly give the same result and, for efficient ethoxylation, low hydroxide and water concentrations must be used.

In industrial ethoxylations, the hydrophobic substrate and catalyst are heated to 120–190°C and the required amount of ethylene oxide is admitted under carefully controlled conditions of temperature and pressure at a rate that just matches its rate of absorption. The reaction is highly exothermic and the concentration of ethylene oxide in the reaction mixture is kept at the minimum consistent with an adequate reaction rate.

(b) *Alkylphenol ethoxylates.* The first non-ionic surfactants to be widely used were ethoxylated alkylphenols. Nonylphenol, produced by BF_3-catalyzed alkylation of phenol with propylene trimer, is the most important raw material for these, but dodecyl- and octyl-phenol, produced by use of propylene tetramer and di-isobutene, respectively, have also been used to a considerable extent. The non-ionic surfactants produced from these branched-chain phenols are biologically hard and are now being replaced by biologically softer materials.

Ethoxylation follows the pattern shown in Figure 13.3. Because the phenol is much more acidic than its first ethoxamer (**17**), the equilibrium in eqn. (10), which corresponds to eqn. (5) in the

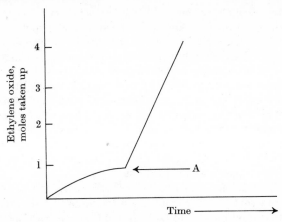

Figure 13.3. Rate of addition of ethylene oxide to an alkylphenol.

scheme on p. 475, is almost completely to the right. Thus the phenoxide ion (**18**) is the only ion present in appreciable concentration. The nucleophilicity of (**18**) is lower than that of (**16**) but

$$RC_6H_4OCH_2CH_2O^- + RC_6H_4OH \rightleftharpoons$$
(**16**)

$$RC_6H_4CH_2CH_2OH + RC_6H_4O^- \quad (10)$$
(**17**) (**18**)

this is far from sufficient to compensate for the difference in concentration between them, so that the first mole of ethylene oxide adds almost exclusively to (**18**). The difference in reactivity between the alkoxides does, however, account for the slow initial addition of ethylene oxide (up to A in Figure 13.3). Once the first mole of ethylene oxide has reacted, consuming all the phenol, the more reactive alkoxide (**16**) takes over, and the ethoxylation accelerates and proceeds normally to give close to a Poisson distribution of products [the equilibria corresponding to eqn. (8) have constants close to unity and the rate constants for the reactions corresponding to eqns. 6, 9, etc. are very similar].

(*c*) *Alcohol ethoxylates.* Long-chain primary alcohols from ethylene telomerization and the hydroformylation process, as well as those produced by reduction of fats, have all been used as raw materials for this type of non-ionic surfactant.

Primary alcohols are only slightly less acidic than their poly-oxyethylene derivatives, and the alkoxides corresponding to the alcohol and these derivatives have very similar nucleophilicities. Thus, base-catalyzed ethoxylation results in approximately a Poisson distribution of polyoxyethylene derivatives. The product, in fact, has a distribution slightly flatter than a Poisson distribution and contains rather more unconverted alcohol than would be required by Poisson statistics. Thus, like alkylphenols, primary alcohols can be treated efficiently with ethylene oxide and basic catalysts to give a narrow distribution of products peaking at polyoxyethylene chain-lengths close to the number of moles of ethylene oxide condensed with the substrate.

Secondary alcohols behave rather differently. They are important raw materials for non-ionic surfactants and are produced by the air-oxidation (5% of O_2 in N_2) of C_{10-15} n-paraffins in the presence of boric acid at 165°C, as follows (R and R' are n-alkyl groups). This is a radical-chain process and substitution occurs

$$3CH_2RR' + 3H_3BO_3 \xrightarrow{\ O_2\ }$$

$$\xrightarrow[H_2O]{} RR'CHOH$$

preferentially at the secondary carbon atoms to give an equimolar mixture of all the possible isomeric secondary alcohols.

These alcohols are substantially less acidic than their poly-oxyethylene derivatives and the equilibrium in eqn. (11) is therefore displaced largely to the left-hand side.

$$RR'CHOH + RR'CH(OCH_2CH_2)_nOCH_2CH_2O^- \; \underset{\longleftarrow}{\text{-----}\rightarrow}$$

(**19**)

$$RR'CHO^- + RR'CH(OCH_2CH_2)_nOCH_2CH_2OH \quad (11)$$

(**20**)

Because of this, under normal base-catalysed conditions (ca. 0.5 mole-% of base), the total alkoxide present contains only a minor proportion of (**20**), even when only a small amount of ethylene oxide has been added. This, coupled with the fact that (**19**) and (**20**) have comparable nucleophilicities, means that the products react in preference to the starting alcohol, and such an ethoxylation gives a wide shallow distribution of products peaking at polyoxyethylene chain lengths substantially higher than the number of moles of ethylene oxide added. A considerable proportion of the starting alcohol remains unconverted even when quite a large amount of ethylene oxide (e.g. 20 moles per mole of alcohol) has been added. Products with such wide distributions have unsatisfactory properties and secondary alcohols have to be ethoxylated by other methods.

One obvious solution to this problem is to convert all the alcohol into alkoxide; this avoids the directive effects of the equilibrium reactions and gives close to a Poisson distribution of products. This cannot be done cheaply, however. The use of a molar amount of sodium hydroxide simply results in the production of polyethylene glycol in the alcohol as solvent (see p. 476) and the use of short-chain sodium alkoxides or direct dissolution of sodium in the secondary alcohol is expensive. A two-stage procedure (described on p. 481) is in fact used.

Alkyl and alkylphenyl ethoxylates are the most important non-ionic surfactants. They are used in household detergents (see p. 483) and in numerous industrial applications such as textile processing, paper making, rubber processing and the manufacture of emulsion paints.

(*d*) *Fatty acid ethoxylates.* Ethoxylation of fatty acids has some features in common with ethoxylation of phenols (acidic substrate which gives an anion of low nucleophilicity) but a complication is that the conditions for base-catalysed ethoxylation are highly suitable for rapid transesterification. Hence the products contain free polyethylene glycol and its diester as well as the mono-(polyethyleneglycol) ester.

These non-ionic surfactants have been used in a variety of household and industrial applications but have not achieved the importance of alkyl and alkylphenyl ethoxylates.

(*e*) *Amine ethoxylates.* Alkylamines are sufficiently basic to react directly with ethylene oxide and they first give diethanolamines by normal S_N2 processes:

$$RNH_2 + 2H_2C\!-\!CH_2 \longrightarrow RN\!\!\begin{array}{l} CH_2CH_2OH \\ CH_2CH_2OH \end{array}$$

Higher ethoxylates can be produced by conventional base-catalysed ethoxylation of the diethanolamines.

These products are mostly used in applications in which they are protonated and function as cationic surfactants (see p. 481).

(*f*) *Acid-catalysed ethoxylation.* The mechanism of Lewis-acid-catalysed ethoxylation is less clear-cut than that of the base-catalysed process. An S_N1-type reaction (*c*) involving opening of

$$H_2C\!\!-\!\!CH_2 \xrightarrow{BF_3} \overset{H_2C\!\!-\!\!CH_2}{\underset{O\!-\!\overset{-}{B}F_3}{\overset{+}{}}} \longrightarrow \overset{O\!-\!CH_2\overset{+}{C}H_2}{\underset{-BF_3}{}} \xrightarrow{RXH}$$

$$\overset{O\!-\!CH_2CH_2\overset{+}{X}HR}{\underset{-BF_3}{}} \longrightarrow \overset{\overset{+}{H}O\!-\!CH_2CH_2XR}{\underset{-BF_3}{}} \xrightarrow{-BF_3} HOCH_2CH_2XR \quad (c)$$

the epoxide ring followed by reaction of the substrate has been proposed. Under most conditions, this seems less likely than S_N2-type attack (*d*) on the epoxide–catalyst complex. The actual

$$\overset{RX:}{\underset{H}{\overset{\curvearrowright}{|}}} \quad \overset{H_2C\!-\!CH_2}{\underset{O\!-\!\overset{-}{B}F_3}{\overset{+}{\curvearrowleft}}} \longrightarrow R\overset{+}{H}XCH_2CH_2O\overset{-}{B}F_3 \longrightarrow$$

$$\overset{RXCH_2CH_2\!-\!\overset{+}{O}H}{\underset{-BF_3}{|}} \xrightarrow{-BF_3} RXCH_2CH_2OH$$

$$(d)$$

mechanism depends on the structure of the substrate RXH, the catalyst and the reaction conditions and is probably somewhere between pure S_N1 and S_N2. As would be expected from these mechanisms, the distribution of products is much less dependent on substrate structure than is the case for the base-catalysed process.

Lewis acids (BF_3, $SnCl_4$, $SbCl_5$) are the preferred catalysts. The reaction is fast and exothermic but despite the fact that it occurs under milder conditions than the base-catalysed process (ca. 80°C compared with ca. 150°C) it gives lower yields and more by-products. The major by-products are polyethylene glycol (presumably produced by dehydration followed by addition of ethylene oxide), and cyclization products such as dioxan, probably produced by reactions such as (*e*). Because of this by-product formation, acid-catalysed ethoxylation is not widely used, except in one major application, namely the commercially important two-stage ethoxylation of secondary alcohols.

$$RX(CH_2CH_2O)_nCH_2CH_2-O: \overset{\underset{\displaystyle |}{CH_2}}{\underset{HO^+-\bar{B}F_3}{}} \overset{\displaystyle H_2C-\overset{H_2}{C}-O}{\underset{CH_2}{\diagdown}} \longrightarrow$$

$$RX(CH_2CH_2O)_nCH_2CH_2-\overset{+}{O} \overset{\displaystyle H_2C-\overset{H_2}{C}-O}{\underset{CH_2}{\diagdown}} \quad + \text{ } HO\bar{B}F_3$$

$$\downarrow \qquad\qquad\qquad\qquad (e)$$

$$RX(CH_2CH_2O)_nCH_2CH_2\overset{+}{\underset{\underset{\displaystyle -BF_3}{|}}{O}}H \quad + \quad \overset{\displaystyle H_2C-\overset{H_2}{C}-O}{\underset{O\diagdown\diagup CH_2}{}}$$

As explained above, base-catalysed ethoxylation of secondary alcohols gives an unsatisfactory wide distribution of products. A product with a narrow distribution (close to the Poisson distribution) and good surfactant properties can be produced by first adding a small amount of ethylene oxide (ca. 1 mole) to the alcohol under acid-catalysed conditions, then distilling off and recycling the unconverted alcohol and adding further ethylene oxide to the distillation residue (mostly monoethoxylated alcohol with a small proportion of higher ethoxylates) under conventional base-catalysed conditions to give the required oxyethylene level.

Cationic Surfactants

Cationic surfactants are composed of a hydrophobic alkyl group attached to a positively charged hydrophilic group. All the industrially important materials of this class are based on quaternary nitrogen compounds or amines.

The most widely used route to them, termed the fatty route, starts from long-chain fatty acids. These are converted by NH_3 at 200–300°C into nitriles which are hydrogenated either to the corresponding primary amines or to amines containing two long-chain alkyl groups, as in (*f*). The primary amine, e.g. (**21**), is

$$n\text{-}C_{17}H_{35}COOH \longrightarrow n\text{-}C_{17}H_{35}CN \overset{\displaystyle \longrightarrow n\text{-}C_{17}H_{35}CH_2NH_2 \text{ } (\mathbf{21})}{\underset{\displaystyle \longrightarrow (n\text{-}C_{17}H_{25})_2NH \qquad (\mathbf{22})}{}} \qquad (f)$$

obtained by using Raney nickel at 100–150°C and $1.5–7 \times 10^6$ Nm^{-2} with ammonia added to suppress formation of secondary amine; the secondary amine, e.g. (22), is obtained by change of conditions to >200°C with venting to release ammonia.

The amines are converted into a number of derivatives that are used technically, the most important being quaternary compounds, e.g. (23), amine acetate (24) and the so-called 'fatty diamines' (25).

$$R_2NH \xrightarrow[\text{NaOH}]{CH_3Cl} (CH_3)_2NR_2^+ \ Cl^- \ \textbf{(23)}$$

$$RNH_2 \begin{cases} \xrightarrow[\text{NaOH}]{CH_3Cl} (CH_3)_3NR^+ \ Cl^- \\[6pt] \xrightarrow{CH_3COOH} {}^+NH_3R \ {}^-OOCCH_3 \ \textbf{(24)} \\[6pt] \xrightarrow{CH_2=CHCN} RNHCH_2CH_2CN \xrightarrow{H_2} RNHCH_2CH_2CH_2NH_2 \end{cases}$$

$$\textbf{(25)}$$

Their production is outlined in the above reactions, where R is usually a mixture of long-chain alkyl groups derived from tallow fatty acids.

Other important materials in this class are quaternized imidazolinium salts such as (26; R derived from hydrogenated tallow), and amine ethoxylates (see p. 479).

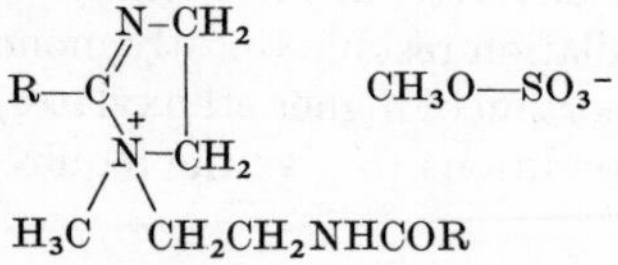

(26)

A variety of other routes to cationic surfactants exist, but the only other method of commercial importance involves the reactions:

$$RCH=CH_2 \xrightarrow[\text{peroxides}]{HBr} RCH_2CH_2Br \xrightarrow{NR'_3} RCH_2CH_2\overset{+}{N}R'_3Br^-$$

where $RCH=CH_2$ represents a long-chain olefin. The radical addition of hydrogen bromide is rapid and efficient, and the resultant terminal bromide can be treated with short-chain amines [e.g. $(CH_3)_2NH$] to give the required nitrogen derivatives.

Because the long-chain bases carry a positive charge and most surfaces are negatively charged in contact with aqueous solutions, these surfactants are substantive to a large number of materials.

Their adsorption tends to nullify the electrical repulsions which aid detergency and, hence, cationic surfactants cannot normally be used for cleaning purposes. However, the substantivity which produces a protective surface film and, in some cases, their bacteriostatic and bactericidal action, makes cationic surfactants valuable for a number of special applications. The most important of these are in fabric softening (quaternary compounds containing two long-chain groups, see p. 487), for surface modification of minerals (quaternary compounds), ore flotation (amine acetates), asphalt emulsions (amine ethoxylates and other derivatives), in the petroleum industry as corrosion inhibitors and as bacterio-stats/bactericides (salts of fatty diamines, and quaternary derivatives).

Household Detergent Products Based on Synthetic Surfactants

There are a very large number of uses of synthetic surfactants, some of which have been mentioned above, and here it is only possible to deal in some detail with the more familiar household detergency products.

Fabric-washing Products. These fall into two major classes: heavy-duty powders, the most important single detergent products, designed primarily for the main wash (white and heavily soiled articles and most coloured items); and light-duty powders and liquids designed for washing woollens, sensitive coloured articles, and the more delicate fabrics. Obviously, the major task of such products is to clean fabrics and clothes, and the main ingredients, surfactant and builder, as well as the anti-redeposition agents that they contain are there to do this. In addition, the products must be convenient and pleasant to use, and must help to retain the new appearance of articles that are repeatedly used and washed. Detergent formulations contain a number of ingredients providing these additional properties. Fabric-washing techniques differ from country to country owing both to social habits and to the different types of washing machines used. Fabric-washing products must also be formulated to take this into account.

(a) *Fluorescent whitening agents.* After repeated wash–wear cycles, many white articles tend to develop a yellow or grey appearance and fabric-washing products usually contain fluorescent whitening agents to combat this. The fluorescers do this by absorbing ultra-violet light (at ca. 360 nm) and re-emitting the absorbed energy by fluorescence at 430–440 nm in the blue region of the visible spectrum. The resultant 'blueing' corrects the tendency of fabrics to look yellow and makes them appear whiter and, since light is absorbed outside the visible range and re-emitted as visible light, the fabrics are in fact brighter. The light absorption must obviously be concentrated at frequencies corresponding to the ultraviolet light present in daylight and the absorption bands must not tail into the visible as this will produce yellowing, especially when the incident light does not contain much in the ultraviolet region. Emission outside this wavelength range can produce undesirable pastel-shade colouring of white fabrics. The fluorescers must, in addition, have high quantum efficiencies, be light-stable and, as they are expensive, be effective at very low concentrations in the product. It is so far impossible to build all these properties into a single molecule that will function for all fabrics and so fabric-washing products usually contain a fluorescer 'cocktail' comprised of compounds designed to be effective for both natural and synthetic fibres. For cotton, materials that function as direct dyes are used; these are usually large, planar conjugated molecules such as (27) solubilized by sulphonate groups, and are usually used at a level of 0.1–0.8% in the detergent formulation. For nylon, materials that function as disperse dyes and diffuse into the nylon fibre are used; because nylon is highly viscous, this diffusion is a relatively slow process; hence, small non-polar molecules, which give the best diffusion rates, are usually used at levels between 0.02 and 0.1%; (28) is a typical nylon fluorescer.

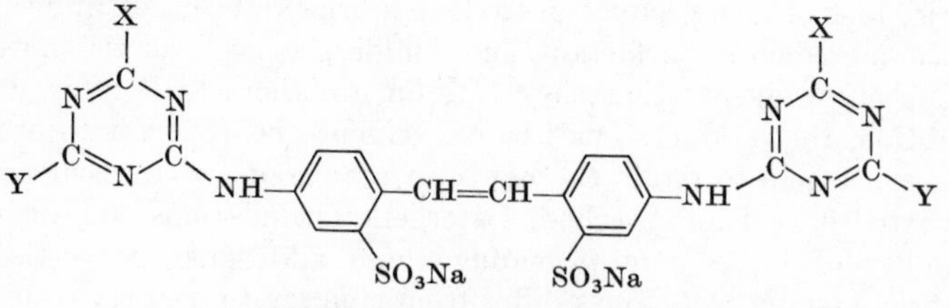

(27) X = Anilino. Y = Morpholino

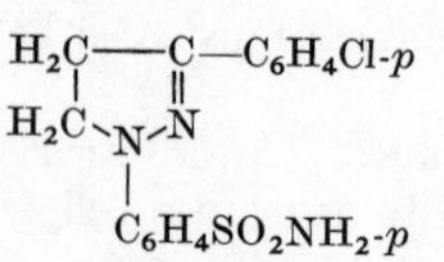

(28)

(*b*) *Bleaches*. In addition to providing general cleaning, a successful fabric-washing process must deal with localized stains from, e.g.

food and drink. Traditionally this is achieved either by incorporating a bleach in the fabric-washing formulation or by adding a separate bleach, usually aqueous sodium hypochlorite, to the wash. The only bleach used widely in fabric-washing powders is sodium perborate ($NaBO_2 \cdot H_2O_2 \cdot 3H_2O$). During the wash, this releases hydrogen peroxide which bleaches effectively above about 70°C. Compounds that generate per-acids in the wash by reaction with the hydroperoxide ion derived from the perborate can also be included in the formulations, e.g.:

$$p\text{-}CH_3COOC_6H_4SO_3Na + HOO^- \longrightarrow p\text{-}^-OC_6H_4SO_3Na + CH_3COOOH$$

Per-acids bleach at temperatures below 60°C and hence such systems are more effective than perborate systems at lower wash temperatures.

(c) *Enzymes.* Although conventional washing powders perform very well, they do not always remove certain difficult proteinaceous stains, e.g. blood. To solve this problem, powders containing proteolytic enzymes have recently been marketed. Such products function best when used in a cold-water soaking operation prior to the hot wash, although they are also very effective when used in the wash itself. They break down proteinaceous material during the soak and the result of the soak–wash cycle is very effective removal of the difficult stains.

(d) *Foaming.* The traditional view that high foaming is synonymous with good cleaning, which is true for the soap powders which were the only fabric-washing powders available before the advent of synthetic surfactants, is still held by housewives in some areas, particularly where manual bowl-washing and top-loading washing machines are used to a significant extent. For products based on synthetic surfactants no such correlation exists and it is possible to formulate products with good detergency properties to give almost any desired foam level.

The first synthetic detergent formulations were high sudsers, and such products are still very important. With the increasing incidence of automatic washing machines, most of which will not tolerate high-sudsing products (owing to overflow of foam) and in which the washing operation is totally enclosed so that the house-

$$n\text{-}C_{11}H_{23}CONHCH_2CH_2OH \qquad\qquad n\text{-}C_{12}H_{25}\overset{+}{N}(CH_3)_2$$
$$\underset{O^-}{\mid}$$

$$(29) \qquad\qquad\qquad\qquad (30)$$

wife does not in any case see the foam level, low-sudsing products are becoming more and more important.

High-sudsing is usually achieved by an anionic surfactant in combination with a lather-booster such as the ethanolamide (**29**)

Table 13.2. Typical composition ranges of heavy-duty fabric-washing powders

Component	Concentration (%) in: low-sudsing formulations	high-sudsing formulations
Sodium alkylbenzenesulphonate[a]	3–12	14–20[f]
Sodium soap[b]	2–10	—
Non-ionic surfactant[c]	2–5	—
Lather-booster[d]	—	2
Sodium phosphates[e]	25–60	30–50
Sodium perborate	0–30	0–20
Sodium silicate	4–8	5–10
Sodium sulphate	4–18	10–15
Sodium carboxymethylcellulose	1–2	0.5–1.5
Fluorescers, perfume, enzymes (if present) and water	to 100%	to 100%

[a] Usually C_{10-15} straight-chain alkyl.

[b] Usually tallow soap.

[c] E.g. C_{12-18} alcohols or C_9-alkylphenol, condensed with 9–18 moles of ethylene oxide.

[d] Usually C_{11-17}-alkyl—$CONHCH_2CH_2OH$.

[e] Usually 80% sodium triphosphate $Na_5P_3O_{10}$.

[f] In one very successful formulation, approximately half the alkylbenzene-sulphonate is replaced by sodium tallow-alcohol sulphate.

or an amine oxide such as (**30**). Low-sudsing is achieved by using a ternary mixture of anionic surfactant, soap and non-ionic surfactant. Table 13.2 gives typical composition ranges for high- and low-sudsing heavy-duty fabric-washing powders. The normal in-use concentration for these products lies in the range 0.1–1%.

Light-duty formulations are designed to give the milder washing conditions required for sensitive coloured articles, woollens and the more delicate fabrics. These formulations are based on the same main ingredients as heavy-duty formulations but contain smaller proportions of bleach and builder and give lower pH values in solution. Some formulations in fact contain neither builder nor bleach.

Fabric-washing powders are usually produced by taking the slurry from the alkylbenzene sulphonation, adding to this the other non-heat-sensitive ingredients of the formulation and pumping the resultant slurry through nozzles in the top of a spray-drying tower. The slurry atomizes and the resultant droplets are dried as they fall in a current of hot air to give the familiar spheroidal particles of the fabric-washing powder. Any heat-sensitive ingredients (perborate, enzymes, etc.) are dry-mixed into this base-powder. The final powder must be crisp and free-flowing and must not 'cake' during storage or use; the sodium silicate helps to provide these properties.

The phosphates used as builders are under pressure in some areas of the world as it is claimed that they accelerate eutrophication, i.e. the excessive growth of algae which leads to fouling of lakes. This is already considered to be a serious problem in some countries, particularly Sweden, the U.S.A. and Switzerland. Although phosphates from detergents account for only a part of the total phosphate reaching natural waters (agricultural fertilizers and human sewage are also major phosphate sources), concern over eutrophication could lead to replacement of detergent phosphates in some countries in the future if environmentally acceptable effective substitutes as safe as phosphate can be found.

Fabric-softening Products. Certain articles such as towels or nappies (diapers) can become harsh through normal washing and drying procedures. If a cationic surfactant is added to the final rinse water in the fabric-washing process, it is adsorbed on the fabric and imparts a pleasant soft feel. It is not possible to produce this effect by adding a cationic surfactant to the anionic detergent formulation used for washing as this would give an inactive cationic surfactant–anionic surfactant complex in the wash solution. Despite the complication of having to use two products, fabric-softening products are now well established and their use is growing rapidly. The best softening is achieved with quaternary compounds, such as (**31**) and (**32**), which possess two long chains.

$$(n\text{-}C_{18}H_{37})_2\overset{+}{N}(CH_3)_2 \quad Cl^-$$

(**31**)

$$n\text{-}C_{18}H_{37}-C\overset{\displaystyle N-CH_2}{\underset{\displaystyle \overset{+}{N}-CH_2}{\big|}} \quad X^-$$

$$H_3C \quad CH_2CH_2NHCOC_{18}H_{37}\text{-}n$$

(**32**)

Typical commercial products are aqueous solutions or dispersions containing 5–8% of quaternary salt, 0–1% of non-ionic surfactant, fluorescent whitening agents and perfume.

Dishwashing Liquids. Dishwashing is obviously more pleasant if the water is hot and foaming throughout the operation. A successful product must therefore both aid removal and suspension of food residues and give a high level of foam that is not readily depressed by fats and other food residues. As dishwashing involves immersion of the hands the products must also be mild to skin. The most important dishwashing liquids are aqueous solutions containing a mixture of surfactants together with additives such as ethanol, sodium xylenesulphonate or urea which help to keep the surfactants in solution and give the product a suitable viscosity. Other ingredients (e.g. perfume, colourants, glycerol) are often present. There is a variety of products on the market with surfactant concentrations ranging from 10% for cheap products to 40% for more expensive ones. Table 13.3 gives compositions of typical high-concentration products.

Table 13.3. Typical composition ranges for dishwashing liquids

Component	Concentration (%) in formulation:	
	Type 1	Type 2
Surfactants		
Sodium alkylbenzenesulphonate[a]	18–26	10–25
Sodium alkyl ether sulphate[b]	10–18	—
Alkylphenol-based non-ionic[c]	—	2–10
N-Alkylethanolamide[d]	1–5	—
Solubilizing agents[e]		
Ethanol	5–15	5–15
Urea	2–8	2–8
Sodium xylenesulphonate	2–8	2–8
Perfume, colourants, other additives and water	to 100%	to 100%

[a] Usually C_{10-12} straight-chain alkyl.

[b] n-$C_{12}H_{25}(OCH_2CH_2)_3OSO_3Na$ or mixture of compounds with similar chain length.

[c] E.g. octyl- or nonyl-phenol condensed with 10–12 moles of ethylene oxide.

[d] E.g. n-$C_{11}H_{23}CONHCH_2CH_2OH$.

[e] One or two of these ingredients may be present. All three are not normally used together.

Bibliography

A. M. Schwarz, J. W. Perry and J. Berch, *Surface Active Agents and Detergents*, Interscience, New York, Vol. 1, 1949; Vol. 2, 1958.

Kirk-Othmer, *Encyclopedia of Chemical Technology*, Interscience, New York, 1963, under headings *Detergency, Soap, Sulfonation and Sulfation*, and *Surfactants*.

K. Durham (Ed.), *Surface Activity and Detergency*, Macmillan, London, 1961.

K. Shinoda, *Solvent Properties of Surfactant Solutions*, Edward Arnold, London, 1968.

M. Sittig, *Practical Detergent Manufacture*, Noyes Development Corporation, Park Ridge, New Jersey, U.S.A., 1968.

M. J. Schick, *Nonionic Surfactants*, Edward Arnold, London, 1967.

Normal Paraffins, Supplement to *European Chemical News*, December 2nd, 1966.

A. K. Sarkar, *Fluorescent Whitening Agents*, Merrow Publishing Co. Ltd., Watford, England, 1971.

W. P. Evans, 'Cationic Fabric Softeners,' *Chem. and Ind.*, **1969**, 893.

(i) Chemistry and Combustion of Petroleum Fuels

(D. M. WHITEHEAD, BP Research Centre, Sunbury-on-Thames)

(ii) Fuels, Explosives and Propellants

(O. A. GURTON, Nobel's Explosives Company Limited)

(i) CHEMISTRY AND COMBUSTION OF PETROLEUM FUELS

'Surely, among the millions of fire worshippers and fire users who have passed away in earlier ages, some have pondered over the mystery of fire.'

William Crookes, 1861

Fire and Heat

Fire, resulting from the combustion of a fuel, was for over 100,000 years a mysterious phenomenon essential to man's survival, broadly under his control, but until comparatively recently incomprehensible to him.

The nature of heat produced by fire was one of the earliest subjects for serious speculation. Greek philosophers classified fire (with air, water and earth) as one of the essential elements and these earlier theories of fire as 'matter' persisted in one form or another until the beginning of the seventeenth century. At this time the beginnings of modern scientific method began to be applied to observable phenomena and Francis Bacon's experiments with the structure of the candle flame possibly began the slow revolution in the understanding of combustion technology and heat with which so many famous names (chemists, physicists and engineers) are associated. These studies were instrumental in the publication in 1777 of the classical *Reflexions sur la Phlogistique* by Antoine Lavoisier which began to lay not only the foundations of a quantitative theory of heat, but indeed the basis for the whole of modern chemistry and has since earned him the title of 'the father of modern chemistry'. Thus chemistry and combustion studies have been closely associated from the earliest times.

From these early beginnings, through the intense activity of the Industrial Revolution in Britain, there has followed an ever increasing interest and fascination with pure and applied aspects of combustion technology. Many people associated with such studies were in fact scientific amateurs who had very real links with the world outside the laboratory and this may have had some bearing on the apparent close coupling between pure and applied research interests, which were so productive during the late eighteenth and nineteenth centuries in this country. One aspirant to an academic career, James Joule, was in fact rejected for a professorship at St. Andrews University because of a slight physical deformity; his well known and important work on the equivalence of heat and energy which led to the first law of thermo-dynamics was carried out while a proprietor of a large brewery and one might question whether a career as an academic even at St. Andrews could have been any more distinguished or productive!

Man's dependence upon energy derived from fuel combustion has increased dramatically since the time of the Industrial Revolution and it is regrettable that during the present century the influence of fundamental combustion studies on practical combustion systems has been small. In the case of continuous combustion technology (flames) as well as with transport (gasoline and diesel engines), the engineer has been astonishingly successful with empirical developments without reference to the considerable research activity in University and Polytechnic Departments.

It is clear, however, that some basic changes in combustion technology are today being pioneered because of environmental pollution pressures, developing energy constraints, and for economic reasons following the continuing rise in fuel prices. It can therefore be foreseen that increasing collaboration between pure research and applied development, apparent during the eighteenth and nineteenth centuries, could become re-established during the latter part of the twentieth century.

But what are the problems which make this co-operation difficult? To enable us to form an opinion, let us first look at the petroleum fuels upon which man is so dependent and then at the basic chemistry which enables these fuels to burn as 'fire' to give 'heat'. Following this review some examples are given of important areas associated with applied combustion technology in which the chemist has contributed significantly, and will continue to do so, by applying the results of fundamental combustion studies. However, very important and significant changes have occurred over the last twenty years in applied combustion practice as a result of many influences, generally non-technical ones, over which the scientist has no control. These trends are briefly reviewed in an attempt to underline some of the problems facing the contribution of the fundamental scientist to the ever developing and increasingly important field of combustion technology.

Common Petroleum Fuels

Liquid petroleum fuels for heating purposes can be classified with coal and natural gas as primary fuels, because they require relatively little pretreatment (apart from the simple physical separation process of distillation) before being utilized. The subject of coal combustion (the main source of energy in the U.K. since the Industrial Revolution) is not covered in this Chapter as it would require a treatise of its own (cf. Lowry[1]).*

The common gaseous and liquid fuels are summarized, together with some typical properties, in Tables 14.1 and 14.2.

Table 14.1. Combustion properties of common gaseous fuels

Property	Hydrogen	Town gas	Methane	Commercial propane	Commercial butane
Calorific value					
($kJ \ m^{-3}$ dry units)	12,100	18,400	37,700	93,900	121,800
($kJ \ kg^{-1}$ 15° dry)	141,900	32,700	55,600	50,400	49,500
Stoichiometric air–fuel requirements (v/v)	2:1	~4:1	9:1	23:1	30:1
Flame speed (max.) (cm/sec)	260	~100	33.8	39.6	36.6
Flammability limits (vol.-%)	4–75	~5–31	5–15	2–11	1.7–9

* For references, see p. 521.

Town gas is a secondary fuel as it does not occur naturally but is manufactured by carbonization of coal or by gasification or reforming of liquid petroleum fuels. It contains a significant proportion of hydrogen.

Table 14.2. Typical properties of common liquid fuels

Property	Gasoline	Kerosine	Gas oil	Fuel oil
Specific gravity at 15.5°C	0.75	0.78	0.84	0.96
Boiling range (°C)	30–180	150–260	180–350	180–>370
Sulphur (wt.-%)	<0.1	0.1	0.6	2.5
Calorific value (kJ/kg)	47,300	46,500	45,600	43,900
Stoichiometric air–fuel requirements (w/w)	14.7:1	14.7:1	i 14.7:1	14.0:1

Natural gas occurs alone or in association with crude petroleum or coal. Its main constituent is methane which varies between about 70% and 95% by volume depending upon the source.

Liquefied petroleum gases (LPG) include both propane and butane and are particularly suitable for portable applications as they are readily liquefied under pressure.

Gasoline (petrol), the most volatile of the liquid petroleum fuels, is used almost exclusively in automotive spark-ignition engines.

Kerosine (colloquially known in the U.K. as paraffin oil) is the main domestic heating fuel in Britain and in some other countries (notably Australia). A similar fuel, with a tighter specification, is used as aviation turbine kerosine (ATK) for sub- and super-sonic jet aircraft.

Gas oil is the main domestic heating fuel used in Europe, the U.S.A. and Canada, and under its automotive name of diesel fuel is used widely in the U.K. and abroad for powering road transport, particularly public and goods vehicles (compression-ignition engines).

Fuel oil, predominantly a non-distillate fuel, comprises the residual material from the crude-oil distillation process, the cracking processes and materials discarded during the production of lubricating oils. Several grades are marketed and are classified primarily according to their viscosity. The sulphur content of these residual fuel oils, formerly of concern because of the corrosive effects of the combustion products, is now attracting renewed attention because of tightening pollutant-emission requirements.

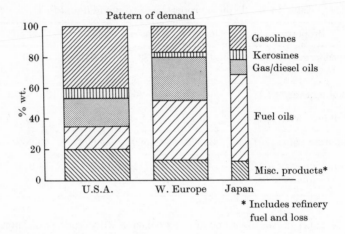

Figure 14.1.   Pattern of fuel demand in U.S.A., W. Europe and Japan in 1970.

The fraction of each ton of crude oil converted into saleable product varies with the demands of the consumer country and three typical patterns of demand for Western Europe, Japan and the U.S.A. are shown in Figure 14.1 (see also Chapter 1). It is not generally realized that as much as 80% of each ton of crude oil in the U.S.A., and 90% in Europe and Japan is used for combustion purposes. A small but significant fraction is utilized for chemical production (solvents, plastics, rubbers, detergents, etc.).

The utilization of petroleum for combustion purposes is clearly a major outlet, so let us look further at the fundamental chemical processes that enable man to use fuel to provide energy for his modern industrial society.

Chemistry of Combustion

Practical Flames

Combustion may be broadly defined as any relatively fast exothermic reaction, and a flame is a combustion reaction that can propagate subsonically through space (by contrast with a detonation, which propagates at supersonic velocity).

The propagation property of a flame enables a flowing fuel–air system to support a stationary flame which can be stabilized by using a burner. Rates of fuel flow can vary from a few ml/min for a domestic gas pilot flame up to a few tons/hour for an industrial fuel-oil flame.

Stationary flames are of two general types:

(*a*) Diffusion flames where both neat fuel and all the air required for combustion 'mix' across the boundary where combustion occurs. These flames may be laminar or turbulent according to the rates of flow and mixing. Practical examples are a Bunsen burner with closed ports (Figure 14.2a), a candle, a simple refinery gas flare and liquid fuel droplet combustion. These flames may therefore range from a few centimetres up to many metres in length.

(*b*) Premixed flames where the fuel and a proportion of the stoichiometric air requirement are mixed (usually within their flammability limits) before combustion takes place. This is known as primary aeration; secondary air is induced into the flame to complete the combustion. A Bunsen burner with open ports is a practical example (Figure 14.2b).

Although most industrial flames are of the turbulent diffusion type, the chemistry of diffusion flames is not well understood.

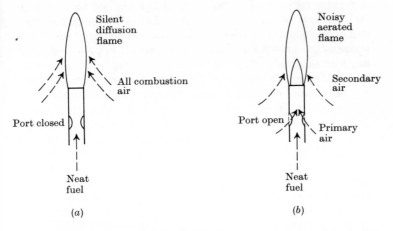

Figure 14.2. The Bunsen burner. (*a*) Diffusion flame. (*b*) 'Premixed' (aerated) flame.

Physical processes of diffusion and turbulence are generally of predominant importance in determining flame stability, shape and luminosity. It has been established that under steady burning conditions the fuel and oxygen do not actually come into contact but are separated by a boundary where the concentration of each falls to zero. Reaction occurs on both sides of this high-temperature boundary and the general mechanism for hydrocarbon fuels appears to be one of carbon formation (via a pyrolysis process) on the fuel side and the formation of reactive radicals (possibly OH) on the oxidant side; thus the species reacting at the boundary are not the initial reactants. The detailed chemical mechanism occurring in diffusion flames is still obscure although much work has been done on the physics of the 'mixing' process. Some problems encountered with practical diffusion flames will be discussed below.

Premixed flames are characterized by their burning velocity or rate of propagation of the flame front into the unburnt premixed fuel–air mixture. This velocity depends primarily on the inlet composition, temperature and pressure of the mixture. When stabilized on a burner, the velocity of the flame front can exceed that of the mixture and hence the problem of flash back into the mixture can occur. This is not possible with a diffusion flame. (A detailed discussion of the structure of flames is given by Fristrom and Westenberg[2].)

In a practical premixed flame most of the enthalpy of reaction is rapidly released in a reaction zone, giving rise to very high temperatures. It has been calculated that in a typical premixed flame this zone will be a few tenths of a mm thick with a maximum temperature gradient approaching 100,000°C per cm and a gas acceleration greater than a thousand times that due to gravity. It is therefore not surprising that molecular transport processes of conduction convection and diffusion assume considerable significance in flames and are responsible for many features of premixed flame structure and stability. These physical factors also affect the chemistry of flames because radical species that initiate flame reactions can easily be transported against the flow of reactants from high-temperature regions back to low-temperature regions by a thermal or concentration diffusion mechanism; hence there is some uncertainty in defining the precise chemical initiation reactions. The physical transport of heat or active species imparts the self-propagating property to a flame which enables a stationary flame to be stabilized on a burner and utilized practically for domestic and industrial purposes.

From the foregoing, it is clear that a flame is a unique chemical system that is strongly dependent upon physical processes, so let us look more closely at the chemistry involved in flame reactions.

Fast Flame Reactions

The majority of gas-phase explosion and flame phenomena can be explained in terms of a branching-chain mechanism. The chain reaction is propagated by reactive species which are consumed but which generate further active intermediates. If two or more active intermediates are generated per reaction then the chain is said to branch, and the rate of reaction is no longer stationary but increases exponentially, giving rise to explosion phenomena or flame with typical reaction times varying from one second down to 10^{-13} second.

This mechanism in which energy is released in the form of active intermediates (atoms and radicals) should be contrasted with a thermal explosion mechanism in which energy is released as heat. If the rate of energy released as heat exceeds the rate of energy dissipation, then a thermal explosion will occur. The number of combustion phenomena that are adequately explained by the thermal theory alone is, however, small.

The behaviour of any flame system cannot be fully quantified until its complete reaction mechanism has been elucidated. So far, the only flames for which a mechanism has been confidently proposed are those supported by the hydrogen–oxygen reaction. Since the mechanism and related thermochemical data are known for this system, continuity equations have now been derived that allow the flame to be expressed as a set of differential equations, and these enable physical flame characteristics to be computed. This procedure has been carried out for the 'simple' H_2–O_2 system, and experimental structure studies of a slow-burning, fuel-rich, H_2–O_2–N_2 premixed flame compare very closely indeed with computed values over wide mixture limits.

The proposed mechanism involves the radical species, $H\cdot$, $\cdot OH$, O and $HO_2\cdot$, and consists of the following steps (1)–(13) (M is a 'third body'):

$$\cdot OH + H_2 = H_2O + H\cdot \qquad \text{(1) propagating}$$
$$H\cdot + O_2 = HO\cdot + O \qquad \text{(2) branching}$$
$$O\cdot + H_2 = HO\cdot + H\cdot \qquad \text{(3) branching}$$
$$H\cdot + O_2 + M = HO_2\cdot + M \qquad \text{(4)}$$
$$H\cdot + HO_2\cdot = HO\cdot + \cdot OH \qquad \text{(5)}$$
$$H\cdot + HO_2\cdot = O\cdot + H_2O \qquad \text{(6)}$$

$$H \cdot + HO_2 \cdot = H_2 + O_2 \qquad\qquad (7)$$

$$HO \cdot + HO_2 \cdot = H_2O + O_2 \qquad\qquad (8)$$

$$O \cdot + HO_2 \cdot = \cdot OH + O_2 \qquad\qquad (9)$$

$$H \cdot + \cdot H + M = H_2 + M \qquad\qquad (10) \text{ recombination}$$

$$H \cdot + HO \cdot + M = H_2O + M \qquad\qquad (11) \text{ recombination}$$

$$H \cdot + O \cdot + M = HO \cdot + M \qquad\qquad (12)$$

The probable initiation reaction is

$$H_2 + O_2 = 2HO \cdot \qquad\qquad (13)$$

but is not shown in the scheme as it has very little effect on the overall kinetics.

The basic chain sequence in the H_2–O_2 reaction is given in reactions (1)–(3), and these (involving atoms and radicals) are the chain-branching reactions referred to above. In flames, however, other steps become important as the flame reactions proceed, and branching ceases when reactions (1)–(3) become significantly reversed or when competing reactions deplete these active intermediates. A dynamic equilibrium therefore exists in a flame (in contrast to a detonation) and a large amount of the available enthalpy is associated with the high radical and atom concentrations existing in the flame. This energy is dissipated by the slow third-order recombination reactions, principally (10) and (11), which persist in space and time well after the propagating and branching reactions are complete and are responsible for bringing the reactive species slowly into thermal equilibrium with the post-flame gases. The surfaces surrounding a flame can influence termination reactions by adsorbing any of the radicals and atoms that diffuse there. Recombination then takes place, giving stable molecular species.

Both early combustion studies of the CO–O_2 reaction and later studies of the basic chemistry of methane combustion show marked similarities to the hydrogen-oxidation mechanism. It is now accepted that an understanding of the hydrogen reaction is also essential to the elucidation of hydrocarbon combustion.

Slow Combustion Reactions

Apart from the fast flame reactions, slow combustion phenomena exist which have been the subject of considerable research work since before the turn of this century.

The slow-combustion temperature region in which hydrocarbons and oxygen react relatively slowly without development

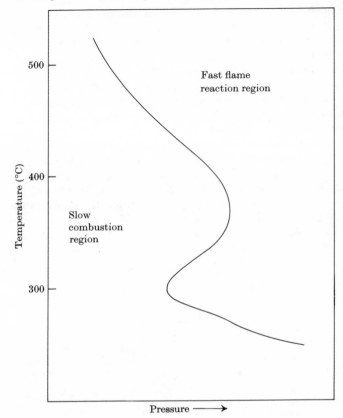

Figure 14.3. Typical shape of a T–P curve for hydrocarbon fuels.

of fast-flame reactions lies between about 200° and 600°C. Both the induction and the reaction times for slow combustion range from a few tenths of a second up to tens of minutes. Slow-combustion reactions, therefore, are more amenable to detailed study than the very fast flame phenomena described above.

A greatly simplified version of a curve (termed a T–P curve) which represents the boundary between hydrocarbon–oxygen slow-combustion reactions and the 'explosive' fast-flame reactions (sometimes termed true ignition) is sketched in Figure 14.3. It can be seen that as the temperature of the mixture is increased the necessary pressure for fast flame reactions to initiate passes through maximum and minimum values (owing to a complex dependence of reaction rates on temperature). The region surrounding a minimum is known as a lobe. The T–P curves for almost all hydrocarbon fuels have been shown to be astonishingly similar in general shape and position relative to the temperature scale. The position of individual curves relative to the pressure scale, however, is very variable. For stoichiometric mixtures of, e.g., normal paraffins with oxygen, the T–P curve moves in the direction of decreasing pressure for increasing chain length of the fuel molecule.

T–P curves have been obtained by using static reactors; these are widely used for slow-combustion studies because the boundary conditions of pressure and temperature, and reaction rate at constant volume, enable accurate kinetic data to be obtained.

In the case of hydrocarbon–oxygen mixtures the slow-combustion region of the T–P curve is in fact considerably more complex than the single region shown in Figure 14.3. The transition from slow-combustion to fast-flame reaction can occur under certain

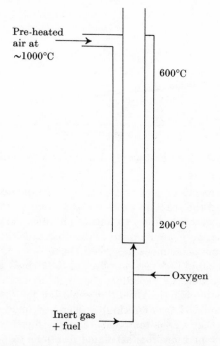

Figure 14.4. Air-heated vertical flow reactor.

conditions of temperature and pressure via what is known as a two-stage ignition mechanism. This transition can be 'observed' in a static reactor (either visually or with the aid of pressure- or temperature-sensing equipment) as a passing sequence of luminosity and periodic light emissions known as cool and second-stage 'flames'. These 'flames' must not be confused with practical flames involving fast chemical reactions. Static-reactor studies are adequately described in the literature[3] and will not be treated in detail here.

Although the static reactor has been widely used in combustion research for the reasons described above, flow reactors afford a more striking visual impression of slow-combustion reactions as the two-stage ignition phenomena may be stabilized in a flowing reactant stream for convenient visual examination. Experiments with *n*-heptane–oxygen mixtures have been reported[4] in which a vertical tube 'separated flame' flow reactor was used, and the present author has operated a similar apparatus; some results are as follows.

Figure 14.4 shows a heated, vertical flow reactor, about 25 mm in diameter, in which a thermal gradient from 200° to 600°C is maintained in a flowing inert-gas stream at atmospheric pressure. If hydrocarbon vapour (e.g. *n*-heptane) is premixed with the inert gas, and oxygen is slowly added, the full sequence of slow-combustion phenomena may be induced and observed in a darkened laboratory.

First a blue 'haze' appears at the top of the reactor; it moves slowly down against the flowing reactant stream and stabilizes after some minutes as a clearly defined blue 'flame' at low intensity near the base of the reactor. This is a cool flame and, surprisingly, only a small proportion of the fuel and oxygen are consumed in this region of the reactor. This is due to a negative temperature coefficient of reaction caused by a change in the basic branching mechanism. A temperature peak of only about 150°C occurs (hence the term cool flame).

If the oxygen concentration is increased the cool flame remains but is followed by a much brighter blue second-stage flame which stabilizes some distance above the cool flame. Considerably more fuel and oxygen are consumed in the second-stage flame, but not all. Under certain reactor conditions a series of cool flames (multiple cool flames) may be observed between the first cool flame and the second-stage flame.

With further increase in oxygen concentration a yellow flame (visually similar to a candle flame) is formed in the wake of the

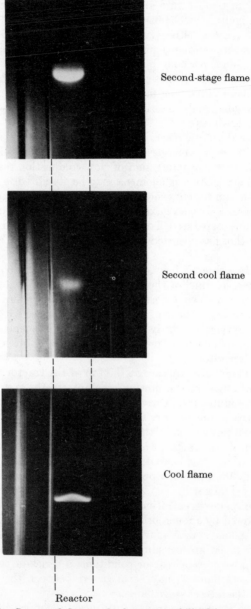

Figure 14.5.   Separated flames of *n*-heptane stabilized in a vertical tube flow reactor.

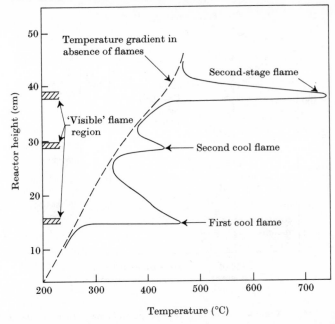

Figure 14.6. Temperature profile through the separated flames of *n*-heptane.

second-stage flame and this completes the consumption of the fuel and oxygen and corresponds to a point on the fast-flame reaction side of the T–P boundary.

Up to this stage the flame sequence described and shown in Figure 14.5 with just two cool flames and a second-stage flame is stable and reversible with oxygen concentration.

With further increase in oxygen the whole structure becomes unstable and finally accelerates (sometimes reaching a detonation velocity) down to the flame trap located at the inlet to the reactor.

In the stable condition, the flow reactor allows temperature and composition profiles to be obtained relative to the flame stages, thus giving additional insight into the likely chemical mechanism associated with two-stage ignition phenomena. Figure 14.6 shows the temperature profile associated with the reactor operating to give the flame sequence shown in Figure 14.5.

Because the proportion of fuel to oxygen changes at each of the flame stages shown (Figures 14.5 and 14.6), and because the flow reactor operates at constant (atmospheric) pressure, it is not

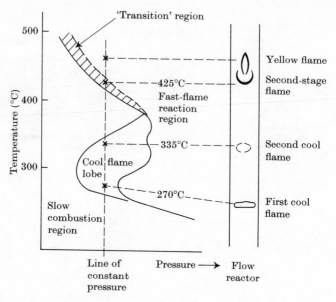

Figure 14.7. Comparison of flow reactor phenomena with a typical T–P curve.

possible to relate the flow-reactor results directly to the static-reactor T–P curves published for *n*-heptane–oxygen mixtures. However, a sketch may be made (Figure 14.7) which indicates the broad connection between a typical T–P curve and those data described above obtained at constant pressure in the flow reactor. Initiation temperatures obtained for *n*-heptane for each of the flame stages shown are given in Figure 14.7.

It may be seen that a cool-flame lobe has been added to indicate the region in which a cool-flame mechanism operates, and a transition region shows the onset of a second-stage flame mechanism which precedes the transition to fast-flame reactions, across the T–P curve boundary.

Additional slow-combustion phenomena observable in a static reactor are periodic cool flames and delayed ignition as well as a considerable variation in the induction period (sometimes tens of minutes) between the mixing of reactants and the onset of slow-combustion reactions.

Although the slow-combustion phenomena are not fully understood it is generally accepted that two-stage ignition results from chemical and thermal modification of the reactant mixture by the cool flame itself, which transfers the system from a slow-combustion mechanism to a fast-flame reaction mechanism (true ignition).

Any theory of slow hydrocarbon combustion must therefore account for the following phenomena: (*a*) two-stage ignition; (*b*) multiple cool flames; (*c*) periodic cool flames (observed in both static and flow reactors under certain experimental conditions); (*d*) delayed ignition (in static reactors); (*e*) the presence of lobes in the T–P curve of all hydrocarbons; (*f*) the negative temperature coefficient of reaction which causes the cool flame 'explosion' to die out after a rise of only about 150°C; and (*g*) the variation in the T–P curve scale of pressure for different hydrocarbon species.

In explaining some of these phenomena the breakthrough occurred when in the period 1925–1935 a number of workers began examining slow-combustion reactions as well as flame processes in terms of chain theory (described above in connection with the H_2–O_2 reaction).

Semenov in particular established that the phenomenology of gas-phase slow combustion of hydrocarbons is explicable in terms of a chain reaction that proceeds with degenerate branching. This term is used to describe a special type of branching-chain reaction in which the multiplication of active intermediates occurs very infrequently, so that the build-up of reaction rate is gradual and eventually falls off either because of competing reactions or because all the reactants are consumed.

Semenov postulated that, for a degenerate branching mechanism, a stable intermediate having an appreciable life-time reacts either to give radicals (leading to chain branching) or a further stable product:

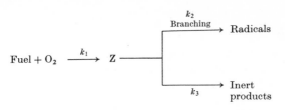

k_1, k_2, k_3 are the relative rate constants. In this way not all the species denoted by Z lead to branching and hence a delay or degeneracy occurs.

Semenov's theory, however, did not take into account any thermal effects, which clearly are present during cool-flame reactions, and many workers have since contributed to this field in an effort to explain all the observed phenomena.

Latterly mathematical models have been constructed, using not a detailed chemical mechanism, but an overall kinetic scheme with hypothetical active intermediates.

A kinetic–oscillation theory, due to Frank–Kamenetszki, in which the concentration oscillation of two intermediate critical products (possibly peroxides and aldehydes) is used, is untenable as the necessary condition of oscillation deduced from the theory is independent of temperature and pressure, a fact that is not supported by experiment.

A thermokinetic theory due to Salinkoff appears to be more acceptable in predicting cool-flame phenomena but does not account for the negative temperature coefficient of reaction in the cool-flame region.

In a more recent theory Yang and Gray (1969) followed the Salinkoff approach by considering the thermokinetic coupling of an intermediate product with temperature, and this approach unifies many of the ideas from previous theories. All the observed cool-flame phenomena may now be predicted by using acceptable values of rate constants and enthalpies, and the computer calculations agree quantitatively with static-reactor experiments.

Although it would appear (as a result of the considerable effort over the last 70–80 years) that the basic underlying mechanism of slow hydrocarbon combustion in static reactors is close to being quantitatively established, these recent mathematical theories throw little light on the species involved, nor can they be rigorously applied to flow-reactor phenomenology. Clearly much detailed work remains to be done before the mechanism of hydrocarbon combustion can be fully elucidated for all fuel–oxygen systems.

Applications of Chemical Knowledge

A few selected areas in which basic chemical data have been, or could be, applied with advantage are next reviewed.

Kinetics

Applied combustion technology has progressed in spite of the lack of fundamental understanding in the systems utilized and developed. One considerable obstacle to the chemist has in the past been a mathematical one. Now that mathematical methods are available for solutions of coupled non-linear differential equations of various degrees of complexity, it is likely that the recent progress in understanding the kinetics of flames will rapidly develop further. As examples, the emission of nitric oxide has been calculated for a spark-ignited engine by use of known kinetic, thermodynamic and physical data. The model proposed may also be used to predict the effect of engine variables and recirculation of exhaust gas on the levels of nitrogen oxides. Similar work is also being carried out on industrial diffusion flames to predict concentrations of nitrogen oxides and methods for their control. The importance of such studies for predicting and reducing future pollution levels is clear.

Knock

The lack of understanding of basic flame reactions was apparent early in this century when the development of the spark-ignited gasoline engine ran into problems with uncontrolled ignition that led to excessive heat transfer, loss of power, occasional engine damage and noise. This problem has been termed knock (Part 1, p. 174) and there is now considerable evidence associating the incidence of knock with the autoignition (via a cool and a second-stage flame mechanism) of the fuel–air mixture ahead of the advancing spark-ignited flame front. A solution to the problem was discovered by an engineer and a chemist (Midgley and Boyd, 1922) who added tetraethyl-lead (TEL) to the fuel. The mechanism of inhibition of cool-flame reactions by TEL was not then understood and even today has not been fully elucidated. It is likely to be associated with the breakdown of TEL during the cool-flame reactions to give a 'fog' of finely divided oxides of lead; these solid particles may provide a surface for hydrocarbon chain-termination reactions and hence prevent the second-stage flame-initiation reactions from building up to give ignition. Lead alkyls appear to be unique antiknock agents, possibly because the cool-flame reactions themselves provide the necessary energy for the inhibiting process. The proposed mechanism is supported by experiments in the flow reactor described above in which a small addition of TEL to the fuel allowed the cool flame to remain intact but caused the second-stage flame to disappear slowly from the reactor.

There has been an increasing trend in Europe towards small engines of higher compression ratio. However, the environmental pressures to prohibit the use of lead-containing additives in gasoline means a reversion to engines of lower compression ratio. A detailed understanding of slow-combustion reactions occurring in spark-ignition engines and possibly the means of controlling them are clearly of growing importance.

Cool Flames

It has been shown that the position at which a cool flame stabilizes in a flow reactor can be correlated with small changes in octane number of the flowing fuel feed. This effect has been utilized in a continuous octane monitor for gasoline process streams in which the pressure in the reactor is varied in order to stabilize the cool flame at one position. A feed-back loop using two sensors is employed (Figure 14.8) and the applied pressure change may be calibrated in terms of variation of octane number over about 10 octane numbers. This technique is being used to replace an expensive and lengthy method for determination of octane number (using a standard engine and a knock-monitoring instrument) and can clearly result in considerable saving in refinery operating costs. The theory of this effect is not known in detail but may be associated with the influence of pressure and fuel structure on the induction period of slow-combustion reactions.

Cool flames have been reported in aircraft fuel tanks if surface temperatures exceed about 200°C. The small pressure rise as a result of the passage of a cool flame in the static vapour space above the fuel could be enough to rupture the fuel tanks and is therefore a potential hazard. Work is being carried out both in static and flow reactors to investigate this problem, which may become particularly critical in supersonic aircraft where fuel is used as a cooling medium to prevent overheating of the aircraft skin.

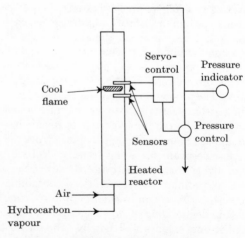

Figure 14.8. Cool-flame octane monitor.

The formation of polycyclic aromatic hydrocarbons in a flow reactor where there is two-stage ignition of gasoline fuels has been compared with combustion deposits removed from a gasoline-engine exhaust. The remarkable similarity in composition and order of occurrence (along the reactor and the exhaust pipe) emphasizes the similarity of hydrocarbon reaction mechanisms at low and high pressures and further underlines the growing importance of laboratory studies in the field of air pollution.

Other direct uses for cool-flame reactions are for the production of 'heavy chemicals' (oxygenated derivatives of the fuel) and as a means of chemical ignition in diesel engines by the direct injection of cool-flame products of reaction into the inlet manifold of the engine. Neither of these applications has yet been commercially exploited but they illustrate the possible outlets for a slow-combustion phenomenon that has traditionally been of academic interest only.

Distillate Fuels

Fundamental combustion studies have not been widely applied to distillate oil-fired domestic burners because the few combustion problems encountered (mainly incomplete fuel combustion) have been easily overcome by empirical means.

An exception is the formation of precombustion deposits (generally termed 'carbon' deposits) in the base of 'pot'-type burners which rely on thermal feed-back from the flame to

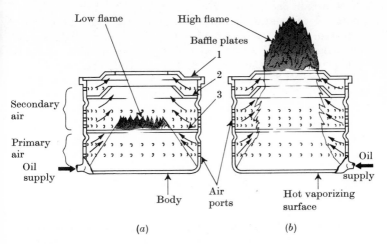

Figure 14.9. Section of pot burner: (*a*) low flame; (*b*) high flame.

vaporize fuel from a pool of heated oil in the base of the pot (Figure 14.9, p. 509). A detailed mechanism has not been established, but it has been proposed that paraffinic radicals produced during reactions of cool-flame type react at about 400°C with the aromatic hydrocarbons present in the fuel, generating aromatic radicals. These are relatively stable and polymerize at the metal wall to produce deposits of high molecular weight. Only small quantities of oxygen are required to initiate these reactions, which, one can argue, can be minimized by reducing the time between vaporization and combustion and by excluding vapour contact with 'cool' metal surfaces.

It should be noted that generic terms such as 'tars' and 'coke' are used because of the complex nature of these deposits. They are, however, primarily carbon, with a few per cent of hydrogen. Thus the C–H ratio and the basic crystallography determine the characteristics of these deposits.

'Droplet' Combustion

The most common method for firing liquid fuels for domestic or industrial use is to atomize the liquid into droplets ($\sim$50–500 $\times$ 10^{-6} m) before combustion. This in effect increases the surface area of the fuel and enables ignition and combustion (primarily by a diffusion flame mechanism) to occur more readily.

Many interdependent processes take place in the vicinity of each droplet and within the drop itself, before complete combustion has occurred. The events of evaporation, distillation, cracking, mixing, soot formation and soot burning have all been extensively studied in model systems and with reference to all types of atomizing burner. Fuel oil contains petroleum fractions of high boiling point and these may undergo partial cracking which delays the combustion and leads to undesirable soot or 'particulate' emissions (hollow, round, coke-like particles called 'cenospheres'). Of particular interest to the chemist is the possibility of rapid chemical reaction within the liquid droplet or coke residue itself as heating and distillation take place. These include the behaviour of ash constituents such as sodium and vanadium because their interactions are of the greatest importance: high-temperature corrosion of heating surfaces in boilers and gas turbines is known to be associated with the formation of complex sodium and vanadium compounds.

Summing up the present status of droplet combustion research work—most studies to date have been concerned with mixing of the droplets with air and hot combustion products, evaporation of the droplets, and general studies of turbulence. The effect of the chemical composition of the fuel on reaction rates is beginning to attract more attention as it may be a limiting factor when fuel is burnt at extremely high rates.

Pollution and Corrosion Problems

A large percentage of all man-made air pollution arises from the combustion of fossil fuels (coal, oil and gas). An understanding of the undesirable pollutant reactions in flames is therefore of prime importance as a prerequisite for their reduction or elimination. Of particular interest at the present time are sulphur dioxide (SO_2), sulphur trioxide (SO_3), nitrogen oxides (NO_x) and soot.

On combustion, most of the sulphur in petroleum fuel is oxidized to sulphur dioxide and at the most 3% of this is further oxidized to SO_3. Although SO_3 is unstable at high temperatures, below about 325°C it combines with water vapour in the flue gases to form sulphuric acid which can either be emitted as a persistent mist or condense on metal surfaces when the metal temperature falls below about 120°C; in the latter case severe metal corrosion can occur.

Nitrogen oxides constitute a pollution problem in some parts of the world as they play a major role in the formation of photochemical 'smog' frequently experienced in some major American cities. The reactions occur in sunlight between NO_x and hydrocarbon vapours to give lacrimatory fogs in which compounds such as peroxyacetyl nitrate have been identified. Air-quality standards for NO_x and SO_2, etc., are being set.

Many studies to minimize the formation of sulphur trioxide and nitrogen oxides have been carried out and most workers agree that these compounds are formed in reactions involving atomic oxygen. Operating burners at low excess air levels is now well documented and does reduce formation of both pollutants, but their complete elimination is clearly desirable.

The mechanism of the formation of soot (also referred to as carbon or particulates) in flames has been the subject of much laboratory and pilot-scale work, and emissions of this kind are already the subject of legislation in many countries. Luminosity is an important property of flames and is due to the spectral emission from solid soot particles. The burning-off of these particles is part of natural combustion in flames (particularly those supported by atomized fuel droplets), but under certain fuel-rich conditions it does not go to completion and undesirable soot emission occurs. Apart from improving burner performance to reduce these emissions, metal additives have been used that are thought to reduce soot by an indirect mechanism involving catalytic dissociation of water vapour to provide hydroxyl radicals; the reaction between OH radicals and solid carbon is extremely fast and gives gaseous products including CO (cf. 'water gas' reaction). The emission of metals as a result of additive treatment of fuels may in time not be acceptable on pollution grounds, so that alternative methods for the reduction of soot formation in flames will clearly be required.[5]

Britain is one of the few countries where ground-level soot concentrations have fallen appreciably (since the Clean Air Acts of 1956), mainly owing to the reduction of coal-burning in domestic open fires and in industry. This one fact emphasizes that much can be done by fuel selection without the aid of the combustion scientist. As pollution legislation becomes stricter, good combustion practice based upon a sound understanding of flame processes must assume greater importance and significance in overcoming future pollution problems.

Summary

From the foregoing it should be apparent that the combustion chemist is potentially able to make a significant contribution to combustion technology over a very wide range of applied topics.

Empirical methods (trial and error) will still continue to feature very significantly in future developments, but the engineer will no doubt need to refer more and more to fundamental combustion data in order to reach an optimum solution to his problems with the minimum of expensive experimental effort.

As a cautionary warning to combustion chemists who would pursue their studies without reference to the real world outside, it may be judicious to follow this Section by reviewing some factors (mainly non-technical ones) that have recently caused a 'revolution' in the world of applied combustion practice, and including some examples of the response of the combustion engineer to this changing and challenging situation.

The World of Applied Combustion Technology

Fuel Influences

Until about 1955 about 90% of the U.K.'s fuel requirements were satisfied by coal. Coal was used widely in open fires for domestic heating, in power-station boilers for electricity generation, industrial boilers for steam-raising and water-heating and in our railway locomotives for propulsion. Clearly these and many other developments had been the subject of considerable work since the advent of the Industrial Revolution in the eighteenth century. Combustion technology associated with these developments has remained largely unchanged in principle and certainly the growing volume of fundamental combustion studies, particularly since the turn of the century, did not have, and possibly could not have had, much impact upon the course of applied combustion technology.

During the 1950's a 'fuel revolution' took place in which the U.K. began to move to a multifuel economy. Nuclear power began to show promise, and the quest for alternative supplies of crude oil (caused partly by political problems in the Middle East) led to a relative abundance of oil fuels which were very competitive in price with the rising costs of coal. As a result the contribution from oil rose rapidly from 15% to 37% of our total U.K. energy utilization over the period 1958–1967 and has since risen more slowly to somewhere near 50% by 1973. In the 1960's a further impetus to the fuel 'revolution' occurred. The coal-carbonizing route for the manufacture of 'high speed' town gas was largely replaced by the steam-reforming of petroleum hydrocarbons over nickel catalysts; this provided a boost for the gas industry and was followed in 1965 by the discovery of commercial quantities of natural gas under the North Sea. By 1967 natural gas was being utilized in both the domestic and the industrial energy sector. With this discovery the whole technology of the domestic gas industry had to change almost overnight to accommodate to this slow-burning hydrocarbon fuel (Table 14.1). Since 1945 almost all U.K. gas appliances (with the exception of gas cookers) had been fitted with burners comprising an array of small diffusion flame 'Bray jet' burners (Fig. 14.10, p. 514) which gave a neat, silent, stable blue flame when burning manufactured town gas (containing up to 50% of hydrogen). Unfortunately the Bray jet burner was unsuitable for burning natural gas because of flame lift, and no other suitable diffusion flame burner was forthcoming in spite of considerable research. The conversion of domestic appliances from town gas to natural gas therefore went ahead with premixed burners (aerated burners) similar in principle to those of the pre-1945 era which were noisy and prone to light-back and to 'fouling' of the burner head by household dust introduced with the primary air. Such burners do not readily suit U.K. gas appliances situated in the kitchen and living room (central-heating boilers and radiant gas fires) principally because of the annoyance from flame noise and

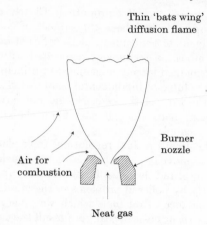

Thin 'bats wing'
diffusion flame

Burner
nozzle

Air for
combustion

Neat gas

Figure 14.10.   Single 'Bray jet' burner.

noise from the injector nozzles. Premixed gas burners are acceptable and widely used in other countries because appliances are
generally not placed within the living area of the house but in a
cellar or outhouse.

It is not therefore surprising that since the 1950's there has been
a marked increase in oil-fired combustion studies applied to
specific areas of industrial technology, and more recently to the
study of the combustion of methane for domestic heating use. Far
from being a monopoly of industry, some studies have been carried
out by University and Polytechnic Departments and have been of
the highest quality. It must be said, however, that important
developments in many areas have occurred because of economic,
environmental or commercial pressures and have therefore not
always waited for the longer-term fundamental studies to mature.
These studies, nevertheless, are leading to a better understanding
of practical flame systems and will therefore be of the utmost
value when future pressures demand further improvements and
changes in applied combustion technology.

A few examples of recent or imminent changes will next be
discussed.

Combustion of Natural Gas for Domestic Use

The research and development activity that took place both
before and during the domestic gas appliance conversion programme in the U.K. aimed at (a) improving the reliability and
acceptability of the premixed (aerated) burner, and (b) designing

and developing a diffusion flame 'conversion' burner for natural gas.

It was envisaged that a 'conversion' burner would be installed to operate on town gas and, at the time of change-over to natural gas, would be simply modified to accommodate the change in calorific value and flame speed of the new fuel. The old advantages of the Bray jet diffusion flame burner were an essential prerequisite for any proposed new burner.

The British Gas Corporation research and test centre at Watson House in London were responsible for co-ordinating the work on new burner designs, and a considerable programme of work was undertaken by them in testing the many prototypes submitted from a wide range of experimental establishments.

Improvements to existing aerated burner technology were made primarily in two design areas: (a) premixed burner heads were improved to prevent flash-back of the flame through the burner head and into the premixed gas–air region; and flame stability was improved by providing small retention flames to 'anchor' the main flames by appropriate design of the burner head; (b) injector nozzles for entraining primary air for premixing with the gas were adapted to reduce noise levels.

In spite of considerable organizational and technical problems the major part of the domestic and commercial conversion programme has been successfully carried out by using the best of the premixed burner designs.

In contrast, the quest for a suitable diffusion flame burner for low flame speed hydrocarbon gases has evoked considerable fundamental and applied research effort and has clearly been, and still is, a problem of considerable magnitude.

Extensive studies of methane combustion in diffusion flames showed that it may not be possible to develop a methane burner having similar characteristics to a town-gas diffusion flame Bray jet burner. The reasons given were the very different burning characteristics of the two gases, and the discovery of a fundamental difference in air-flow pattern at the base of the town-gas flame, giving increased mixing of gas with air and hence greater stability of the faster-burning fuel.

A recent review by Watson House on the progress of diffusion flame burner developments for natural gas described three burners that they have finally selected, after thorough laboratory evaluation, for district trials.[6]

The first burner, developed at Watson House, uses a principle that directs neat gas through a slot at the base of a surface along which the gas flows, entraining air before the combustion. The second, a Dutch 'pinhole burner', incorporates a burner head with holes for main burner flames and auxiliary ports for smaller

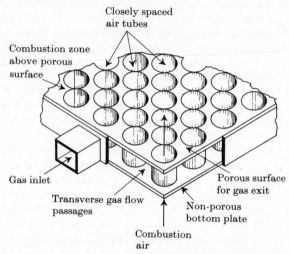

Figure 14.11. Porous plate construction of BP 'matrix' burner.

retention flames, rather similar in practice to aerated burner technology. The third burner, developed at the BP Research Centre,[7] is a complete departure from known gas practice and is illustrated in Figure 14.11. It has been reported by Watson House that this burner concept may in time enable a domestic central heating boiler no bigger than a 20-cm cube to be developed, complete with control system, with a remote fan for forced draught. None of these three types of burner is yet on the market.

The design of appliances for domestic heating has received much attention in recent years and the now familiar trend of gas heating appliances to compact wall-mounted units began in about 1967 with exciting new ideas using powered-flue (or fan-assisted) boilers. A good summary of developments in this field is given by Sugg.[8]

The Industrial Packaged Boiler

One of the most striking developments of the 1960's linked with the growing use of petroleum fuels was the advent of the liquid-fuel-fired, industrial, packaged boiler for steam-raising, to replace the old dual-fuel oil–coal boilers. These new boilers were pre-assembled by a single manufacturer and delivered to the site complete and (with the exception of the plumbing) ready for firing. The burner and boiler system were designed to give maximum efficiency (up to about 80%) over a wide range of boiler loads and resulted in cleaner combustion, greater flexibility and lower operating and installation costs.

Excluding power generation, steam-raising is at present the most important outlet for residual fuel-oil sales. In some urban areas legislation requires the use of distillate fuels in small industrial boilers in order to reduce emission of sulphur. Reduced soot deposits and maintenance requirements can offset a proportion of the higher fuel cost.

Figure 14.12 is a diagrammatic representation of a packaged fire-tube boiler, showing the fire tube, in which heat transfer is mainly by radiation from the flame, and several banks of relatively narrow-bore tubes, in which heat transfer is by convection from the combustion gases.

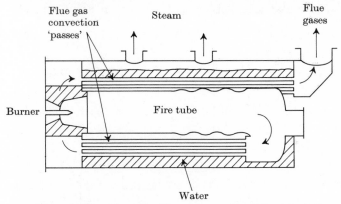

Figure 14.12. Diagram of three-pass packaged fire-tube boiler.

Research and development work continues, particularly into the design and into automatic and low-polluting steam-raising and water-heating boilers. The use of heat-transfer oils instead of water or steam for heat exchanging has been suggested to increase boiler-surface temperatures and hence reduce the incidence of SO_3 corrosion.

Iron-making

The blast-furnace process (Fig. 14.13, p. 518) accounts for over 95% of the world's annual production of iron, and research into improving the economy of the process began in the 1950's.

With the shortage and therefore increasing costs of 'coking' coal (used in the furnace to generate heat and CO for reducing the ore) the injection of hydrocarbons, particularly fuel oil, into the hot-air blast to replace a portion of the coke burden has been increasingly used since the 1950's and is now a well-proven technique that has led to a rise in productivity and smoother furnace operation. It is in fact an outstanding example of an advance in technology in which the development and capital costs have been recovered in a comparatively short time. Incentives for an even greater level of fuel-oil injection to replace coke have been considerable, both because of economic factors and because sulphur in the fuel is retained in the lime flux used in the process and not emitted as sulphur dioxide. The blast furnace therefore represents a non-polluting outlet for fuel oil of high sulphur content.

Technically fuel-oil injection rates higher than those currently used raise some complex problems of droplet combustion within the 1000°C near-sonic hot-air blast, and experience in a parallel

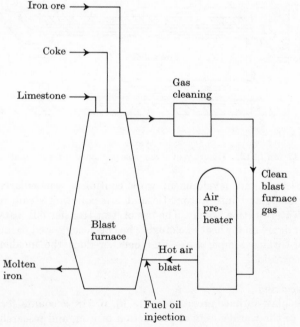

Figure 14.13.   Simplified blast furnace flow sheet.

field of study, possibly that of rocket engine or gas-turbine technology, may perhaps provide some clues for improving the quality of combustion in the hot-air blast.

High-intensity Burners

The technology of compact flames having heat transfer rates above those normally experienced in conventional practice is termed high-intensity combustion.

The advantages of high-intensity combustion burners in boilers are two-fold: (*a*) increased heat transfer results from the higher velocity of impingement of flame gases, possibly augmented by the latent heat of recombination (termination) of active flame species on the 'cool' heat-exchanger walls; and (*b*) economic advantages of smaller combustion chambers.

The problems arising from high-intensity combustion technology are associated with an increase in the normally small conversion of SO_2 into SO_3 in the flame, and with fixation of nitrogen to form nitrogen oxides at high flame temperatures as discussed above.

A recent method for achieving high-intensity combustion with little formation of nitrogen oxides is that of two-stage combustion. The burner is fired with substoichiometric quantities of air in the first stage, with air injection into the second stage to complete the combustion. This form of burner has not so far been widely adopted in industry.

Fluid-bed Combustion

Another promising industrial (and possibly domestic) 'reactor' for the non-polluting combustion of fuel, coupled with a high intensity of combustion is the fluid bed.

Bone pioneered much of the early work on 'surface' and 'catalytic' combustion, but this particular development began in the U.K. about 1965 and was selected as the most promising new technique for the combustion of coal. Basically, coal ash is used as an inert bed material and is mixed with sized coal particles as fuel. Air is blown up through the relatively shallow bed, which fluidizes and can be ignited by initial gas firing. The coal burns in the fluidizing air, and heat is transferred directly to heating coils within the bed itself. The 'spent' particles are removed, charged with more fuel, and returned to the bed. Currently the possibility of firing with fuel oil is being investigated by injecting the oil into beds of fluidized refractory particles.

Experimental reactors are now operating in both the U.S.A. and the U.K. with both coal and oil firing, and an important future is predicted for this process.

The following results and possible future trends for fluid-bed combustion have been cited: (*a*) Complete combustion can be achieved in the bed itself. With the addition of limestone to the bed, up to 100% of the sulphur can be retained when firing with sulphur-containing fuels. (*b*) At the operating bed temperature of 800°C very high rates of heat transfer have been confirmed. Heat-release rates three times those for normal boilers are obtained. At these low bed temperatures the formation of nitrogen oxides is reduced by about 50%. (*c*) Further benefits may be gained by operating the combustor at elevated pressures. It has been calculated that a fluid-bed steam generator working at 2×10^6

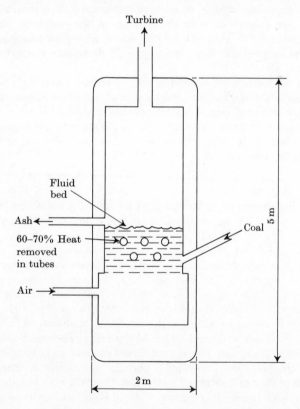

Figure 14.14. Pressurized fluid-bed combustor.

Nm^{-2} would be only about 1/25 the size of a conventional plant of similar output. A considerable saving in capital costs could therefore be made. In combination with a gas turbine for electricity generation, such a plant would be more efficient than conventional power-station electricity generation. An experimental pressurized fluid-bed combustor is shown schematically in Figure 14.14.

It is now generally considered that engineering problems rather than combustion or pollution ones remain to be solved before development to a commercial scale can be confidently predicted for the fluid-bed combustor.

Postscript

It has been suggested that a 'river of ignorance and prejudice' divides the fundamental scientist and the practical combustion engineer. This sentiment was certainly true ten years ago, but today, with far greater economic and environmental pressures and the advent of a world energy shortage, a 'wind of change' is blowing. There are signs that research is becoming more accountable, that Universities and Polytechnics are co-operating with Industry (through Government-sponsored schemes, etc.), and applied degree courses are gaining wider recognition and support. This greater awareness of the need for practical innovation founded upon basic scientific knowledge comes at a time when real progress, outlined in several Sections above, is being made in the fundamental understanding of flames and combustion systems. Thus the note of cautious optimism sounded in this Chapter regarding the role of the chemist in present and future technological advance is not a hollow one in spite of the acknowledged complexities of combustion chemical phenomena.

For the future, chemistry together with other disciplines will need to explore the reactions occurring, not only in flames, but at considerably higher temperatures (above about 4000°C) where ions and atoms predominate and a 'new chemistry' exists. That, however, is another story which, when written, may recount the synthesis of valuable chemicals that cannot easily be made at ordinary temperatures. Will this be the next revolution?

Bibliography

[1] H. H. Lowry (Ed.), *The Chemistry of Coal Utilisation*, Wiley, New York, 1963.

[2] R. M. Fristrom and A. A.Westenberg, *Flame Structure*, McGraw-Hill, New York, 1965.

[3] B. Lewis and G. von Elbe, *Combustion, Flames and Explosions of Gases*, Academic Press, New York, 1961.

[4] J. E. Johnson, K. G. William and H. W. Carhart, *The Vertical Tube Reactor—A Tool for Study of Flame Processes*, 7th Symposium (International) on Combustion, Butterworth, London, 1959, p. 392.

[5] *Combustion Science and Technology*, 1972, **5**, 187–272, Special Issue, 'Soot in Flames'.

[6] B. C. Dutton, J. A. Harris and B. J. Kavarana, 'Non-Aerated Burners for Natural Gas,' *J. Inst. Gas Engrs.*, 1971, **11**, 476.

[7] D. H. Desty and D. M. Whitehead, 'New Burners for Old', *New Scientist*, **1970**, 147.

[8] P. C. Sugg, 'Opportunities in Appliance Development for Natural Gas', *J. Inst. Gas Engrs.*, 1969, **9**, 601.

(ii) FUELS, EXPLOSIVES AND PROPELLANTS

Although explosions of petroleum fuel–air mixtures are valuable in driving engines and tools, they are not of much practical use as propellants for guns or rockets, as military high explosives, or for blasting rock. The reason is that air density is too low. This can be overcome by using a solid oxidizing agent and probably the first useful propellant and explosive was gunpowder, a mixture of charcoal and sulphur as fuels with potassium nitrate as oxidizing agent. This was a stable mixture which burned when heated locally, releasing energy and gases, but it had several disadvantages that were to be overcome as organic chemistry developed.

For most practical purposes, fluorine is too expensive and the best oxidizing element is oxygen; and the least energetic products of its chemical reactions are oxides. For the organic chemist these products are water, carbon dioxide and some nitrogen oxides. The nitrogen oxides are, however, easily reduced by carbon and hydrogen, and the energy difference between nitrogen and its oxides is small.

We might therefore say that the minimum energy is needed in C—O, H—O and N—N bonds. More energy is needed for C—C, C—H and N—O bonds. To build a CHNO molecule capable of rearrangement with energy release, we should therefore use groups such as —NO, —NO$_2$ or —ONO$_2$ to hold oxygen, and couple them to alkyl or aryl groups.

For example, nitromethane, CH_3NO_2, is an explosive, but in reaction it produces gases such as carbon monoxide and hydrogen which, if oxygen were available, could react to give more energy.

Trinitromethane, $CH(NO_2)_3$ ('Nitroform'), also is explosive, but this has an excess of oxygen and produces in its products nitrogen oxides and free oxygen which, if other fuels were present, could give more energy. The ideal organic explosive is one which is 'oxygen balanced' and has all its oxygen in N—O bonds. There are a few such substances, e.g. 'Nitroglycol' ($—CH_2ONO_2)_2$:

$$\begin{array}{c} CH_2ONO_2 \\ | \\ CH_2ONO_2 \end{array} \longrightarrow 2CO_2 + 2H_2O + N_2$$

Few nitroso-compounds are of interest as practical explosives because they are unstable. Nitro-compounds are much more stable and find their use in military high explosives, e.g. 2,4,6-trinitrotoluene (TNT), N-methyl-2,4,6-trinitrophenylnitramine

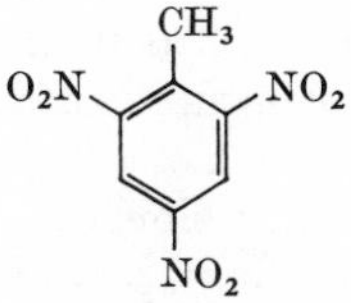

TNT 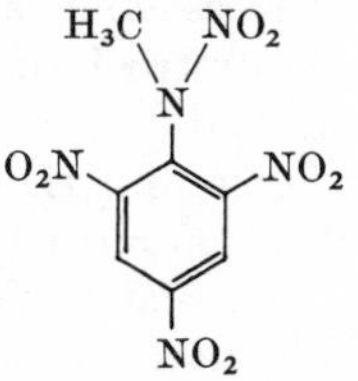   Tetryl 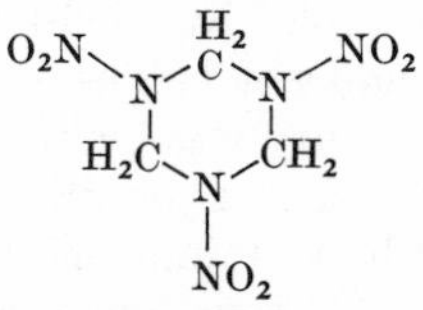   RDX

(tetryl) and 'cyclotrimethylenetrinitroamine' (hexahydrotri-N-nitro-s-triazine) (RDX). Almost without exception, military high explosives are lean in oxygen, but this is not serious because carbon monoxide and hydrogen are as effective as carbon dioxide and water as working gases and the toxicity of CO is unimportant.

For propellants and commercial explosives, organic nitrates have proved to be much more useful. They are less stable than nitro-compounds but, when mixed with stabilizing additives, they can be kept for many years at almost any terrestrial ambient temperature.

Glycerol trinitrate (nitroglycerine) (Part 1, p. 35) was the first explosive organic compound to be used on a large scale. Its chief limitation was its liquid form, but cellulose nitrate (Part 4, p. 93) swells in the liquid to give, according to the type and quantity of the cellulose derivative, anything from a jelly to a brittle 'solid'.

So the mixture was used for propellants (cordite and 'Ballistite'), and for 'Blasting Gelatine', a most successful mining explosive.

Propellants could be prepared from this mixture in very precise shapes. When ignited they burned over the surfaces, the rate of burning depending on the concentration of products near the burning surfaces and therefore on the pressure. It now became possible to design a propellant so that it burned or decomposed into simpler molecules at just the right rate to maintain acceleration of a projectile over the full length of a gun barrel. It was also possible to design rocket charges that would burn at a constant and predictable rate to give a constant thrust. Today propellants do not all contain nitroglycerine. Some (single-based) contain only nitrocellulose and modifying agents. Others are mixtures of organic resins and ammonium perchlorate, and some are bubble-free high-explosive mixtures.

The most important use of nitroglycerine has been in mining, quarrying and tunnelling. For this work it had marked advantages over gunpowder. It was much more energetic, shattered the hardest rock and gave much less flame (a useful safety advantage in coal mines). But it was not nitroglycerine alone that made this possible. Nobel discovered that the explosion of nitroglycerine could not be ensured by flame, but that if a little fulminate of mercury was ignited its effect on nitroglycerine was devastating. Nobel used this discovery in his 'Patent Detonator', a blind copper tube containing mercury fulminate. The explosion Nobel produced was a new type—detonation—and all modern high explosives, military and civil, depend on this phenomenon. Until this discovery, explosives had reacted by thermal decomposition which spread to all available surfaces by the movement of hot gaseous or solid products, and then the chemical reaction continued through thermal conduction or diffusion, increasing the temperature until deeper layers were brought to burning. The mechanism of reaction could vary and the kinetics could have a marked effect. Often the controlling step was, however, a diffusion rate rather than a rate of chemical reaction.

In detonation a wave of initiation spread through the explosive. The front of this wave was an intense shock, giving a sudden increase in pressure. Behind this the compressed explosive underwent the chemical change, remaining denser than the original explosive, and only at the rear of this reaction zone did the products begin to expand. The products of reaction were in fact accelerated towards, and not away from, the explosive and reached a steady state when the difference in the velocity of the

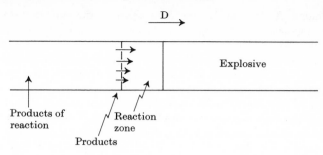

Figure 14.15. Infinite high-explosive reaction. Products moving towards explosive at speed W so that $D - W =$ local velocity of sound, where $D =$ velocity of the detonation shockwave.

wave and the mass movement of the products was equal to the local velocity of sound. In other words, detonation was a shock wave continuously supported by the chemical reaction. The speed of the wave was determined by the energy released in the reaction, and the mechanism was unimportant.

Why then did gunpowder not detonate? The answer lay in the rate of reaction. Although for an infinite quantity of material the speed of a detonation wave is independent of the rate or mechanism of reaction, in practice all explosive charges are finite and as detonation proceeds in one dimension expansion of the products is possible in the other two.

Figure 14.15 shows the infinite case for a cylinder where the reaction zone simply expands until reaction is complete, while Figure 14.16 shows the practical case where the reaction zone length is small and certainly less than the diameter of the explosive

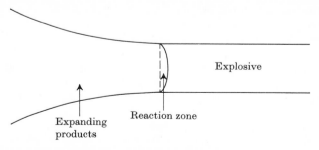

Figure 14.16. Practical case of a high-explosive reaction.

cylinder. A prerequisite for detonation is therefore a rate of reaction rapid enough to yield a substantial amount of energy within the 'supersonic' reaction zone.

For nitroglycerine in a cylinder of 10 mm in diameter, the speed of detonation will be about 8000 m sec^{-1}. The reaction zone length is less than the diameter (about 2 mm) and therefore the reaction must be completed in much less than one microsecond. Of course, the detonation temperature approaches 6000°K but even so the time is so short that diffusion from one vaporized particle to another would take too long. Therefore gunpowder would not detonate at this diameter, whereas nitroglycerine does so readily. Basically this is because all reactants are intimately linked in one molecule.

Why, then, do most high explosives not detonate from flame? Professor Bowden with many co-workers studied the initiation of detonation by impact, friction, heat, etc. Briefly, his conclusions were that all initiation is basically thermal and that burning or deflagration builds up to detonation when confinement conditions cause rapid acceleration or the development of shock waves. There are some explosives in which this build-up is very rapid, but for most organic explosives the transition is difficult. This is because the first steps in thermal ignition are endothermic, releasing molecules which then react exothermally. In the early stages the rates of reaction are controlled by the endothermic steps and energy is lost to the surroundings by conduction, convection, etc., rapidly enough to prevent acceleration.

Primary explosives are those that detonate from flame, impact or friction. They probably differ from secondary or high explosives more in degree than kind. They are nearly all the simple but unusual salts, fulminates or azides. Of these, mercury fulminate, $Hg(ONC)_2$, and lead azide, $Pb(N_3)_2$, are the most valuable commercially: they are thermally stable, but as soon as decomposition starts it accelerates rapidly. It seems probable that for these substances the first step in the reaction is an exothermic one, but there are several theories to account for the behaviour of these primary explosives.

Organic primary explosives include azides, e.g. cyanuric triazide, heavy-metal salts of nitrophenols (lead picrate, lead trinitroresorcinate) and some unusual compounds of high nitrogen content, e.g. tetrazene.

The lead salts of nitrophenols and tetrazene are, however, usually used as sensitizing additives to azides since they are not so energetic and need confinement to ensure burning to detonation.

Modern Explosives

Nobel's concept that explosives much safer to handle than gunpowder but much more energetic could be set off in a devastating way by a small amount of a sensitive explosive has been carried still further with the most modern explosives. Figures 14.17 and 14.18 show a typical chain of explosives used in a modern blast. One small detonator contains all the dangerous explosive safely encapsulated. It is set off by an electrically heated bridge wire which causes a flash of flame to ignite 0.1 g of lead azide. This reacts so rapidly in its simple fission to lead and nitrogen that a detonation shock wave is set up. The shock detonates an organic nitrate (pentaerythritol tetranitrate). This detonation propagates in the 6-mm diameter of the detonator and so through the even smaller diameter of the same substance in a detonating fuse (see Figure 14.18). The flexible fuse passes down a shot hole bored in rock (Figure 14.18). The detonation wave meets primers attached to the fuse; these are made from organic nitro-compounds some-

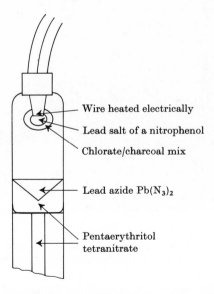

Figure 14.17. An electric detonator. The lead–nitrophenol salt flashes, the chlorate–charcoal burns with a hot flame, the simple molecule lead azide decomposes to lead and nitrogen by detonation, and the pentaerythritol tetranitrate, $C_5H_3(NO_3)_4$, detonates from shock although it would decompose only slowly if heated.

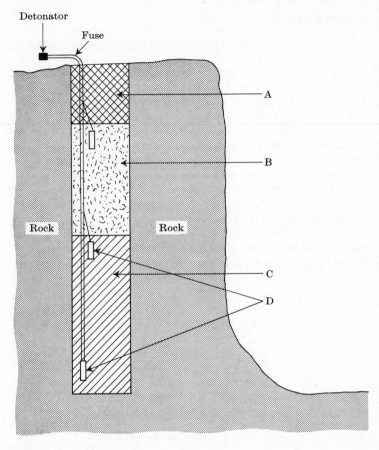

Figure 14.18. Blasting in rock. A, Rock stemming. B, Ammonium nitrate and diesel oil. C, Base charge of slurry or nitroglycerine–ammonium nitrate explosive. D, Primers containing TNT and PETN; less sensitive than PETN itself but detonated readily by shock from fuse.

times containing an organic nitrate, and, although insensitive, are made from explosive molecules and detonate in diameters of 23 mm or more.

The diameter of the bore hole will be over 100 mm and at its base a dense explosive charge C is placed: either a mixture of nitroglycerine, nitrocellulose, ammonium nitrate and combustibles, or a slurry explosive. Ammonium nitrate is an explosive but it is not energetic and, unless in a fine low-density form, will not detonate in 100-mm diameter. In reacting it yields free oxygen which, if burned with cellulose, doubles the energy output. By mixing the ammonium nitrate and combustible with a nitroglycerine–nitrocellulose jelly, an explosive is obtained that detonates to yield most of the available energy.

Explosive slurries are of two types, sensitized and unsensitized. The sensitized ones are thickened aqueous solutions of ammonium nitrate containing TNT or other 'secondary' detonating explosive; the latter is detonated by a primer, and energy from the ammonium nitrate is liberated behind this detonation. Unsensitized slurries are thickened aqueous solutions of ammonium and other nitrates containing an excess of solid ammonium nitrate and combustibles either in solution or in suspension; if the combustible is soluble it is thoroughly mixed with the ionized solution of ammonium nitrate, but not as intimately as it would be if chemically combined. These solutions have proved to be detonating explosives of much lower sensitivity than organic nitro-bodies but they can be sensitized by aeration. Bubbles incorporated in any detonating explosive provide local centres of high temperature and ignition as the pressure wave reaches them. One method of incorporating the bubbles is by the use of paint-grade aluminium; the hydrophobic surface of this material holds the bubbles and when they become hot spots they immediately ignite aluminium and become even hotter. Here, then, is a mixture which, in spite of the diffusion necessary for reaction, is able to react rapidly enough for detonation because of rapid ignition, high temperature of reaction and intimacy of content of the reactants.

Higher up the shothole in Figure 14.18 is a low-density mixture B of ammonium nitrate and diesel oil and air. Again air-sensitization helps detonation of the ammonium nitrate, but the intimacy of contact between the hydrocarbon and the oxidizer is also of great assistance. The ammonium nitrate used for this explosive is made by pouring a melt of the salt with about 5% of water down a cooling 'shot tower' and drying the granules so formed. Because of the drying these granules are absorbent and low-viscosity oils penetrate them to maximize intimacy of contact.

Not all blasting involves the full chain of detonator–fuse–primer–explosive. For most underground work the explosive must detonate in small diameter and is usually sensitive to initiation by a detonator. The explosives used are gelignites (mixtures of about 30% of nitroglycerine with nitrocotton, ammonium nitrate and fuels) or powders (similar mixtures of lower nitroglycerine content). Some of these are specially designed

Table 14.3. Explosive substances and their chief modern uses

Explosive	Manufacture	Uses
Primary explosives		
Mercury fulminate, $Hg(ONC)_2$	Mercury dissolved in nitric acid and alcohol added. After fuming, the salt crystallizes and is washed.	Filling for copper detonators and in some percussion caps. Less used and may soon be obsolete.
Lead azide, $Pb(N_3)_2$	Sodium + ammonia → sodamide; sodamide + nitrous oxide → sodium azide. Sodium azide solution + Pb nitrate give precipitate of lead azide.	Filling for detonators and percussion caps. The explosive is usually made in the presence of colloids (dextrin, gelatin, cellulose ethers) to reduce its sensitivity.
Tetrazene, $C_2H_8N_{10}O$	Aminoguanidine sulphate reacts with sodium nitrate to give 1-guanyl-4-nitrosoaminoguanyl-tetrazone.	Sensitizer in percussion caps.
Lead styphnate, $PbO_2C_6H(NO_2)_3 \cdot H_2O$	Resorcinol nitrated to 2,4,6-trinitro-derivative. Dissolved in water with MgO. Magnesium salt solution reacts with Pb nitrate.	Sensitizer with lead azide in detonators. Also used in percussion caps.
Diazodinitrophenol, $C_6H(NO_2)_2ON_2$	Reaction of nitrous acid with picramic acid.	Detonator filling (mainly in U.S.A.).

Secondary explosives

Nitroglycerine, $C_3H_5(NO_3)_3$ (Glycerol trinitrate)	Nitration of glycerol with mixed nitric and sulphuric acids.	Blasting explosive mixtures. Double-based propellants (with cellulose nitrates).
Nitroglycol, $C_2H_4(NO_3)_2$ (Glycol dinitrate)	Nitration of glycol with mixed nitric and sulphuric acids.	Used in conjunction with nitroglycerine to give low freezing properties.
Cellulose nitrates, $C_6H_7O_2(NO_3)_3$	Nitration of cellulose (cotton or wood cellulose) with mixed nitric and sulphuric acids.	Thickens nitroglycerine to make 'gelatines' and in propellants.
Ammonium nitrate, NH_4NO_3	Neutralization of ammonia gas in nitric acid.	With nitroglycerine in gelignites and coal-mining explosives. In aqueous slurries with TNT or aluminium + air.
Pentaerythritol tetranitrate, $C_5H_8O_4(NO_2)_4$	Nitration of pentaerythritol with nitric acid.	Base charge for detonators, charges for detonating fuse. Mixed with TNT for primers and special charges.
Trinitrotoluene, $C_7H_5(NO_2)_3$	Nitration of toluene with mixed nitric and sulphuric acids.	For slurries, primers, etc., as above and for military purposes.
Cyclotrimethylene trinitramine (RDX), $C_3H_6N_6O_6$	Nitration of hexamine.	Mainly for military purposes, but used commercially where temperature is unusually high, e.g. in steel and iron furnaces.

for coal-mining and contain cooling salts in sufficient quantity to prevent ignition of firedamp in rigorous detonation trials.

Although explosives have military applications, by far the greatest quantity is used today for mining, quarrying, road building, tunnelling, etc. For these purposes the chief chemical is ammonium nitrate because it is cheap, easily made, and insensitive. It has been shown how other explosives are used as aids to the release of energy from ammonium nitrate. There are some interesting attempts to introduce alternatives to ammonium nitrate which may yet change the pattern of explosives manufacture. Nitric acid forms powerful explosive mixtures with organic substances. Liquid oxygen and cellulose form a sensitive, detonatable mixture, and hydrazine nitrate solutions have been used.

For specialist applications, e.g. use at high temperatures under high pressures, special explosives have to be designed. High pressure can usually be tolerated by normal secondary explosives but their sensitivity drops as the gas bubbles are compressed; this can be overcome by introducing finely powdered substances of high density, which become hot spots when detonation occurs. High temperatures (above 200°C) cannot be tolerated by the usual commercial explosives. After a careful study of the effects of structure on stability some unusual organic explosives have been synthesized and some of these have found uses in space exploration. Examples are hexanitrostilbene and 'tetranitrobenztetrazene' (dibenzo[*ae,g*]-[1,2,5,6]-tetrazocine). Their main fea-

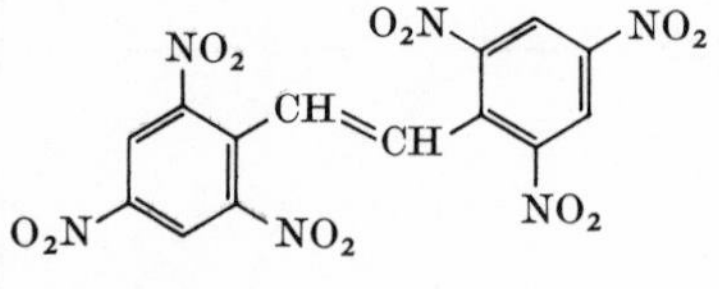

Hexanitrostilbene

Tetranitrobenztetrazene

tures are high melting point, symmetrical structure and all the oxygen in NO_2 groups.

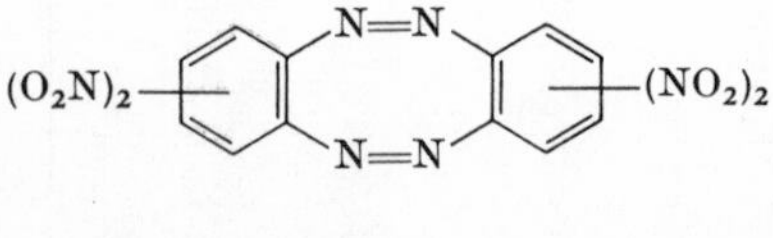

The methods of manufacture and chief uses of explosives in current use are summarized in Table 14.3 (pages 530–531).

Finally, we should mention nuclear explosives. Eventually organic explosives may be replaced by nuclear ones, but there is a long road to travel. Present nuclear devices are complex, far too large for most civil work and give radioactive products. Even their use for blasting out new harbours is too dangerous a source of pollution.

CHAPTER 15

Food Chemistry

(A. Crossley, Unilever Research, Port Sunlight Laboratory)

Most foods are based largely on natural products and a wide variety of chemical types is therefore available. The subject is further complicated by the fact that we have only a limited knowledge of many of the macromolecular species present which include complex lipoproteins, glycolipids, mucopolysaccharides and mucoproteins. Moreover, many of these materials are highly labile as they originate in the living cell and alter considerably during processing, so that only in a few areas can complete understanding of the chemical nature of foodstuffs and the changes that occur during processing be achieved. In most cases information must be gathered from a variety of sources, and foodstuff studies are therefore essentially multidisciplinary; they range from organic chemistry and biochemistry to physics, biochemical engineering and biology.

It is an over-simplification to categorize foodstuffs in terms of carbohydrate, lipid and protein, and foods must be considered in terms of origin of raw material: meat, fish or vegetable.

A number of industries has grown around the processing of these raw materials, such as flour milling, sugar refining, extraction of oilseeds, and industries based on preservation involving freezing, curing, canning and drying. Apart from these 'basic' industries, there is further processing of these semi-processed raw materials to provide such products as bread, confectionery, chocolate, margarine and a variety of dairy products.

Some details of various foods in the U.K. diet are shown in Table 15.1.

Within the limits of this Chapter it is not possible to discuss all the various food industries in detail and so two examples have

Table 15.1. Food consumption in the U.K. per head

Food	Weight (g) consumed per week per head (U.K. 1969)	Joules	Composition (g), per 100 g of uncooked		
			protein	fat	carbo-hydrate
Milk	3700	285	3.2	4.0	5
Cheese	112	1800	26	35	—
Fats, butter and margarine	320	3460	—	89	—
Eggs	250	680	12	13	—
Meat products	1100	500–1800	18–23	2.5–43	—
Fish	140	328	18	0.36	2.9
Sugar and preserves	530	1100–1700	0.36	—	72–107
Vegetables	2460	104–374	0.36–2.1	—	5–16
Fruit	840	60–149	0.36–1.1	—	2.9–20
Bread, flour, baked goods	1620	1040–1800	6.4–8.2	1.4–21	54–82
Miscellaneous	280	—	—	—	—

been chosen: oils and fats processing, and bread making. In addition, two topics of general relevance will be discussed: food spoilage and its prevention, and new raw material sources.

The general topic of food additives deserves a special mention and is an area of particular interest to the organic chemist. Increased sophistication in utilizing food raw materials has increased demand for 'additives', such as colours (see Chapter 10), flavours (see Chapter 16), preservatives, gelling and emulsifying agents, sweeteners, vitamins, etc. (Table 15.2, p. 536). Frequently, such materials are developed by identification of naturally occurring materials, but the object is commonly to provide cheaper and more versatile synthetic additives. Whether or not such additives occur naturally, it is essential that their metabolism in the human body is studied thoroughly, because there is an increasing awareness of the hazards of some compounds to living processes whether they are naturally occurring or synthetic organic chemicals.

Oils- and Fats-based Industries

More than 10×10^6 tons of oils and fats are produced world-wide for edible purposes. Extracted oils and fats are processed (refined) to remove a variety of impurities. Lecithin, a mixture of phospholipids derived largely from soyabean oil, and fatty acids are

Table 15.2. Examples of food additives

Class	Examples
Antioxidants	Butylated hydroxyanisole, butylated hydroxytoluene, various gallate esters, ascorbic acid, tocopherols
Preservatives	Benzoic acid, sorbic acid, biphenyl-2-ol
Emulsifiers and stabilizers (structuring agents)	Stearoyl tartrate, fatty monoglycerides, aceto-, lacto- and citro-glycerides, certain polyglycerol esters, fatty sorbitan esters, cellulose esters, sodium (carboxymethyl)cellulose, agar, alginic acid, carrageenan, lecithin, Na and Ca caseinate, starch and modified starches
Sweeteners	Saccharin, cyclamate, peptides
Flour improvers	Benzoyl peroxide
Colours	Various azo dyes, β-carotene
Food acids	Acetic, citric, lactic, malic, succinic acids
Humectants	Sorbitol
Glazing agents	Beeswax, spermaceti
Antifoaming agents	Sodium stearate, silicones
Anticaking agents	Magnesium stearate
Release agents	Sperm oil, butyl stearate, silicone
Sequestrants	Glycine, salts of ethylenediaminetetra-acetic acid (EDTA)
Propellants	Freons (fluorinated hydrocarbons)
Flavours	Lactones, acetic acid, hydrolyzed protein, esters
Flavour enhancers	Sodium glutamate, inositate and guanylate

important by-products of these purification processes. The refined oils contain largely triglyceride esters (Part 4, p. 150), together with small amounts of di- and mono-glycerides. A few typical fatty acids are listed in Table 15.3.

Table 15.3. Some important natural fatty acids

Palmitic acid	$CH_3(CH_2)_{14}COOH$
Stearic acid	$CH_3(CH_2)_{16}COOH$
Oleic acid	$CH_3(CH_2)_7CH{=}CH(CH_2)_7COOH$
Linoleic acid	$CH_3(CH_2)_4CH{=}CHCH_2CH{=}CH(CH_2)_7COOH$
Linolenic acid	
	$CH_3(CH_2)_2CH{=}CHCH_2CH{=}CHCH_2CH{=}CH(CH_2)_7COOH$

CH$_2$OOCR	CH$_2$OH	CH$_2$OH
CHOOCR'	CHOOCR	CH$_2$OH
CH$_2$OOCR"	CH$_2$OOCR'	CH$_2$OOCR
Triglyceride	Diglyceride	Monoglyceride

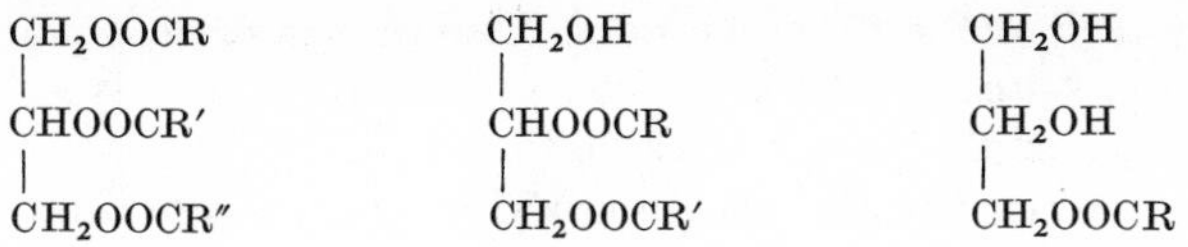

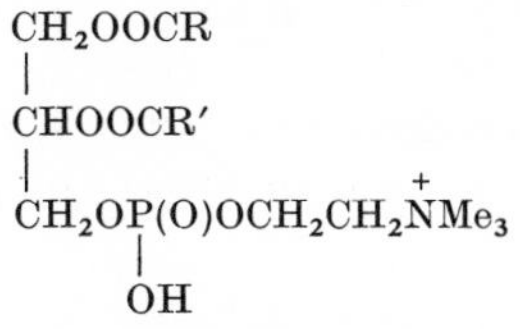

CH$_2$OOCR

CHOOCR'

CH$_2$OP(O)OCH$_2$CH$_2$NMe$_3^+$

OH

Lecithin

The glyceride esters of natural fats are very specific in their composition. Thus the unsaturated acids in vegetable species have double bonds almost exclusively in the *cis*-configuration, and the various fatty acyl groups are found in quite specific positions in the glycerol esters. Oils from vegetable sources have typically an unsaturated fatty acid residue in the 2-position; for instance, cocoa butter contains 40% of 1-palmitoyl-2-oleoyl-3-stearoylglycerol with virtually none of its isomers present. The physical properties of oils and fats are derived not only from the fatty acids present but also from the specific glyceride isomers characteristic of each particular fat.

The needs of various processed foods demand a greater range of properties than can be obtained from natural fats, and various means of modification are widely used. Thus the distribution of acids among the glycerides can be modified by transesterification, a process catalyzed by traces of sodium or sodium alkoxides. Transesterification leads to a statistically random distribution of

Table 15.4. Products of transesterification of 2-oleoyl-1,3-distearoylglycerol (SOS)

Glyceride		%
Tristearoylglycerol	(SSS)	29.6
2-Oleoyl-1,3-distearoylglycerol	(SOS)	14.8
1-Oleoyl-2,3-distearoylglycerol	(OSS)	29.7
1-Stearoyl-2,3-dioleoylglycerol	(SOO)	14.8
2-Stearoyl-1,3-dioleoylglycerol	(OSO)	7.4
Trioleoylglycerol	(OOO)	3.7

fatty acids on the glycerol residue. Thus the glyceride, 1-palmitoyl-2-oleoyl-3-stearoyl (POS) yields a mixture of 14 glycerides. A simple example is given in Table 15.4 (p. 537).

In practice various oils are mixed before transesterification, to give very complex mixtures of glycerides leading to considerable modification of product properties. The exact mechanism and the nature of the active catalyst species are still a matter of some dispute.

Partial glycerides (mono- and di-esters) are even more labile and can be intra-esterified under very mild conditions [see reaction (a)].

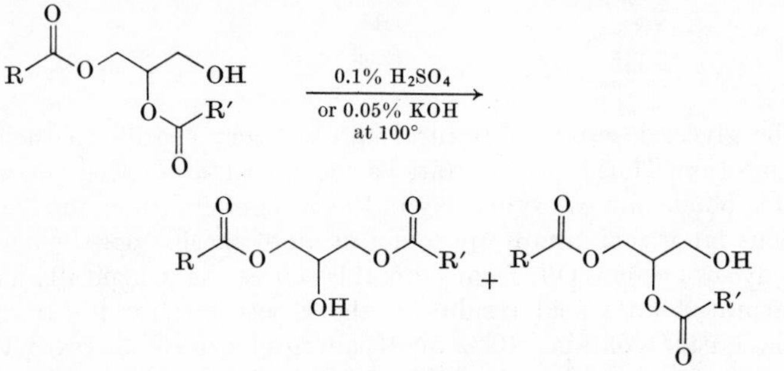

Perhaps the most interesting method of modification is hydrogenation (Part 1, p. 156). Its importance can be judged by the fact that more than a million tons of oil are hydrogenated for edible purposes each year. This process is used to raise the melting point of fats, and also to lower the concentration of the more unsaturated glycerides such as linolenic acid which give rise to by-products by autoxidation on storage.

Hydrogenation is normally carried out in the absence of a solvent between $100°$ and $180°C$, commonly using a finely divided nickel catalyst on kieselguhr or silica at pressures between 100×10^3 and 500×10^3 Nm^{-2}. Only a partial hydrogenation of double bonds present is carried out, which leads to a great diversity of products. The products from a simple linoleate ester are shown in Table 15.5 for two levels of hydrogenation.

It can be seen that not only are double bonds saturated in the process but *trans*-isomers and a variety of positional isomers are produced. Hydrogenation is a selective process depending on the fact that the rate of hydrogenation of a fatty acid is greater the more double bonds in the fatty-acid chain. In an ideally selective

Table 15.5. Partial hydrogenation of methyl linoleate at 100°C with Ni/SiO$_2$ catalyst under selective conditions

	Hydrogenation time: 4 min		Hydrogenation time: 7 min	
Diene (%)				
cis-cis (remaining linoleate)	72.5		37.2	
cis-trans and *trans-cis*	0.5		1.0	
Monoene (%) ($\Delta \equiv$ double-bond position)	*cis*	*trans*	*cis*	*trans*
$\Delta 4$	0.02	0.02	0.05	0.07
$\Delta 5$	0.05	0.09	0.20	0.23
$\Delta 6$	0.08	0.12	0.28	0.30
$\Delta 7$	0.20	0.30	0.58	0.64
$\Delta 8$	0.35	1.00	1.10	2.3
$\Delta 9$	6.1	1.25	13.5	3.7
$\Delta 10$	0.78	3.7	1.9	7.2
$\Delta 11$	0.90	3.4	2.0	7.0
$\Delta 12$	5.0	1.30	11.6	3.2
$\Delta 13$	0.25	1.00	0.75	2.3
$\Delta 14$	0.16	0.38	0.42	0.87
Saturated (methyl stearate) (%)	0.7		1.6	

hydrogenation no fully saturated species would be formed before polyunsaturated esters are hydrogenated to monounsaturated esters. In fact, selectivity is very dependent on conditions at the catalyst surface, and the ratio of saturated to unsaturated species for any degree of hydrogenation can be modified considerably by choice of catalyst and hydrogenation conditions. The ratio of *cis*- to *trans*-isomer formed can be regulated similarly. This is important as the *trans*-esters have considerably higher melting points than the *cis*-isomers (e.g. the melting point of triolein is −5°C and of its *trans*-isomer, trielaidin, is 40°C). It is clear that the properties of hydrogenated fats are governed largely by the amounts of saturated and *trans*-unsaturated glycerides formed. The variety of industrial processing conditions employed has been developed largely to regulate the amounts of these components. In this way cooking fats, frying and salad oils, and fats suitable for margarine, ice-cream, and confectionery purposes may be obtained.

One major influence of lipid in the foods area deserves further mention, namely, the importance of linoleic acid in human nutrition. Linoleic acid cannot be synthesized in the human body and man's intake, therefore, depends solely on his diet. There is growing evidence that relatively high levels of this acid are necessary to reduce the incidence of arteriosclerosis, a major cause of death. Such levels cannot normally be achieved in conventional diets and there is considerable current activity by food scientists to maintain linoleic acid in its *cis-cis*-form as far as possible during processing, e.g. by using moderate conditions for the hydrogenation. In addition, one could obtain a product of similar physical kind by interesterification of linoleic-containing glycerides with more saturated types.

Natural unsaturated fatty acids have a close resemblance to a family of hormone-like substances, the prostaglandins (Part 4, p. 161), which exhibit a wide range of potent physiological effects and are currently the subject of intensive study. The relationship is seen in the following formulae which illustrate one proven

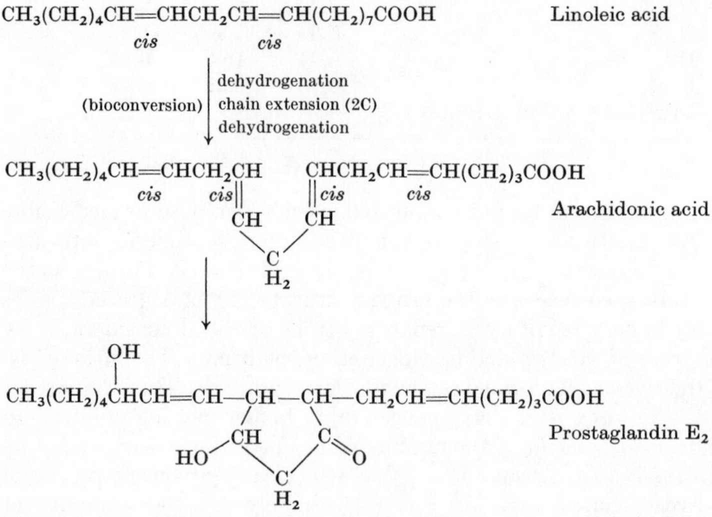

$$CH_3(CH_2)_4CH{=}CHCH_2CH{=}CH(CH_2)_7COOH \qquad \text{Linoleic acid}$$
$$\qquad\quad cis \qquad\qquad\quad cis$$

(bioconversion) | dehydrogenation
chain extension (2C)
dehydrogenation

$$CH_3(CH_2)_4CH{=}CHCH_2CH \qquad CHCH_2CH{=}CH(CH_2)_3COOH$$
$$\quad cis \qquad cis\| \qquad\quad \|cis \qquad cis$$
$$\qquad\qquad\qquad CH \qquad CH \qquad\qquad\qquad\qquad \text{Arachidonic acid}$$
$$\qquad\qquad\qquad\qquad C$$
$$\qquad\qquad\qquad\qquad H_2$$

$$\qquad OH$$
$$\qquad |$$
$$CH_3(CH_2)_4CHCH{=}CH{-}CH{-}CH{-}CH_2CH{=}CH(CH_2)_3COOH$$
$$\qquad\qquad\qquad\quad CH \qquad C$$
$$\qquad\qquad HO \qquad\qquad C{=}O \qquad\qquad \text{Prostaglandin E}_2$$
$$\qquad\qquad\qquad\quad C$$
$$\qquad\qquad\qquad\quad H_2$$

pathway from linoleic acid *in vivo*. The influence that these discoveries will have on human nutrition and foodstuff composition is not yet clear but is likely to be of major importance.

The Bread-making Industry

Flour and derived foodstuffs are so basic to the Western diet that they deserve a special mention. Furthermore, a study of these materials serves to illustrate the complex effects and inter-actions of the protein, carbohydrate and lipid constituents.

Wheats vary widely in properties. 'Strong' wheats normally contain 13–14% of protein and 'weak' wheats as little as 6%. In the U.K., bread flours are produced by milling a mixture of strong and weak wheats in such a way that a considerable amount of the bran and wheat kernel is removed to provide an 'extraction rate' of ca. 72% of flour containing 11–12% of protein. Bread flour is bleached by treatment with chlorine dioxide which decolourizes carotenoids; a small proportion of potassium bromate may also be added. White flours are then fortified with calcium, iron and nicotinic acid.

There are three stages in bread making: dough development, dough leavening and baking. The traditional process involves mixing a dough from the basic ingredients of flour, yeast, salt and water, plus other optional materials, and allowing it to ferment for 1–3 hours before processing it further into loaves. As fermentation proceeds, the dough becomes extensible and begins to show improved ability to hold entrapped gas in the form of finely dispersed bubbles.

The maturing of a dough and the development of viscoelastic properties is generally attributed to the development of a three-dimensional gluten–protein network by cross-linking of the protein molecular species found in the individual flour particles. The protein molecules are present in the flour in a tightly coiled form and are held in this state by physical forces, e.g. intramolecular covalent disulphide bonds between cysteine residues. Mixing the dough ingredients results in the disruption of some of the weaker cohesive forces such as hydrogen bonds, so that the proteins can then hydrate, swell and uncoil in the salt solution of the dough. This initiates a series of reactions both within and between proteins which leads to the stable network of the mature dough. It is widely held that probably the most important of these reactions are thiol–disulphide and disulphide–disulphide interchanges.

In modern high-speed continuous processes, cross-linking is facilitated by addition of oxidizing agents (ascorbic acid and potassium bromate). A small amount of fat is an essential ingredient although its role is not clear.

After the dough has been developed, it is divided into portions and held under conditions that are favourable to the multiplication of the yeast. By a series of enzymic actions, some of the flour starch is hydrolysed to shorter-chain saccharides, and finally glucose is broken down to ethanol and carbon dioxide. The carbon dioxide aerates the dough and is necessary for a final good loaf volume and crumb structure. Yeast and enzyme activity is finally destroyed in the baking process. The loaf structure is set by the rupture of some of the starch granules (i.e. gelatinization of some of the starch) in the cooking process, the gluten-protein forming the continuous matrix of the crumb.

The flavour of bread is chemically complex but can be considered as consisting of crumb and crust components. The crumb flavour is due to by-products of yeast action on the starch, whereas the crust flavour is a result of the Maillard reaction (browning of crust) between various reducing sugars and protein amines.

The major deterioration taking place on storage is staling. Thus, wheat starch consists of amylose and amylopectin molecules and both are involved in staling, though it is thought that the change in the amylopectin is the more important of the two.

Amylose, which is removed from the starch granules during baking, forms a soft gel which 'retrogrades' very rapidly by molecules 'lining up', forming hydrogen bonds between glucose units on adjacent chains. The amylopectin left in the granule has more room and tends to be in 'open-tree' configuration. Over a period of days this tends to close up, and again intrachain hydrogen

Open-tree form of amylopectin   Closed form

bonds are formed, but this time the chains are components in a single molecule (amylose is a linear molecule). The coascervation of amylopectin can be reversed to some extent by heating to 50–60°C. Hence the 'freshening' of a stale loaf by reheating.

The staling of bread occurs over a period of days and is temperature-dependent, the maximum rate occurring at about 2°C. At deep-freeze temperatures (−20°C) molecular motion is restricted so that no change occurs. The use of emulsifiers, e.g. mono-glycerides, which form complexes with starch, can increase shelf-life by retarding the changes in the starch molecules.

Spoilage and Its Prevention

Foods are subject to continuous attack by all forms of life. Some 30% of all food produced is destroyed by insects and further depredations are made by fungi. It is scarcely surprising, therefore, that loss and damage continue after harvest and slaughter. Apart from such obvious pests as beetles and rodents, much damage is done by fungi and micro-organisms. To this must be added modification and degradation by enzymes from the food organisms themselves and by autoxidation and other spontaneous chemical reactions.

The history of civilization is intimately bound up with the solution to these problems. Dry and carefully stored grain, however, keeps remarkably well and samples of considerable antiquity have been found almost intact. Damp grain, on the other hand, is subject to attack by fungi even when it is still in the ear. Bunts, rusts and smuts are a familiar sight in cornfields and the rye fungus *Clariceps purpurea* produces a devastating family of toxins which are amides of lysergic acid (Part 4, pp. 344, 389). An oil-seed fungus, *Aspergillus flavus*, found chiefly on groundnuts, produces compounds such as aflatoxin B_1 which are highly carcinogenic in mammals and gave rise to 'Turkey X disease' in Britain in 1960. Pride of place, however, goes to a series of deadly peptides produced by the anaerobic bacterium *Clostridium botulinum* which are among the most poisonous substances known.

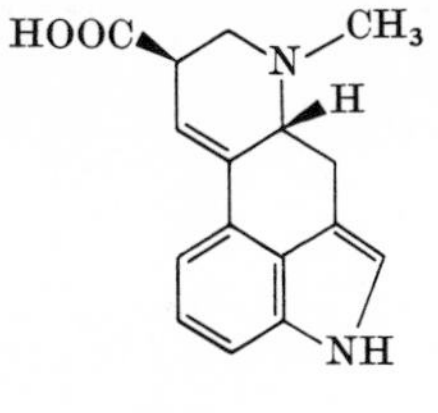

Lysergic acid 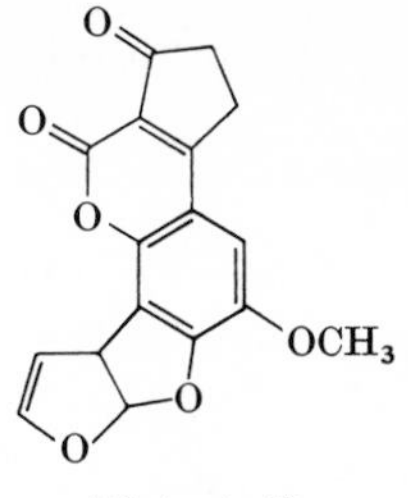   Aflatoxin B_1

Long before the existence of micro-organisms was suspected, methods had been devised for eliminating them from foodstuffs. Thus it was noted that the products of souring were still quite useful as foodstuffs and kept better than the original materials. Indeed, pickling in vinegar is one of the oldest methods of food preservation. Smoking, dehydration, salting, preserving (with sugar) and freezing are all ancient techniques which owe much of their success to lowering of the 'water activity'. Sterilization of food in airtight containers was developed during the Napoleonic wars: in glass bottles in France (N. Appert, 1809) and in tin-coated steel cans in England (P. Duranc, 1810), although it was not until 1839 that the latter came into widespread use in America. In his famous experiment (1864) Pasteur showed that beef broth would keep for long periods with free access of air if precautions were taken to exclude microbes.

The fact that canned goods have to be sterilized is a disadvantage, adding to the cost and frequently detracting from the flavour or texture of the products. Consequently there have been numerous attempts to improve older methods of preservation. The introduction of the plate freezer (Clarence Birdseye, 1929) enabled food to be quick-frozen so that large ice crystals did not form and spoil the texture. Freeze-drying of liquids was first described in 1906 (M. d'Arsonval and F. Bordasm) and more recently vegetables have been dehydrated without loss of texture and flavour by first pricking them so that moisture can be removed from the interior (S. Gunson, J. P. Savage, 1957).

$$\text{EtCH} \overset{c}{=} \text{CHCH}_2\text{CH} \overset{c}{=} \text{CHCH}_2\text{CH} \overset{c}{=} \text{CH(CH}_2)_7\text{COOR} \qquad \xrightarrow{\text{Initiation}}$$

Linolenate

$$\text{EtCH} \overset{c}{=} \text{CHCH}_2\text{CH} \overset{c}{=} \text{CH}\overset{\bullet}{\text{C}}\text{HCH} \overset{c}{=} \text{CH(CH}_2)_7\text{COOR} \qquad \longrightarrow$$

Free radical

$$\text{EtCH} \overset{c}{=} \text{CHCH}_2\overset{\bullet}{\text{C}}\text{HCH} \overset{t}{=} \text{CHCH} \overset{c}{=} \text{CH(CH}_2)_7\text{COOR} \qquad \xrightarrow{\text{O}_2}$$

$$\underset{\underset{\text{O}-\text{O}\bullet}{|}}{\text{EtCH} \overset{c}{=} \text{CHCH}_2\text{CHCH} \overset{t}{=} \text{CHCH} \overset{c}{=} \text{CH(CH}_2)_7\text{COOR}} \qquad \xrightarrow[\substack{\text{(H from another} \\ \text{molecule)}}]{\text{Propagation}}$$

$$\underset{\underset{\text{OOH}}{|}}{\text{EtCH} \overset{c}{=} \text{CHCH}_2\text{CHCH} \overset{t}{=} \text{CHCH} \overset{c}{=} \text{CH(CH}_2)_7\text{COOR}} \qquad \xrightarrow[\text{scission}]{\text{Chain}}$$

Hydroperoxide

$$\text{EtCH} \overset{c}{=} \text{CHCH}_2\text{CHO} + \cdot\text{CH} = \text{CHCH} \overset{c}{=} \text{CH(CH}_2)_7\text{COOH} + \cdot\text{OH}$$

cis-Hex-3-enal   Radical by-product

Autoxidation of linolenates

The foregoing techniques are well able to control microbial growth, but hydrolysis and autoxidation of fats remain serious problems. Even pasteurized butter soon becomes rancid owing to hydrolysis with the formation of some short-chain fatty acids, including butyric acid (threshold concentration for taste: 25 ppm); and 'desiccated' coconut is not notably more stable than fresh. Other animal fats and vegetable oils can also become rancid; in this case, although their triglycerides do not contain short-chain fatty acids, such acids can be produced by autoxidation and chain scission; a more immediate result of autoxidation, however, is the production of aldehydes for which a typical reaction sequence is shown on p. 544.

Although hexenal is only one of a number of aldehydes and other compounds formed during the autoxidation of oils containing linoleate, it has a pronounced 'green bean' odour (threshold concentration for taste: 0.11 ppm) which is characteristic of such oils in general and of soyabean oil in particular. Several other unsaturated aldehydes with even lower taste thresholds have been identified in autoxidized oils, and indeed many characteristic food flavours and 'off' flavours appear to be in part due to them, for example, fish, chicken, lean beef, tallow, and hydrogenated fish oil.

Various antioxidants have been added to foods with the object of reducing autoxidation, with limited success in some cases, such as with refined lard. Most vegetable oils already contain natural antioxidants (tocopherols) which confer protection, provided that pro-oxidants are absent, by themselves becoming oxidized to

α-Tocopherol

innocuous by-products such as quinones. It is thought that under certain circumstances autoxidation proceeds by free-radical attack on the methyl groups with production of dimers and, ultimately, of formaldehyde, but the precise sequence of reactions occurring at ambient temperature is still not fully understood. Certain heavy metals, e.g. copper or iron, can be highly pro-oxidant and these can often be removed or converted into less active forms by use of chelating agents such as citric acid or ethylenediamine-tetra-acetic acid.

New Sources of Food Raw Materials

Of the three major constituents of human food, carbohydrates are the most plentiful; edible fats are in short supply in some developing countries, but there is a much wider and more serious deficiency of proteins. Therefore, current interest centres mainly around ways and means to augment the supply of edible proteins.

Proteins

Although chemical synthesis may provide ultimately a useful source of proteins, at present there are many natural sources that are not fully utilized. Considerable progress is being made in up-grading a number of agricultural protein sources previously unused, and in converting inorganic materials, carbohydrates waste and petroleum hydrocarbons into usable proteins by microbiological means.

Protein 'flour' derived from extracted oilseeds is rich in high-quality proteins, and problems associated with its use have been largely overcome, namely the destruction of growth inhibitors in soyabean, the removal of the toxic gossypol from cottonseed, and the control of toxic moulds on groundnuts. The flour can be incorporated into a variety of products and even spun into meat-like filaments.

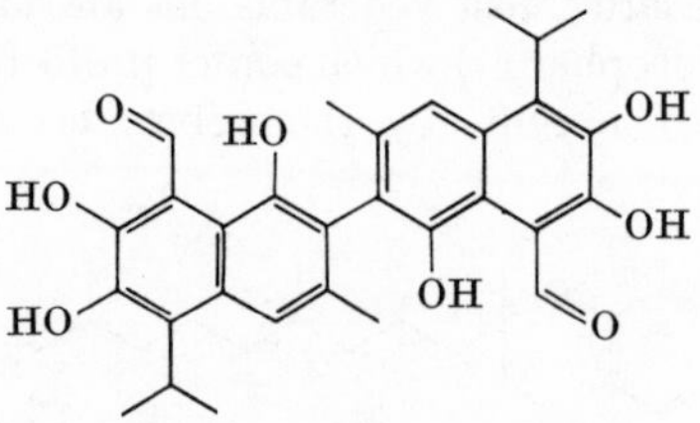

Gossypol

The green alga, *Chlorella*, which occurs in Nature as a green scum on ponds and waterways, has been given serious consideration as a food source. It shows a two-fold increase in protein-rich biomass in 2–6 hours' growth on inorganic nutrients. Unlike proteins from higher plants, the protein from *Chlorella* is not deficient in lysine, methionine or tryptophan.

Micro-organisms, especially yeasts, have been grown on a large scale from carbohydrate nutrients. Both molasses (a by-product from the production of sugar cane) and sulphite pulp liquor (waste effluent from the paper industry) have been used as nutrients.

Although proteins with physicochemical properties similar to that of milk protein (casein) have been produced, the products have been used entirely as animal feeds. It is likely, however, that intensive studies currently under way on the culture of a micro-fungus grown on hydrolyzed starch will result in large-scale production of proteins suitable for human food.

Several soil bacteria and yeasts can utilize aliphatic hydro-carbons, if provided with nitrogen from inorganic sources such as ammonium salts, to produce proteins suitable for use as animal feedstuffs. In a recent development, natural gas is first chemically converted into methanol, which is used as the feedstock for a specially selected micro-organism; the amino-acid composition of the product varies with the organism, the carbon source and the process, but all are said to be good as animal feeds.

Protein synthesis involves the linkage of amino-acids in a predetermined sequence and is currently limited to laboratory preparation of relatively simple species. Manipulation becomes easier if the amino-acid is covalently linked to a water-insoluble polymer and this may form the basis of future commercial processes (see Part 4, p. 338). Random polymerization to form 'proteinoids' of molecular weight over 8000 has been reported to provide potentially nutritious material.

$$CH_2{=}CHCHO + CH_3SH \longrightarrow$$

$$CH_3SCH_2CH_2CHO \xrightarrow[\text{KCN}]{NH_3-} CH_3SCH_2CH(NH_2)COOH$$

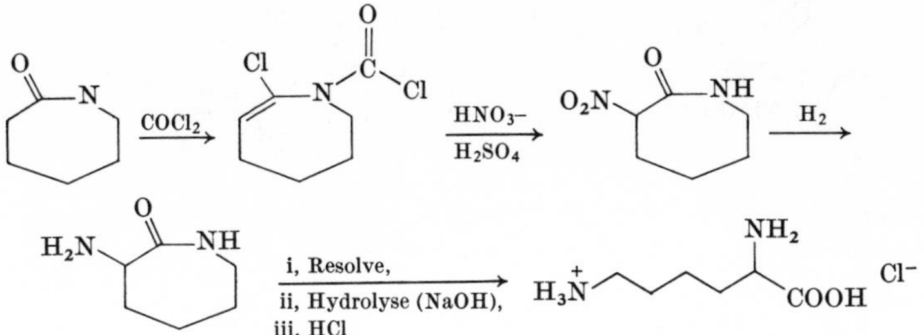

Methionine

Any synthetic processes must depend on the availability of amino-acids. These materials are also used for supplementing materials deficient in particular amino-acids. Fortunately many of them are available from natural sources. Commercial processes for the preparation of lysine and methionine are illustrated in the above formulae.

Carbohydrates

Carbohydrates in the form of grain crops are generally abundant, and sucrose extraction from cane or beet forms the basis of a major industry. D-Glucose is obtained commercially by hydrolysis of starch by acid or various enzymes (Part 1, p. 158), as illustrated.

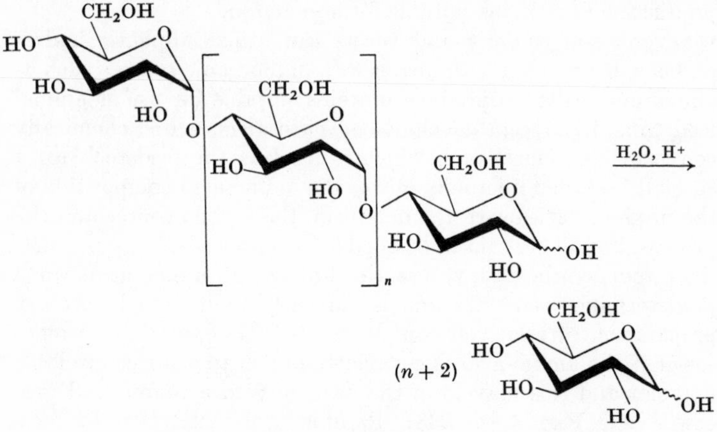

Hydrolysis of starch to glucose   D-Glucose

Purely synthetic commercial processes are not currently used. Recent work using vinylene carbonate derived from petrochemical intermediates, however, has aroused interest in synthetic processes.

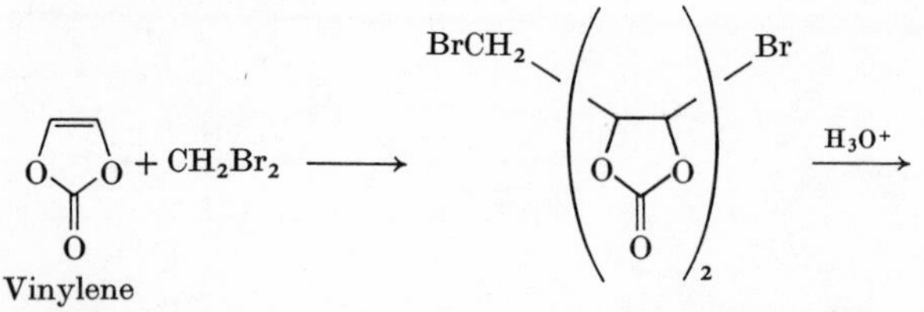

Vinylene
carbonate

$BrCH_2$—$[CH(OH)]_3$—CHO

Fats

Significant quantities of synthetic fatty acids were manufactured for the first time in Germany during World War II. The process involved the oxidation of alkanes (from Fischer–Tropsch hydrogenation of coal) by air in the presence of aqueous potassium permanganate as catalyst (see Chapter 6). After fractionation and purification, the C_{10}–C_{20} fraction contained oxygenated derivatives of straight- and branched-chain hydrocarbons with both even and odd numbers of carbon atoms. This fraction was esterified with glycerol and the resulting synthetic fat was refined and used in margarine.

Large quantities of synthetic fatty acids are currently manufactured in Eastern European countries by similar oxidation of high-boiling paraffins from petroleum. Before purification, the crude acids are heated to dehydrate hydroxy-acids. Technology is also available for polymerization of ethylene, followed by oxidation or carbonylation, but the products are more expensive. Although not normally used for edible purposes, these synthetic acids help to release large quantities of natural fats from non-edible to edible uses.

The Future of Foodstuffs

In more prosperous countries there will be an increasing demand for a greater variety of foodstuffs and also for foods that are convenient to use and have extended keeping qualities. On the other hand, the rapid increase of population will place a considerable strain on world resources. Scientific advances must attempt to cater for both of these conflicting demands.

Nutritional requirements will increase in importance. For many countries the provision of a balanced and adequate diet must be the first objective. At the same time, advances in medical nutritional knowledge are likely to require changes of composition of foodstuffs. Already there are strong indications of a need for larger amounts of linoleic acid in the diet, and many processed foods must by law be fortified by addition of vitamins and essential minerals.

Scientific advances across a wide field will advance food technology. Thus, knowledge of protein structures is developing rapidly after the α-helix discoveries, and similar advances are being made in the carbohydrate field. Techniques are also being developed to study complexes such as the lipoproteins. These and similar studies should help us to understand the physical nature of foodstuffs and how they can be modified. Another area of progress is in understanding enzyme-catalyzed transformations; simple synthetic systems are now being developed that simulate these enzyme systems, and continuous processes based on enzyme action are envisaged. Advances are to be expected, but it will be a long time before scientists can entirely displace the 'art' from food technology.

CHAPTER 16

Perfumes and Flavours

(E. COWLEY and W. D. FORDHAM, Bush Boake Allen Ltd., London)

Probably the first flavours used were aromatic spices, and the Chinese are believed to have used clove buds a thousand years before Christ. The growth of the spice trade was one of the major economic developments during the Middle Ages. The use of terpenoid compounds for medical purposes dates back to antiquity but it was not until the fifteenth and sixteenth centuries that the use of flowers for preparing agreeably fragrant mixtures originated in Southern France. Lavender and rosemary oils were widely produced in the seventeenth century and the famous 'Eau de Cologne' first appeared in 1725. A very high proportion of the original perfumery compounds and a substantial number of the original natural flavours are terpenes or terpenoid compounds (see Part 4, Chapter 5).

The chemistry of perfumes and flavours has been revolutionized by the development of modern analytical techniques, of which gas chromatography is one of the most important. Most perfumes and many flavours are extremely complex mixtures of compounds; we shall see below that lemon oil contains no less than 60 different, identified compounds and, although the lemon taste may be characteristic of a mixture containing only a small number of them, taste and smell are both sensitive to extremely small amounts of material and it is not necessarily the major components in a mixture which are responsible for its particular odour or taste.

We shall not be concerned in this Chapter with the physiological (and psychological) aspects of smell and taste but only with the organic chemistry. Taste and smell are necessarily to some extent subjective, and we shall discuss only the types of naturally occurring compound used in perfumes and flavours and the preparation and application of synthetic and 'unnatural' compounds.

PERFUMERY COMPOUNDS

(W. D. FORDHAM, Bush Boake Allen Ltd., London)

Essential Oils

Most essential oils contain a wide variety of monoterpenoids such as (+)-limonene, α-pinene, citral, geraniol, carvone and borneol; and in recent years many sesquiterpenes such as α-cedrol, vetiverol and α-santalol have also been isolated. In addition, many benzenoid compounds occur as, for example, phenethyl alcohol, eugenol and methyl anthranilate. Aliphatic sulphur and nitrogen compounds also occur in some oils, particularly those employed for culinary purposes. A few oils are rich in one compound only and these are invariably used as a commercial source of these compounds. Thus, the oil of *Eucalyptus globulus* contains 75% of cineole; lemongrass oil contains 75–80% of citral; and the oil expressed from the peel of the citrus fruit contains 90% of (+)-limonene. However, these are exceptions and we shall consider here a few of the more important essential oils that are the usual complex mixtures.

Rose oil is produced in Bulgaria primarily from two types of rose: *Rosa damascena* Mill and *Rosa alba* Linn. France and Morocco cultivate predominantly *Rosa centifolia* Linn. Different methods of extraction are practised commercially to produce three types of rose oil: 'Rose Concrete', 'Rose Absolute' and 'Rose Otto'.

'Rose Concrete' is produced by extraction of freshly harvested flowers with a volatile solvent such as butane or purified petroleum ether. The solvent is removed by distillation, leaving a dark waxy mass known as concrete. It contains approximately 60% of non-volatile fatty substances and 35–45% of volatile compounds comprising phenethyl alcohol (60%), (−)-citronellol (20%), geraniol and nerol (12%), and traces of aromatic compounds. The concretes have a limited perfumery application in cosmetic preparations and are used primarily to produce rose-absolutes.

Extraction of the rose concrete with ethyl alcohol separates most of the odorous compounds from the insoluble material. After removal of the alcohol by distillation a yellow viscous oil remains, known as 'Rose Absolute', which is used extensively in perfumery formulations. Absolutes are rich in phenethyl alcohol, (−)-citronellol, geraniol and nerol.

Steam-distillation of the flowers of *Rosa damascena* Mill gives a pale yellow liquid having a very floral, spicy and honey-like odour, which is known as 'Rose Otto'. It is rich in (−)-citronellol and geraniol and contains small quantities of linalool, phenethyl alcohol, citral, eugenol, carvone and sesquiterpenes.

Sandalwood oil, one of the earliest-known perfumery materials, is produced from the trees *Santalum album* L., indigenous to East India, and *Eucarya spicata* Sprag. et Summ of Western Australia. East Indian sandalwood oil is produced by steam-distillation of the powdered wood and roots of the tree, whilst the Australian oil is obtained by a combination of solvent-extraction and steam-distillation of the wood. These oils are used in perfumery for their powerful and lasting odour and their ability to blend well with a large number of other essential oils. Both oils contain the sesquiterpene alcohol, α-santalol, as the major constituent.

Geranium oils are commercially produced from several varieties of *Pelargonium* and the name 'geranium oil' is thus really a misnomer. The greater part of the world's supply of geranium oil comes from Reunion, known formerly as Bourbon, and the oil is often referred to as geranium Bourbon oil. It has a powerful rose-like odour and, because of its low cost, is extensively employed in perfume compounds, particularly for use in soaps and cosmetics. It consists mainly of citronellol (40–50%), geraniol (20–25%), and small quantities of linalool, α-terpineol, phenethyl alcohol, eugenol, (−)-isomenthone, hex-3-en-1-ol, and sesquiterpene hydrocarbons and alcohols.

Several citronella oils are produced from grasses of the *Cymbopogon* genus which grow extensively in the Ceylon and Java areas. They contain geraniol and citronellol as their major constituents and are used extensively in cheap perfumes for detergents, cleansers and other household products. They have also been used as commercial sources of geraniol and (+)-citronellal, which are important perfumery chemicals in their own right. (+)-Citronellal is also used as an intermediate in the synthesis of (−)-menthol. However, these oils are now relatively little used as sources of geraniol and citronellal, which can be produced more economically on an industrial scale by synthetic methods.

Odour and Molecular Constitution

The quality and intensity of the olfactory stimulus evoked by a chemical compound is determined by its molecular structure.

However, the situation is confused because of the following circumstances: (1) the lack of objective methods of odour assessment; (2) our inability to quantify odour effects; (3) the dubious purity of many compounds under test; and (4) the low sensitivity of chemical or physical analysis compared with the sensitivity of olfactory analysis. Most odours are complex, although we identify them as single entities, and there are no means of communicating these perceptions except by simile. Thus, professional perfumers address each other in an intuitive language involving such terms as 'green notes', 'floral top-notes', 'heavy undertones', etc. The most mysterious quality of odours is that frequently they do not combine in an understandable manner. Many compounds in the pure state give an odour impression very different from that which they exhibit in admixture with other odoriferous chemicals. A knowledge of the interplay of odour phenomena of this type forms the basis of the art of perfumery.

It is generally accepted that the perception of odour arises by contact of the chemical concerned with the olfactory receptors, although the nature of the interaction is not known. Certain atoms or groups of atoms, known as 'osmophores' have been found from experience to impart strong odour effects. For example, the introduction of a sulphur atom into an organic molecule invariably gives rise to very strong, and usually objectionable, odours; the carbonyl group imparts a characteristic odour quality, which is further modified if the carbonyl group is conjugated with an olefinic bond. However, this is only part of the story and it is necessary also to consider the stereochemistry of the molecule; this affects its olfactory character in a manner somewhat analogous to its influence on chemical reactivity. A particularly striking example is that of the two isomers of 3-(2,2-*exo*-3-trimethyl-*exo*-5-norbornyl)cyclohexanol wherein the epimer (**1**) with the *trans*-(axial)-hydroxyl group possesses a very strong sandalwood odour, whilst the isomer with the *cis*(equatorial)-hydroxyl group is completely odourless.

(**1**)

Many other cases are known of such effects of stereoisomerism in ring structures as, for example, with borneol and isoborneol, the menthols and the carvomenthols. Y. R. Naves has built up a strong case for supposing that a clear and intense odour, or even the absence of odour, is associated with conformational rigidity which favours the development of stable intramolecular interactions, particularly in compounds of relatively high molecular weight.[1]* This line of reasoning is said to explain the well known fact that ketones of high molecular weight only develop their full odour characteristics at high dilution. The macrocyclic ketones having a musk odour are excellent examples of this phenomenon.

The influence of chirality on odour is still somewhat confused, and conflicting statements appear in the literature. For example, α-ionone (2) has been resolved by a number of workers and the conclusions reached are contradictory. One school states that the odours of the (+)- and (−)-forms differ, whilst the other maintains that they are the same but only one-tenth as strong as the odour of the racemate prepared by mixing the two enantiomers. In general, the odours of optical isomers appear to differ when they are rigid molecules, but this generalization does not seem to hold for aliphatic molecules. It is reported, for example, that it is not possible to distinguish olfactively between the (+)-, (−)- and racemic forms of citronellal (3) or citronellol (4). In the case of linalool (5) opinions are divided between those who maintain that the enantiomorphs are indistinguishable and those who consider that the isomers have different odours. The ability of an expert panel to distinguish between the two enantiomers can always be said to be attributable to the presence of trace impurities derived from either the starting materials or the process involved rather than to inherent differences between the (+)- and (−)-forms.

(2)   (3)   (4)   (5)

Theimer and McDaniel have attempted to overcome this difficulty in the following ingenious manner.[2] Five compounds with definite fragrance characteristics were synthesized in both (+)-

* For references, see p. 577.

and (−)-forms and were compared by panels. Each compound was made by two different routes so that there were obtained in each case, two pairs, namely (+)-1 and (+)-2, and (−)-1 and (−)-2. The synthetic route was the same for each (+)-1 and (−)-1 compound and differed from the preparative method for (+)-2 and (−)-2. This procedure means that any build-up of impurities due to the chemistry involved would automatically bring the pairs [(+)-1, (−)-1] and [(+)-2, (−)-2] close together, at the same time putting the pairs [(+)-1, (+)-2] and [(−)-1, (−)-2] further apart. The olfactive effect of impurities in the end-products thus would have to be subordinate to the olfactive effect of the chirality if the (+)- could still be distinguished from the (−)-enantiomer with statistical significance. By this means it was demonstrated that the (+)- and (−)-forms of myrtenal diethyl acetal (**6**), *trans*-pinocarvyl propionate (**7**), myrtenal (**8**), pinoacetaldehyde (**9**) and *trans*-pinocarveol (**10**) could be distinguished in every case.

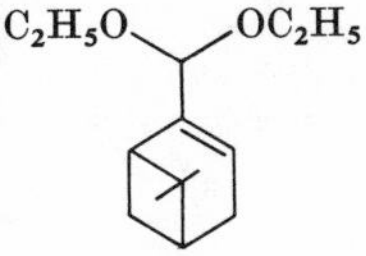

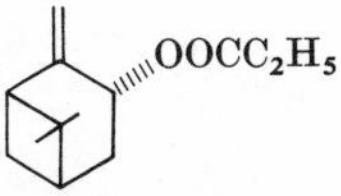

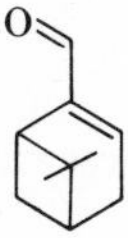

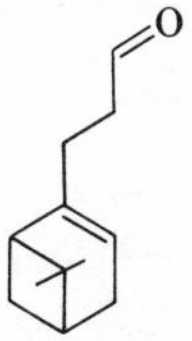

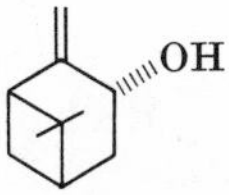

There seems little doubt that the majority of ethylenic *cis–trans*-isomers differ in odour, sometimes very markedly. The well-known difference in odour between geraniol (**11**) and nerol (**12**) is a case in point. The 2′-*trans*-isomer of α-ionone (**2**) is reported to have a violet-like odour whilst the *cis*-isomer possesses a cedarwood-like bouquet. *cis*-Hex-3-en-ol (**13**), known as leaf alcohol, possesses an odour that is more highly prized by perfumers than that of the *trans*-isomer.

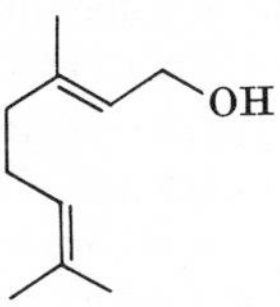

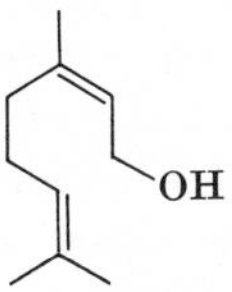

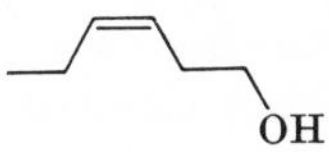

From the few examples quoted it is clear that the diversity of chemical structures gives rise to a wide variety of odour effects.

However, it does not follow that compounds of different chemical structure need necessarily possess different odours. In fact, it is well recognized that compounds of widely differing chemical types may possess closely similar odours. This is well demonstrated with the musk-odoured compounds, which are to be found among indane, tetralin, androstane, macrocyclic and nitrobenzene structures (which are discussed below). The subject has been fully reviewed by Beets.[3]

It is also reported by Beets,[4] a little surprisingly perhaps, that the odour quality of the three structures (A), (B), (C) is the same.

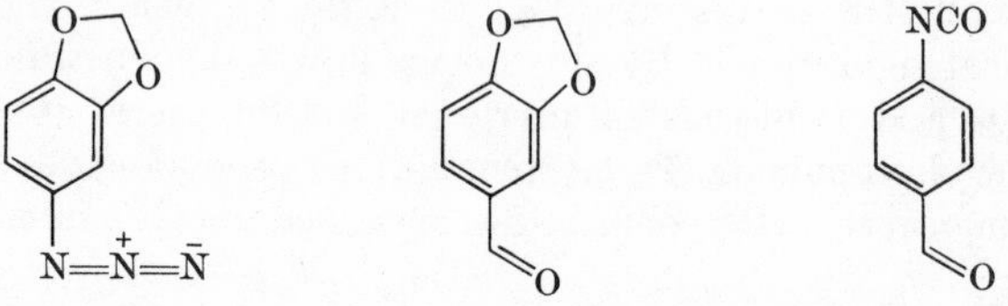

| (A) | (B) | (C) |

Mechanism of Odour Perception

Ottoson[5] showed that direct contact must occur between the olfactory receptors and the odorous material in order that odour sensation may be experienced. He demonstrated that the olfactory receptors cannot be excited when they are separated from the odorous particles by a thin membrane. Several workers have proposed the theory that molecules will cause odour sensation if they possess a steric form that allows them to fit into positions on the nasal receptors having corresponding profiles. Amoore has enlarged on this theme and attempted to correlate the size and shape of molecules with their odour from a study of six hundred or so chemicals.[6] He suggests that even primary types of odour occur which he describes as ethereal, camphoraceous, musky, floral, pepperminty, pungent and putrid, and the theory attempts to correlate these odours with molecular shape and size. Thus, the camphoraceous odour is said to be exhibited by spherical molecules of diameter 0.7 nm; the musky odour by disc-shaped molecules about 1.0 nm across; the floral odour by kite-shaped molecules; the pepperminty odour by wedge-shaped molecules; ethereal odours by small or thin molecules; pungent odours by electrophilic molecules; and putrid odours by nucleophilic molecules. All odours are considered to be complexes of these basic seven, caused by the molecules fitting two or more different odour-reception sites.

Some biological support for this type of hypothesis has been supplied by Gesteland *et al.*[7] who found evidence for eight different kinds of olfactory receptors in the olfactory area of the frog.

The theory has met with considerable criticism. For example, the almond-like odour of hydrocyanic acid seems inconsistent with the theory since its profile structure is quite unlike that of other compounds having a similar odour, e.g. benzaldehyde. Objections to the theory have also been raised on the grounds that Amoore's literature survey of odoriferous chemicals can only provide a record of odour descriptions which are subjective, frequently vague, sometimes contradictory and often incorrect, if only because of lack of purity of the substances under test. Lack of attention to purity has contributed greatly to the confusion that exists in the field of structure and odour correlation.

The 'functional profile group' concept introduced by Beets[8] suggests that the profile of the molecule and the position of its functional groups play the central role in the process of olfactory interaction. This theory leads to the following generalizations:

(1) Details of the molecular profile can be replaced without influencing the general character of the odour, provided that this substitution does not introduce polar groups that may change the order of the orientations.

(2) The functional groups can also be replaced by other groups of similar polarity, without changing the olfactory character.

(3) Molecules that do not possess a polar group, e.g. saturated hydrocarbons, or those having functional groups that are not easily accessible sterically and are consequently not very active chemically, have not very characteristic odours, or are even odourless.

(4) Polar groups in neighbouring positions in the same molecule are able to co-operate with a single functional group and thus contribute to a uniform organization, whilst when they are in positions distant from one another their coexistence contributes to disorientation and thus weakens the odour.

In 1928 Dyson proposed that a relationship existed between molecular vibration and odour and Wright[9] has more recently correlated odour with low-frequency infrared vibrations (below 800 cm^{-1}). He has reported statistically significant correlations in the Raman spectra of sixteen compounds having a nitrobenzene odour. Similarly he has correlated the far-infrared spectra of ten musks, ten non-musk compounds and thirteen compounds having a 'green' odour. Wright has also pointed out that, in the case of optically active compounds, no known example exists wherein one enantiomer has an odour and the other not, from which he deduces that the primary olfactory process is physical rather than chemical.

The problem of mechanism of olfactory perception is thus still open to speculation.

Commercial Synthesis of Some Perfumery Compounds

Geraniol, Nerol, Linalool, Citral

Until comparatively recently these important acyclic mono-terpenoid compounds were produced solely from natural sources such as Citronella, Bois de Rose and Lemongrass Oils. Industrial synthetic methods have been developed in the past five to ten years, however, yielding products that compete in quality, price and volume of production with the materials of natural origin. It is estimated that at present, over 5000 tons p.a. of this family of compounds is produced synthetically. Two routes are currently employed, one starting from natural products and the other from acetylene.

The natural product is β-pinene, most of which is isolated by fractional distillation of sulphite turpentine, a by-product of the kraft paper-manufacturing industry. On pyrolysis at 400°C β-pinene gives a high yield of myrcene (15), which is then hydro-chlorinated under anhydrous conditions in the presence of a catalytic quantity of cuprous chloride to give a mixture containing linaloyl (16), neryl (17) and geranyl chloride (18). It is interesting

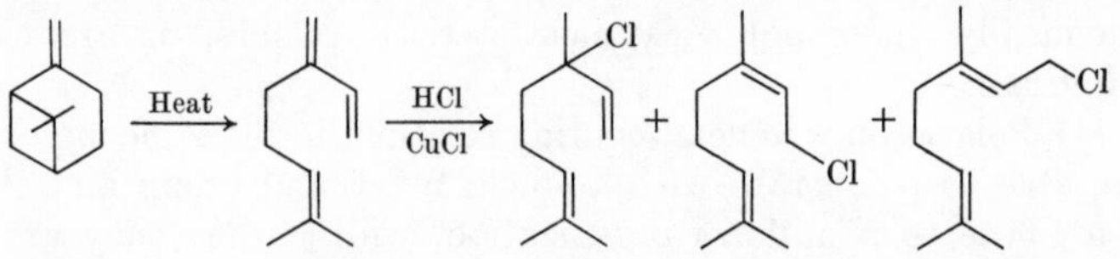

(14)   (15)   (16)   (17)   (18)

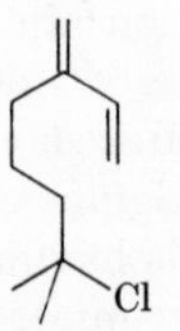

(19) 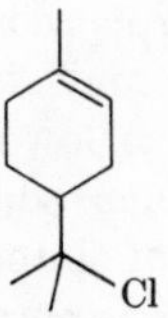   (20)

that in the absence of cuprous chloride these allylic chlorides are produced only in minor quantities and the main products are myrcenyl chlorides (**19**) and α-terpinyl chloride (**20**). The mechanism of the catalytic action of cuprous chloride in this reaction is obscure. It is known that cuprous chloride forms labile complexes with conjugated dienes, including myrcene, and it is reasonable to assume that this complex is an intermediate in the catalysed reaction; but its structure and the nature of its reaction with HCl have not yet been elucidated.

The proportions of the allylic chlorides (**16**), (**17**) and (**18**) formed depend on the reaction conditions and may change subsequently on storage owing to equilibration processes. The ease with which such allylic isomerizations occur is, in fact, the basis of the flexibility of this synthetic route. Thus, when the mixture of chlorides reacts with anhydrous sodium acetate in the presence of triethylamine as catalyst, a typical S_N2 mechanism operates to give a mixture of acetates wherein geranyl and neryl acetate predominate (ca. 90–95%). Under conditions such that an S_N1 (solvolysis) reaction predominates as, for example, in the reaction of the chlorides with acetic acid in the presence of an acid acceptor, the product is almost exclusively linaloyl acetate. This possibility of varying the product composition by proper selection of reaction conditions is put to practical industrial use. Saponification of the respective acetates yields the required linalool or geraniol/nerol mixture. The geraniol and nerol are separated by fractional distillation for sale as pure chemicals or the mixture may be employed without separation as an intermediate for the synthesis of other perfumery products. Thus, it may be oxidized to a mixture of the two geometrical isomers of citral by passing its vapour in admixture with oxygen over a heated copper catalyst.

Citral is employed extensively in the perfume and flavour industry and as an intermediate for the manufacture of α-ionone, an important perfumery product, and of β-ionone which is employed for vitamin A manufacture. The nerol/geraniol mixture may also be hydrogenated to citronellol (**21**), another important perfumery alcohol, which is also an intermediate for synthesizing 'hydroxydihydrocitronellal' (**22**). This compound has not been found to occur naturally but is employed in tonnage quantities by the perfumery industry because of its lily-of-the-valley type odour. The route is as shown.

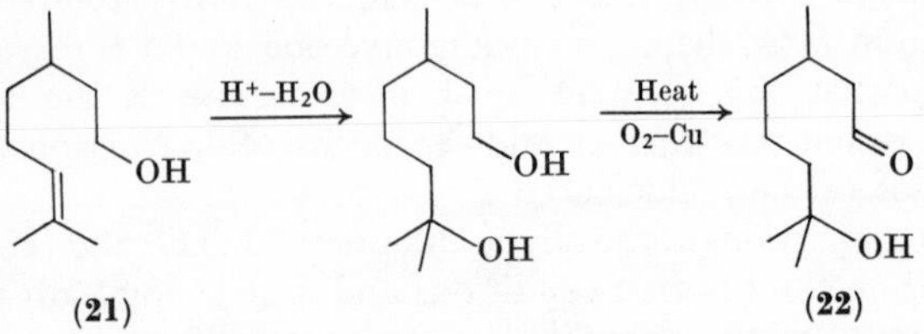

(**21**)   (**22**)

The purely synthetic route to monoterpenoid compounds employs acetylene and acetone as its basic raw materials and is particularly valuable for the manufacture of linalool and citral. In outline, the steps are as follows.

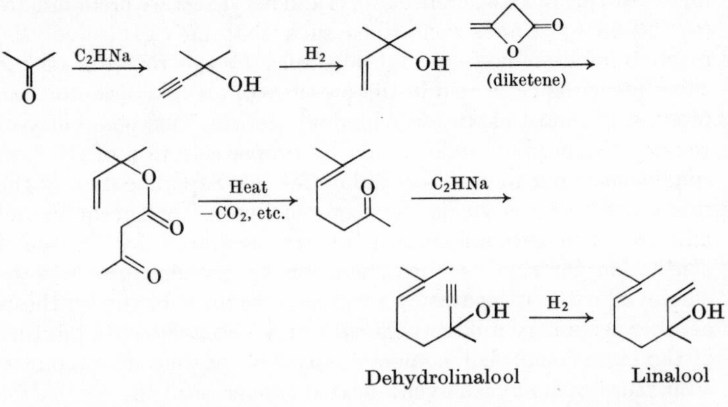

Dehydrolinalool   Linalool

Dehydrolinalool is also a versatile intermediate since citral may be obtained in high yield by heating its acetate with a catalytic quantity of copper. Also, pyrolysis of the acetoacetate of dehydrolinalool (the Carroll reaction[10]) yields pseudoionone (**23**), the precursor of the ionones.

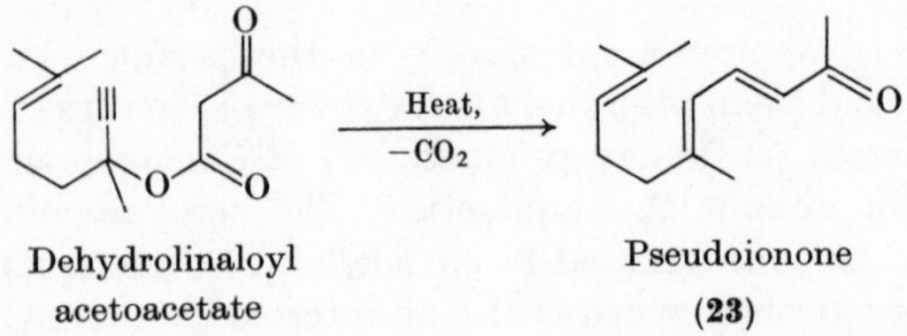

Dehydrolinaloyl   Pseudoionone
acetoacetate   (**23**)

The Ionones

α- and β-Ionone are commercially important compounds and are usually manufactured from citral. The first step involves an aldol condensation of citral (**24**) with acetone under alkaline conditions to give pseudoionone (**23**), which is then cyclized by acidic reagents to give a mixture of isomeric ionones. If phosphoric acid or boron trifluoride is employed, then α-ionone (**2**) is the major product, whereas when strong sulphuric acid is used β-ionone (**25**)

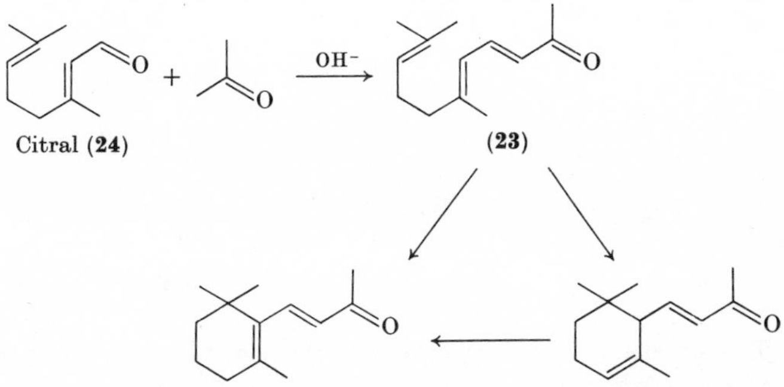

Citral (**24**)   (**23**)

β-Ionone (**25**)   α-Ionone (**2**)

predominates. Russian workers have demonstrated that α-ionone is first formed under both conditions and that under the influence of sulphuric acid the α-ionone isomerizes to the fully conjugated β-ionone.

α-Ionone is used extensively in the perfumery industry because its odour resembles that of violet flowers. β-Ionone, however, is used in perfumery to a limited extent only, and its main outlet is as an intermediate in the synthesis of vitamin A. Both ionones are employed in the flavour industry, particularly in imitation berry flavours.

Condensation of citral with ethyl methyl ketone gives rise to two 'methylpseudoionones' (**26**) and (**27**), each of which may be cyclized to the α- or β-methylionone. There are thus four structurally different methylionones, known as α-isomethylionone (**30**), α-n-methylionone (**28**), β-isomethylionone (**31**) and β-n-methylionone (**29**). From the perfumer's point of view this multiplicity of isomers creates a complex situation which is outside the scope of this Chapter. Suffice it to say that all commercial methylionones contain varying proportions of these isomers although the odour of α-isomethylionone (**30**) is generally considered to be the most attractive.

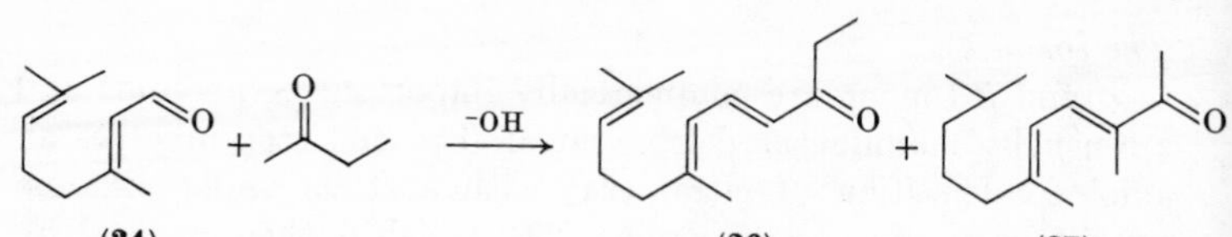

(24)   (26)   (27)

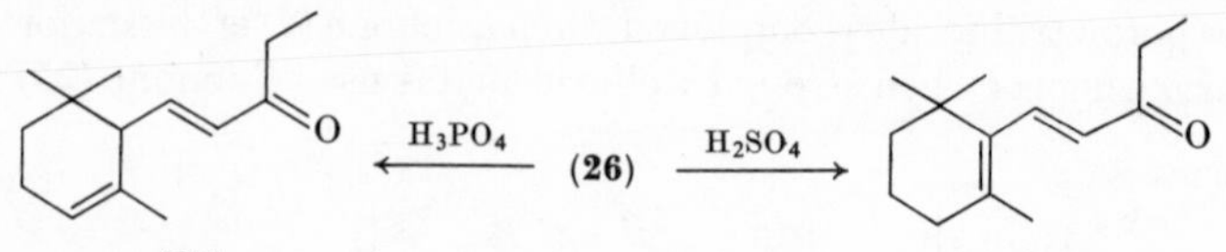

(27)   (26)   (28)

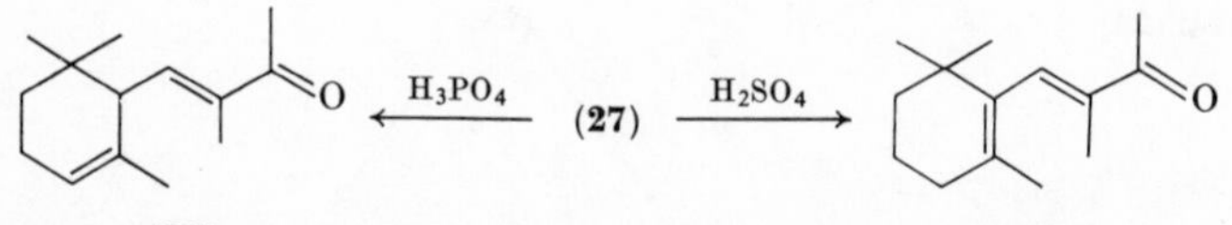

(30)   (27)   (31)

The 6-methylionones, α-irone (**32**), β-irone (**33**) and γ-irone (**34**), are very costly. Their odour is extremely powerful and is of the orris-violet type. They occur naturally in orris rhizomes wherein the α-isomer generally predominates. Here again the effect of structural and stereochemical features on the odour is complex and there is still disagreement on the issue. However, most perfumers agree that α-irone, particularly the *cis*-isomer, possesses a highly desirable odour. The irones are isolated from the volatile oil of orris rhizomes and are also synthesized from 5,6-dimethyl-hept-5-en-2-one by an acetylene route similar to the linalool synthesis mentioned on p. 560.

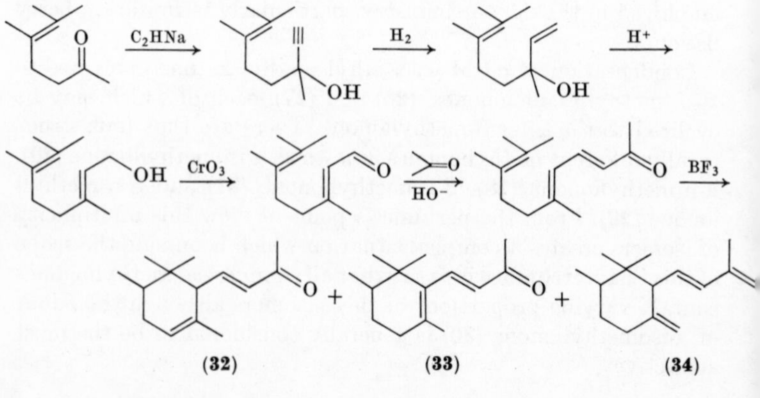

(32)   (33)   (34)

Some Musk Compounds

The macrocyclic ketones muscone (**35**), civetone (**36**), dihydro-civetone (**37**) and exaltone (**38**) occur naturally in certain animal secretions, whilst the lactones ambrettolide (**39**) and exaltolide (**40**)

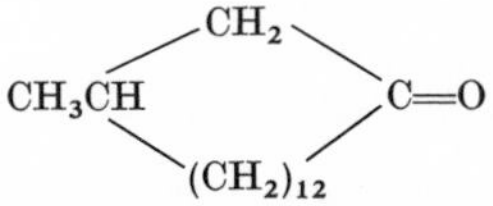

Muscone (**35**) 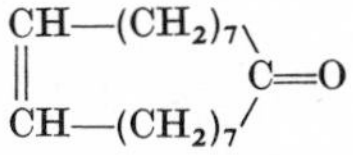 Civetone (**36**) 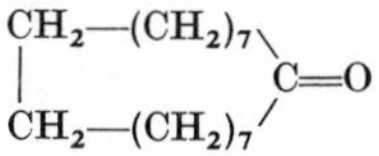 (**37**)

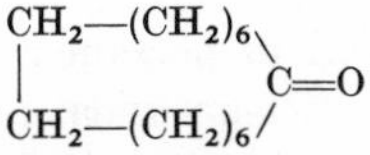

Exaltone (**38**) 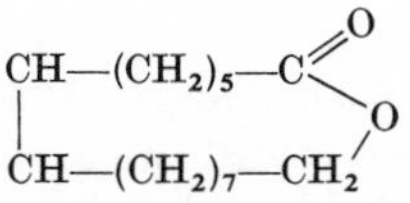 Ambrettolide (**39**) 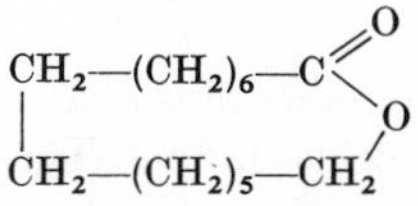 Exaltolide (**40**)

are of vegetable origin. The constitution of a number of these macrocyclic compounds, particularly civetone and muscone, was elucidated by Ruzicka during the 1920's.

Considerable academic work has been carried out on this range of compounds since that time, but few commercially viable syntheses have been developed, with the result that these chemicals are invariably costly. The majority of the synthetic routes (see Part 3, Chapter 3, pp. 190–194) start from dicarboxylic acids or near derivatives, and most of the earlier syntheses relied on preparation of the cyclic ketone either by heating a salt of elements of the third and fourth groups of the Periodic System or by cyclizing an intermediate bromo-ester. The first method gives very poor yields of large-ring ketones and the second method suffers the commercial disadvantage that very high dilution is required in order to obtain good yields of the desired products. One of the most useful, from a preparative point of view, is the acyloin condensation. This method depends upon the fact that α,ω-dicarboxylic acids, when treated with finely divided sodium, furnish cyclic acyloins, a reaction that appears to take place on the surface of the metal. The acyloins are converted into macrocyclic ketones by reduction with zinc and hydrochloric acid.

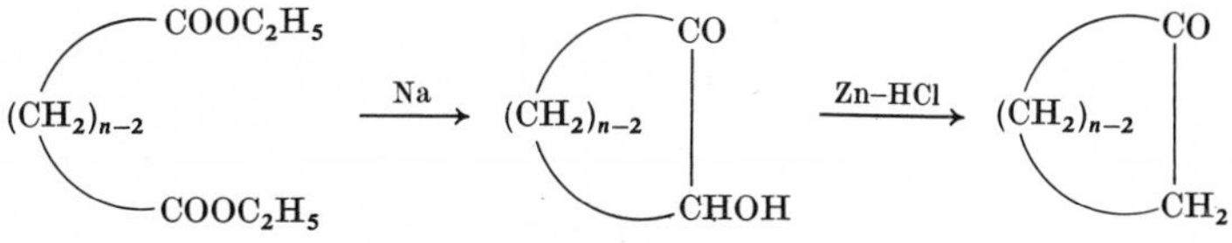

The macrocyclic ketones may be oxidized by Caro's acid (H_2SO_5) to the corresponding macrocyclic lactones and many of these compounds are prepared in this manner.

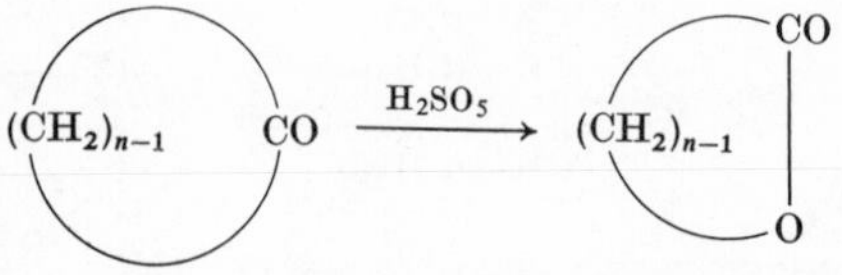

Both the macrocyclic ketones and the lactones possess a very intense musk odour which finds wide application in perfumery compounding. They also retard the evaporation of more volatile constituents of a perfume and frequently show synergic and odour-amplifying effects on perfumes and flavours. Their effects are in some cases perceptible at concentrations as low as 0.01 ppm.

Several aromatic nitro-compounds are known that have a musk-like odour. The more commercially important compounds include musk xylene, musk ambrette and musk ketone. Carpenter

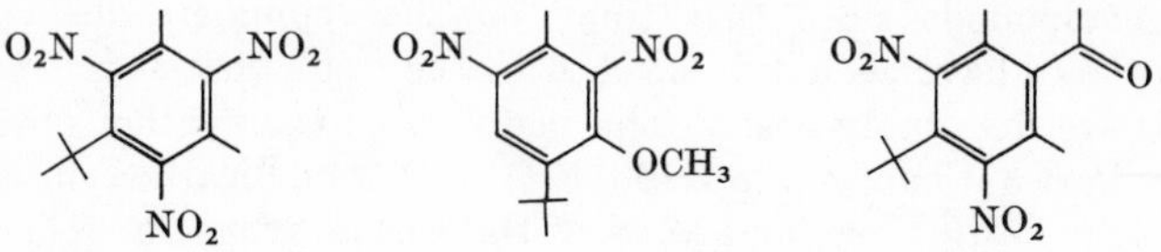

Musk xylene   Musk ambrette   Musk ketone

and his co-workers[11] have made a detailed study of the relation between structure and odour of musk ambrette isomers and homologues and conclude that the most important requisite for musk odour is that the tertiary alkyl group should be *ortho* to the alkoxy-group. The presence of the *tert*-butyl group also appears essential since its replacement by *sec*-butyl destroys the musk odour.

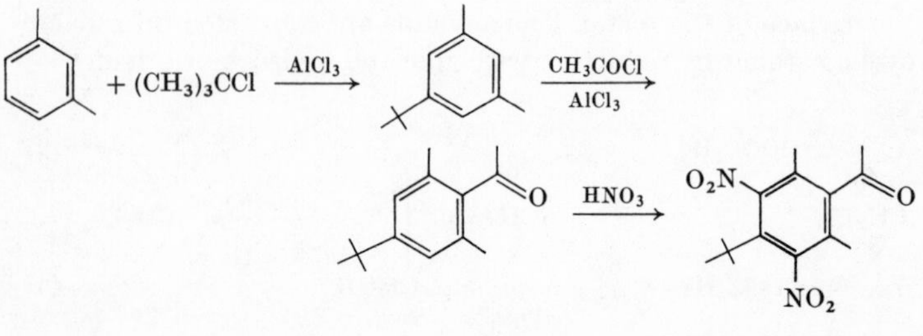

The aromatic nitro musks are prepared commercially by a Friedel–Crafts reaction followed by nitration. The synthesis of musk ketone from *m*-xylene is shown.

A number of tetrahydronaphthalenes are known to have a musk odour, although at present only versalide (**41**) and tonalid (**42**) are of commercial importance. In this series of compounds the

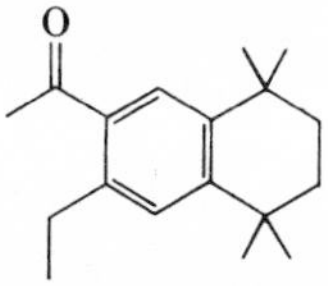

(41) 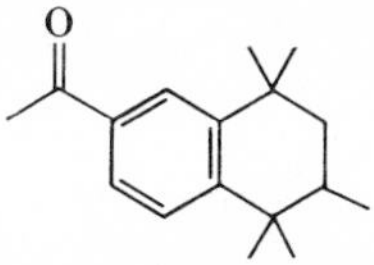   **(42)**

development of a musk odour appears to depend on the presence of a carbonyl group attached to the benzene nucleus. The tetralin musks are usually prepared from 2,5-dichloro-2,5-dimethylhexane as shown.

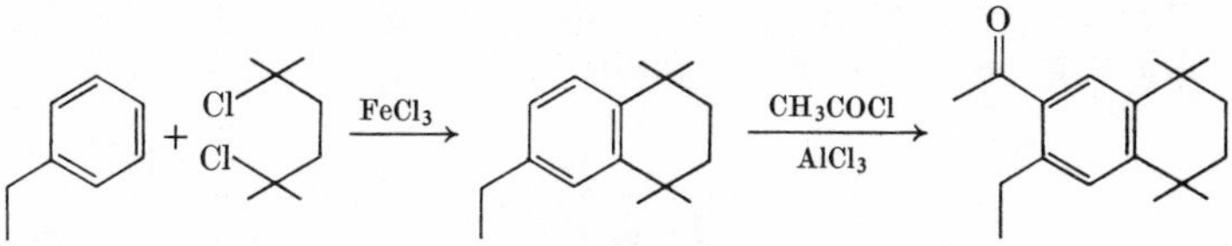

(41)

Beets and his co-workers[12] have synthesized a large number of indan derivatives in order to determine the influence of substitution on the musk odour of this class of compound. The basic structure (**43**) is odourless but the introduction of an acetyl or formyl group produces (**44**) and (**45**) having strong musk odours.

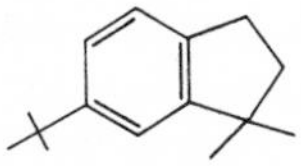

(43) 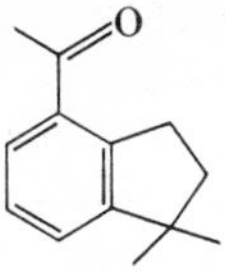   **(44)** 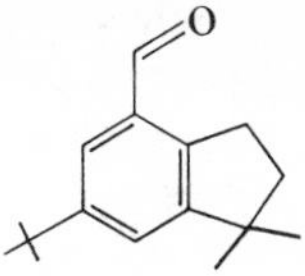   **(45)**

Substitution in the five-membered ring also appears necessary for the appearance of the musk odour since (**46**) is odourless and the introduction of alkyl substituents as, for example, one methyl group in (**47**) gives a moderate musk odour and the two methyl groups in (**48**) give a strong musk odour.

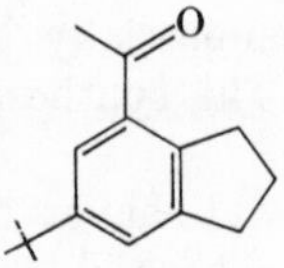

(46) 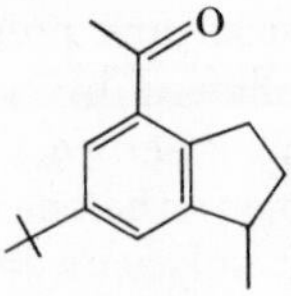 (47) 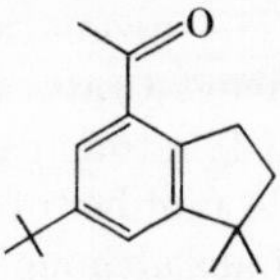 (48)

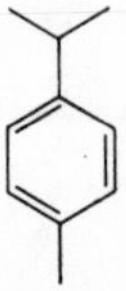

p-Cymene

The compound (48) is known under the trade-name 'Celestolide'. The majority of indan musks are manufactured by the condensation of *p*-cymene with a tertiary alcohol under acid conditions, followed by a Friedel–Crafts reaction with an acid chloride.

The indan musks are stable and do not discolour soaps and cosmetic products and, in view also of their relatively low price, appear to have a promising future in the perfumery industry.

FLAVOURS

(E. COWLEY, Bush Boake Allen Ltd., London)

Analysis and Identification of Flavouring Materials

The advent of sophisticated analytical techniques, especially gas chromatography, mass spectrometry and infrared spectroscopy has made it possible to identify minor constituents in flavours. Compounds present in minute quantities are often key ingredients in the organoleptic qualities of the material. One of the earliest analyses was of the 'aromatic' principles of raspberry fruit.[13] The 'aromatic' principles of raspberry fruit were extracted and concentrated; separation of the carbonyl compounds from the concentrate by reaction with 2,4-dinitrophenylhydrazone left a residue that still had a characteristic odour which was found to be due to some 10 different alcohols and to methyl acetate. The fractions isolated and identified are listed in Table 16.1.

As a second example, Table 16.2 (p. 568) shows the ingredients discovered by W. G. Jennings[14] in a study of the volatile constituents of the Bartlett pear by the full battery of modern analytical techniques.

Table 16.1. Volatile flavour constituents of the raspberry

More than 1.0 ppm	0.1–1.0 ppm
Acetaldehyde	Acetoin
cis-Hex-3-enal	Hex-2-enal
	Hexanal
	α-Ionone
	Geraniol
	cis-Hex-3-en-1-ol

0.01–0.1 ppm	Less than 0.01 ppm
Acetone	Biacetyl
β,β-Dimethylacrolein	Acrolein
Pent-2-enal	Propanal
Pent-1-en-3-ol	Pentan-2-one
Hexan-1-ol	β-Ionone
3-Methylbut-3-en-1-ol	Methanol
Butan-1-ol	Pentan-1-ol
Ethanol	*trans*-But-2-en-1-ol

Chemical Structure and Organoleptic Qualities

We saw, when discussing the odorous properties of compounds that minor changes of structure—indeed even different chiralities —were sufficient to make a substantial change in the compound's smell. In a similar way we find flavour characteristics vary widely within a range of similar compounds. The 'pear-like' flavour of isopentyl acetates is well known. Table 16.3 (p. 569) shows how a range of similar esters vary in their taste.

Geraniol (3,7-dimethylocta-2,6-dienol), so important in perfumery compounds, is also of importance as a flavouring and the flavour characteristics of some of its esters are listed in Table 16.4.

The comparison of chemical structure with organoleptic properties does not reveal any uniform pattern. In the esters listed in Tables 16.3 and 16.4 the flavour's strength remains virtually constant but the flavour characteristics change substantially on alteration of the carbon skeleton. A quite different trend is found in vanillin (**49**), the chemical constituent of vanilla, and 'ethyl-vanillin' (**50**) where the basic characteristics are the same but

Table 16.2. Volatile constituents of Bartlett pears*

Ethanol[a]
Propanol[a]
Butanol[a, b]
Pentanol[a, b]
Hexanols[a, b, f]
Heptanol[a]
Octanol[a, b]
Me acetate[a]
Et acetate[a, b]
Pr acetate[a]
Bu acetate[a, b]
Pentyl acetate[a, b]
Hexyl acetate[a, b]
Heptyl acetate[a, b]
Octyl acetate[a, b]
cis-Hexenyl acetate[b]
Me octanoate[a]
Et octanoate[a, b]
Me decanoate[a, b]
Et decanoate[a, b]
Et dodecanoate[a]
Et tetradecanoate[a, b, c]
Me hexadecanoate[a, b, c]
Et hexadecanoate[b, c]
Me octadecanoate[a, b, c]
Me *trans*-oct-2-enoate[b]
Et *trans*-oct-2-enoate[a, b]
Et *trans*-dec-2-enoate[a, b]
Me *cis*-dec-4-enoate[a, b, f, g]
Et *cis*-dec-4-enoate[b, c, f]
Me *trans*-dodec-2-enoate[a, b, c]
Et *trans*-dodec-2-enoate[a, b, c]
Et *cis*-dodec-6-enoate[b, c]
Me *cis*-tetradec-8-enoate[b]
Et *cis*-tetradec-8-enoate[b]
Me *cis*-octadec-9-enoate[a, b, c]
Me deca-*trans*-2,*cis*-4-
   dienoate[a, b, d]
Et deca-*trans*-2,*cis*-4-
   dienoate[a, b, d, g]
Pr deca-*trans*-2,*cis*-4-
   dienoate[a, b, c, g]

Bu deca-*trans*-2-*cis*-4-
   dienoate[b, g]
Hexyl deca-*trans*-2-*cis*-4-
   dienoate[a, c, g]
Me deca-*trans*-2,*trans*-4-
   dienoate[b, g]
Et deca-*trans*-2,*trans*-4-
   dienoate[b, g]
Pr deca-*trans*-2,*trans*-4-
   dienoate[b, g]
Hexyl deca-*trans*-2,*trans*-4-
   dienoate[a, b, c, g]
Me deca-*cis*-2,*trans*-4-
   dienoate[b, g]
Et dodecadienoate[a, b, c]
Me dodeca-*trans*-2,*cis*-6-
   dienoate[a, b, c]
Et dodeca-*trans*-2,*cis*-6-
   dienoate[a, b]
Me tetradeca-*cis*-5,*cis*-8-
   dienoate[a, b, c, h]
Et tetradeca-*cis*-5,*cis*-8-
   dienoate[a, b, c, h]
Et hexadecadienoate[a, b]
Et deca-*trans*-2,*trans*-4,*cis*-7(?)-
   trienoate[a, b, c]
Et dodeca-*trans*-2,*cis*-6,*cis*-9(?)-
   trienoate[a, b]
Me tetradeca-*trans*-2,*trans*-4,*cis*-
   8(?)-trienoate[b, c]
Et tetradeca-*trans*-2,*trans*-4,*cis*-
   8(?)-trienoate[b, c]
Me 4-hydroxy-*trans*-
   butenoate[b, d, e, f]
Et 4-hydroxy-*trans*-
   butenoate[b, d, f]
Me 3-hydroxyoctanoate[b, c, f]
Et 3-hydroxyoctanoate[b, c]
Sesquiterpenes,
   triunsaturated[a, b, d]

---

* Compounds were identified by the following techniques:

   [a] Retention time.
   [b] Infrared spectrum.
   [c] Mass spectrum.
   [d] UV spectrum.

   [e] Melting point.
   [f] Derivatives.
   [g] Synthesis and comparison.
   [h] Nuclear magnetic resonance spectrum

Table 16.3. Variation in flavour of some structurally related esters (cf. p. 567)

Compound	Flavour
Variation of acid group	
Isopentyl acetate	Pear
Isopentyl propionate	Apricot-plum
Isopentyl butyrate	Plum
Isopentyl cinnamate	Fruity, spicy
Isopentyl salicylate	Strawberry
Variation of alkyl group	
Methyl butyrate	Apple
Ethyl butyrate	Pineapple
Butyl butyrate	Fruity, buttery
Isopentyl butyrate	Plum

Table 16.4. Flavours of geraniol and some of its esters (cf. p. 567)

Compound	Flavour
Geraniol	Rose-like
Geranyl acetate	Mild rose
Geranyl butyrate	Floral but sharp
Geranyl formate	Raspberry-like
Geranyl phenylacetate	Honey-rose-like
Geranyl propionate	Bitter grape-rose-like

where ethylvanillin is found to have $2\frac{1}{2}$ times the flavour value (or strength) of vanillin itself. A similar change in flavour strength is

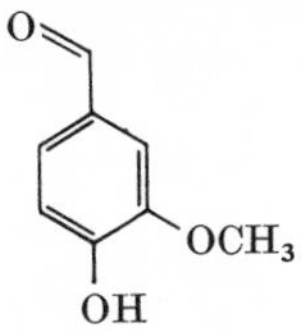

Vanillin (**49**)

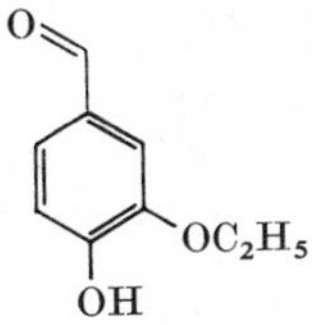

Ethylvanillin (**50**)

noted with maltol (3-hydroxy-2-methyl-4-pyrone) (**51**) and ethylmaltol (**52**) (p. 570). Again the flavour characteristics are similar but the increase in flavour with the ethyl derivative is ten-fold.

Maltol is used commercially to improve the taste of bread and cakes, being added at a rate of 50–100 ppm. It occurs naturally in roasted malt and in caramel, and is found in bread crust.

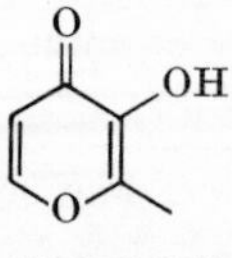

Maltol (**51**)

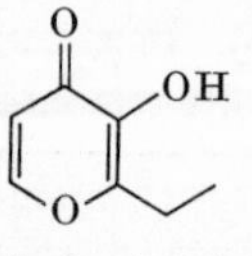

Ethylmaltol (**52**)

Just as with odours so with flavours, a great effort to understand the mechanism of taste has been made. The organoleptic qualities of a particular compound depend on a subjective interpretation of the response of the olfactory nerve ending for a given individual, and the taste of a substance is very often dependent on its odour. However, apart from the complex mingling of taste and odour responses, it is generally accepted that there are four basic tastes: acidity, sweetness, saltiness and bitterness. The physiological interpretation of the first two sensations can be ascribed to the interaction of ions with the taste buds of the tongue. Sodium chloride, bromide and iodide, and potassium and ammonium chloride, all taste salty, and this salty taste can be thought to be due to interaction of both the cation and the anion of a completely dissociated salt with the tongue. A bitter taste accompanies the salty taste of potassium bromide and ammonium iodide, whereas the heavier Group I metal salts, caesium chloride, caesium and rubidium bromide, and potassium iodide, are predominantly bitter. In a similar simple fashion the fact that most acids in aqueous solution taste sour leads to the conclusion that the sensation of a sour taste is due to the interaction of H_3O^+ ions with the tongue.

The sensations of bitterness and sweetness are less obviously classified by a particular chemical feature. Sweet-tasting sub-

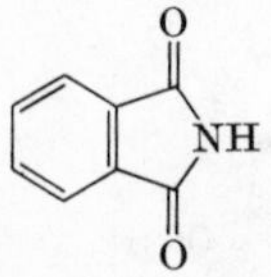

Saccharin (**53**)

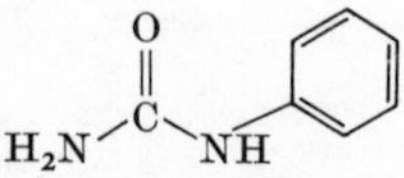

Dulcin (**54**)

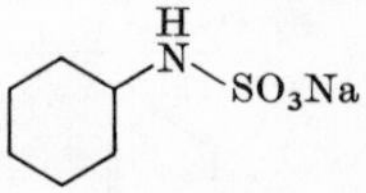

Sodium cyclamate (**55**)

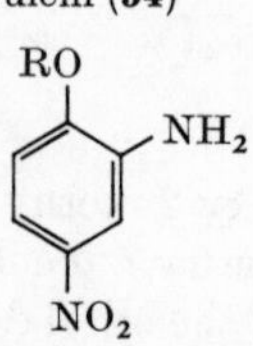

(**56**)

stances have a variety of molecular structures, ranging from polyhydroxy-compounds such as the mono- and di-saccharides, through chloroform and some amino-acids (where R is larger than ethyl), to imides and ureas such as saccharin (**53**) and dulcin (**54**) and some 2-amino-4-nitrophenyl alkyl ethers (**56**). It has been suggested[15] that a common feature of this wide variety of compounds is the presence in the sweet-tasting molecule of a weakly acidic proton such as is present in the alcoholic OH group, the imide or amide hydrogen, the proton adjacent to the nitro-group (**56**) or the single hydrogen in chloroform, together with a proton acceptor site (Lewis base) about 0.3 nm from the acidic proton. The oxygen of the alcoholic OH group, the oxygen of the carbonyl, sulphonyl or nitro-group, or even the electronegative chlorine atom, could function as a Lewis base.

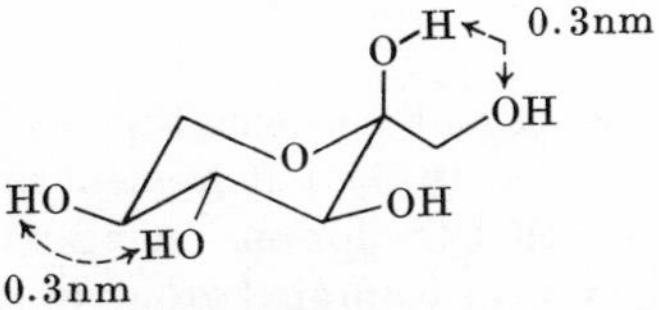

α-D-Fructopyranose

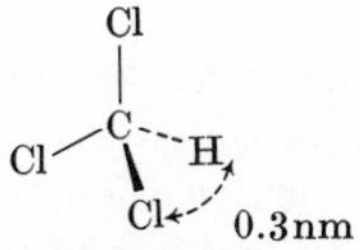

Chloroform

To be sweet-tasting, a compound must also be soluble in water, and the taste will thus obviously be influenced by the presence of groups that affect solubility. If the compounds are very soluble in water and thus less lipid-soluble as, for example, the mono-saccharides, they will be less sweet than compounds such as saccharin.

It is then assumed that, if these two groups—an acidic proton and a proton acceptor—are at the correct distance from each other in the molecule, the taste bud responsible for detecting sweetness has a pair of complementary proton-donor/Lewis-base

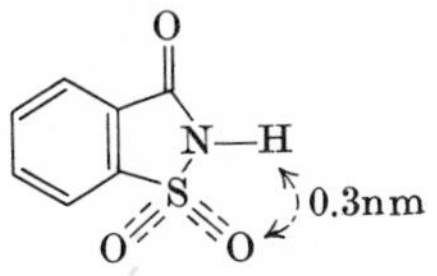

Saccharin

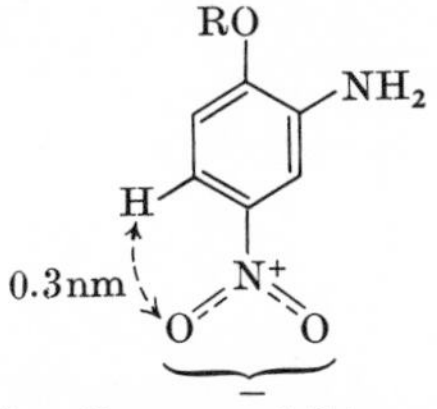

When R = propyl this compound is 4000 times as sweet as sucrose

sites, for example the —NH—C(=O)— peptide link in a protein chain, which forms a stronger hydrogen bond with the sweet-tasting molecule at the proton-donor/Lewis-base sites, for example OH and O of a sugar or glycerol. The enantiomers of the simple

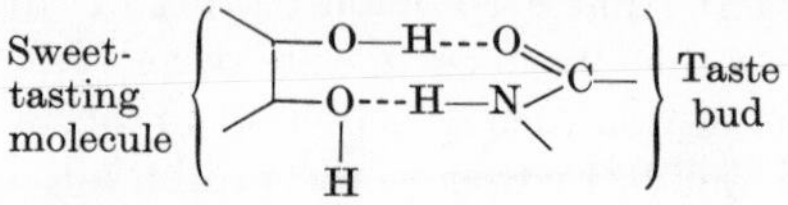

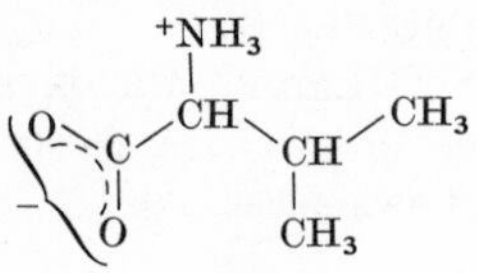

Valine 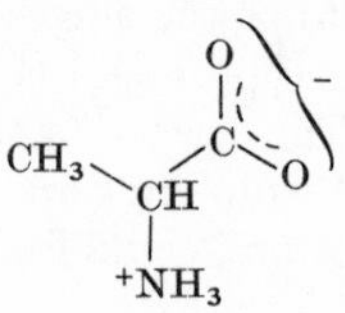   Alanine

amino-acid alanine are equally sweet; but the enantiomers of valine differ in sweetness, possibly owing to the requirement of a particular orientation on the protein that fits for one enantiomer but is subject to steric interference by the isopropyl group in the other, and this effect would not be so marked in alanine with its smaller side chain.

It is thought[16] that the structural feature responsible for bitterness is similar to the sweet-tasting proton-donor/proton-acceptor pair, except that the distance between the proton and the Lewis base site is 1.5 Å. Again it is suggested that the proton-donor and -acceptor sites of the proteins on the surface of the taste bud act as a receptor site and form hydrogen bonds with the bitter-tasting molecule. This idea is demonstrated by comparing the molecular structure of compounds such as (**57**) and (**58**).

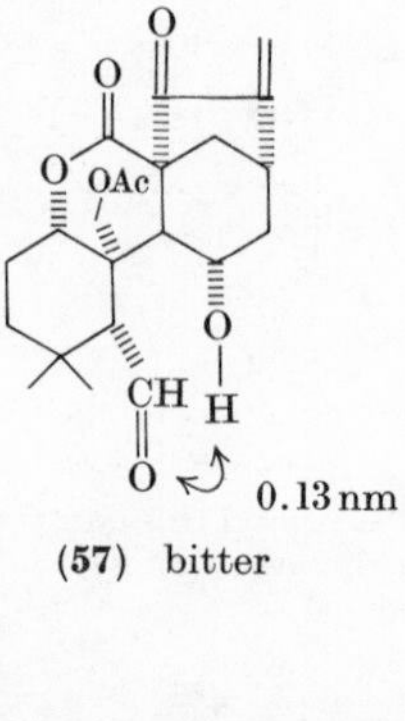

(**57**) bitter 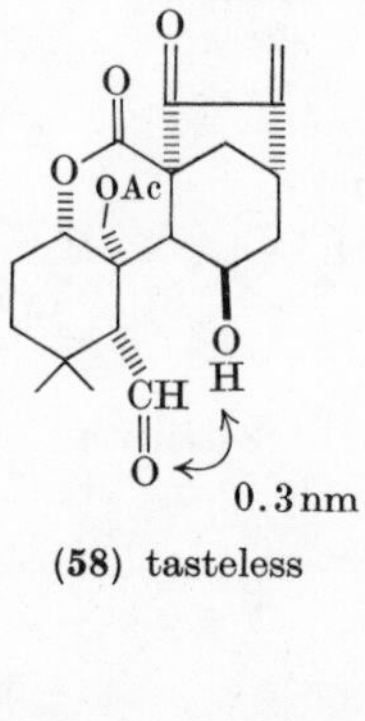   (**58**) tasteless

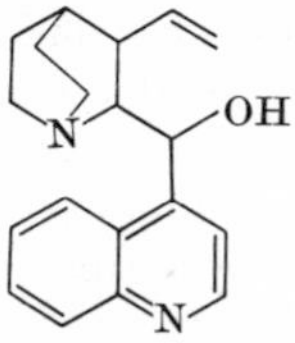

(59)

Alkaloids have pronounced bitter tastes. In quinine **(59)**, for example, the OH and nitrogen of the quinuclidine ring may act as proton donor and acceptor, respectively.

Manufacture of Some Typical Flavouring Materials

Vanillin must have been observed as crystals on the surface of vanilla beans long before it was reported in chemical literature. Its structure was determined by Tiemann and Haarmann in 1874, and two years later Reimer synthesized it from guaiacol **(60)**.

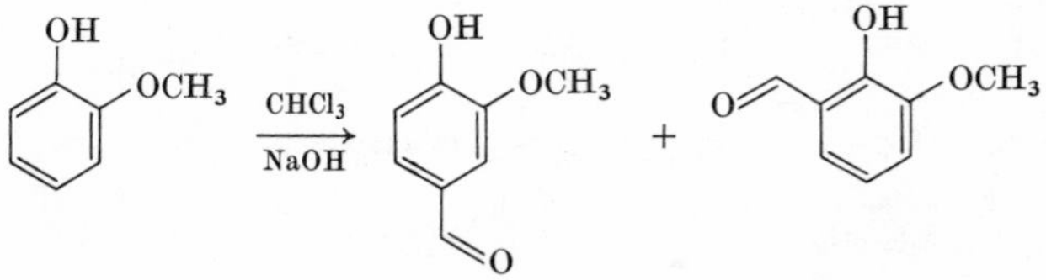

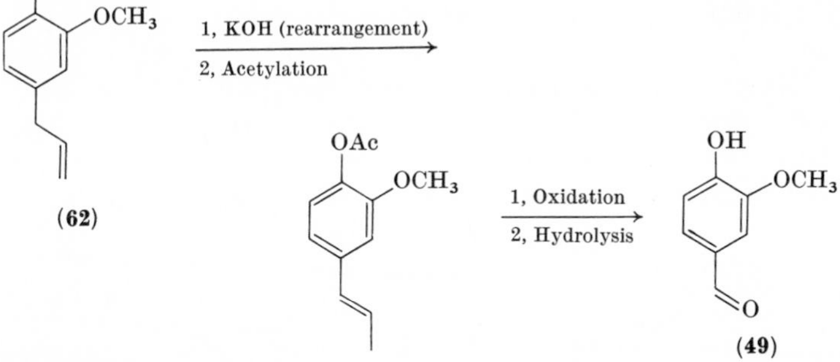

The mechanism of the Reimer–Tiemann reaction is discussed in Part 2, p. 290. The 2-hydroxy-3-methoxybenzaldehyde **(61)** produced as a by-product can easily be separated by steam-distillation. Another starting point for vanillin (**49**) was eugenol (**62**), which is readily obtainable from a number of essential oils, notably oil of cloves. Eugenol was first isomerized by treatment with alkali, the isoeugenol produced was acetylated, then oxidized with dichromate, and the resulting aldehydo-ester was hydrolysed to vanillin.

In the 1930's yet another approach was developed, utilizing the guaiacyl groups in lignin (see Part 4, p. 143). This process involves treating sulphite waste liquor with alkali and extracting the alkali-metal salt of vanillin with an alcohol. Vanillin is at the present time manufactured at the rate of thousands of tons p.a.

So far we have been mainly concerned with chemicals that produce an 'aromatic' flavour. Sweetening agents are in great demand because many people wish to reduce the carbohydrate intake of their diet either for dietetic reasons or because of diabetes. The most important sweetener is saccharin (*o*-sulphobenzoic imide) (**53**).

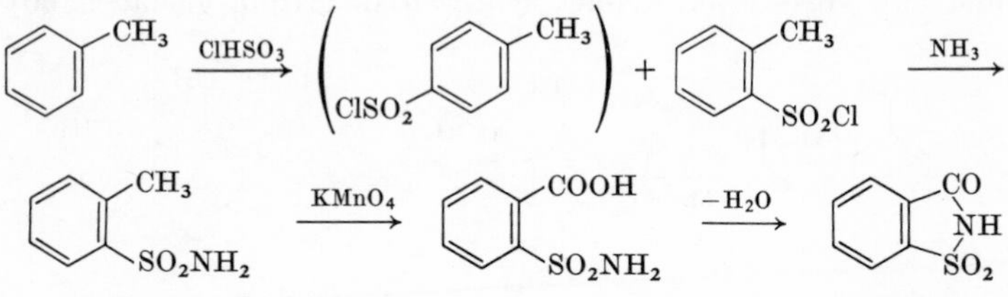

(**53**)

Saccharin is manufactured from toluene as indicated, the *p*-toluenesulphonyl chloride being separated from the *o*-isomer by crystallization. Saccharin is some 550 times sweeter than sugar (sucrose). Recently there has been a world-wide ban on sodium cyclamate (**55**); it is 30 times as sweet as sugar but its importance lay in that it has a synergic effect on saccharin: a blend of sodium cyclamate and saccharin produced sweetness without the bitter after-taste that is characteristic of saccharin itself.

Sorbitol (**63**) is used as a sweetening agent in foods for diabetics, but suffers the disadvantage of providing nearly as many calories in the diet as sucrose; thus it cannot be used in slimming diets. It can be manufactured by hydrogenation of glucose.

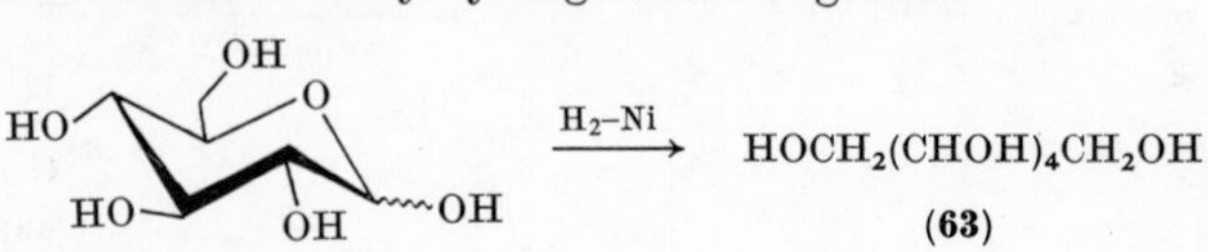

$$\xrightarrow{\text{H}_2\text{-Ni}} \text{HOCH}_2(\text{CHOH})_4\text{CH}_2\text{OH}$$

(**63**)

At the present time there is also much interest in the use as perfumes and flavour components of the pyrazines discovered in heat-treated foods. For example, when added at a level of 10–100 ppm, 2-ethyl-3,5,6-trimethylpyrazine (**64**) enhances the cocoa flavour of chocolate products. This pyrazine can be prepared from biacetyl and propylenediamine by the reaction sequence illustrated; or it can be made from tetramethylpyrazine by radical bromination (*N*-bromosuccinimide and benzoyl peroxide) of one of the methyl groups, followed by alkylation of the corresponding Grignard reagent with methyl iodide.

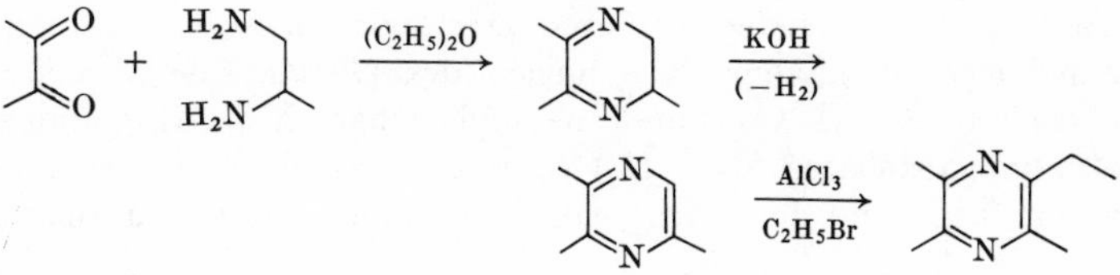

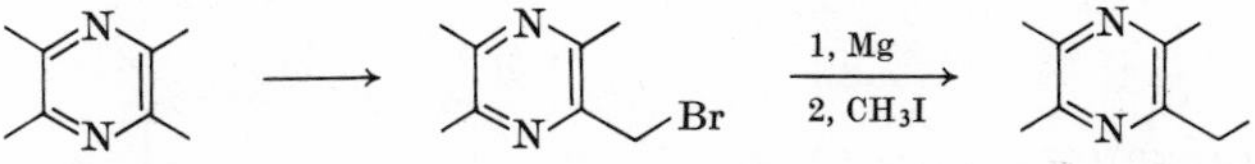

Ethyltrimethylpyrazine (**64**) has been detected in roasted coffee and is thought to be produced in cooked food by heat-induced reactions of the sugars and amino-acids naturally present.

The simpler ethylpyrazine (**65**) has been used to improve the flavour of tobacco; and the odour of green peas has been ascribed to a mixture of the pyrazines (**66**)–(**68**). The odour threshold of these pyrazines is as low as 2 parts per 10^{12}.

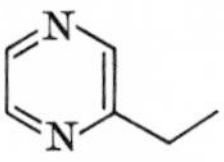

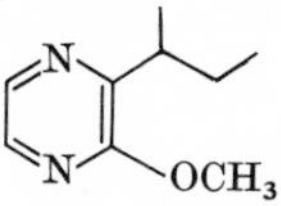

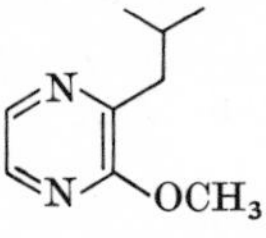

The pyrazine (**66**) can be synthesized by chlorination (Cl_2 and H_2O at 400°C) of methylpyrazine, followed by replacement of the chlorine by methoxyl and alkylation with, successively, methyl iodide and ethyl iodide by way of the sodium salt in liquid ammonia.

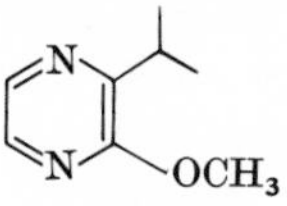

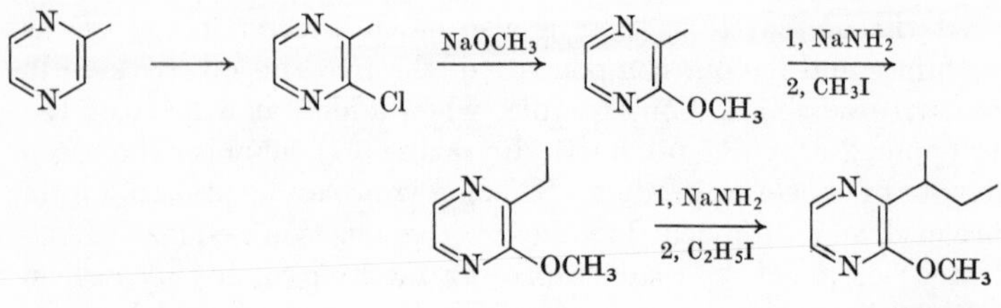

(66)

Total flavour potentiation was first discovered with monosodium glutamate (69) which has an intense meat-like taste. There is also considerable interest in mononucleotides for enhancing meat flavour. The nucleotides (70; inosine 5'-monophosphate, X = H; xanthine 5'-monophosphate, X = OH; guanine 5'-monophosphate, X = NH_2) have a threshold taste level of about 0.1% and act synergically with monosodium glutamate.

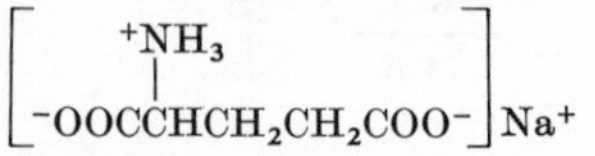

(69)

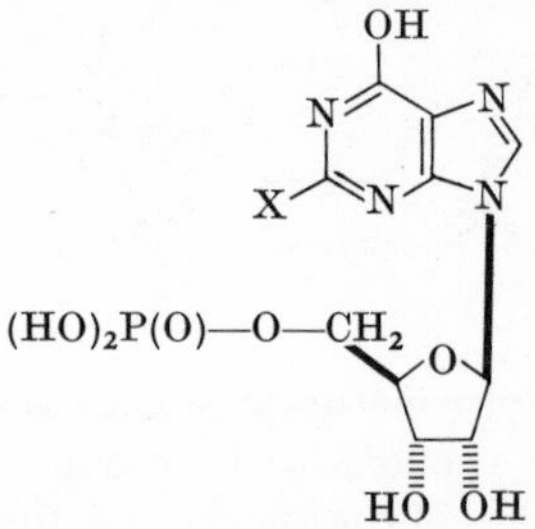

(70)

With these nucleotides, taste cells in human beings are probably stimulated only when the compound fits the active structure or site of the taste cell.

Basic work is progressing with regard to the physiology of taste but results from this are very much in the future. How interesting it would be to explain fully the bite of pepper or the coolness of menthol!

For the future development of the flavour industry it is possible to visualize greater sophistication in building synthetic flavours as more 'key' chemicals are discovered, growth in the use of the flavours as the trend to convenience foods progresses and as the 'farming' of natural flavours becomes more expensive, some new breakthrough with regard to total flavour potentiation and basic taste sensation and, finally, an understanding of the physiology of flavour and taste perception.

Bibliography

[1] Y. R. Naves, 'Relationship between Stereochemistry and Olfactory Properties,' VIth Mediterranean Symposium on Smell, *La France et ses Parfums*, 1970, **13** (**68**), 160.

[2] E. T. Theimer and M. R. McDaniel, *J. Soc. Cosmet. Chem.*, 1971, **22**, 15–26.

[3] M. J. Beets, *Molecular Structure and Organoleptic Quality*, Soc. Chem. Ind. Monogr., 1957, No. 1, p. 38.

[4] M. J. Beets, *Amer. Perfumer Cosmet.*, 1961, **76**, 54.

[5] D. Ottoson, *Acta Physiol. Scand.*, 1956, **35**, Suppl. 122.

[6] J. E. Amoore, *Proc. Sci. Sec. Toilet Goods Ass.*, 1962, No. 37, Suppl. 1, p. 13.

[7] R. C. Gesteland, J. Y. Lettvin, W. H. Pitts and A. Rojas, *Olfaction and Taste* (Y. Zotterman, Ed.), 1963, Macmillan, New York, p. 19.

[8] M. J. Beets, *Perfums, Cosmet Savons*, 1962, **5** (4), 167.

[9] R. H. Wright, *J. Appl. Chem.*, 1954, **4**, 611, 615.

[10] M. F. Carroll, *J. Chem. Soc.*, **1940**, 704.

[11] M. S. Carpenter, W. M. Easter and T. F. Wood, *J. Org. Chem.*, 1951, **16**, 589.

[12] M. J. Beets, H. Van Essen and W. Meerburg, *Rec. Trav. Chim.*, 1958, **77**, 854.

[13] R. W. Moncrieff, *The Chemical Senses*, 3rd edn., Leonard Hill, London, 1967.

[14] W. G. Jennings, Flavour Symposium, Paris, 1971.

[15] R. S. Schallenberger and T. E. Acree, *Nature*, 1967, **216**, 480.

[16] T. Kubota and I. Kubo, *Nature*, 1969, **223**, 97.

CHAPTER 17

(i) Photographic Materials

(G. F. DUFFIN, Minnesota 3M Research Limited, Harlow)

(ii) Chlorofluorocarbons—Production and Applications

(B. D. JOYNER, I.S.C. Chemicals Limited, Avonmouth)

(i) PHOTOGRAPHIC MATERIALS

Silver Halide Processes

While there has been much attention to chemical methods of imaging, mostly of an organic nature, in the majority of processes used commercially today a silver halide is still employed as the light-sensitive element. This halide, usually bromide with a small proportion of iodide in the more sensitive films or a mixture of chloride and bromide in the less sensitive materials, has nevertheless to be substantially backed up by organic chemistry.

Gelatin

The silver halide is present as a dispersion of small crystals ('grains') of 0.1–1.5-μm diameter, in a colloid binder that consists always of gelatin, sometimes with other polymers added. This suspension of silver-halide grains in gelatin is termed the 'photographic emulsion'. The gelatin suspending agent plays an important part in the photographic process and it is here that the properties of gelatin are so excellently suitable. For example, gelatin, amongst other things, suspends the crystals in warm aqueous solution during the grain growth (Ostwald ripening) and sets to a jelly on cooling, thus providing a solid matrix for the final film. It swells readily and reversibly to permit access and egress of the aqueous processing solutions, sustains the final image of black silver or dyes for colour and contains reactive groups

578

permitting excellent hardening of the layer to give good, permanent images.

Gelatin is a polypeptide of molecular weight about 100,000; it can be obtained in a substantial homogeneous form although the commercial product is usually degraded to some extent, giving a mean molecular weight about 70,000. Hydrolysis of the fibrous protein collagen, isolated from cattle bones or the skins of cattle or pigs, is achieved by treatment with lime or mineral acid. Apart from breaking the collagen into separate chains, these treatments hydrolyze amide and other groups and thus the final balance of the all-important terminal or side-chain reactive groups depends on the method of preparation. The soluble gelatin is finally extracted by hot water and isolated by evaporation.

The common amino-acids present in gelatin include glycine (**1**), alanine (**2**), proline (**3**) and hydroxyproline (**4**), the last two being characteristic of collagen. The guanidino-group in arginine (**5**) is

H_2NCH_2COOH

(**1**)

$CH_3CH(NH_2)COOH$

(**2**)

$H_2C{\diagdown}CHCOOH$
$H_2C{\,}|$
$H_2C{-}NH$

(**3**)

$HO{-}H_2C{\diagdown}CHCOOH$
$H{\diagup}C{\,}|$
$H_2C{-}NH$

(**4**)

believed to be important in the gelling process which involves reversible hydrogen bonding.

$H_2N{\diagdown}$
${\diagup}CNH(CH_2)_3CHCOOH$
HN

with NH_2 above CH

(**5**)

CH_2O

(**6**)

CHO
$|$
CHO

(**7**)

$H{\diagdown}{\diagup}OH$
$Cl{-}C{\diagup}C{\diagdown}O$
${\diagdown}C{-}CO$
Cl

(**8**)

Hardening, usually carried out by adding aldehydes, such as formaldehyde (**6**), glyoxal (**7**) or mucochloric acid (**8**), to the liquid emulsion just before coating on film or paper, involves particularly the ϵ-amino-group of lysine (**9**) or the hydroxyl groups of serine (**10**) or hydroxypyroline (**4**).

$H_2N(CH_2)_4CH(NH_2)COOH$
(**9**)

$HOCH_2CH(NH_2)COOH$
(**10**)

Stabilizers

The emulsion is prepared by forming and growing the silver halide by double decomposition between a soluble silver salt and mixed alkali halides and, apart from the gelatin's function, organic chemistry plays little part in this stage. The crystals, even when grown to the desired size (potential speed is approximately proportional to crystal volume), are still of low sensitivity and must be sensitized in a critical process in which small specks of silver sulphide are formed on the surface. The optimally sensitized state is an unstable one. Further growth or redistribution of these specks can cause loss of speed or formation of fog, that is, a developable image in the non-exposed regions. The only satisfactory stabilizers of this optimally sensitized state are silver-complexing heterocyclic organic compounds, of which easily the best are the tetra-azaindenes, such as 4-hydroxy-6-methyl-1,3,3a,7-tetra-azaindene (**11**). Others include 5-mercapto-1-phenyl-

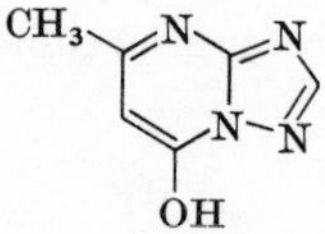

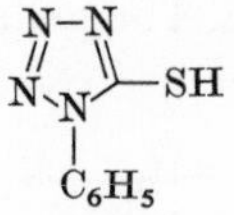

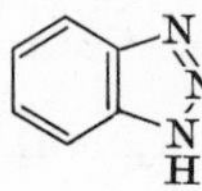

tetrazole (**12**) and benzotriazole (**13**). These stabilizers are therefore added, in suitable solvents, to the optimally sensitized emulsion.

Sensitizing Dyes

The silver-halide crystals, being colourless or yellow, are naturally sensitive only to blue and violet light. There are, however, certain classes of dye which, when added to the emulsion, extend the sensitivity over the whole of the visible spectrum (panchromatic films) or even into the near-infrared region. All the good spectral sensitizing dyes belong to the class of cyanines, of which there are two main types. The first are the true cyanines containing the resonance structures (**A**) and (**B**), where Z and Z′

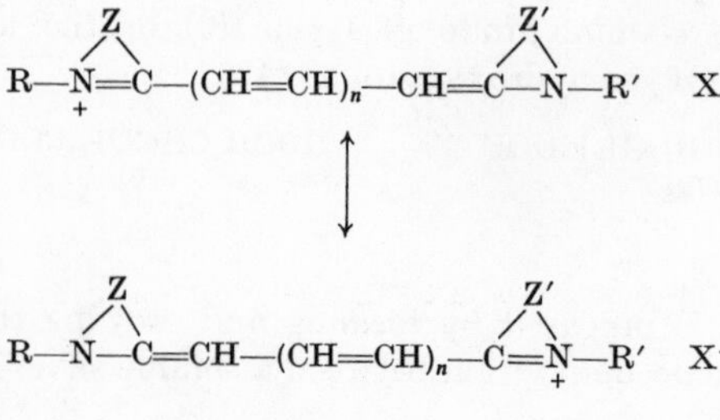

are the residues of heterocyclic rings, frequently containing hetero atoms other than N, X^- is an anion, R and R′ are alkyl or similar groups, n is zero or an integer, and additionally the atoms of the polyene chain may be substituted. Examples of this class are (**13**), (**14**) and (**15**).

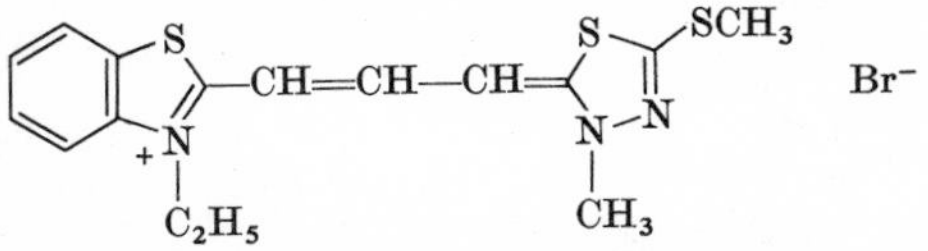

(**13**)

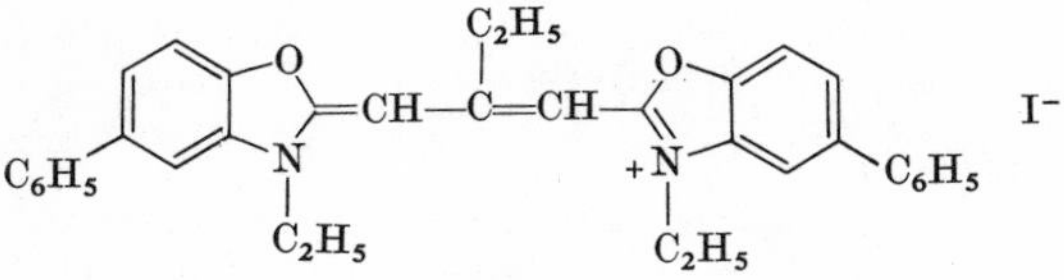

(**14**)

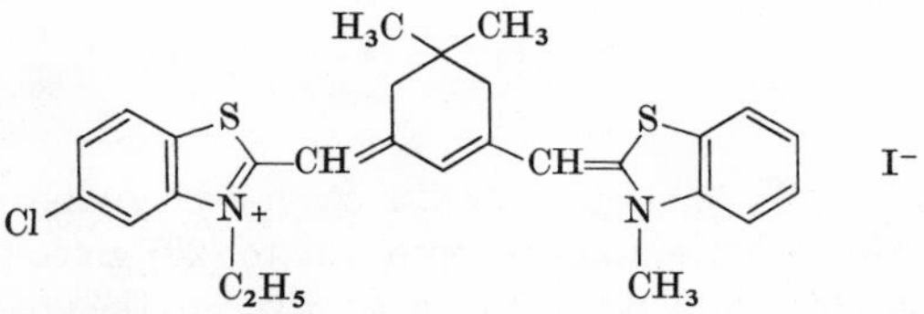

(**15**)

The second class is that of the merocyanines with the vinylogous amide charge-separation resonance, (**C**) and (**D**), where R, Z, n

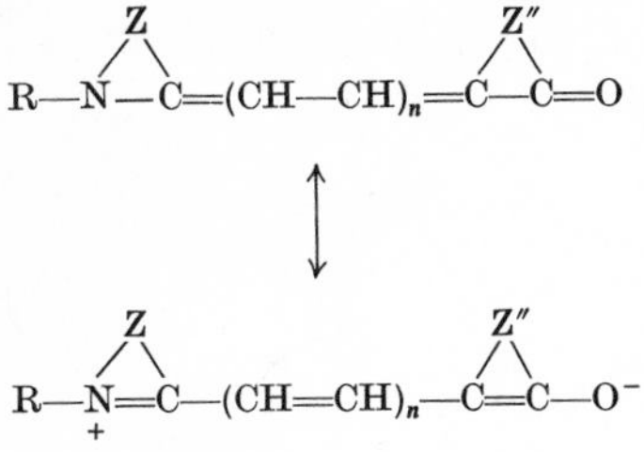

$$R-N-C=(CH-CH)_n=C-C=O \quad (\mathbf{C})$$

$$R-N=C-(CH=CH)_n-C=C-O^- \quad (\mathbf{D})$$

have the same significance as above and Z″ is the residue of a ketomethylene heterocyclic nucleus, such as pyrazolin-5-one or thiazolid-4-one.

The synthesis of the polymethine compounds utilizes the nucleophilic reaction of a reactive methyl or methylene heterocycle with the suitably substituted heterocyclic salt or similar reaction intermediate (Part 3, p. 69). Thus, the reaction of quinaldine ethiodide (16) with 2-methylthiobenzothiazole methotoluene-*p*-sulphonate (17) gives the dye (18).

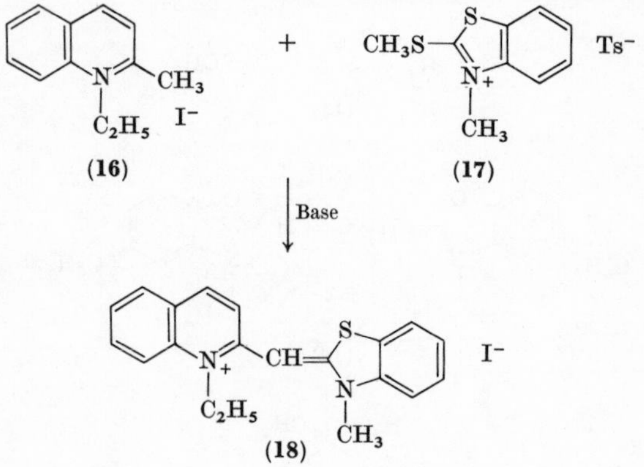

Reaction of two molecules of 2-methyl-5-phenylbenzoxazole ethiodide (19) with triethyl orthopropionate (20) gives the symmetrical dye (14) which is a common sensitizer to the green region of the spectrum.

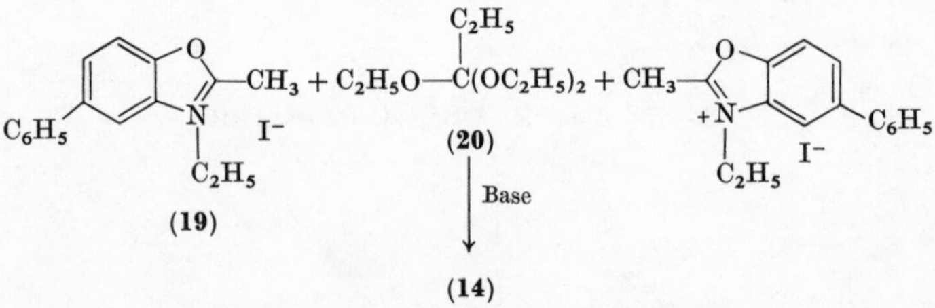

Finally, the merocyanine dye (23) is prepared by a two-stage process, first from the benzothiazole salt to the vinyl intermediate (21), which is then treated with the triazolidone (22).

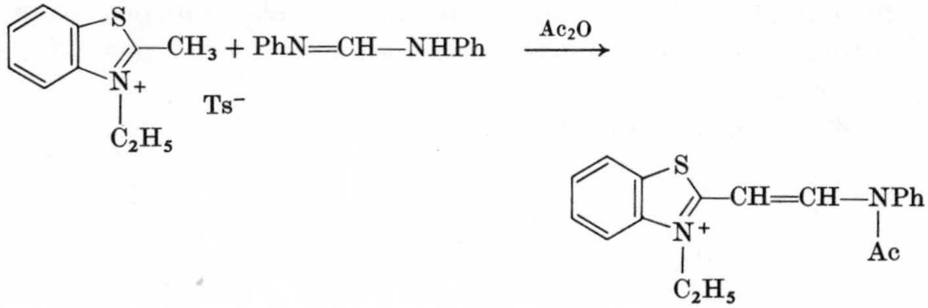

(21)

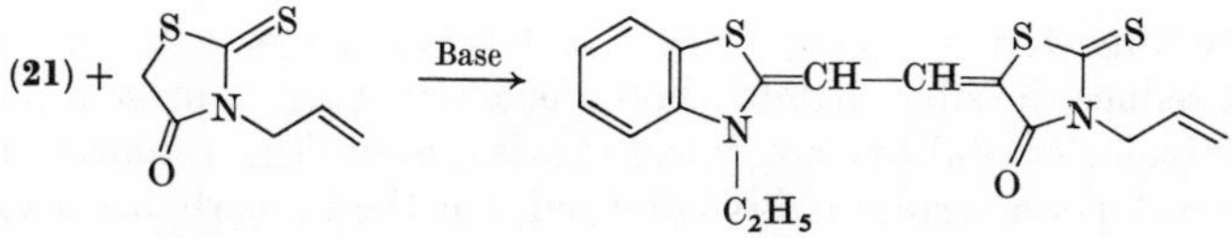

(22)   (23)

Developers

The light-exposure of the silver-halide crystal, prepared as above in its state of desired sensitivity, requires that only a few photons per grain be absorbed. By consecutive electron and ionic stages, the photoelectrons so produced yield a permanent image, on each exposed crystal, of a small group (3–4) of silver atoms called the latent image. This is then converted into the final image by development in which each light-struck crystal is converted completely into silver, a quantum gain of up to 10^9.

All present developing agents used in silver halide photography are organic. For black-and-white images, the almost universal developer is hydroquinone (24), usually employed in synergic combination with metol (25) (*p*-methylaminophenol) or 'Phenidone' (26) (1-phenylpyrazolidin-3-one).

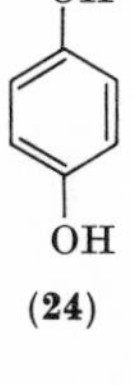

(24)    (25) 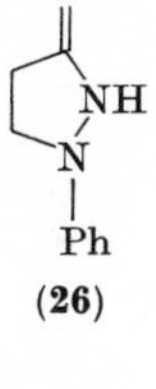   (26)

These agents are used in mild alkaline solution, most commonly carbonate, with sulphite and some halide added. The sulphite is added as a preservative, the halide to provide an initial concentra-

tion of halide so that the properties of the developer do not change
drastically as halide ion is released during the development of the
first film processed.

The overall reaction, if hydroquinone is the developing agent,
is (a).

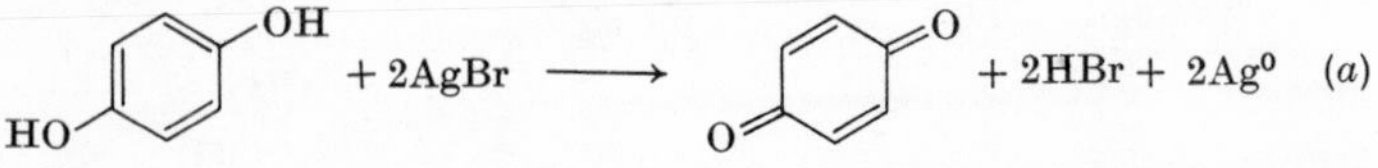

$$\text{(structure)} + 2AgBr \longrightarrow \text{(structure)} + 2HBr + 2Ag^0 \quad (a)$$

The important property of such a developing agent is that it
must reduce the silver halide where the latent image is present on
a particular crystal but not where this is absent. This requires of
the developer a certain redox potential, but there clearly are also
other necessary conditions for a truly effective differentiation
between the light-struck and non-light-struck grains. A difference
in rate of up to 10^5 is achievable in practice.

The quinone (27) produced in the above reaction reacts with
sulphite to give the monosulphonate (28).

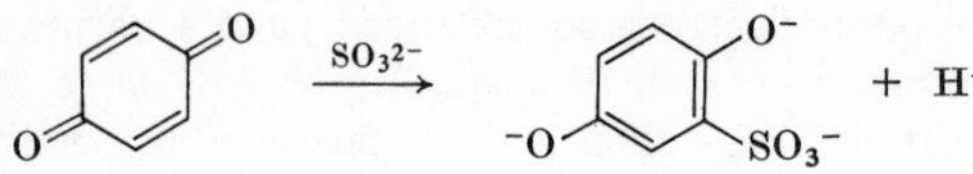

$$\text{(27)} \xrightarrow{SO_3^{2-}} \text{(28)} + H^+$$

(27)   (28)

When the synergic combination of hydroquinone (24) with
metol (25) or 'Phenidone' (26) is employed, the primary reduction
of the silver halide is produced by this second agent, usually
present in much smaller quantity, e.g. 1/50 to 1/10 molar relative
to the hydroquinone. Thus 'Phenidone' behaves as shown in

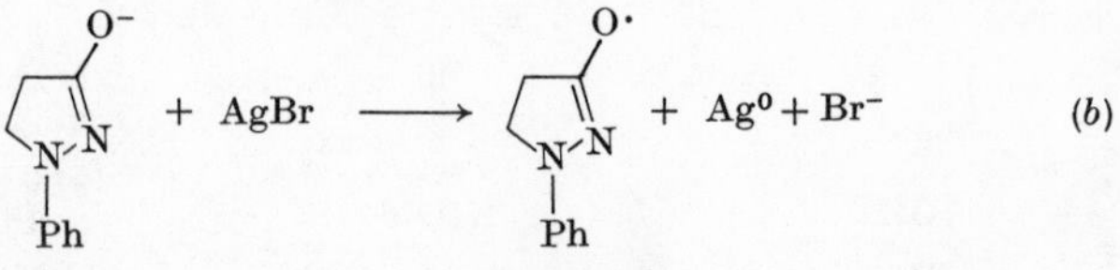

$$\text{(29)} + AgBr \longrightarrow \text{(30)} + Ag^0 + Br^- \quad (b)$$

(29)   (30)

reaction (b); the free radical (30), which is orange–red and relatively
stable, is then reduced back to the 'Phenidone' anion (29) by the
action of the hydroquinone [reaction (c)].

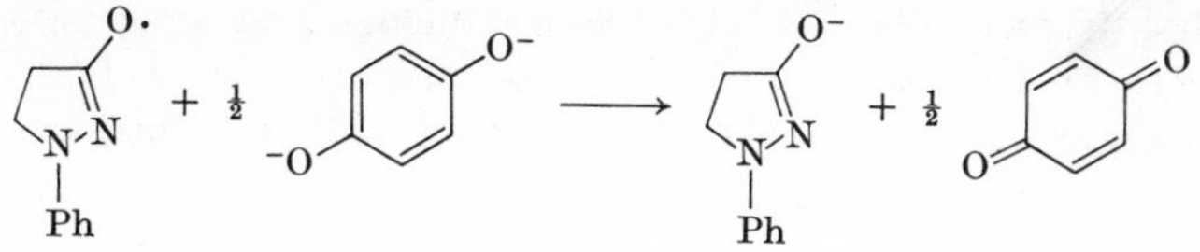

$$(c)$$

A similar process occurs with metol. An important difference, however, conferring great utility on the 'Phenidone'-based process, is that the 'Phenidone' radical does not react with sulphite and is very efficiently regenerated, whereas the oxidation product of metol is partly removed by reaction with sulphite and thus lost to the cyclic process. 'Phenidone' is therefore effective in very small quantities and the developer shows no loss of activity until the hydroquinone reserve is depleted, an important point technologically.

Colour Processes

So far consideration has been only to processes where the final image is a black one in developed silver. Colour processes are today very important and more common than black-and-white. All silver halide colour processes employ the basic elements described above but additionally employ further organic chemistry to create a dye image, the silver image being then removed, e.g. by bleaching, or discarded in some way, for example by being left in the negative sheet of the Polacolor method.

The colour image reactions may be classified as (1) dye-forming processes, (2) dye-destroying processes, and (3) dye-migrating processes. The commonest of these is the first and provides the basis for Agfacolor, Kodachrome and other well-known methods.

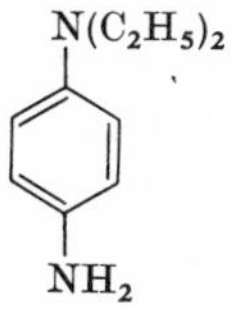

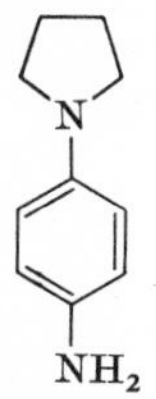

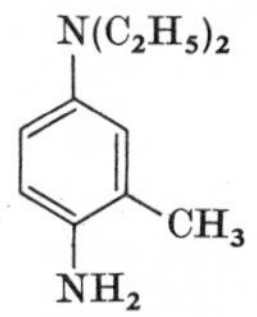

Turning therefore to dye-forming processes: those in commercial use all employ chromogenic development. This involves the use of a developing agent which is a *p*-dialkylaminoaniline or simple derivative such as (**31**), (**32**) or (**33**), in which the development of the light-struck silver grains, while producing silver, gives also a reactive oxidized product from the developer. Thus, *p*-diethyl-aminoaniline (**31**) gives the quinodi-imine (**34**) and this, being

fairly unstable, reacts rapidly in the film layer to give the leuco-form of the dye which then by further reaction, e.g. with silver, gives the final dye. The reagents for this are commonly an acylacetanilide to give the yellow image, a pyrazolone to give the magenta and a naphthol to give the cyan (blue-green).

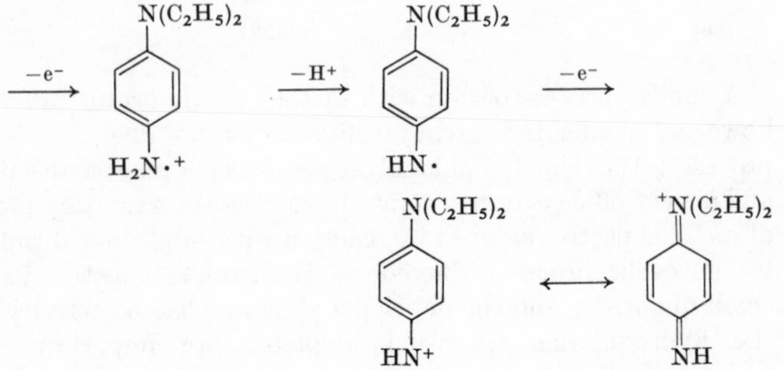

The physics of the overall process require that, for a reversal film such as Ektachrome where the exposed piece of film is to be developed to give a transparency of the original, the comple-mentary colour be formed in each layer. Thus, a blue-sensitive layer (no sensitizing dye) will be developed to give a yellow image, a green-sensitive layer to yield a magenta image and a red-sensitive layer to give a cyan (blue-green) image.

It will be appreciated that this system requires that only the correct colour be developed in a given layer. The commonest way to ensure this is to incorporate the appropriate colour coupler in the desired layer. This may be done by two means. The first, as employed in the original Agfacolor film, uses substantive couplers, that is those containing a long-chain ballasting group such as C_{18} chain combined with a solubilizing acid group such as SO_3H. Each coupler is then dissolved in alkali and added to its complementary emulsion when interaction with the gelatin—again the side-chain amino-groups come into play—holds the coupler firmly in the layer after coating; this process may be done consecutively or in one stage by multiple-slot application to the moving web of film or paper. Couplers of this type are (**35**), (**36**) and (**37**).

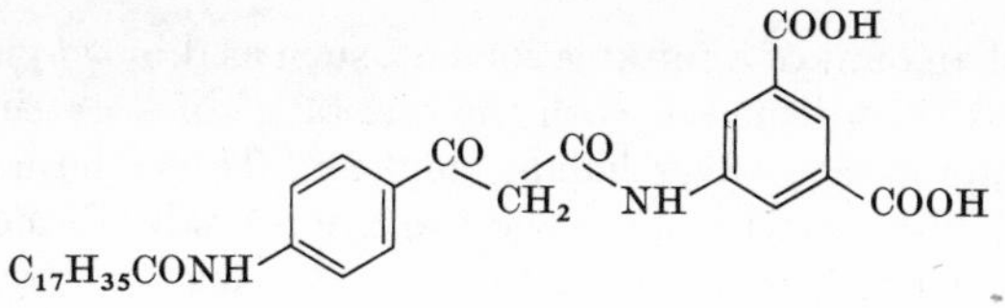

(35)

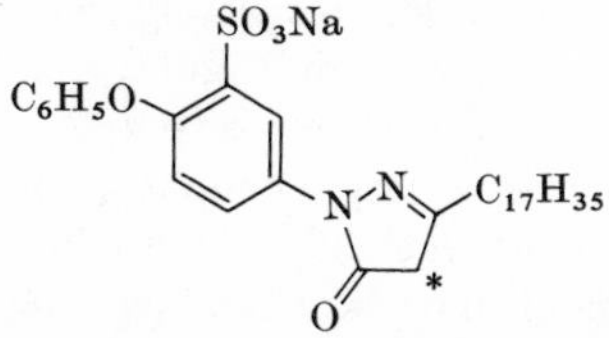

(36) 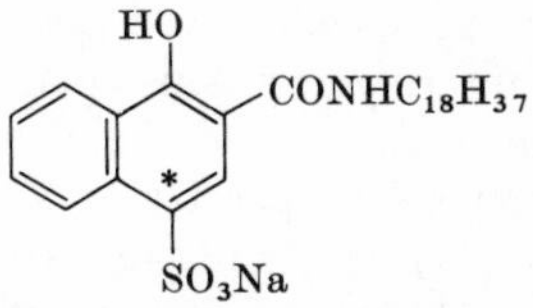   (37)

* Coupling position

Alternatively, the method used in most modern films, such as Kodacolor, is to incorporate an oil-solubilizing group in each coupler, such as (38), (39) and (40). The coupler is then dissolved

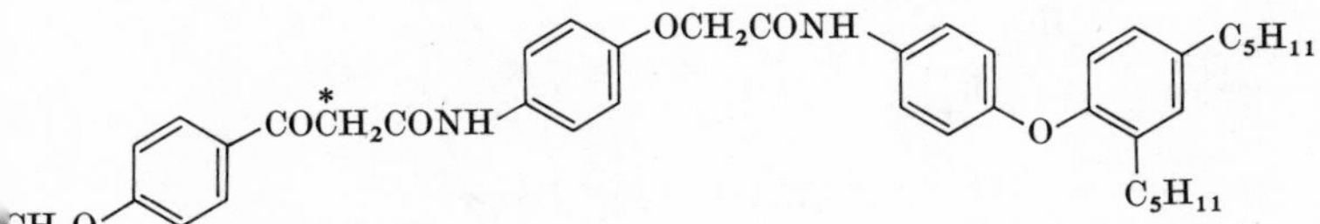

(38)

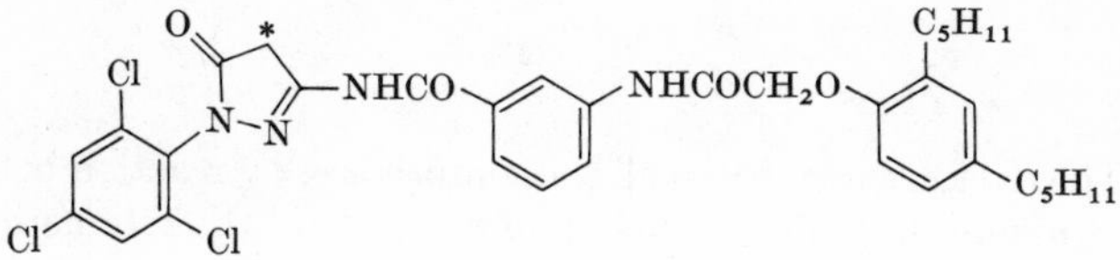

(39)

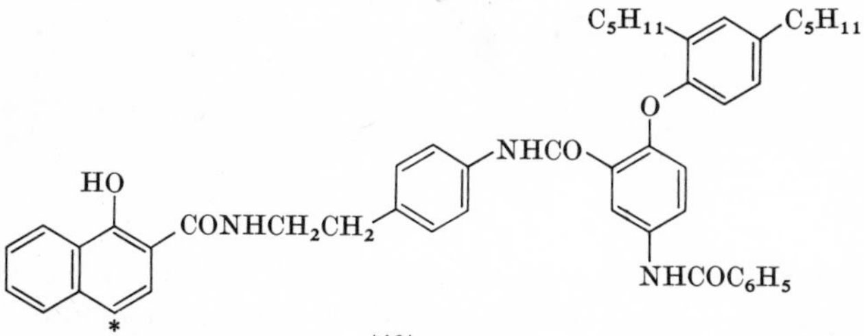

(40)

in a small amount of a suitable solvent, such as dibutyl phthalate, and this is then dispersed with the assistance of a wetting agent in the appropriate silver halide emulsion. Thus, the coupler is dispersed oil-in-water fashion and the dye is finally formed in the oil phase, where it remains fixed.

In both these cases, the rapid reaction of the oxidized developer ensures broadly that only the desired dye is formed by development in a specific layer. Sometimes, in order completely to ensure this, interlayers are also placed between the emulsion layers.

The other method of chromogenic development, as employed in Kodachrome, operates quite differently. Simple couplers, such as those (**41**), (**42**) and (**43**), are dissolved in the developers but it is arranged that the colour development of each of the layers occurs in turn such that, although the dye is formed from reagents in solution, only the desired dye is formed in each layer.

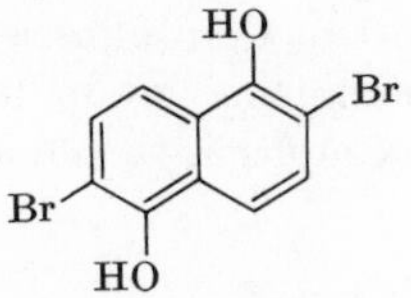

(**41**)

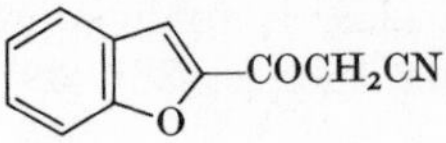

(**42**)

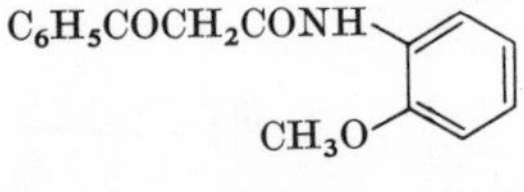

(**43**)

The dyes are formed, in all these processes, by reaction between the quinone di-imine, e.g. (**34**), and the anion of the ketomethylene coupler [reaction (*d*)].

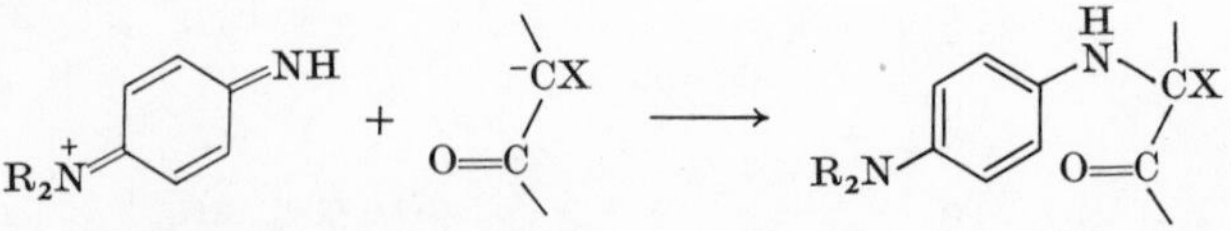

(**34**)   (*d*)

In the examples given above, where $X = H$, the intervention of another two silver ions then oxidizes this leuco-dye into the final dye [reaction (*e*)].

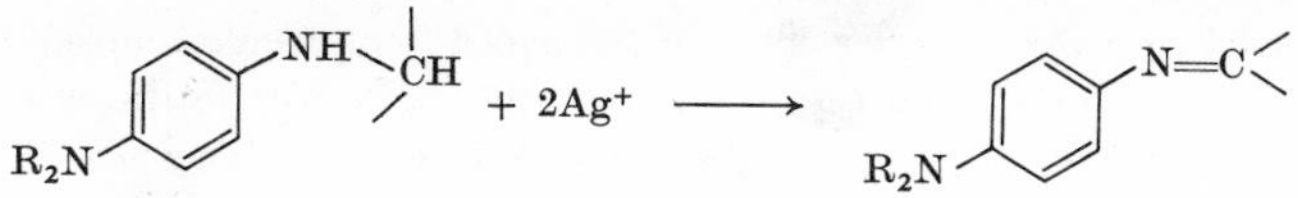

$$R_2N-\text{(ring)}-NH-CH< \quad + 2Ag^+ \longrightarrow \quad R_2N-\text{(ring)}-N=C< \quad (e)$$

For example, a pyrazolone dye would be (**44**).

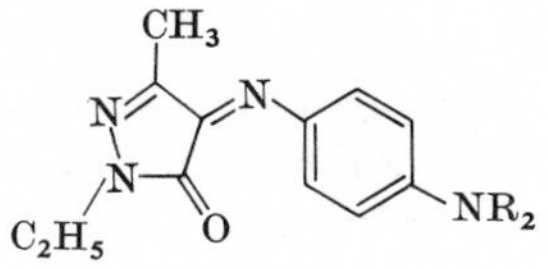

$$\text{(structure 44)}$$

(**44**)

In some cases, X can be such that the elimination of HX gives the dye directly without further oxidation by silver. Couplers substituted in the reactive position with chlorine are used in some products, while in others use of phenylazo for X results in couplers that are themselves coloured and are used for colour-correction in negative films.

It should also be stressed that there are two broad classes of colour material, those reversal materials giving a positive rendering directly and those yielding first a colour negative which must then be printed on to another colour material to give the final positive print. The reversal films may use substantive, oil-soluble or developer-soluble couplers but the negative-working material can employ only the fixed couplers.

Turning now to the dye-destroying colour processes: the only commercial materials of this type are based on the silver-dye-bleach method. In this process, each layer contains a carefully chosen azo-dye of colour complementary to spectral sensitivity of the emulsion. A first black-and-white development gives a silver image in the light-struck areas. The film is then placed in a bath of high acidity containing a silver-complexing agent, such as thiourea, and a substance known as a dye-bleach catalyst of which

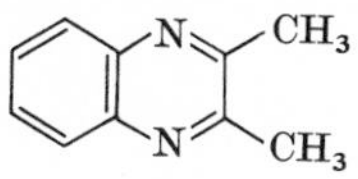

$$\text{(structure 45)}$$

(**45**)

2,3-dimethylquinoxaline (**45**) is a common example. The reaction is then:

$$A\!-\!N\!=\!N\!-\!B + 4Ag + 4HX \longleftarrow A\!-\!NH_2 + BNH_2 + 4AgX$$

where A and B are the residues of the azo-dye molecules chosen so that the fragments are easily washed out of the film. Subsequent processing then removes all the silver halide, etc., from the film, leaving only the original azo-dye in those areas where the light did not fall in the exposure. As with the other method discussed above, the process is fully stoichiometric, thus giving good gradation.

This dye-bleach system has been commercialized in 'Ilford Colour Print' and, more recently, in 'Cibachrome'. Azo-dyes of the type used are exemplified by (46), (47) and (48).

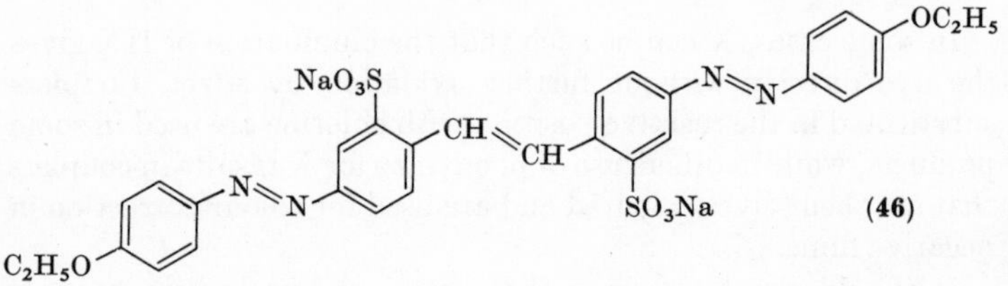

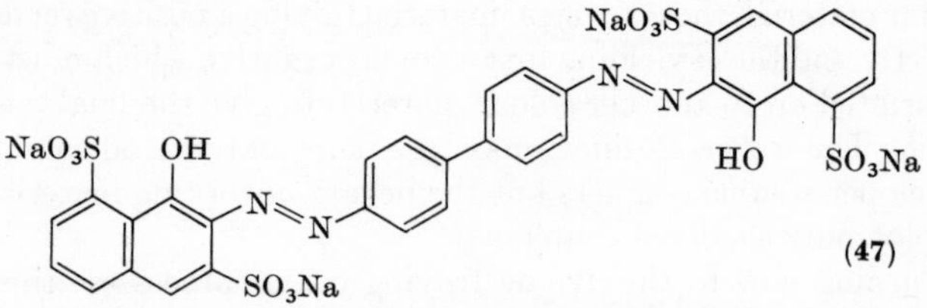

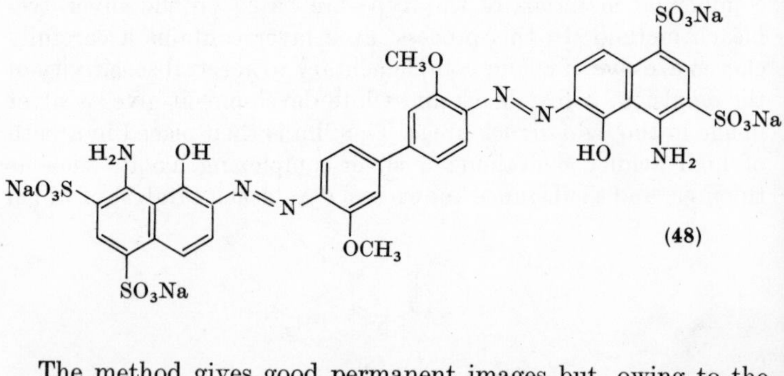

The method gives good permanent images but, owing to the presence of all the dyes in the light-sensitive layers, it is only applicable to slow materials such as print papers and the processing is less straightforward than that of chromogenic development.

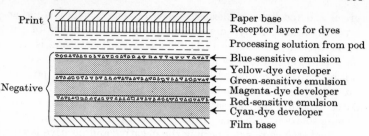

Print { Paper base
Receptor layer for dyes
Processing solution from pod

Negative {
← Blue-sensitive emulsion
← Yellow-dye developer
← Green-sensitive emulsion
← Magenta-dye developer
← Red-sensitive emulsion
← Cyan-dye developer
Film base

Figure 17.1. Schematic representation of the Polacolor process. The print, negative and processing solution are brought together after exposure.

The Polacolor process provides at the present the sole example of a commercial material based on dye diffusion. The light-sensitive element contains the useful sensitive layers responding to blue,

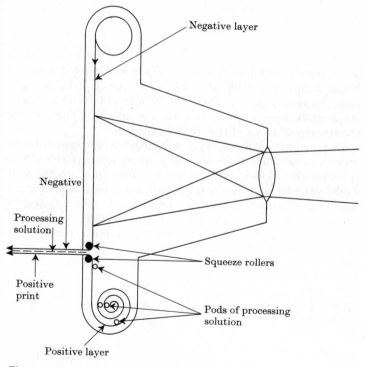

Negative layer

Negative

Processing solution

Positive print

Squeeze rollers

Pods of processing solution

Positive layer

Figure 17.2. Schematic representation of a Polaroid camera and the process. (Reproduced, by permission, from G. F. Duffin, *Photographic Emulsion Chemistry*, Focal Press Ltd., London, 1966.)

green and red light; underneath each of these layers is another layer containing a dye developer, the essential element of the whole system (see Figure 17.1). This composite emulsion is held on one roll in the camera and constitutes the negative; a second roll holds paper coated with a receptor layer and having pods of processing solution attached at intervals to the coated paper. After exposure the two layers are squeezed together, with processing solution in between from a broken pod (see Figure 17.2). The final step consists of peeling the negative and positive layers apart. Development of the silver halide occurs in the light-struck areas, oxidation of the dye developer converting it from a soluble into an insoluble molecule. Such a dye developer is shown

(49)

schematically in (49). In this case, the quinone produced would no longer possess the alkali-solubility of the quinol. Thus, in those areas where no light fell, the dye developer will diffuse across to the receptor layer where it is mordanted to give a final, direct-positive reproduction of the original exposure.

More recently Polaroid have announced a new system based upon a revolutionary camera and a novel arrangement of the light-sensitive element and the receptor layer in which the final dye image is formed. This enables the film and print to be a single entity which is expelled from the camera via a squeeze system to break the pod automatically as the exposure is made. As indicated in Figure 17.3, exposure occurs *through* the receptor layer on to the film beneath it. The dye image is formed by a substantially similar diffusion process, but the whole thing remains together in one piece. The white background from the print is formed by a dispersed material which is spread, from the pod, in the processing solution between the receptor and the element which was originally light-sensitive.

Miscellaneous Organic Components

There are many other organic aspects of silver halide photography. Wetting agents of the long-chain alkylsulphonate type are used to assist even coating of the final formulation on the film or

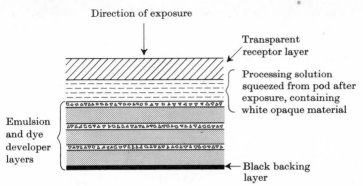

Figure 17.3. Schematic representation of the most recent (1973) Polacolor material.

paper base and, of course, this base is itself of interest. For many years, after the abandonment of the dangerous cellulose nitrate (celluloid), films have been made of cellulose acetate, usually the triacetate—although this designation is not quite correct. The fully acetylated material is partially hydrolysed to a 80–85%-acetylated polymer which is more conveniently soluble for casting from solvents such as dichloromethane.

Of recent years, polyester base has become increasingly popular, especially for industrial films such as those for printing and medical X-ray usage. This base is chemically identical with 'Terylene', consisting of poly(ethylene terephthalate), and is manufactured by the hot extrusion of the polymer into sheets which are then oriented by elongation and sideways stretching, very much as is the fibre.

Poly(ethylene oxide) derivatives, either the polyglycols themselves or their aryl end-terminated derivatives (**50**) where $R = H$

$$R(OCH_2CH_2)_nOH$$

(50)

or aryl, and $n = 10$–50, are incorporated in some films for their effect on development: increased speed is produced in the fastest camera film, this being vital in the high speed of present-day products. Certain other films, called lith films because of their use in lithographic half-tone processes, employ these poly(ethylene oxides) to achieve the very high contrast required for good half-tone dots.

Polymers are used for certain special purposes. These include conferment of better flexibility, for which polyethylacrylate latexes are used, and physical improvements such as matting of the film surface or providing protection against static. Toning of the colour of the silver image to give a good black is effected by some heterocyclic mercapto-compounds, such as (51), while ultraviolet absorbers, for example (52), give good permanence to colour prints.

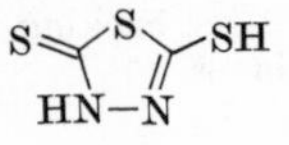

(51) 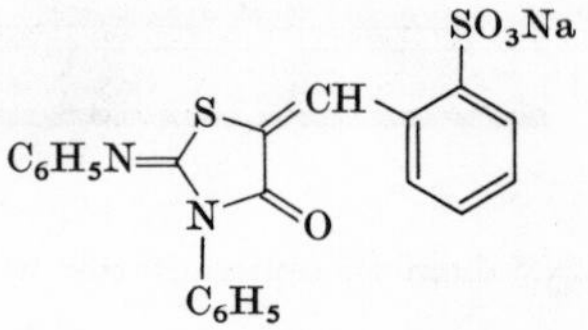   (52)

Non-Silver Processes

Diazo

In spite of the enormous range of organic processes devised and described in the patent literature, that based on the light-sensitivity of diazonium salts remains the most popular. Exposure of a coating of a diazonium salt to ultraviolet light (the mercury line at 365 nm is commonly used) causes the reaction:

$$ArN_2X \xrightarrow[\text{H}_2\text{O}]{h\nu} ArOH + N_2 + HX$$

The diazonium compound is destroyed in the exposed areas. The remaining salt is then caused to couple with a phenol to give a dye image, leaving the exposed areas substantially colourless.

The simplest diazonium salts, such as that from aniline itself, are unstable, but those with electron-donating (+R) substituents are thermally much more stable and may readily be isolated and handled, especially as their double salts, e.g. with zinc chloride. Three common materials are (53), (54) and (55).

Also, such salts possess high sensitivity to ultraviolet light and, because the minimum amount of light required depends on the coating weight of the salt, have the additional advantage of giving dyes of high tinctorial power. Thus, higher speed is attained from them.

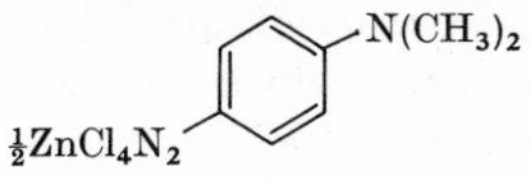

(53)

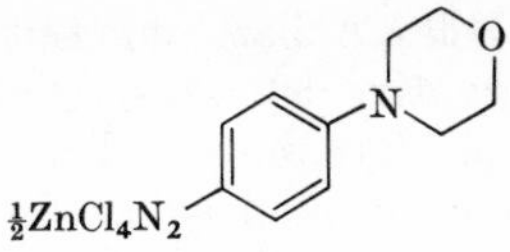

(54)

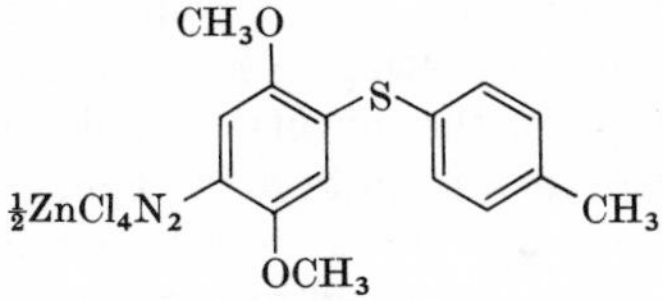

(55)

Broadly, there are two different modes of producing the final image. The phenol may be incorporated in the coated layer, along with diazonium salt, and then merely introduction of a base will cause coupling; this may be done by using ammonia vapour, the 'ammonia' papers. Alternatively, only the diazonium salt is coated and the coupling achieved by applying an alkaline solution of a phenol, called the semi-dry or semi-wet process. The ammonia papers usually employ a naphthol, such as (56) or (57), in com-

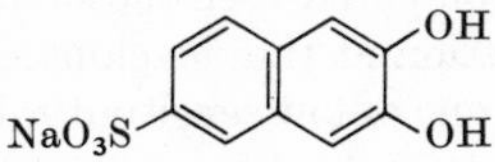

(56)

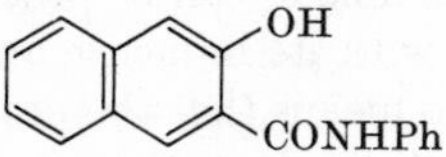

(57)

$$CH_3COCH_2CONHPh$$

(58)

bination with acetoacetanilide (58). The naphthol gives a blue dye and the anilide a yellow one, the combined result being a fairly good black image.

In the semi-dry process, the most commonly used coupling component is phloroglucinol [$1,3,5-C_6H_3(OH)_3$], which reacts with the diazonium compound to mono-, bis- and tris-azo dyes that are yellow, brown and blue, respectively, and the combination of which gives the desired good black image.

The Kalvar Process

While basically also dependent on diazonium salts, the Kalvar process is very different in that the image consists of nitrogen bubbles. A diazonium salt such as the common p-dimethylamino-derivative (**53**) is dispersed in an acrylic ester–acrylic acid polymer layer. Exposure to light generates nitrogen which, by careful choice of the polymer, forms micro-bubbles and also polymerizes the macromolecule further to ensure that the bubbles remain intact. The image thus appears as a white cloudiness in the otherwise clear polymer layer but, owing to scattering by refraction at the solid–vapour interface, the density to well-collimated light, such as on a slide projector, is quite high. Thus, this gives good black-and-white images when viewed by projection.

Alternatively, the nitrogen formed by exposure to light may be allowed to escape slowly; the remainder of the diazo salt is then thermally decomposed, the nitrogen forming bubbles that are stabilized by the continued polymerization. The former method gives a negative image, the latter a positive.

Photochromic Systems

Another process, which has arisen to a position of importance recently, involves a pronounced colour change occurring on exposure to light, i.e. photochromism. This has been applied particularly to micro-images because, being a molecular process in which one photon causes the change in only one molecule, very fine images can be produced compared with the silver-halide process with its immense amplification at the development step. This also means that the process is one of low sensitivity, but that is no real disadvantage and readily accepted in view of the high information-storage capacity of the system.

Many compounds show photochromism but the most commonly mentioned class are indoline–pyran spiro-compounds. Condensation between an o-hydroxy-aldehyde (**60**) and an indolenine salt

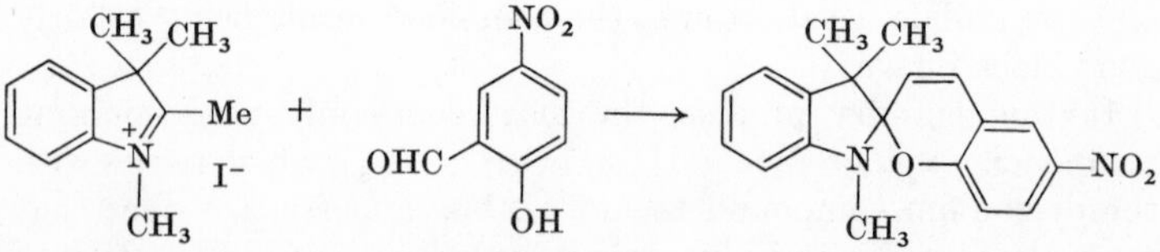

(**59**), for example, gives a product (**61**) that is colourless and in the excited state passes readily into the valence isomer (**62**) which has an intense magenta colour.

The reverse process can also be achieved by exposing strongly to light absorbed by (**62**), but the image is fairly stable to normal handling and storage.

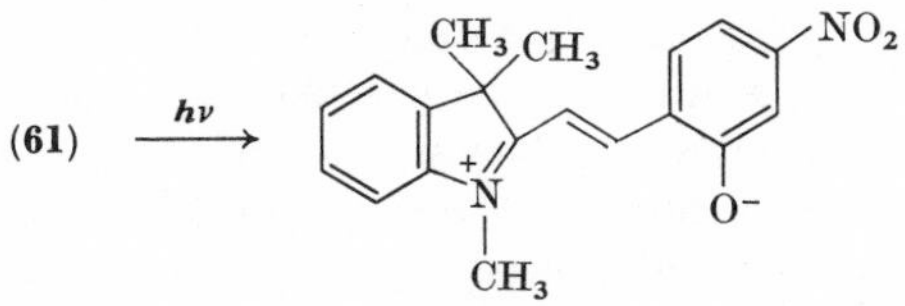

(**61**) $\xrightarrow{h\nu}$

(**62**)

(ii) CHLOROFLUOROCARBONS—PRODUCTION AND APPLICATIONS

World production of chlorofluorocarbons now exceeds 800×10^3 tons p.a.; the uses include aerosol propellants, refrigerants, solvents, foam blowing agents, fire extinguishers and chemical intermediates. This important sector of the chemical industry developed from the need for safer refrigerants some 40 years ago. The term 'chlorofluorocarbons' is a misnomer, implying, as it does, compounds consisting solely of carbon, chlorine and fluorine. This is the case with most of the largest-tonnage compounds but others,

Table 17.1.   Chlorofluorocarbons produced commercially

Country	Company	Trade name
U.S.A./Canada	Allied Chemical	Genetron
	E. I. du Pont	Freon
	Kaiser Chemicals	Kaiser
	Pennwalt	Isotron
	Union Carbide	Ucon
U.K.	ICI	Arcton
	I.S.C. Chemicals	Isceon
France	Pechiney	Flugene
	Ugine Kuhlmann	Forane
W. Germany	Hoechst	Frigen
	Kali Chemie	Kaltron
Italy	Montecatini Edison	Algofrene
Japan	Daikin Kogyo	Daiflon
	Mitsui Fluorochemicals	—
	Asahi Glass	Asahiflon
Australia	Australian Fluorine Chemicals	Isceon
Holland	Zinc Organon	FCC

containing elements such as hydrogen or bromine, are also important and 'halogenated hydrocarbons' is a more correct expression. The refrigeration industry originally depended upon gases such as ammonia, sulphur dioxide and methyl chloride with their attendant problems of toxicity and flammability. Ammonia is still very widely used for large-scale plants, but the trend towards domestic and retail-shop refrigeration in the 1920's stressed the need for non-toxic, non-flammable refrigerants. The use of chlorofluorocarbon compounds as refrigerants was pioneered by Henne and Midgely, to be developed by E. I. du Pont de Nemours and Co. Inc. under the name 'Freon'. These compounds are now manufactured by a number of companies around the world as Table 17.1 (p. 597), which is by no means comprehensive, shows.

Chemical Composition

All the commercial chlorofluorocarbons are halogenated derivatives of common alkanes and all contain at least one atom of fluorine. Fluorine is the most reactive element known and combines violently with many other materials unless the reaction conditions are controlled. However, fluorinated alkanes are characteristically unreactive, and the greater the fluorine content of the molecule the less reactive it is.

There are three basic starting materials, each giving rise to one series of chlorofluorocarbons, formed by reaction with hydrogen fluoride, in essence as follows:

(i) Carbon tetrachloride
$$CCl_4 \longrightarrow CCl_3F \longrightarrow CCl_2F_2 \longrightarrow CCl_3F \longrightarrow CF_4$$

(ii) Chloroform
$$CHCl_3 \longrightarrow CHCl_2F \longrightarrow CHClF_2 \longrightarrow CHF_3$$

(iii) Hexachloroethane
$$CCl_3{-}CCl_3 \longrightarrow CClF_2{-}CCl_2F \longrightarrow$$
$$CClF_2{-}CClF_2 \longrightarrow CClF_2{-}CF_3$$

Other compounds of commercial importance contain bromine as well as fluorine, viz.

(iv) $CBrF_3$; $CBrF_2{-}CBrF_2$

or are cyclic fluorocarbons, viz.

(v) $F_2C{-}CF_2$

     | | octafluorocyclobutane

    $F_2C{-}CF_2$

In all cases the successive replacement of chlorine by fluorine results in a lowering of the boiling point, and the majority of the chlorofluorocarbons are gases under normal conditions of temperature and pressure. However, with the exception of those few compounds whose critical temperatures are below normal ambient temperature, they can all exist as liquefied gases under pressure, and it is this property that gives rise to many of their applications.

Nomenclature

One feature immediately apparent from even a brief consideration is the similarity and considerable length of the names. This could lead to confusion and a common numbering system was devised which would be prefixed either with manufacturers' trade names or with the letter 'R' in general literature. This system is based upon a three-digit number, XYZ, where Z is the number of fluorine atoms in the molecule and Y is one more than the number of hydrogen atoms in the molecule, while X is one less than the number of carbon atoms in the molecule and is omitted if this makes it zero.

Any discrepancy in satisfying the valency of carbon is assumed to be made up by chlorine.

This nomenclature is best explained by examples. Thus dichlorodifluoromethane CCl_2F_2 is designated R12; here X = 0, Y = 1 and Z = 2. Chlorodifluoromethane, where X = 0, Y = 2 and Z = 2 is designated R22. 1,2-Dichloro-1,1,2,2-tetrafluoroethane is designated R114, X = 1, Y = 1 and Z = 4.

The presence of bromine in the molecule is indicated by the letter 'B' and a terminal figure indicating the number of bromine atoms, e.g. $CBrClF_2$ = R12B1 and $CBrF_3$ = R13B1.

There are further complications to this system of nomenclature in order to take isomerization into account but they will not be dealt with here.

Production

The basic step in the production of these chlorofluoro-compounds is the replacement of chlorine by fluorine, a reaction that can be carried out in either the liquid or the vapour phase.

In the liquid-phase process, the appropriate chlorocarbon (e.g. carbon tetrachloride) and anhydrous hydrogen fluoride are pumped into an autoclave reactor containing an antimony chlorofluoride catalyst at 100–150°C and 1–3×10^6 Nm^{-2}. The vapour-phase process operates at atmospheric pressure with a higher temperature, 250–350°C, and an impregnated carbon catalyst. In this case, carbon tetrachloride and anhydrous hydrofluoric acid are separately vaporized, mixed, preheated in a heat-exchanger using exhaust vapours from the reactor, and passed over the catalyst.

When carbon tetrachloride is used, the product from both processes is a mixture of R11 (CCl_3F) and R12 (CCl_2F_2), viz.

$$2CCl_4 + 3HF \longrightarrow CCl_3F + CCl_2F_2 + 3HCl$$

The product ratio can be varied within fairly wide limits by varying reaction conditions, feed rates, etc. $CClF_3$ is also produced in traces but the reaction is generally controlled to minimize this compound for which there is only a small demand.

Purification consists of removal of by-product HCl and traces of unused HF, either by distillation under pressure or by an aqueous scrubbing system. This gives a crude product mix which is then separated into its pure components by distillation.

The production of $CHClF_2$ from $CHCl_3$ is very similar but there are slight differences in the production of two-carbon compounds. Although hexachloroethane was quoted earlier as the starting material, this is actually a solid (m.p. 187°C) and is usually prepared within the reactor with a feed of tetrachloroethylene and chlorine. Hydrogen fluoride is also added at the same time, and feed rates and reactor conditions are so controlled as to ensure that the addition of chlorine is complete before fluorination commences. There are six possible products from the fluorination of hexachloroethane, with fluorine content ranging from CCl_3—CCl_2F to CF_3—CF_3, but only three are really of commercial significance, namely CCl_2F—$CClF_2$ (R113), $CClF_2$—$CClF_2$ (R114) and $CClF_2$—CF_3 (R115).

Purity

It has been claimed that the chlorofluorocarbons are the purest large-tonnage chemicals produced and the annual output of over 800×10^3 tons tends to support this view. With one or two very specialized exceptions, the refrigeration application demands the most stringent specification and, since almost all the chlorofluorocarbons are used in refrigeration, the general practice is to produce all material to that quality.

The usual specification for refrigeration quality is shown in Table 17.2. These limits are sufficient to ensure that any particular

Table 17.2. Specification of refrigeration-quality fluorocarbons

Maximum water content (ppm)	10
Boiling range (95%–5%) (°C)	0.25
High-boiling residue (vol.-%)	0.01
Non-absorbable gas content (vol.-%)	1.5
Maximum acidity, as Cl⁻ (ppm)	0.5

compound will perform its expected refrigeration duty satisfactorily but, in addition, gas chromatography is used to check on chemical purity and to determine the composition of mixtures of chlorofluorocarbons required for some applications.

Applications

Table 17.3 lists the applications for the main fluorocarbons in commercial use (see p. 602).

Refrigeration continues to be a major outlet, but R11 and R12 are more used as aerosol propellants than as refrigerants and the refrigeration industry comes in second place in terms of consumption of chlorofluorocarbons, as shown in Table 17.4 (p. 603).

Refrigerants

The choice of refrigerant is decided by the temperature reduction needed. Where only modest reductions in temperature are required, one of the higher-boiling compounds would be used, whereas a low-temperature application would require one of the compounds with a low boiling point, although there are large areas of overlap in the useful working ranges of specific refrigerants.

Dichlorodifluoromethane (R12) is the most widely used refrigerant, suitable for centrifugal, rotary and reciprocating compressors. It is used in large-scale air-conditioning systems as well as in domestic refrigerators. Trichlorofluoromethane (R11) is often used with centrifugal compression systems in industrial air-conditioning systems. Chlorodifluoromethane (R22) is popular for low-temperature work in both domestic and commercial freezers. Azeotropic mixtures

$$(CCl_2F_2 + CH_3CHF_2 \text{ and } CClF_2CF_3 + CHClF_2)$$

are also important, and other chlorofluorocarbons have special applications.

Table 17.3. Properties and applications of chlorofluorocarbons

Refrigerant no.	Formula	Boiling point (°C)	Vapour pressure at 21°C (p.s.i.a.)	Applications
113	CCl_2F—$CClF_2$	47.6	6	Solvent, refrigerant, chemical intermediate.
114B2	$CBrF_2$—$CBrF_2$	47.3	6	Fire extinguishant.
11	CCl_3F	23.8	13	Propellant, refrigerant, blowing agent, solvent.
21	$CHCl_2F$	8.9	23	Possible refrigerant, also considered as working fluid.
114	$CClF_2$—$CClF_2$	3.8	28	Propellant, refrigerant.
12B1	$CBrClF_2$	−4.0	38	Fire extinguishant.
C-318	C_4F_8	−5.8	40	Propellant.
12	CCl_2F_2	−29.8	85	Propellant, refrigerant, blowing agent.
115	$CClF_2$—CF_3	−38.7	113	Refrigerant, propellant.
22	$CHClF_2$	−40.8	137	Refrigerant, chemical intermediate.
13B1	$CBrF_3$	−57.8	213	Fire extinguishant, refrigerant.
13	$CClF_3$	−81.4	473	Refrigerant.
14	CF_4	−128.0	543[a]	Refrigerant, but very little used.

[a] Critical pressure.

Table 17.4. Applications (as %) of chlorofluoro-carbons in U.S.A. and U.K.

Application	U.S.A.	U.K.
Propellants	50	70
Refrigerants	28	10
Blowing agents	5	10
Solvents	5	5
Chemical intermediates	10	3
Miscellaneous	2	2

Although a limited amount of work has been done on the use of chlorofluorocarbon compounds in absorption cycle refrigeration they are, for all practical purposes, confined to vapour compression systems. The chilling effect in a compression cycle system is obtained by the evaporation of liquid refrigerant in the low-pressure area of a closed circuit. Mechanical suction, provided by a compressor, recovers the low-pressure vapour and returns it to the high-pressure part of the circuit. This process is best explained by reference to the operation of a domestic refrigerator:

(*a*) Liquefied refrigerant, exerting the vapour pressure appropriate to the temperature of the liquid, passes through an expansion valve into a larger volume (the low-pressure side of the system). The liquid evaporates, expanding to form vapour; this causes a drop in temperature, heat being taken from the surrounding metalwork which forms the cold plate or ice-box of the refrigerator.

(*b*) The vapour is drawn into the compressor and the process of compression to high-pressure vapour raises its temperature.

(*c*) Hot compressed vapour is cooled as it passes through a radiator and condenses to liquid. This radiator is the grille of fine tubes found at the back of a refrigerator, while in industrial refrigeration it is more likely to be a water-cooled heat exchanger.

(*d*) Liquid refrigerant is passed on to the expansion valve to continue the cycle.

This is the basic vapour compression system and multi-stage units or cascade systems, in which hot refrigerant vapour is cooled by the evaporation of a second refrigerant in a separate circuit, are merely elaborations of this basic system.

Aerosol Propellants

The strict definition of an aerosol is an ultrafine suspension of solid or liquid particles in gas or air, e.g. smoke or fog. However, the term is more commonly used to describe any package whose contents are dispensed by means of internal pressure. This includes such diverse products as insecticides, hair sprays, paints and shaving foams. Despite the modern image of the aerosol pack, the basic principles of packaging products under pressure were appearing in patents and other technical literature at the beginning of this century and, indeed, a controllable valve for use on bottles was patented in 1862. Nevertheless, it is generally accepted that the real fathers of the true aerosol package were a Norwegian, Eric Rotheim, who patented a pressurized packaging system in 1933, and Goodhue and Sullivan, who obtained a patent for an aerosol insecticide in 1943.

In the early 1940's, U.S. troops fighting in the Pacific areas were suffering more casualties from tropical insect bites than from enemy action. Goodhue and Sullivan, scientists at the U.S. Bureau of Entomology working on the problem of the most effective and convenient way of dispensing insecticides, dissolved the insecticidal agents in a small amount of oil and mixed this solution with dichlorodifluoromethane (R12) to give a self-pressurized pack. The vapour pressure of CCl_2F_2, 500×10^3 Nm^{-2} at $21°C$, ensured that when the solution was discharged into the atmosphere, the liquefied CCl_2F_2 evaporated immediately and violently and broke up the droplets of insecticide into an extremely fine mist or 'aerosol'. This insecticide mist remained suspended in the atmosphere for a considerable length of time and thus was much more effective against flying insects than the previous coarse wet sprays provided by pump-action dispensers. Something like 40 million 'bug bombs' were produced between then and 1945 and provided the initial introduction to aerosols for a large number of servicemen.

These early products were not suitable for civilian use since the pressure demanded the use of an expensive and heavy steel cylinder, but it was quickly realized that trichlorofluoromethane (R11) could be mixed with the dichlorodifluoromethane to give a propellant mixture of lower vapour pressure (250×10^3 Nm^{-2} for a 50/50 mixture), suitable for safe handling in light-weight tinplate cans of the type already in use for packaging beer. That was the real starting point of the aerosol industry as we know it today, an industry which in 1972 produced over 2700 million packs in the U.S.A. and 360 million in the U.K. A sketch of a pack is given

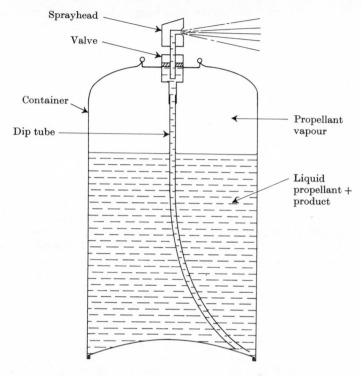

Figure 17.4. Principles of an aerosol pack.

in Figure 17.4. Aerosol technology has made a great number of advances and several textbooks[1–4] have been published, but the basic principles are essentially the same. The operating pressure in an aerosol pack is supplied either by the vapour pressure of a liquefied gas or by means of a compressed gas. Both chlorofluorocarbon and hydrocarbon liquefied gases are used and the choice is determined by a number of factors which include the product formulation, compatibility, safety and economics. Compressed gases, such as nitrogen, carbon dioxide or nitrous oxide, are much less commonly used, because the operating pressure decreases as product is dispensed from the container and the vapour space increases.

This decrease in pressure does not occur with liquefied-gas propellants since, although liquid propellant gradually evaporates to fill the increasing vapour space, the vapour pressure over a liquefied gas remains constant so long as the slightest drop of liquid remains.

In general the basic propellant is dichlorodifluoromethane and the vapour pressure is kept down to an acceptable level either by means of trichlorofluoromethane or by using other volatile solvents such as ethyl alcohol, methylene chloride, methyl chloroform, etc. However, some formulations contain significant amounts of water, such as shaving lathers, colognes, perfumes and some pharmaceutical products. Trichlorofluoromethane and the chlorinated solvents are prone to hydrolysis and so it is customary to use 1,2-dichlorotetrafluoroethane (R114) to reduce the pressure of CCl_2F_2 in such cases.

CCl_2F_2, CCl_3F and $CClF_2$—$CClF_2$ are the only chlorofluorocarbons commonly used as aerosol propellants, but others are used for specialized applications. $CHClF_2$ (R22) has found a very limited use in products requiring a particularly high pressure, while $CClF_2$—CF_3 (R115) and cyclo-C_4F_8 have FDA approval in America for use in aerosol foodstuffs such as whipped cream.

Some idea of the variety of aerosol products and the major product categories that make up an industry whose output is currently at a level of 7 aerosol packs per head of population in the U.K. is shown in Table 17.5.

Table 17.5. Uses of chlorofluorocarbons in the U.K.

Product category	Millions of packs	% of total
Hair sprays	117	32.2
Deodorants/antiperspirants	53	14.6
Perfumes/colognes	16	4.4
Shaving foam	11	3.0
Other personal products	3.5	1.0
Polishes	34.5	9.5
Insecticides	26	7.8
Air fresheners	23	6.4
Oven cleaners	7	1.9
Starches	5.5	1.6
Other household products	13	3.6
Pharmaceutical	11	3.0
Automotive products	9.5	2.6
Paints and other coatings	15.5	4.3
Industrial products	8	2.2
Miscellaneous	7	1.9
	360.5	

These broad product categories are made up of over 300 separate product types with around 2000 brands.

Solvents

1,1,2-Trichlorotrifluoroethane (R113) is a very mild solvent, often referred to as a 'selective' solvent, suitable for removing oil, grease and similar contaminants from most plastic materials. Hence it is a suitable solvent for use in the electrical and electronics industries where metals/plastics combinations provide many cleaning problems. Another area of application is precision engineering, particularly that concerned with aerospace and military equipment, where solvents are required that are particularly stable, non-corrosive and easily removed without leaving any residue. In general, the refrigeration quality of trichlorotrifluoroethane is adequate but, in the U.S.A., very considerable quantities of an even purer grade ('white room' quality) are used in the space programme.

1,1,2-Trichlorotrifluoroethane is also used in dry-cleaning where its mild solvent action and low boiling point make it very suitable for delicate or difficult cleaning operations such as furs, suedes, and heat-sensitive fabrics. In this context R11 is also used since its solvency power is rather higher, making it suitable for replacement of perchloroethylene which is much in use in general dry-cleaning work. Additionally, trichlorofluoromethane is commonly used for cleaning refrigeration systems, generally during servicing or after repairs, before charging with fresh refrigerant (commonly CCl_2F_2 or $CHClF_2$). The advantages of CCl_3F for this application are that it adequately removes oil and grease while it does not affect the plastics used in hermetic motor construction and is easily removed by evacuation.

Foam-blowing Agents

In the production of flexible polyurethane foam, the reaction between isocyanate and polyol is exothermic and this heat can be used to evaporate a volatile liquid mixed into the reactants, thus forming a foam. Trichlorofluoromethane (R11) is used for this application and is either premixed with the polyol or incorporated into the reaction mixture at the point of discharge on the foaming machine. The foaming reaction mass then passes through a heated tunnel to complete the reaction. The CCl_3F plays no part in the chemistry of the reaction; it is purely a physical blowing agent.

Rigid polyurethane foams, generally used for thermal insulation, have a closed cell structure and derive an additional benefit from the use of chlorofluorocarbon blowing agents since the vapours have a lower thermal conductivity than that of carbon dioxide, the other common foaming medium. Trichlorofluoromethane (R11) used with reactants chosen to give a closed-cell rigid foam, fulfils this role. For *in situ* cavity foaming, dichlorodifluoromethane (R12) is used in the 'frothing' process, where its very rapid evaporation expands the liquid reaction mass to something like the consistency of shaving lather before it enters the cavity. This ensures thorough filling of narrow areas and reduces foaming pressures by virtue of the reduced expansion that is required. Dichlorodifluoromethane is also used in the production of poly-(vinyl chloride) foam by the plastisol process. In this application, CCl_2F_2 under pressure is compounded into the PVC plastisol which is kept under pressure until it is discharged on to the foaming bed. This release to atmospheric pressure allows expansion of the CCl_2F_2 vapour and foams the plastisol. Further expansion and curing of the PVC occurs as the foam passes through a heated tunnel.

Polystyrene is another polymer that is being foamed in sheet and board form by means of CCl_2F_2. The process is somewhat similar to that for poly(vinyl chloride) in that CCl_2F_2 is mixed into the liquid polymer under pressure before release into atmospheric pressure through an extrusion die. Foamed polystyrene produced in this way can range from 3-inch board used for constructional insulation to thin sheet of a few millimetres thickness used for thermoforming into packaging material.

Fire-extinguishing Agents

All the chlorofluorocarbons are non-flammable and have some degree of fire-extinguishing ability but the bromine-containing compounds are most usually employed in this application. These are:

'R' Number	Formula	B.p. (°C)	Trade name
114B2	$CBrF_2$—$CBrF_2$	47.3	Halon 2402, Fluobrene
12B1	$CBrClF_2$	−4.0	Halon 1211, BCF
13B1	$CBrF_3$	−57.8	Halon 1301, BTM

The extinction efficiency of a fire-extinguishing agent is very dependent on the total extinguisher system and on details such as operating pressure, rate of discharge and nozzle design. For this reason it is not easy to say which extinguishing agent alone is best for a particular application. All the bromofluorocarbon compounds listed above function as vaporizing liquid extinguishing agents but there are differences in mode of application.

By virtue of its low boiling point and consequent high vapour pressure, bromotrifluoromethane (R13B1) reaches the fire as a cloud of vapour. This, coupled with its virtually non-toxic properties, makes it very suitable for total flooding systems where extinguishing equipment is fixed permanently in place and operates automatically in response to an increase in temperature beyond a set level. The original application for this system was in aircraft engine nacelles, but it is being increasingly chosen for civilian uses such as art galleries, museums, libraries and computer installations. The most important factor in such situations is that the use of water on a fire would cause almost as much damage as the fire itself, whereas a vapour extinguishing agent has no deleterious effect. Additionally, there is always a risk of leakage or accidental discharge from a fixed system and here the non-toxic nature of the extinguishant is important.

$CBrClF_2$ (R12B1) is the most common vaporizing liquid extinguishant currently used in the U.K.; it is more generally known as 'BCF' (Bromo-Chloro-di-Fluoro-methane). It does not evaporate so rapidly on discharge and tends to reach a fire in the form of a cloud of vapour and droplets of liquid. This provides a greater chance that the extinguishing agent will penetrate to the heart of a fire, and such agents are generally more suitable for hand or mobile operation and particularly for tackling petrol and other liquid-fuel fires. In addition to its value for hand and mobile extinguishers it is also used in fixed installations for aircraft, electrical equipment, engine compartments, etc.; this reinforces the point that it is the complete extinguisher system that determines suitability for a specific situation rather than the particular extinguishing agent.

Chemical Intermediates

Chlorodifluoromethane (R22) is the starting material for the production of tetrafluoroethylene:

$$2CHClF_2 \xrightarrow{\quad 600^\circ \quad} CF_2{=}CF_2 + 2HCl$$

The polymerization of tetrafluoroethylene is described in Chapter 8. Another fluoropolymer is obtained from chloro-trifluoroethylene: the monomer is obtained from 1,1,2-trichlorotrifluoroethane (R113) by treatment with zinc powder and ethanol:

$$CCl_2FCClF_2 \longrightarrow CClF{=}CF_2 + Cl_2$$

Handling

The very low chemical reactivity of the chlorofluorocarbons, and the stringent specification to which they are produced, allow them to be handled perfectly satisfactorily in mild-steel containers.

Bulk handling of chlorofluorocarbons is largely restricted to the aerosol industry, in storage tanks of, say, 20 tons capacity, supplied by road tankers or by rail tank cars; the latter are not used in this country except for export business, but are common on the Continent and in the U.S.A.

Chlorofluorocarbons are chemically stable and consequently there are few limitations on materials of construction for handling equipment. Of the metals, only zinc, galvanized steel, magnesium, and alloys containing significant levels of zinc or magnesium are not advised and, although some plastics and elastomers are affected by some chlorofluorocarbons, an adequate range of such materials is available for use as gaskets, seals, etc.

The aim of this Section has been to give some idea of the industrial developments that have grown out of the discovery of the first chlorofluorocarbon compounds some 40 years ago. The amounts used in all the applications introduced so far are still growing and new outlets are being developed. Apart from the essential physical properties that determine the choice of compounds for specific applications, the main attributes of the chlorofluorocarbons—non-flammability, very low toxicity, chemical stability and low chemical reactivity—have been the basis of developments so far and will doubtless be a major factor in future applications.

Bibliography

[1] A. Herzka and J. Pickthall, *Pressurised Packaging (Aerosols)*, Butterworth, London, 1961.

[2] H. R. Shepherd, *Aerosols: Science and Technology*, Interscience, New York, 1961.

[3] A. Herzka, *International Encyclopaedia of Pressurised Packaging (Aerosols)*, Pergamon Press, Oxford, 1966.

[4] *The Aerosol Handbook*, Wayne and Dorland and Co., 1972.

Index